MECHANICAL FOUNDATIONS OF ENGINEERING SCIENCE

ELLIS HORWOOD SERIES IN ENGINEERING SCIENCE

STRENGTH OF MATERIALS
J. M. ALEXANDER, University College of Swansea.
TECHNOLOGY OF ENGINEERING MANUFACTURE
J. M. ALEXANDER, R. C. BREWER, Imperial College of Science and Technology, University of London, J. R. CROOKALL, Cranfield Institute of Technology.
VIBRATION ANALYSIS AND CONTROL SYSTEM DYNAMICS
CHRISTOPHER BEARDS, Imperial College of Science and Technology, University of London.
COMPUTER AIDED DESIGN AND MANUFACTURE
C. B. BESANT, Imperial College of Science and Technology, University of London.
STRUCTURAL DESIGN AND SAFETY
D. I. BLOCKLEY, University of Bristol.
BASIC LUBRICATION THEORY 3rd Edition
ALASTAIR CAMERON, Imperial College of Science and Technology, University of London.
STRUCTURAL MODELLING AND OPTIMIZATION
D. G. CARMICHAEL, University of Western Australia
ADVANCED MECHANICS OF MATERIALS 2nd Edition
Sir HUGH FORD, F.R.S., Imperial College of Science and Technology, University of London and J. M. ALEXANDER, University College of Swansea.
ELASTICITY AND PLASTICITY IN ENGINEERING
Sir HUGH FORD, F.R.S. and R. T. FENNER, Imperial College of Science and Technology, University of London.
INTRODUCTION TO LOADBEARING BRICKWORK
A. W. HENDRY, B. A. SINHA and S. R. DAVIES, University of Edinburgh
ANALYSIS AND DESIGN OF CONNECTIONS BETWEEN STRUCTURAL JOINTS
M. HOLMES and L. H. MARTIN, University of Aston in Birmingham
TECHNIQUES OF FINITE ELEMENTS
BRUCE M. IRONS, University of Calgary, and S. AHMAD, Bangladesh University of Engineering and Technology, Dacca.
FINITE ELEMENT PRIMER
BRUCE IRONS and N. SHRIVE, University of Calgary
PROBABILITY FOR ENGINEERING DECISIONS: A Bayesian Approach
I. J. JORDAAN, University of Calgary
STRUCTURAL DESIGN OF CABLE-SUSPENDED ROOFS
L. KOLLAR, City Planning Office, Budapest and K. SZABO, Budapest Technical University.
CONTROL OF FLUID POWER, 2nd Edition
D. McCLOY, The Northern Ireland Polytechnic and H. R. MARTIN, University of Waterloo, Ontario, Canada.
TUNNELS: Planning, Design, Construction
T. M. MEGAW and JOHN BARTLETT, Mott, Hay and Anderson, International Consulting Engineers
UNSTEADY FLUID FLOW
R. PARKER, University College, Swansea
DYNAMICS OF MECHANICAL SYSTEMS 2nd Edition
J. M. PRENTIS, University of Cambridge.
ENERGY METHODS IN VIBRATION ANALYSIS
T. H. RICHARDS, University of Aston, Birmingham.
ENERGY METHODS IN STRESS ANALYSIS: With an Introduction to Finite Element Techniques
T. H. RICHARDS, University of Aston, Birmingham.
ROBOTICS AND TELECHIRICS
M. W. THRING, Queen Mary College, University of London
STRESS ANALYSIS OF POLYMERS 2nd Edition
J. G. WILLIAMS, Imperial College of Science and Technology, University of London.

MECHANICAL FOUNDATIONS OF ENGINEERING SCIENCE

H. G. EDMUNDS, B.Sc.(Eng.), Ph.D., A.C.G.I., F.I.Mech.E.
Professor and Head of Department of Engineering Science
University of Exeter

ELLIS HORWOOD LIMITED
Publishers · Chichester

Halsted Press: a division of
JOHN WILEY & SONS
New York · Brisbane · Chichester · Toronto

First published in 1981 by

ELLIS HORWOOD LIMITED
Market Cross House, Cooper Street, Chichester, West Sussex, PO19 1EB, England

The publisher's colophon is reproduced from James Gillison's drawing of the ancient Market Cross, Chichester.

Distributors:

Australia, New Zealand, South-east Asia:
Jacaranda-Wiley Ltd., Jacaranda Press,
JOHN WILEY & SONS INC.,
G.P.O. Box 859, Brisbane, Queensland 40001, Australia

Canada:
JOHN WILEY & SONS CANADA LIMITED
22 Worcester Road, Rexdale, Ontario, Canada.

Europe, Africa:
JOHN WILEY & SONS LIMITED
Baffins Lane, Chichester, West Sussex, England.

North and South America and the rest of the world:
Halsted Press: a division of
JOHN WILEY & SONS
605 Third Avenue, New York, N.Y. 10016, U.S.A.

British Library Cataloguing in Publication Data
Edmunds, H. G.
Mechanical foundations of engineering science –
(Ellis Horwood series in engineering science)
1. Mechanics, Applied
I. Title
620.1 TA350

Library of Congress Card No. 81–6566 AACR2

ISBN 0–85312–281–4 (Ellis Horwood Limited, Publishers – Library Edn.)
ISBN 0–85312–354–3 (Ellis Horwood Limited, Publishers – Student Edn.)
ISBN 0–470–27253–8 (Halsted Press)

Typeset in Press Roman by Ellis Horwood Limited
Printed in Great Britain by R. J. Acford, Chichester

Table of Contents

Author's Preface

From a perusal of the popular student texts it would appear that the advent of departments of Engineering Science, as opposed to the more traditional departments devoted to one or other of the specialised branches of engineering, has, over wide areas, had little impact on either the content or the treatment of the various courses of which undergraduate curricula are comprised. This suggests, that although the newer foundations have effected some liberalisation of the curriculum, there has been little move toward the greater objective of a more generalised treatment of the science of engineering as a whole. Yet the promise of such a development is obvious, and the means for its realisation are available, and it is to be presumed that it is not pursued with greater vigour for pedagogic reasons only.

Certainly, it must be confessed that the concepts which reveal the affinities between situations which appear physically disparate are necessarily of the nature of mathematical abstractions, and it would be idle to deny that this is a direction in which students of engineering are not always highly motivated. However, experience has shown that it is all too easy both to exaggerate the difficulty of the more modern methods, and greatly to underestimate both the degree of simplification to which they finally lead, and the ability of present-day students to accommodate them, and this book presents the text of a first-year course in Engineering Science, which, over the years has probed in a purely pragmatic way along the road from a traditional introduction to Newtonian Mechanics as a subject in its own right, to a course which also aims particularly to lay a common foundation for the development along modern and reasonably sophisticated lines of the many subsequent studies which stem from it. In particular it aims to prepare for subsequent courses in Rigid Body Mechanics, in Stress Analysis and Strength of Materials, in Fluid Mechanics and Hydraulics, in Soil Mechanics and Plasticity Theory, and in Structural Analysis. Additionally, for those students whose subsequent studies will chiefly lie in other directions, the course aims to give as broad an oversight of the field of mechanics as can reasonably be accommodated within a single course.

Of the topics which usually appear in a first book on Newtonian Mechanics, this includes Point Kinematics in Chapter 3, Particle Dynamics in Chapter 4, Moments of Forces and the equilibrium of distributed systems in Chapter 5, and Moments of Scalars in Chapter 10. Though these are topics to which the great majority of engineering students have previously been introduced, it is necessary to expand and generalise a range of concepts, which, in A-level courses appear to be developed in a form which is suitable for application only in special cases, and here, this is effected by adopting the vector methods that are presented in Chapter 2.

But instead of the more usual Rigid Body Mechanics, we then introduce a study of the mechanics of a deformable continuum. This comprises studies of the stress tensor and its resolution in Chapter 6, the strain tensor and its resolution in Chapter 7, and the strength and stress/strain properties of Hookean Solids and Newtonian Fluids in Chapter 8. Though this section is developed on the tensor concept, it adopts neither the terminology nor the symbolism of formal tensor algebra. Rather, a tensor is treated as a generalised vector which can be resolved for a vector component in exactly the same way as a vector is resolved for a scalar component. In this way we are able to take advantage of sophisticated concepts and methods with a minimum of mathematical commitment.

Chapter 9 presents a brief review of the method of formal stress analysis that is pursued in the Theory of Elasticity, whilst Chapter 11 presents a fuller account of the treatment of long, slender members in Strength of Materials, including tension, cylindrical torsion, bending and transverse shear. These are topics which are obviously of interest in their own right, and are again commonly to be found in an introductory course in mechanics. However, taken together, the two chapters are particularly designed as an introduction to the concept of 'redundancy', and to illustrate the alternative differential and integral formulations by which it is resolved. These are fundamental principles of crucial importance, far beyond the range of subjects which this course specifically aims to serve.

In the SI system of units, the permitted alternative of the N/mm^2 is preferred to the basic unit of pressure, because of the obvious impediment of a unit which is not commensurate with everyday quantities.

List of Symbols

Scalars are indicated by symbols printed in italic.

Geometrical vectors, that is, vectors which are described by a rectangular set of scalar components, are indicated by symbols printed in bold. Except for the symbols **i**, **j** and **k** which are reserved solely to indicate unit vectors in the directions of a set of rectangular coordinates, the symbols which represent unit vectors are further distinguished by the addition of a circumflex.

Matrices, including non-geometrical row vectors and column vectors, are indicated by bold symbols which are enclosed within square brackets.

Time-rates of change of scalars are indicated in the dot convention.

A	Superficial area.
a	Scalar acceleration-along-the-path.
$\mathbf{a}$	Acceleration, (vectorial). Area vector.
$\mathbf{a}_{\mathrm{O/PA}}$	Acceleration as seen by the observer O, of the point P relative to the point A.
B	Number of ball joints in a space truss.
b	Breadth of the section of a beam normal to an axis of symmetry.
C	Centre of Curvature. Centroid of a distributed scalar. Points on the centroidal axis of a set of distributed vectors. Centre of Gravity. Constant of integration.
$\mathrm{D}^n_{\mathrm{O}}$	The operator $\mathrm{d}^n/\mathrm{d}t^n$, as seen by the observer O.
E	Young's Modulus.
$\mathbf{E}$	Electrostatic field.
e	Eccentricity of a conic section.
F	Scalar force.
F_x	Tensile (or compressive) force on a transverse section of a long, slender member.
F_y, F_z	Shear forces on a transverse section of a long, slender member.
f	Flexibility Coefficient (general).
f_{AB}	Tensile force in the member AB of a truss.
$\mathbf{f}$	Force vector.

G	Universal gravitational constant. Shear Modulus.
g	Gravitational acceleration.
$\mathbf{i}, \mathbf{j}, \mathbf{k}$	Unit vectors in the directions of given rectangular coordinates.
I_{ii}	Second moment of a scalar about x_i.
I_{ij}	Product moment of a scalar about x_i and x_j.
$[\mathbf{I}]$	Tensor of second degree moments. The unit identity matrix.
K	Bulk Modulus. Constant of Integration.
K_{ii}	Radius of gyration of a scalar about x_i.
L, l	Length.
$\mathbf{l}, \mathbf{l}'$	Unit vectors (general). The set of direction cosines of a line in given rectangular coordinates.
$\mathbf{l}_\mathrm{n}$	Unit normal vector.
$\mathbf{l}_\mathrm{t}$	Unit tangent vector.
$(\mathbf{l}_\mathrm{r}\ \mathbf{l}_\phi\ \mathbf{l}_\mathrm{z})$	The unit vectors of cylindrical coordinates.
$[\mathbf{l}_{x'x}]$	The matrix of direction cosines of (x'_1, x'_2, x'_3) in (x_1, x_2, x_3).
M	Number of members in a truss.
M_x	Torsional moment on a transverse section of a long, slender member.
M_y, M_z	Bending moments on a transverse section of a long, slender member.
m	Mass.
m_{AB}	Scalar moment about $\overrightarrow{\mathrm{AB}}$.
$\mathbf{m}_\mathrm{A}$	Vector moment about A.
P	Number of pin joints in a plane truss.
p	Principal stress (scalar).
p_T	Thermodynamic pressure.
R	Support reaction.
(r, ϕ, z)	Cylindrical coordinates.
$\mathbf{r}$	Radius vector.
$\mathbf{r}_{\mathrm{PA}}$	Position vector of P relative to A.
s	Arc distance along a line.
T	Periodic time. Temperature.
t	Time.
u_i	Scalar component of point displacement in the direction of x_i.
$\mathbf{u}$	Displacement. Displacement field.
V	Volume.
v	Speed.
$\mathbf{v}$	Velocity. Velocity field.
$\mathbf{v}_{\mathrm{O/PA}}$	Velocity as seen by observer O of P relative to A.
W	Point load.
w	Distributed load per unit length. Loading equation.
$\langle w \rangle$	Loading equation in terms of singularity functions.
$(X, Y, Z), (X_1, X_2, X_3)$	Principal axes.
$(x, y, z), (x_1, x_2, x_3)$	Rectangular coordinates (general).

α	Coefficient of thermal expansion.
α_{pq}	Flexibility coefficient of rotation at $s = p$ per unit couple at $s = q$.
γ_{ij}	Angle of shear of the right-angle between x_i and x_j. Shear strain of the right-angle between x_i and x_j – small strains only.
Δ	Volumetric strain. Deflection of a beam.
Δ_{pq}	Flexibility coefficient of displacement at $s = p$ per unit load at $s = q$.
δ_{pq}	Flexibility coefficient of displacement at $s = p$ per unit couple at $s = q$.
$\partial \mathbf{u}_{ij}$	The partial derivative of the component of point displacement in the direction x_i with respect to x_j.
$[\partial \mathbf{u}]$	The matrix of partial derivatives of the components of point displacements.
$\partial \mathbf{v}_{ij}$	The partial derivative of the component of point velocity in the direction x_i with respect to x_j.
$[\partial \mathbf{v}]$	The matrix of partial derivatives of the components of point velocities.
ϵ_p	Principal strain.
ϵ_{ii}	Linear strain in the direction x_i.
ϵ_{ij}	Shear element for the directions x_i and x_j $(= \frac{1}{2}\gamma_{ij})$.
ϵ_{oo}	Linear strain along the octahedral axes.
θ	Slope of a beam in plane bending.
θ_x	Angle of rotation of a transverse section of a long slender member about its normal, torsional axis.
θ_y, θ_z	Angles of rotation of a transverse section of a long slender member about its in-plane, bending axes.
θ_{pq}	Flexibility coefficient of rotation at $s = p$ per unit load as $s = q$.
λ	Lamé's elastic constant. The second coefficient of viscosity.
μ	Coefficient of friction. Coefficient of viscosity.
μ_v	Coefficient of bulk viscosity.
v	Poisson's ratio.
ρ	Radius of curvature. Density.
σ_{ij}	Scalar component of stress on the section normal to x_i resolved in the direction of x_j.
$\boldsymbol{\sigma}_i$	Stress vector for the section normal to x_i.
$[\boldsymbol{\sigma}]$	Stress tensor.
σ_{oo}	Normal stress on the octahedral sections. Hydrostatic stress.
τ	Shear stress in an unspecified direction.
τ_o	Shear stress on the octahedral sections.
ϕ_{AB}	Tensile force per unit length in the member AB of a truss.

1

Method and Measures

1.1 THE SCIENTIFIC METHOD

In studies of the physical world, it is invariably found that the effects which stem from any one event are indefinitely extended in both time and space, so that the effects of different events are liable to interact in a manner that is, strictly speaking, incomprehensible. Yet if attention is focused on some limited part or aspect of the universe, it is generally found that only a few, dominant effects are significant, and it is this alone that makes scientific investigation possible and which determines the scientific method.

In this, we first identify some comprehensible part or aspect of the universe that is isolated in our imagination as the object of study. This we call 'the system'; but a system is not necessarily conceived to be either tangible or finite, for it may well include some abstraction like a point particle that is indefinitely small, or a force field that is indefinitely large. However it is more commonly identified as a finite physical entity, though this may be defined either as a specific quantity of matter (the control mass), or as the changing contents of some specific region of space (the control volume).

The remainder of the universe is called the surroundings, and the first crucial step of analysis is so to comprehend a given situation that we can adequately identify and quantify, not only those features of the system which determine its essential configuration or its 'state', but also those features of its interactions with the surroundings which have a significant bearing on its state. In effect, we thus define an idealised mathematical model of both the system and the interactions to which it is liable, and by studying the relationships between the state of the system and the interactions, we aim to describe the relationships in natural laws which are simple in form, accurate in interpretation, and comprehensive in scope.

But in their details, the properties of even the simplest systems tend to be somewhat complex, so that laws which seek directly to describe 'real' experience with 'real' systems cannot generally cover any great range of events without sacrificing both simplicity and accuracy. The basic laws of science are therefore usually expressed, not as descriptions of perceived effects in real systems, but rather in terms of abstract concepts such as point masses and forces. These may be no more than direct extrapolations from common experience, or, like the concepts of relativity and quantum theory, they may far transcend that experience, but in either case they are so defined that the idealised model fairly reflects those features of the situation that are essential to the issue whilst excluding those that lead to excessive complication. In this way we succeed in defining simple conceptual situations for which we can formulate simple natural laws.

Of course, the fundamental laws cannot then be applied directly to real situations: indeed, it is not even possible directly to test their validity by experiment. They are therefore more accurately to be regarded as axioms than as verifiable laws, and hence in order to effect their application to real situations it is necessary to associate each axiom with some means of bridging from the conceptual world in which it is expressed to the perceptual world in which it is to be applied. This we do by identifying some means of modelling the features of the real systems in terms of their conceptual counterparts.

Here we are concerned with the axioms and models of **Newtonian Mechanics.** Broadly this is concerned with the motion and equilibrium of finite, tangible bodies, in cases in which the effects of any change of internal energy or thermodynamic state are negligible, and in which we are essentially concerned only with the effects of interactions on the motions of bodies, and not with the transmission in time and space of the interactions themselves. Thus we now distinguish Newtonian Mechanics from **Thermodynamics** on the one hand, and from **Electromagnetic Field Theory** on the other.

1.2 SCALAR QUANTITIES

In all science, we are particularly concerned with relationships that are expressed in quantitative form. The first, essential step is therefore that of quantifying both the state of the system, and those of its interactions which significantly affect its state, and such descriptions are made in terms of quantities each of which has two attributes – **magnitude**, and **dimension.**

Dimensions

Contrary to common usage, the term 'dimension' refers not to the size of a quantity but to its nature. Some of these dimensions, such as length (l), time(t) and mass (m) we recognise intuitively, and from these we choose an arbitrary independent set which we regard as basic. Other dimensions are then determined as functions of the basic set, either by definition or through mathematical or physical laws. Thus, volume acquires the dimension of length cubed, $(l)^3$, speed the dimension of length divided by time, (l/t), and force the dimensions of mass times length divided by the time squared, (ml/t^2).

Magnitude

The magnitude M, of any quantity is determined simply by proportional scaling of an agreed, unit prototype, and it is therefore expressed as a simple product of its measure n, (a pure number), and the appropriate unit, u. Thus

$$M = n.u. \tag{1.1}$$

Evidently the measure n of a given quantity will vary in inverse proportion to the size of the unit u, so that if u_1 and u_2 are alternatives units of a given dimension, then the magnitude M of a given quantity may be equated either to n_1u_1 or to n_2u_2, and it follows that

$$M = n_1u_1 = n_2u_2 \tag{1.2}$$

whence $n_2 = (u_1/u_2)n_1$.

Fundamental Units

In each system of measurement, one set of units is regarded as basic and fundamental. The members of the set are chosen to be independent of one another, and these are the units that are defined directly by particular physical objects or other accepted standards. It is convenient, though not essential, to select as fundamental, the units of those quantities that are also regarded as basic in dimensional analysis.

Derived Units

Derived units are defined in terms of the basic units either through explicit relationships that exist between the corresponding physical quantities, or through relationships based on physical laws. Thus, the geometrical relationships between areas and lengths determine the derived unit of area in terms of the fundamental unit of length, and the relationship between force and rate of change of momentum expressed in **Newton's Second Law of Motion** determines the derived unit of force in terms of the fundamental units of mass, length and time.

Coherence

A system of units is said to be **coherent** if all the derived units are so chosen that they are equal to products and quotients of the basic units. Thus in a coherent system the unit of area is the square of the unit of length, the unit of velocity is equal to the unit of length divided by the unit of time, and so on.

It should be recognised that a mathematical relationship between physical quantities is more than an equation relating pure numbers. Suppose that physical quantities Q_1, Q_2 and Q_3 are related such that

$$M_1 = 3M_2M_3 \ . \tag{1.3}$$

Then in view of equation (1.1)

$$n_1u_1 = 3(n_2u_2)\,.\,(n_3u_3)$$
$$= 3(n_2n_3)\,.\,(u_2u_3)\;. \qquad (1.4)$$

In this equation, n_1, n_2, n_3 and 3 are pure numbers, whilst u_1, u_2 and u_3 are unit quantities. But if the system of units is coherent $u_1 = u_2u_3$, and equation (1.4) reduces to

$$n_1 = 3n_2n_3\;. \qquad (1.5)$$

Equation (1.5) is a mathematical relationship between numbers which has precisely the same form as equation (1.3) has between physical quantities. This complete formal equivalence would not generally be achieved with incoherent units for which $u_1 \neq u_2\,.\,u_3$.

1.3 SYSTÈME INTERNATIONAL

The current, internationally agreed coherent system of units is the **Système International d'Unités**, which is universally abbreviated to **SI**.

Base SI Units

The basic SI units are listed, with their symbols in Table 1.1, and the manner of their definition is briefly outlined in the following:

The meter (*m*) is defined in terms of a number of wavelengths of a particular radiation.

The kilogram (*kg*) is equal to the mass of the international prototype (a cylinder made of a platinum-iridium alloy lodged in the laboratories of the Bureau International des Poids et Mesures at Sévres).

The second (*s*) is the duration of a specific number of periods of a particular radiation.

The ampere (*A*) is the current that would flow in two straight, parallel conductors under specified conditions. Since these involve the derived unit of force it will be convenient to defer a fuller description till later.

The kelvin (*K*) is the fraction 1/273.16 of the thermodynamic temperature of the triple point of water.

The candela (cd) is the luminous intensity of a portion of surface of a black body at a defined temperature and pressure.

The mole (mol) is the amount of substance containing as many elementary units (which may be molecules, atoms, electrons, etc., and must be specified), as there are atoms in 0.012 kg of carbon-12.

Table 1.1

Physical quantity	Name of unit	Symbol
length	metre	m
mass	kilogram	kg
time	second	s
electric current	ampere	A
temperature	kelvin	K
luminous intensity	candela	cd
amount of substance	mole	mol
Supplementary Units:		
plane angle	radian	rad
solid angle	steradian	sr

The radian (rad) is the angle subtended at the centre of a circle by an arc of the circle equal in length to the radius.

The steradian (sr) is the solid angle subtended at the centre of a sphere by an area on the surface of the sphere equal in magnitude to the square of the radius.

Derived SI Units are all defined as multiples and quotients of the basic units. However, it is often convenient to give the quantity a special name, and in this way we arrive at a wide variety of units such as the newton, coulomb, watt, joule, farad, ohm, pascal, and so on. However, their status is not changed by their being given a name, so that, like derived units without special names, they are all expressible in terms of two or more of the seven basic units listed previously.

It should be noted that, for units which are named after people, the symbol which represents the unit is written as a capital letter, but the name itself is written with the initial letter in lower case.

In general, derived units will be introduced as they arise naturally throughout the course; here, we mention only one.

The Unit of Force is defined so that Newton's Second Law of Motion may be written without a factor of proportionality; (or more precisely, with a unit factor of proportionality). Thus

$$\text{force} = \text{time-rate of change of momentum},$$

or, on the assumption of constant mass

$$\text{force} = \text{mass} \times \text{acceleration}\ .$$

It follows that the unit of force is 1 kg m/s^2, but as that is somewhat cumbersome to pronounce and write, it is given a special name – the Newton (N). Thus 1 N = 1 kg m/s^2, and it is that force which, when applied to a mass of one kilogram, gives it an acceleration of 1 metre per second. Note that a space is left between the symbol of a unit and the preceding number.

Multiples of Units

In general, descriptions in terms of the SI units themselves are to be preferred, but this is not always convenient, so multiples and submultiples of the units may be formed by means of the prefixes listed in Table 1.2. However, to avoid errors in calculation it is important that all quantities should be expressed in terms of the units themselves before they are entered into calculations.

Table 1.2

Factor	Prefix	Symbol	Factor	Prefix	Symbol
10^{12}	tera	T	10^{-2}	centi	c
10^{9}	giga	G	10^{-3}	milli	m
10^{6}	mega	M	10^{-6}	micro	μ
10^{3}	kilo	k	10^{-9}	nano	n
10^{2}	hecto	h	10^{-12}	pico	p
10	deca	da	10^{-15}	femto	f
10^{-1}	deci	d	10^{-18}	atto	a

The symbol of a prefix is to be combined with the unit symbol to which it is directly attached, forming with it a new unit symbol which can be raised to a positive or negative power and which can be combined with other unit symbols to form symbols for compound units; for example,

$$1 \text{ cm}^3 = (10^{-2}\text{ m})^3 = 10^{-6}\text{ m}^3,\ 1\ \mu\text{s}^{-1} = (10^{-6}\text{ s})^{-1} = 10^6\text{ s}^{-1}\ .$$

N.B. (i) Compound prefixes must not be used (e.g. mμm).

(ii) Prefix symbols and unit symbols should not be written with space between them, e.g. 1 millinewton is 1 mN *not* 1 m N – the latter, in fact, means the product of metre and newton, the space implying multiplication. Products of unit symbols involving the metre should be written with m as the last term of the product, thus: kg m/s^2 *not* m kg/s^2, etc.

(iii) Plurals must *never* be used; e.g. 5 centimetres must not be abbreviated as 5 cms – otherwise the plural s may be confused with the symbol for the second.

(iv) Note that, although the kilogram is an SI base unit, one does not use mkg, kkg or the like.

When expressing the magnitude of a quantity, use units resulting in numerical values between 0.1 and 1000; for example

express	12000 N	as	12 kN
	0.00394 m		3.94 m
	0.0003 s		0.3 ms or 300 μs

1.4 DIRECTED QUANTITIES

Reflection will show that the fundamental units were all defined as scalars, since it was specified that the magnitude of a quantity is to be determined simply by the proportional scaling of an appropriate unit prototype, and when situations are described in these terms, they may be analysed by the algebra of numbers. However, there are many situations in which whole sets of scalar quantities can be considered to describe different aspects of a single quantity, and this is particularly true of a wide range of quantities which are associated with a direction as well as a magnitude. In mechanics, it is true of a force, for example, and in such a case it is commonly found that the task of assessment can be greatly eased by treating the set of scalars that together determine both the magnitude and the direction of the quantity as a single entity. But this advantage can be realised only if we first establish the principles of an algebra that is appropriate to this kind of quantity, and this is the subject of the next chapter on vector algebra. But vector algebra is itself largely concerned with the relative positions of the terminal points of straight lines, and since this concept of position is central to all mechanics, we preface our approach to vector algebra by a review of the concept of position, with particular regard to the frame of references in which the property is recognised, and to the systems of coordinates in which it is measured.

1.5 POSITION, FRAMES OF REFERENCE, AND COORDINATES

It is assumed that the general import of the word 'position' is understood intuitively. It incorporates the two ideas of distance and direction, and is essentially relative in that we can conceive its meaning only in a manner which relates the position of one thing to the position and orientation of another. Position is therefore a concept which assumes the existence of some appropriate frame of reference, and this raises the question as to what constitutes a frame that is adequate for the purpose.

Frames of Reference

We suppose a frame of reference to comprise a set of 'fixed' points O, X, Y etc. (i.e., a set of points between each pair of which the distance is assumed fixed), and we suppose that the position of any general point P relative to (or 'in'), the frame is to be determined by measuring its modular distance from each of the datum points.

With a single datum point O we could measure only the distance of P from O, so we could determine only that P lies at some unknown position on the surface of a particular sphere centred on O.

With two datum points O and X, we could measure the distance of P from each, and we could infer that P must lie simultaneously on the surfaces of two particular spheres, one centred on O, and the other on X. Excluding the unreal case in which the spheres do not meet, and the special case in which they just touch, this intersection would be a circle which lies in a plane normal to the straight line between the datum points, with its centre on the line.

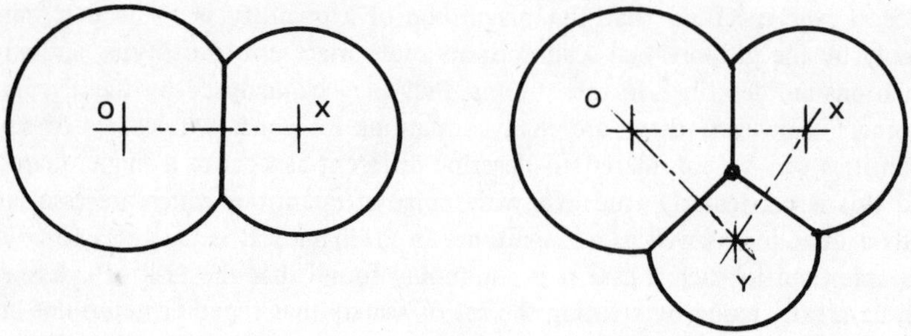

Fig. 1.1

With three datum points O, X and Y we could determine that P must lie simultaneously on the surfaces of three particular spheres, and again ignoring the special cases, this would imply that P must lie on the intersection of the surface of one of the spheres with the circle of intersection of the other two. Now reflection will show that this would restrict P to one of only two possible positions which lie symmetrically on either side of the plane through the three datum points, and to distinguish between them, it would be necessary to add a fourth datum point Z. But we note that the two possible positions would be equidistant from any point in the plane through O, X and Y, so it is necessary that the fourth datum point should not be coplanar with the other three.

We therefore conclude that the position of a point relative to a given frame of reference can be determined uniquely, provided that the frame of reference provides four non-coplanar datum points, between each pair of which the distance can be assumed to be fixed, and in practice, a frame of reference might typically be based on a set of 'fixed' stars or on four points in some substantial body such as the earth, which is supposed to be rigid.

But given the four 'fixed' points of the frame of reference, we are not restricted to measurements that are made to the points themselves. Rather, by drawing lines and planes through the datum points in various sets, we could construct a great variety of frames from which an appropriate set of distances and angles might be measured, and we shall represent a frame of reference notionally by a plane (which is supposed to be drawn through three of the datum points) together with a normal straight line (which is supposed to be dropped from the fourth datum point) as in Fig. 1.2.

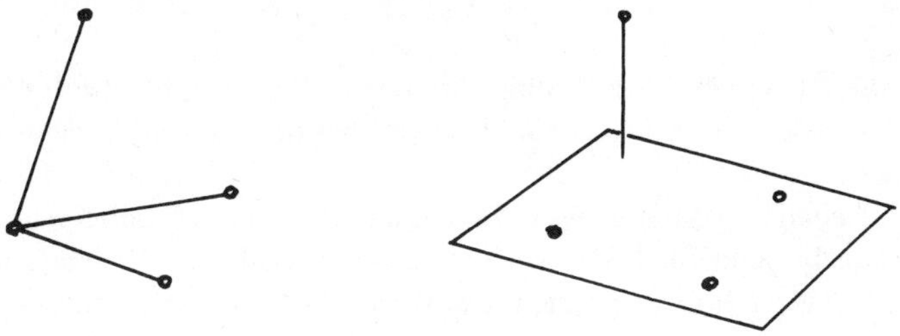

Fig. 1.2

But although a frame of reference must provide at least four datum points, the position of any point in the frame can, in fact, be determined by only three measurements. For we have seen that with only three datum points we are effectively in doubt, only as to which side of their plane the required point lay. Therefore, drawing a plane through each trio of datum points, we use the fourth only to distinguish the two domains that lie on either side of the plane. Then one domain is reckoned positive, and the other negative, and by thus ascribing a sense to each of three measurements, we obviate the need for a fourth.

Coordinate Systems

A coordinate system can be regarded as a formal scheme for the determination of position. It employs a set of coordinate planes and axes which are drawn in the frame of reference, and for each system these have a particular, agreed form, and are used to define a particular, agreed scheme of measurement.

The four most common systems of coordinates are the **Cartesian**, the **Rectangular Cartesian**, the **Cylindrical Polar**, and the **Spherical Polar**, as indicated in Fig. 1.3. It is assumed that these are systems with which you are already acquainted, but we note two features of particular interest.

First we note that the Cartesian systems are unique in that they alone make use solely of measurements of length that are all made in directions that are fixed in the frame of reference. In the cylindrical system, for example, there is only one fixed direction of measurement (that of z), whilst in the spherical

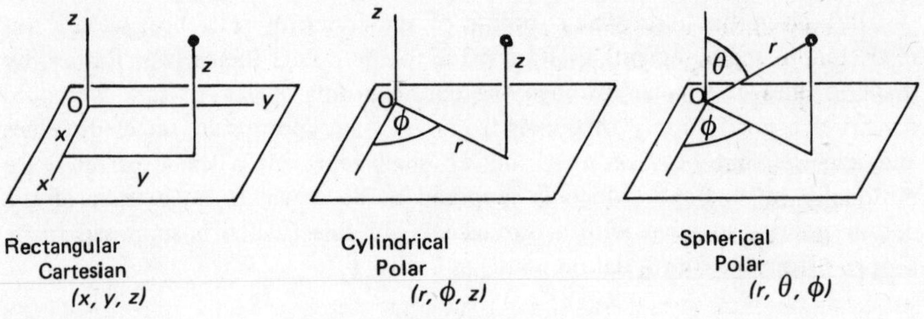

Fig. 1.3

system there is none. When dealing with quantities which themselves have directional properties, we can reasonably expect that this distinction may be of some consequence.

The second is concerned with the sense of any triad of directed lines, though we are largely concerned only with rectangular triads of coordinate directions. With these, we require to assign a positive sense to one of the two possible directions along each axis, and in principle this is a matter of arbitrary choice. Thus, with a given set of axes, we are free to assign the coordinate symbols at will, and in each case we are free to specify either direction along the axis as the positive, but Fig. 1.4 shows that in practice, the various sets of axes that may thus be generated by arbitrary choice fall into two groups. Within each group, one set of axes may be generated from any other, simply by rotation, but no set from one group may be so generated from any set of the other. Of itself, this distinction is not significant, but it will be necessary to define sign conventions for various kinds of quantities, and it falls out, that through an interaction between conventions, algebraic relationships for a set of axes from one of the

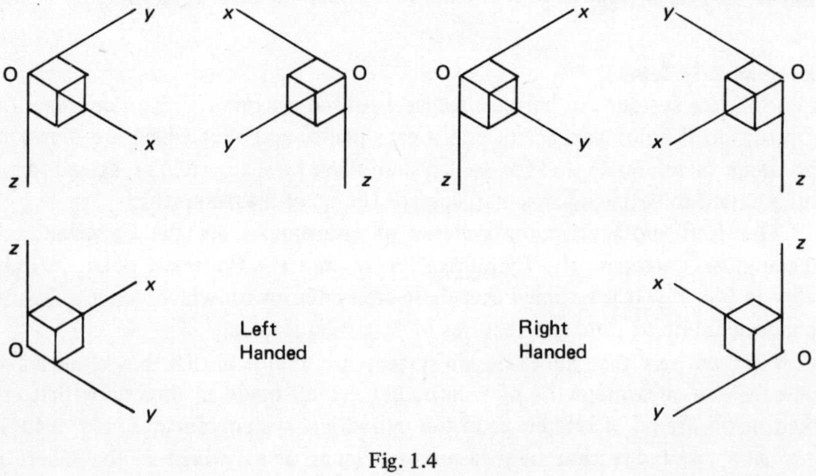

Fig. 1.4

groups may differ from the corresponding relationship for a set of axes from the other group. To avoid this issue we therefore always choose a set of axes from the same group – that which is described as right-handed.

Right-Handed Axes

To distinguish right-handed axes from left, we pay attention to the sequence in which the coordinate system requires the coordinates to be represented. Thus representing this coordinate sequence by (p, q, r), we may ascribe positive senses to the first two, p and q arbitrarily, but the sense of the third coordinate

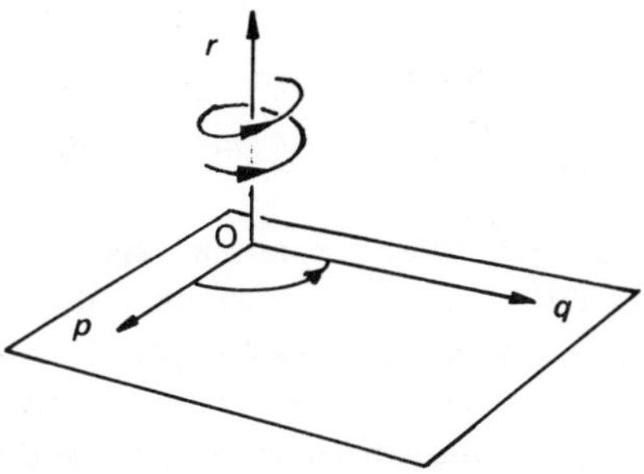

(p, q, r)

Fig. 1.5

r must then be determined by the rule of the right-hand screw. In this we imagine a right-handed screw with its axis along the r axis, when the sense of rotation that would bring the positive direction of the first axis p into coincidence with the positive direction along the second axis q after turning through an angle less than π would cause the screw to advance in the positive direction along the third axis r (Fig. 1.5).

2

Vector Algebra

2.1 INTRODUCTION

There are few directed quantities that are not amenable to vector algebra, but a statement which in one case can be unequivocal, in another may require a measure of qualification in order to take account of any effect of the location of the quantity, and to avoid the intrusion of this complication into the development of the basic algebra, it is convenient to develop the scheme, not in application to directed quantities in general, but only to a particular, geometrical quantity. Then, when we meet other kinds of directed quantities that are generally amenable to the same scheme, they will be referred to, not simply as 'vectors', but rather as **'vector quantities'**, and it will then be necessary to determine whether or not the effect of the location of the quantity requires separate consideration.

In the following development of this topic there are three distinct stages. In the first we define both vectors themselves, and a range of operations that may be performed upon them. Of course, these operations are chosen because they evaluate quantities that are generally found to be significant in a wide range of different situations, but they are still, in principle, arbitrary. It is therefore not to be expected that their significance will be immediately apparent, but will only emerge, case by case, in particular applications. These definitions are, of themselves sufficient to secure for the algebra of vectors a number of important algebraic advantages, but at the end of the algebraic analysis, it is usually necessary to revert to the algebra of numbers in order to reduce the result to scalar terms, and it is a second, important advantage of vector algebra that these computations also are reduced to particularly simple routines.

In the second stage we therefore first identify a standard means of describing a vector as a set of scalars. Then we describe the salient features of a vector as functions of these 'scalar elements', and we determine what may be called the scalar equivalents of the vector operations. These are simple routines of scalar operations on the scalar elements of vectors which evaluate the results of applying the more awkward vector operations, without actually performing them.

The third stage continues throughout the remainder of the text, where the vector algebra is applied to various situations, and it is only here that the significance of the quantities determined by the vector operations will emerge.

2.2 PROPERTIES OF DIRECTED STRAIGHT LINES

A directed straight line is one that is drawn in a specific sense, from a given initial point to a given final point. However, the identification of the terminal points themselves is awkward, particularly in a diagram in which the initial point of one line may coincide with the final point of another, so instead, the sense of a directed line is indicated by an arrowhead which is pointed away from the initial point and toward the final (Fig. 2.1).

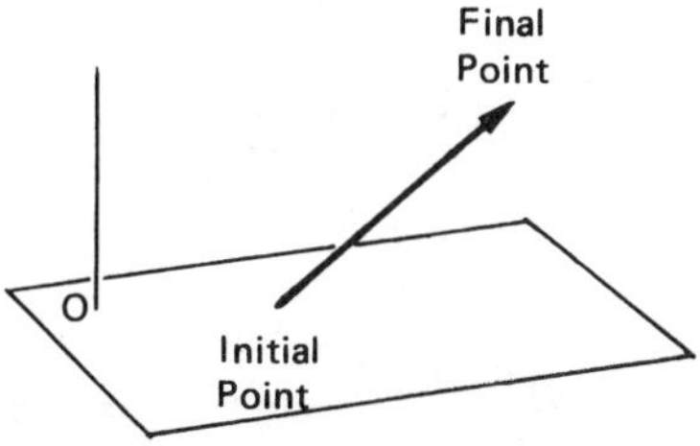

Fig. 2.1

Direction

The meaning of the word 'direction' is taken to incorporate the idea of sense as well as inclination. Parallel lines are therefore said to have the same direction only when they are pointed in the same sense, whilst parallel lines whose senses are opposed are said to have opposite directions.

Modulus

The modulus of a quantity describes its size, without regard to its sense or inclination. It is therefore a scalar which is taken to be essentially positive, and the modulus of a quantity is indicated by enclosing the symbol which represents the quantity within vertical strokes.

Coefficients

If a directed line is parallel to a given reference direction, the sense of the one may be compared with that of the other, and the quantity that associates the modulus of the line with its sense, as thus determined, is called the coefficient of the line in the given direction. The coefficient of a directed line can therefore be either positive or negative, according to whether its sense agrees or disagrees with that of the parallel reference direction. In Fig. 2.2, for example, $\vec{AB}$ and $\vec{CD}$ are both parallel to the reference direction $\vec{OX}$, but whereas $\vec{AB}$ points in the same sense as $\vec{OX}$, $\vec{CD}$ points in the opposite sense, so that whereas the coefficient of $\vec{AB}$ in the direction $\vec{OX}$ is equal to its modulus, that of $\vec{CD}$ is equal to the negative of its modulus.

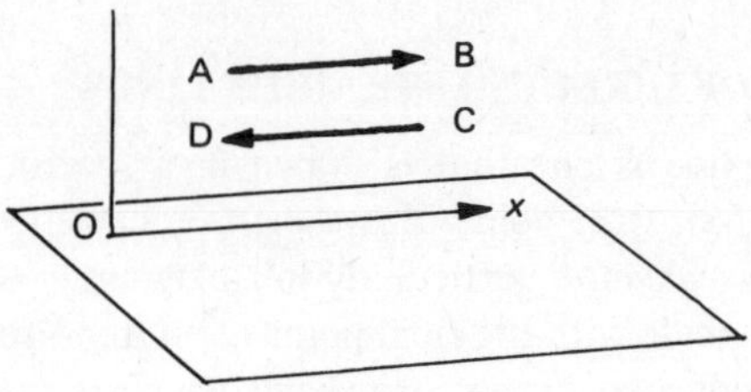

Fig. 2.2

2.3 VECTORS

The geometrical properties of a directed line in a given frame of reference are determined by its modulus, its direction and its location, but of these, the vector of the line is defined to include only the modulus and the direction.

Vector Equivalence

Since the location of a line is specifically excluded from its vectorial properties, it follows, that where the location of the line is relevant to the issue of interest, it will be necessary to bring it into account as a separate consideration. In the event, we shall find that this is easy to do because location can itself be treated as a vector, but that is a question for the future, and for the present it is sufficient to note that the modulus and direction alone comprise the vector properties. The vectors of two lines are therefore equivalent if they share the same modulus and direction, irrespective of their locations, and it follows, first, that the vector of a line is unaltered by a parallel displacement, and second, that the value of a vector can as well be varied by a change of direction as by a change of modulus. A vector can therefore be treated as a constant only when it is fixed in both modulus and direction.

Null or Zero Vectors

A **null** or **zero vector** is a vector of zero modulus; it is one in which the final point effectively coincides with the initial point.

Unit Vectors

A **unit vector** is a vector of unit modulus, and unit vectors are of particular importance because they afford the standard means of defining directions.

Symbols

Just as symbols can be used to represent numbers and operations in the algebra of numbers, so can they be used to represent vectors and operations in the algebra of vectors. However, we shall find, that although it comprises a quite different scheme, the algebra of vectors has adopted many of the terms and operational symbols that are already in use in the algebra of numbers. For example, we talk of the addition of vectors, just as we talk of the addition of scalars, and both these operations are indicated by the same symbol; yet vector addition is different from scalar, and since we do not distinguish between the symbols which indicate the different sets of operations, it is essential that we clearly distinguish between the symbols which represent the quantities to which they are applied.

Symbols which represent vectors are indicated in bold type, with the addition of a circumflex to distinguish unit vectors. Thus if **a** represents a general vector, its modulus is indicated by $|\mathbf{a}|$, and the unit vector in the same direction by $\hat{\mathbf{a}}$.

2.4 OPERATIONS ON VECTORS

It is taken as understood, that just as operations on scalars are always to be interpreted in the scalar manner, so are operations on vectors always to be interpreted in the vectorial manner, as defined in the following.

Vector Addition

In the vector sum $(\mathbf{a} + \mathbf{b})$, the vector **b** is added to vector **a** by first giving it a parallel displacement which bring its initial point into coincidence with the final point of vector **a**, as indicated in Fig, 2.3(a). Then the sum $(\mathbf{a} + \mathbf{b})$ is defined as the vector which is drawn from the initial point of the vector **a** to the final point of the vector **b**.

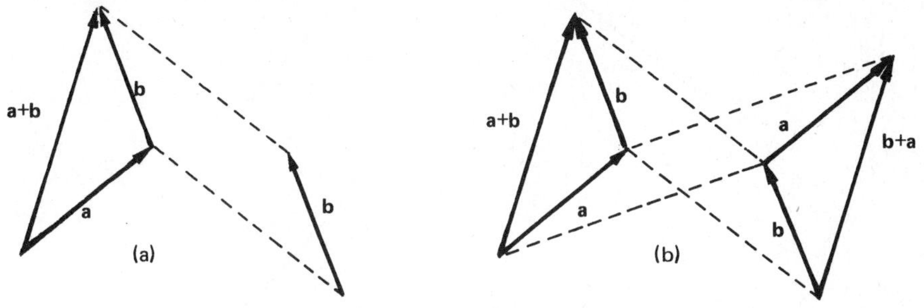

Fig. 2.3

Fig. 2.3(b) shows that if we now form the sum $(\mathbf{b} + \mathbf{a})$ by adding **a** to **b**, the result is vectorially equivalent to $(\mathbf{a} + \mathbf{b})$. The addition of vectors, like the addition of scalars is therefore seen to be commutative.

that is $$\mathbf{a} + \mathbf{b} = \mathbf{b} + \mathbf{a} \tag{2.1}$$

The Negative of a Vector

The negative of a quantity can be defined as the quantity that must be added to the positive to give zero result. But with vectors, the negative is added to the positive by first bringing its initial point into coincidence with the final point of the positive, and for zero result, the final point of the negative must then

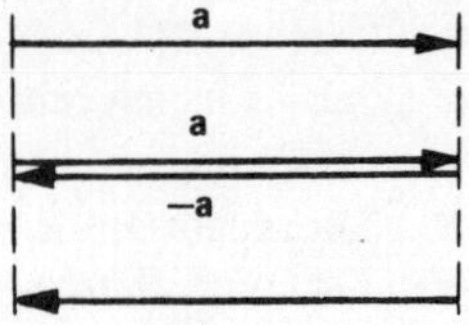

Fig. 2.4

close back to the initial point of the positive. The negative of a vector is therefore defined to be equal but opposite to its positive, and it can evidently be formed from the positive either by interchanging its terminal points, or, what is equivalent, by reversing its arrowhead.

Vector Difference

To evaluate the vector difference $(\mathbf{a} - \mathbf{b})$, the vector $\mathbf{b}$ is subtracted from the vector $\mathbf{a}$ by adding its negative (Fig. 2.5(a)).

that is $$\mathbf{a} - \mathbf{b} = \mathbf{a} + (-\mathbf{b}) \tag{2.2a}$$

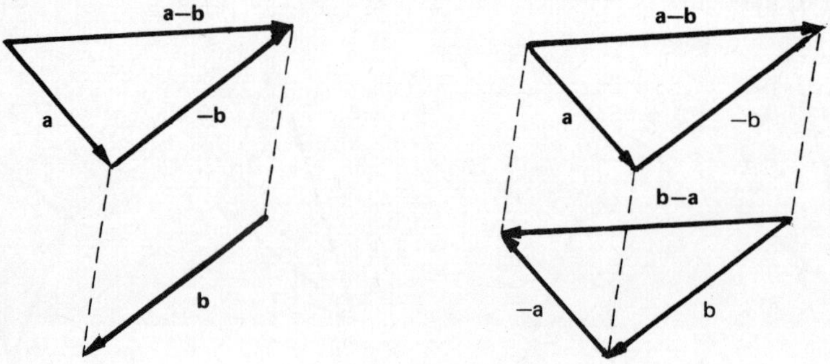

Fig. 2.5

However if $\mathbf{a}$ is now subtracted from $\mathbf{b}$ to form the difference $(\mathbf{b} - \mathbf{a})$, the result is the negative of $(\mathbf{a} - \mathbf{b})$ (Fig. 2.5(b)). A vector difference is therefore not commutative; rather

$$(\mathbf{a} - \mathbf{b}) = -(\mathbf{b} - \mathbf{a}) \tag{2.2b}$$

Multiplication by a number

If a vector a is added to itself, the sum has the same direction as a but twice the modulus, and it could reasonably be called the vector 2a. On the other

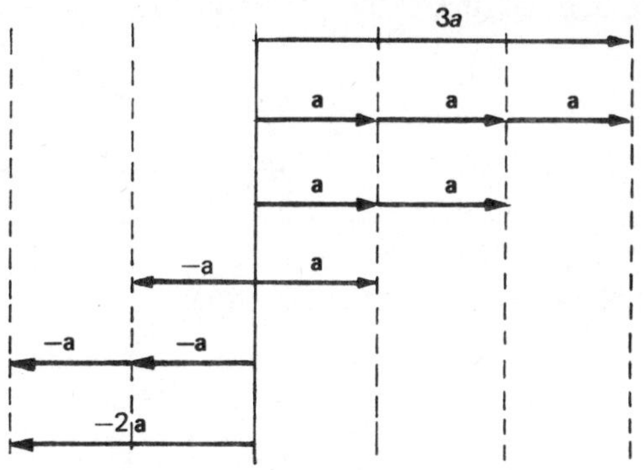

Fig. 2.6

hand, the negative of a vector differs from its positive only in respect of their senses, so recognising that a negative multiplier can be equated to −1 times its positive, we specify as follows:

In multiplying a vector, a positive scalar multiplier operates only on the modulus of the vector, whilst the negative of unity operates only to reverse the sense of the vector.

Alternatively, but equivalently, a scalar multiplier (whether positive or negative) can be considered to operate on the coefficient of the vector in its own direction.

The rule implies that a vector can be equated to its modulus times the unit vector in its own direction, that is, that $\mathbf{a} = |\mathbf{a}|.\hat{\mathbf{a}}$.

Scalar Products of Vectors

A scalar product of vectors is indicated by a dot which is placed, like a decimal point, between the vectors to which the operation is applied. It is defined as the

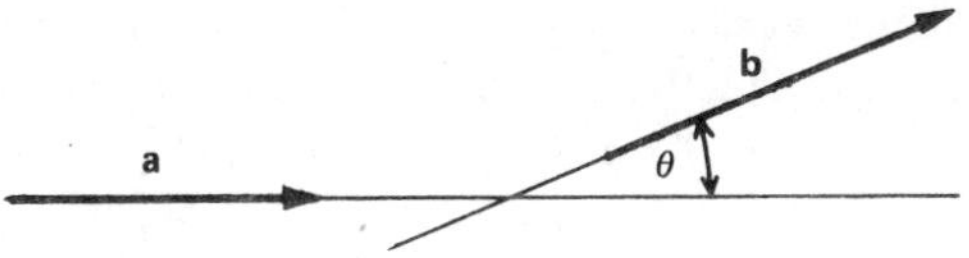

Fig. 2.7

product of the moduli of the vectors and the cosine of the angle between, and being a product of scalars, the result is itself a scalar, and the operation is commutative.

$$\text{Thus } \mathbf{a} \cdot \mathbf{b} = |\mathbf{a}|\,|\mathbf{b}|\cos\theta \qquad (2.3a)$$

$$\text{And } \mathbf{a} \cdot \mathbf{b} = \mathbf{b} \cdot \mathbf{a} \qquad (2.3b)$$

It can be shown that the scalar product is also distributive, that is, that the product of a vector **a** with the sum (**b** + **c**) of two others is equal to the sum of the products, of **a** with **b**, and of **a** with **c**. Thus in Fig. 2.8, let θ_b, θ_c and θ_{b+c} be the angles which **b**, **c** and (**b** + **c**) make with the vector **a**.

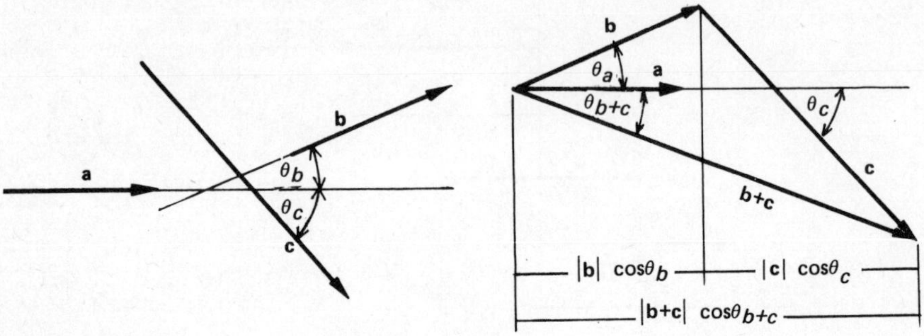

Fig. 2.8

By definition $\mathbf{a} \cdot (\mathbf{b} + \mathbf{c}) = |\mathbf{a}|\,|\mathbf{b} + \mathbf{c}|\cos\theta_{b+c}$

But it is obvious from Fig. 2.8 that

$$|\mathbf{b} + \mathbf{c}|\cos\theta_{b+c} = |\mathbf{b}|\cos\theta_b + |\mathbf{c}|\cos\theta_c$$

Therefore $\mathbf{a} \cdot (\mathbf{b} + \mathbf{c}) = |\mathbf{a}|\,|\mathbf{b}|\cos\theta_b + |\mathbf{a}|\,|\mathbf{c}|\cos\theta_c$

that is $$\mathbf{a} \cdot (\mathbf{b} + \mathbf{c}) = (\mathbf{a} \cdot \mathbf{b}) + (\mathbf{a} \cdot \mathbf{c}) \tag{2.3c}$$

Scalar Product of Rectangular Vectors

Since the cosine of $\pi/2$ is zero, the scalar product of any pair of rectangular vectors is zero

Scalar Product of Parallel Vectors

Parallel vectors may have directions which are either equal or opposite, and we note that whereas the cosine of zero is unity, the cosine of π is the negative of unity. The scalar product of a pair of parallel vectors will therefore be the scalar quantity equal either to the product of their moduli or to its negative, according as the directions of the vectors are either equal or opposite.

The Square of a Vector

The scalar product of a vector **a** with itself is called the square of **a**, and is indicated by $\mathbf{a}^2$. The result is evidently the scalar quantity equal to the square of the modulus of **a**, and in particular, the square of any unit vector is equal to scalar unity.

Thus $\mathbf{a}^2 = \mathbf{a} \cdot \mathbf{a} = |\mathbf{a}|^2$ (2.4a)

and $\hat{1}^2 = \hat{1} \cdot \hat{1} = 1$ (2.4b)

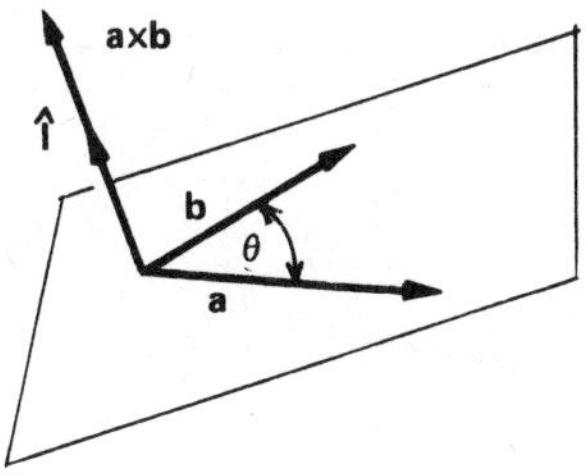

Fig. 2.9

The Vector Product of Vectors

A vector product of two vectors is indicated by the same cross as is used to indicate multiplication in the algebra of numbers. Thus the vector product of **a** with **b** is indicated by **a** × **b**, which is read as **a** 'cross' **b**, in order to avoid confusion with the scalar product **a** · **b** which is read as **a** 'dot' **b**.

The vector product **a** × **b** is defined as the vector whose modulus is equal to the product of the modulus of **a**, the modulus of **b** and the sine of the smaller angle between their positive directions. Its direction is at right-angles to both **a** and **b**, in the sense for which (**a**, **b**, **a**×**b**) constitutes a right-handed set (Fig. 2.9).

Thus if $\hat{1}$ represents the unit vector at right angles to both of the vectors **a** and **b** in the sense such that (**a**, **b**, $\hat{1}$) comprises a right-handed set, and if θ represents the smaller of the two angles between the positive directions of **a** and **b**

By definition: $\mathbf{a} \times \mathbf{b} = |\mathbf{a}| . |\mathbf{b}| . \sin\theta . \hat{1}$ (2.5a)

We note that whereas the cosine of $(2\pi - \theta)$ is equal to the cosine of θ, the sine of $(2\pi - \theta)$ is the negative of the sine of θ. In defining the vector product it is therefore necessary to distinguish between the larger and the smaller of the two angles subtended by the vectors concerned, and although in principle, it is not necessary to draw the same distinction in defining a scalar product, in practice it is probably safer always to use the smaller of the two angles.

It follows from the definition, that, **b** × **a** has the same modulus as **a** × **b**, but the opposite direction. A vector product is therefore not commutative, but rather

$$(\mathbf{b} \times \mathbf{a}) = -(\mathbf{a} \times \mathbf{b}) \qquad (2.5b)$$

On the other hand, a vector product is distributive, and to show that this is so, we first show that the product of one vector **a** with another **b** is equal to the product of **a** with the projection of **b** on the plane normal to **a**. Thus in Fig. 2.10, **b**$'$ is the projection of **b** on the plane normal to **a**, and if θ represents the

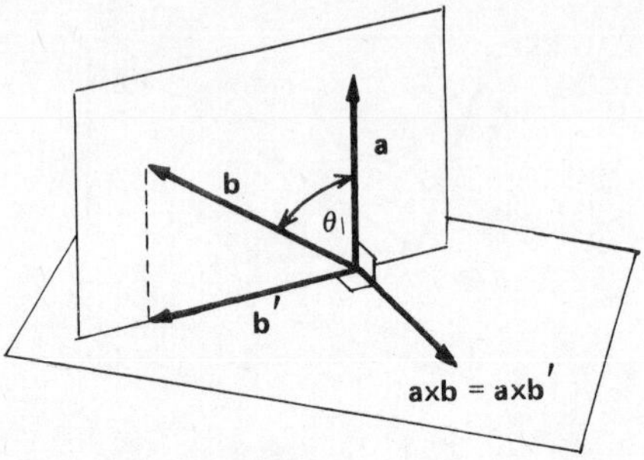

Fig. 2.10

inclination of **b** to **a**, it is obvious that $|\mathbf{b}'| = |\mathbf{b}|.\sin\theta$. Now **b**$'$ is at right-angles to **a**, so the modulus of $\mathbf{a} \times \mathbf{b}'$ is given by $|\mathbf{a}|.|\mathbf{b}'| = |\mathbf{a}|.|\mathbf{b}|.\ \sin\theta$, and this is evidently equal to the modulus of $\mathbf{a} \times \mathbf{b}$. Furthermore, since **a**, **b** and **b**$'$ are co-planar, the direction of $\mathbf{a} \times \mathbf{b}'$ is also equal to that of $\mathbf{a} \times \mathbf{b}$, and we conclude that $\mathbf{a} \times \mathbf{b}' = \mathbf{a} \times \mathbf{b}$.

Now consider Fig. 2.11, where **a**, **b** and **c** are arbitrary vectors, and **b**$'$, **c**$'$ and $(\mathbf{b}' + \mathbf{c}')$ are the projections of **b**, **c** and their sum $(\mathbf{b} + \mathbf{c})$ on the plane normal to **a**. By definition, the vector products of **a** with **b**$'$, **c**$'$ and $(\mathbf{b}' + \mathbf{c}')$ all lie in the plane normal to **a**, in directions normal to **b**$'$, **c**$'$, and $(\mathbf{b}' + \mathbf{c}')$

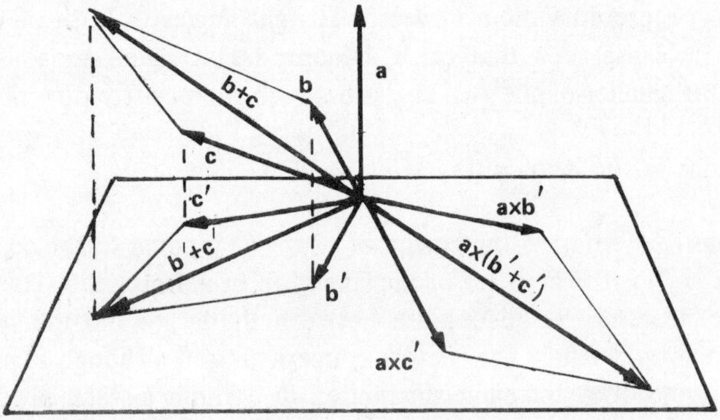

Fig. 2.11

themselves. Therefore, just as $(\mathbf{b}' + \mathbf{c}')$ constitutes the diagonal of the parallelogram formed on **b**$'$ and **c**$'$, so $\mathbf{a} \times (\mathbf{b}' + \mathbf{c}')$ constitutes the diagonal of the paralellogram formed on $(\mathbf{a} \times \mathbf{b}')$ and $(\mathbf{a} \times \mathbf{c}')$, and it follows from the rule of vector addition that the product of **a** with the sum $(\mathbf{b}' + \mathbf{c}')$ may be equated to the sum of its several products with $\mathbf{b}'$ and with $\mathbf{c}'$. But we have seen that the products of $\mathbf{a}$ with the projections are equal to its products with $\mathbf{b}$, $\mathbf{c}$ and $(\mathbf{b} + \mathbf{c})$ themselves, and we conclude that a vector product is distributive over a sum generally, that is, that

$$\mathbf{a} \times (\mathbf{b} + \mathbf{c}) = (\mathbf{a} \times \mathbf{b}) + (\mathbf{a} \times \mathbf{c}) \tag{2.5c}$$

Vector Product of Parallel Vectors

We note that the sine of π, like the sine of zero, is zero, so the cross product of any pair of parallel vectors is zero, irrespective of whether their directions are equal or opposite.

Vector Product of Unit Rectangular Vectors

Since the sine of $\pi/2$ is unity, it follows that the modulus of the cross product of any rectangular pair of unit vectors will also be equal to unity, and we recognise that the cross product of any rectangular pair of unit vectors determines the third member of a right-handed, rectangular triad.

Components

Any set of vectors whose sum is equal to a given vector are said to comprise a **set of components** of the given vector. However, in practice, we are seldom concerned with any but sets which are mutually rectangular, and such sets may include either two components or three (Fig. 2.12).

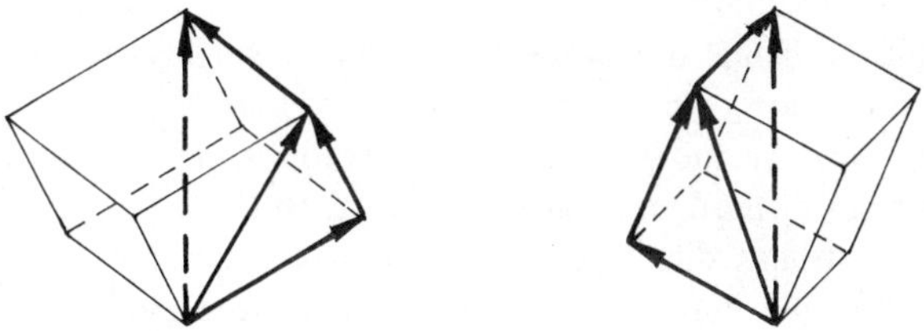

Fig. 2.12

Resolved Components

Evidently, a given vector $\mathbf{a}$ can be equated to an indefinite number of different sets of rectangular components, but consider only those that include one component in a given direction which we may conveniently suppose to be defined by a unit vector $\hat{\mathbf{l}}$. The components of each set being mutually rectangular, it follows in each case that the components other than the one in the fixed directions must lie in a plane which is normal to the direction. Furthermore, since the sum of each set is equal to the given vector **a**, it follows that the terminal points of the sum of each set must coincide with those of **a**. If, therefore, the component in the given direction is drawn with its initial point on the initial point of **a**, its final point must lie at its intersection with the plane through the final point of **a**, normal to the given direction. But this is evidently a unique point, and we conclude that the component in the given direction has a unique value which is common to all the sets.

This is called the component of **a**, resolved in the given direction.

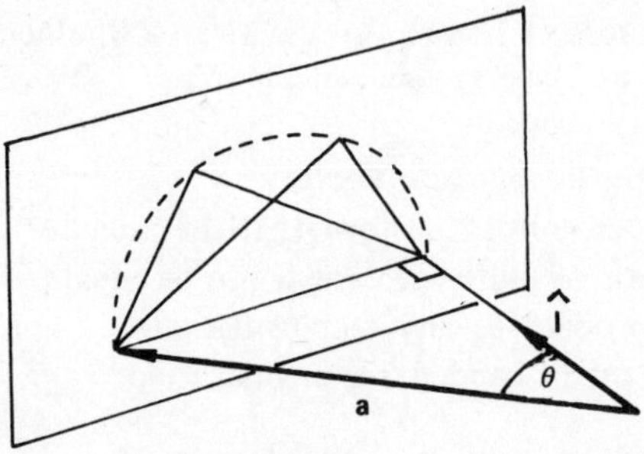

Fig. 2.13

Scalar Components

The coefficient of a resolved component in the direction of resolution, is called the **scalar component** of the vector in that direction.

Fig. 2.13 shows that a scalar component of a vector is equal to the modulus of the vector itself times the cosine of the angle which it makes with the direction of resolution. Thus, if θ is the angle which a vector **a** makes with a given direction of resolution, the scalar component of **a** in that direction is equal to $|\mathbf{a}|.\cos\theta$. But since the modulus of a unit vector is unity, it follows from the definition of a scalar product, that this is equal to the scalar product of **a** with the unit vector in the direction of resolution, so the scalar component of a vector **a** in a direction defined by a unit vector $\hat{\mathbf{l}}$ is given by $\mathbf{a} \cdot \hat{\mathbf{l}}$, the scalar product of the two.

2.5 SCALAR EQUIVALENTS IN FIXED, RECTANGULAR COORDINATES

In applying the vector algebra, we are free to choose any system of coordinates that best reflects the general form of the situation under investigation, but no matter what the form of the coordinates, the vector is always described as a set of scalars by specifying its scalar components in a set of rectangular directions. Of course, these directions are chosen to accord with the directions of the coordinates themselves, and since these are not generally fixed in the frame of reference, it follows that it is generally appropriate to describe different vectors in terms of their components in sets of rectangular directions whose orientations vary with the directions of the vectors themselves. However, the rectangular system is a special case which is of great vectorial convenience, because it makes use solely of coordinate directions which are fixed in the frame of reference, and with these it is obviously appropriate to describe all vectors in terms of their components in the same set of fixed directions, as defined by the coordinates.

We also note that any change in a vector would appear different to different observers, depending on their relative rotation. That this is so is obvious, for if two observers suffer a relative rotation, then a vector which was fixed relative to one, would rotate relative to the other.

For the present we shall therefore consider only the special case of rectangular coordinates which, being drawn in a frame of reference which is fixed relative to the observer, are themselves fixed relative to the observer. Thus we now defer a consideration of two cases: of non-rectangular coordinates which are drawn in the observer's frame of reference, and of any coordinates which are drawn in a frame of reference which moves relative to the observer, and we take it as understood that throughout this chapter we refer only to the special case.

The Unit Vectors i, j and k

Unit vectors in the directions of fixed rectangular coordinates (x, y, z) are indicated by **i**, **j** and **k** respectively, and since these symbols are specifically reserved for this purpose, it is not necessary to add the circumflex that is otherwise used to distinguish unit vectors.

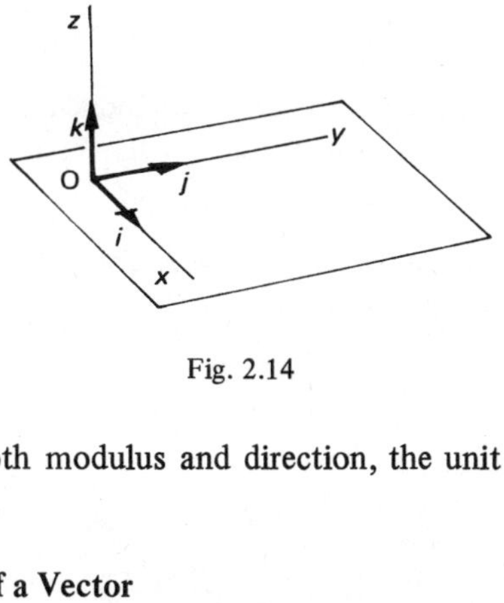

Fig. 2.14

Being fixed in both modulus and direction, the unit vectors **i**, **j** and **k** are constants.

The Scalar Elements of a Vector

The scalar component of a vector in a given direction has a unique value. Therefore, when all vectors are described in terms of their scalar components in a fixed set of coordinate directions, each vector is uniquely associated with a particular set of scalar components and these are called the scalar elements of the vector in these coordinates. Thus if a_x, a_y and a_z are the coefficients of the components of a vector **a** in the directions of x, y and z coordinates in turn, as indicated in Fig. 2.15, then (a_x, a_y, a_z) comprise the scalar elements of **a** in those coordinates.

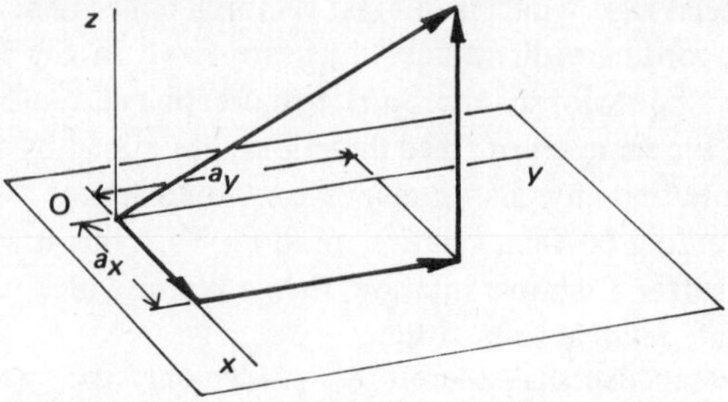

Fig. 2.15

By the definition of multiplication by a number, a vector can be equated to its modulus times the unit vector in its own direction. So just as **a** can be

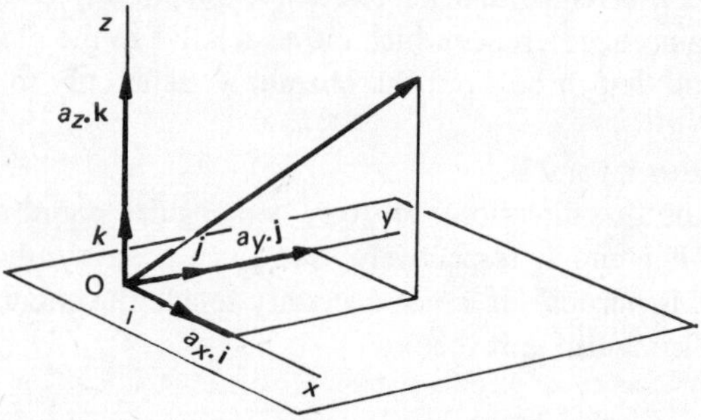

Fig. 2.16

equated to $|\mathbf{a}|.\hat{\mathbf{a}}$. so can the vectors of its components be equated to $a_x.\mathbf{i}$, $a_y.\mathbf{j}$ and $a_z.\mathbf{k}$, and equating the vector to the vector sum of its components, we have

$$\mathbf{a} = |\mathbf{a}|.\hat{\mathbf{a}} = a_x.\mathbf{i} + a_y.\mathbf{j} + a_z.\mathbf{k} \tag{2.6}$$

Evidently, the scalar elements of the vector of a line can be determined by subtracting the coordinates of its initial point from those of its final point.

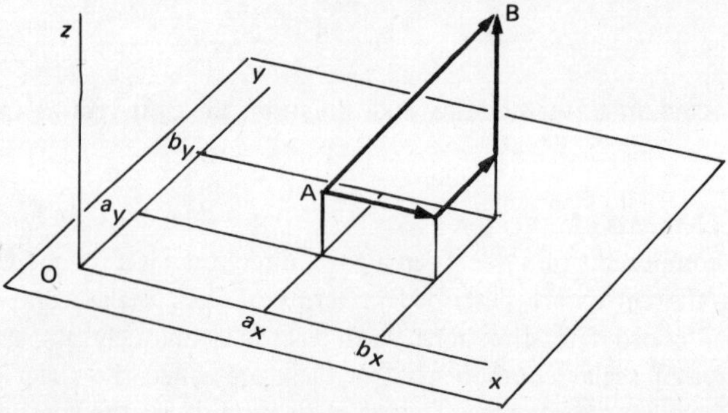

Fig. 2.17

However a vector which is drawn from the origin of coordinates is called a **radius vector**, and the scalar elements of a radius vector can evidently be equated simply to the coordinates of the point to which it is drawn.

The Modulus of a Vector, and the Coincident Unit Vector

By Pythagoras' Theorem, the square of the modulus of a vector can be equated to the sum of the squares of any of its sets of rectangular components, and in particular to the sum of the squares of its scalar elements in any rectangular coordinates

Thus $$|\mathbf{a}| = \sqrt{(a_x^2 + a_y^2 + a_z^2)} \tag{2.7}$$

But $$\mathbf{a} = |\mathbf{a}|.\hat{\mathbf{a}}$$

Therefore $$\hat{\mathbf{a}} = |\mathbf{a}|^{-1}.\mathbf{a}$$

or $$\hat{\mathbf{a}} = \frac{a_x}{|\mathbf{a}|}.\mathbf{i} + \frac{a_y}{|\mathbf{a}|}.\mathbf{j} + \frac{a_z}{|\mathbf{a}|}.\mathbf{k} \tag{2.8}$$

We therefore conclude that the modulus of a vector can be equated to the square root of the sum of the squares of its scalar elements, and that the scalar elements of the unit vector in its own direction can be determined by dividing each of the elements of the given vector by the square root of the sum of their squares.

The Direction Cosines of a Line and the Coincident Unit Vector

The direction of a line in a given system of coordinates could be defined by the angles θ_x, θ_y and θ_z which it makes with each of the coordinate axes in turn (Fig. 2.18). However, it is more convenient to specify, not the angles themselves

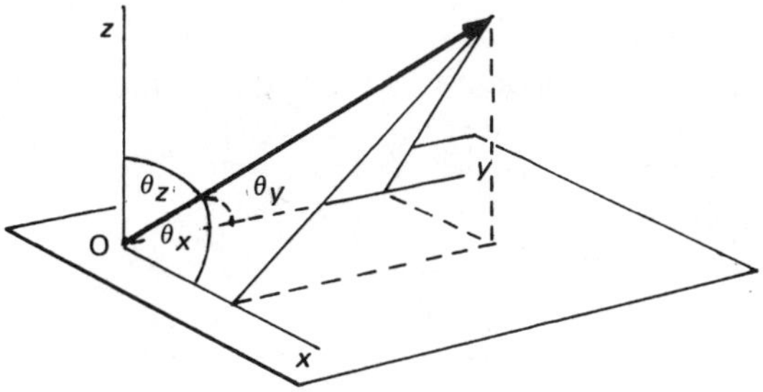

Fig. 2.18

but their cosines, and these are called the **direction cosines** of the line in the given coordinates, and are represented by l_x, l_y and l_z respectively.

We note that a scalar component of a vector can be equated to its modulus times the cosine of the angle which it makes with the direction of resolution, and since the modulus of a unit vector is unity, it follows that the scalar elements of a unit vector are equal to the cosines of the angles which the vector makes with the coordinate axes. The scalar elements of a unit vector in given rectangular coordinates are therefore equal to its direction cosines in the same coordinates.

The Scalar Equivalent of Vector Addition

From Fig. 2.19 it is obvious that when two vectors are added, each of the

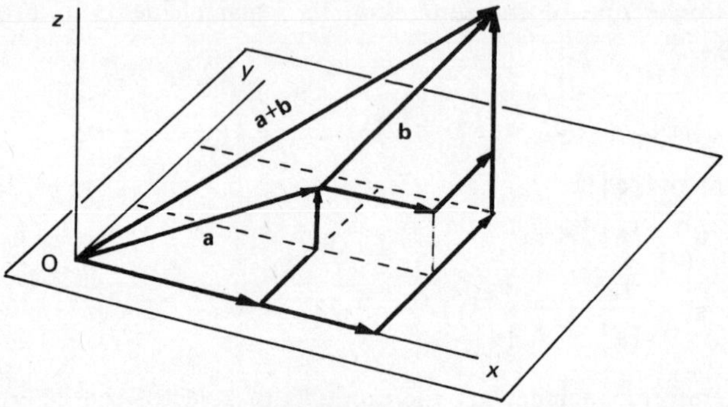

Fig. 2.19

scalar elements of the sum (**a** + **b**) is equal to the sum of the corresponding elements of **a** and of **b**.

Thus if $\mathbf{a} = a_x.\mathbf{i} + a_y.\mathbf{j} + a_z.\mathbf{k}$

and if $\mathbf{b} = b_x.\mathbf{i} + b_y.\mathbf{j} + b_z.\mathbf{k}$

Then $$\mathbf{a} + \mathbf{b} = (a_x + b_x).\mathbf{i} + (a_y + b_y).\mathbf{j} + (a_z + b_z).\mathbf{k} \quad (2.9)$$

The Negative of a Vector

The scalar elements of the negative of a vector are equal to the negatives of the scalar elements of the given vector.

Vector Difference

The scalar elements of the vector difference (**a** − **b**) may be determined by subtracting each scalar element of **b** from the corresponding scalar element of **a**.

Multiplication by a Number

When a vector is multiplied by a number, the multiplier operates on the coefficient of the vector but leaves its inclination unaltered. The operation is therefore one which leaves the proportions of the components of the vector unaltered, and a vector may therefore be multiplied by a number, by multiplying each of its scalar elements by the number.

Scalar Equivalent of a Scalar Product

Describing each vector as the sum of its components in the directions of the coordinates, we have

$$\mathbf{a} \cdot \mathbf{b} = (a_x.\mathbf{i} + a_y.\mathbf{j} + a_z.\mathbf{k}) \cdot (b_x.\mathbf{i} + b_y.\mathbf{j} + b_z.\mathbf{k}) \ ,$$

But the scalar product is distributive over a sum, and taking the product of each component of **a** with each component of **b** in turn we have

$$\begin{aligned} \mathbf{a} \cdot \mathbf{b} &= a_x b_x.\mathbf{i}\cdot\mathbf{i} + a_x b_y.\mathbf{i}\cdot\mathbf{j} + a_x b_z.\mathbf{i}\cdot\mathbf{k} \\ &+ a_y b_x.\mathbf{j}\cdot\mathbf{i} + a_y b_y.\mathbf{j}\cdot\mathbf{j} + a_y b_z.\mathbf{j}\cdot\mathbf{k} \\ &+ a_z b_x.\mathbf{k}\cdot\mathbf{i} + a_z b_y.\mathbf{k}\cdot\mathbf{j} + a_z b_z.\mathbf{k}\cdot\mathbf{k} \end{aligned}$$

But we know that the scalar product of rectangular vectors is zero, and that the square of a unit vector is scalar unity.

Thus $\quad \mathbf{i}\cdot\mathbf{j} = \mathbf{j}\cdot\mathbf{i} = \mathbf{i}\cdot\mathbf{k} = \mathbf{k}\cdot\mathbf{i} = \mathbf{j}\cdot\mathbf{k} = \mathbf{k}\cdot\mathbf{j} = 0,$

and $\quad \mathbf{i}\cdot\mathbf{i} = \mathbf{j}\cdot\mathbf{j} = \mathbf{k}\cdot\mathbf{k} = 1,$

and substituting accordingly gives

$$\mathbf{a}\cdot\mathbf{b} = a_x . b_x + a_y \cdot b_y + a_z . b_z \tag{2.10}$$

The scalar product $\mathbf{a}\cdot\mathbf{b}$ can therefore be equated to the sum of the products of the corresponding pairs of scalar element of **a** and **b**.

Scalar Equivalent of the Vector Product

The scalar equivalent of a vector product is again determined by making use of its distributive property.

Thus $\quad \mathbf{a} \times \mathbf{b} = (a_x.\mathbf{i} + a_y.\mathbf{j} + a_z.\mathbf{k}) \times (b_x.\mathbf{i} + b_y.\mathbf{j} + b_z.\mathbf{k})$

that is

$$\begin{aligned} \mathbf{a} \times \mathbf{b} &= a_x b_x.\mathbf{i}\times\mathbf{i} + a_x b_y.\mathbf{i}\times\mathbf{j} + a_x b_z.\mathbf{i}\times\mathbf{k} \\ &+ a_y b_x.\mathbf{j}\times\mathbf{i} + a_y b_y.\mathbf{j}\times\mathbf{j} + a_y b_z.\mathbf{j}\times\mathbf{k} \\ &+ a_z b_x.\mathbf{k}\times\mathbf{i} + a_z b_y.\mathbf{k}\times\mathbf{j} + a_z b_z.\mathbf{k}\times\mathbf{k} \ . \end{aligned}$$

But since $(\mathbf{i}, \mathbf{j}, \mathbf{k})$ form a right-handed set of rectangular unit vectors.

$$\mathbf{i} \times \mathbf{i} = \mathbf{j} \times \mathbf{j} = \mathbf{k} \times \mathbf{k} = 0$$

and

$$\left.\begin{aligned} \mathbf{i} \times \mathbf{j} &= -(\mathbf{j} \times \mathbf{i}) = \mathbf{k}. \\ -(\mathbf{i} \times \mathbf{k}) &= (\mathbf{k} \times \mathbf{i}) = \mathbf{j} \\ \mathbf{j} \times \mathbf{k} &= -(\mathbf{k} \times \mathbf{j}) = \mathbf{i} \end{aligned}\right\}$$

Substituting accordingly then gives

$$\mathbf{a}\times\mathbf{b} = (a_y b_z - a_z b_y).\mathbf{i} - (a_x b_z - a_z b_x).\mathbf{j} + (a_x b_y - a_y b_x).\mathbf{k} \tag{2.11a}$$

However, the right-hand side can be recognised as the determinant of a matrix in which **i**, **j** and **k** appear in the first row, with the scalar elements of **a** and **b** in the second and third rows respectively.

Thus

$$\mathbf{a} \times \mathbf{b} = \begin{vmatrix} \mathbf{i} & \mathbf{j} & \mathbf{k} \\ a_x & a_y & a_z \\ b_x & b_y & b_z \end{vmatrix} \tag{2.11b}$$

2.6 VECTOR ANALYSIS

The principles of infinitesimal calculus can as well be applied to vectors as to scalars, but this is a calculus of variations, so it is particularly significant, that whereas a variation in a scalar is independent of the observer, a variation in a vector may appear different to different observers, depending on their relative rotation. Moreover, with all systems of coordinates but one it is appropriate to describe a vector in terms of its components in a set of directions which vary with the direction of the vector itself, and for both these reasons, the procedures by which the calculus is applied to vectors are generally different from those that apply to scalars. However, when all operations are referred to an observer who uses fixed rectangular coordinates, the operations of differentiation and integration can be applied to a vector simply by applying the scalar operations to each of the scalar elements of the vector in turn, and we now review the principles of vector analysis, chiefly to justify this manner of application, but partly to define the terminology that this text adopts.

Constants, Variables and Parameters

In analysis, a system is conceived as an aggregation of parts which have specific characteristics, and which interact with one another and the surroundings in specific ways. Some of these interactions stem from the operation of the natural laws, whilst others are only contrived in the manner in which the parts of the particular system are assembled, but in either case they determine the behaviour of the system and limit the range of configurations that are open to it, and the object of analysis is to relate the characteristics of the system as a whole to the characteristics of its parts, and the interactions to which they are liable.

In this, the crucial step is that of reflecting the essential features of the situation in a mathematical model. This is couched in terms of quantities of which some describe features which are constant, whilst others describe features which may vary from one configuration or 'state' of the system to another. However, we commonly require to refer to a special case in which a quantity that is generally variable is temporarily fixed at a particular value, and this fixed value of the variable is called the **parameter** of the special case that is thus determined.

Independence, Degrees of Freedom and State Coordinates

In general, a given system may allow the arbitrary adjustment of several of its variables simultaneously, but with each system there is a characteristic limit to the number of variables that can be treated in this way. When the value of each of a set of variables can be assigned at will, irrespective of the values of the others, the members of the set are said to be **independent**, and the greatest number of scalar variables that can be included in an independent set is called the **number of degrees of freedom** of the system. In other words, an independent set of scalar variables equal in number to the degrees of freedom of a system are sufficient to determine the configuration of the system completely, and by analogy with the coordinates of position, such a set is called a **set of coordinates of the state** of the system.

Functions

A function specifies a set of operations on a set of quantities of which the variables are called the **arguments** of the function. In general, the arguments need not comprise a set of quantities of the same kind, and even when they do, the quantity determined by the function may yet be of a different kind. So as well as scalar functions of scalars, we can also define scalar functions of vectors, (e.g. a scalar product of vectors), vector functions of vectors (e.g., a vector product of vectors), and vector functions of scalars. Here we are concerned only with the latter, that is, with vector functions of scalar arguments.

Equations of State

Let $(p, q, r\,..)$ represent a set of state coordinates of a system, and let u represent any other of its variables. From the definition of state coordinates it follows that a unique value of u will be associated with each combination of values of the coordinates, and this property of the system can be described mathematically by an equation that is drawn between u on the one hand, and on the other, a function of the coordinates whose value for each combination of values of its arguments is equal to the corresponding value of u in the system. Formally this equation may be represented symbolically as follows

$$u = u(p, q, r\,..)\ .$$

Here u alone, as on the left of the equation, represents the quantity u itself, whilst the expression on the right is taken, as a single whole, to represent some function of the arguments listed inside the brackets, whose value in every configuration is equal to the quantity indicated before the bracket. However, we shall follow the common practice of using the same symbol u to indicate either the quantity u itself, or some function of unspecified arguments whose value is equal to u, on the assumption that the intention will be obvious from the context. However, the arguments being unspecified, it will be particularly important to remember that the standard differentials and integrals are determined only for operations that are performed on a function in respect of its own argument.

2.7 THE CALCULUS OF A VECTOR FUNCTION OF A SINGLE SCALAR ARGUMENT

The calculus is applied to functions of several arguments through the concept of 'partial' operations in which the single function of several arguments is effectively treated as a succession of functions, each of which has only one argument. For the present, it is therefore sufficient to restrict consideration to vector functions of a single scalar argument, on the understanding that the results may subsequently be applied to functions of several arguments by the methods of partial calculus.

Increments

Let a represent a vector function of a single scalar argument s, and supposing that the value of the independent variable suffers a finite variation from s to $(s + \delta s)$, let the corresponding values of the dependent variable be represented by $\mathbf{a}$ and $(\mathbf{a} + \delta\mathbf{a})$. Then the corresponding increments in the values of the variables are $(s + \delta s) - s$, and $(\mathbf{a} + \delta\mathbf{a}) - \mathbf{a}$, though these expressions are usually abbreviated by cancelling the s's and the $\mathbf{a}$'s to δs and $\delta\mathbf{a}$. But operations on vectors must always be interpreted in the vectorial manner, so that even though an increment of a vector is indicated briefly by a symbol such as $\delta\mathbf{a}$, it must still be interpreted as the vector difference between the final and initial values of the vector, or, what is equivalent, as the vector which, added to the initial value $\mathbf{a}$, gives a sum equal to the final value $\mathbf{a} + \delta\mathbf{a}$, as indicated in Fig. 2.20.

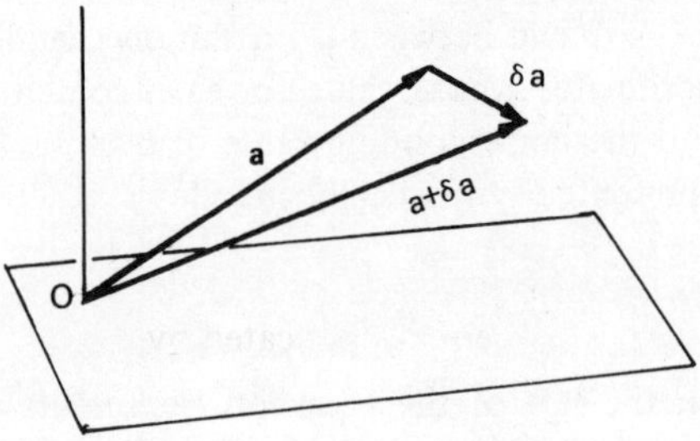

Fig. 2.20

Differential Coefficients of a Vector Function

Choosing a parametric value of the independent variable s as a fixed starting value, let δs represent a variable increment in its value, and, representing the corresponding increment in the value of the dependent variable by $\delta\mathbf{a}$, consider the ratio of the variations in the dependent and independent variables, $\delta \mathbf{a}/\delta s$. In general, the ratio would vary with δs, but commonly, as δs tends to zero, the ratio tends to a specific limiting value, and where such a limit exists it is called the **differential coefficient** of **a** with respect to s for the particular parametric value of s. A differential coefficient therefore describes the ratio of the changes in the dependent and independent variables for indefinitely small changes from a particular starting point, and it is therefore commonly referred to as an **instantaneous rate of change** of the dependent variable with respect to the independent. However, the word 'instantaneous' does not necessarily refer to an instant of time, but to an 'instant' of the independent variable, whatever that may be.

Symbolically, an indefinitely small increment is indicated by the letter d, so, representing the parametric value of the independent variable to which it applies by $s = S$, the corresponding value of the differential coefficient of **a** with respect to its argument s, and its derivation (by the operation known as differentiation) as the limiting ratio of finite variations, are indicated as follows

$$\left(\frac{\mathrm{d}\mathbf{a}}{\mathrm{d}s}\right)_{s=S} = \underset{\delta s \to 0}{\text{Limit}} \frac{\delta \mathbf{a}}{\delta s} = \underset{\delta s \to 0}{\text{Limit}} \frac{(\mathbf{a} + \delta \mathbf{a}) - \mathbf{a}}{(S + \delta s) - S} \tag{2.12}$$

Since the ratio $\delta \mathbf{a}/\delta s$ can be equated to the scalar $1/\delta s$ times the vector $\delta \mathbf{a}$, it is itself a vector parallel to $\delta \mathbf{a}$ whose coefficient in that direction is equal to the ratio of scalars $|\delta \mathbf{a}|/\delta s$. The differential coefficient $\mathrm{d}\mathbf{a}/\mathrm{d}s$ is therefore the vector parallel to the limiting direction of the increment $\delta \mathbf{a}$, whose coefficient in that direction is equal to the limiting value of the ratio $|\delta \mathbf{a}|/\delta s$.

The Derivative of a Vector Function

Where each value of the argument is associated with a unique value of the differential coefficient, a function is said to be smoothly continuous, and the differential coefficients of such a function can evidently be regarded as a second function, called the **derivative** of the original, which derives from the original by the operation of differentiation. The derivative of a vector function **a** with respect to its scalar argument s is indicated by $\mathrm{d}\mathbf{a}/\mathrm{d}s$.

A derivative has the same argument as the function from which it derives, so as long as the result does not reduce to zero, the operation can evidently be repeated so as to define derivatives of progressively higher orders. Thus the derivative of the first derivative is called the **second derivative**, and so on, and the nth derivative of **a** with respect to s is indicated by $\mathrm{d}^n\mathbf{a}/\mathrm{d}s^n$. In this the placement of the indices is determined by dimensional considerations, for reflection will show that the dimensions of the nth derivative are equal to the dimensions of **a** divided by the nth power of the dimensions of s.

Differentiation of a Sum

It can be shown that for vectors as for scalars, the operation of differentiation is distributive over a sum, so that if **a** and **b** are vector functions of the same scalar argument s, the derivative of the sum of **a** and **b** is equal to the sum of the derivatives of **a** and **b** severally. Referring to Fig. 2.21, suppose that the value of the independent variable changes from s to $s + \delta s$, and let the corresponding values of the functions be represented by **a** and $\mathbf{a} + \delta\mathbf{a}$, and by **b** and $\mathbf{b} + \delta\mathbf{b}$. In Fig. 2.21 we add **b** to **a** to form the initial sum $(\mathbf{a} + \mathbf{b})$, and we add

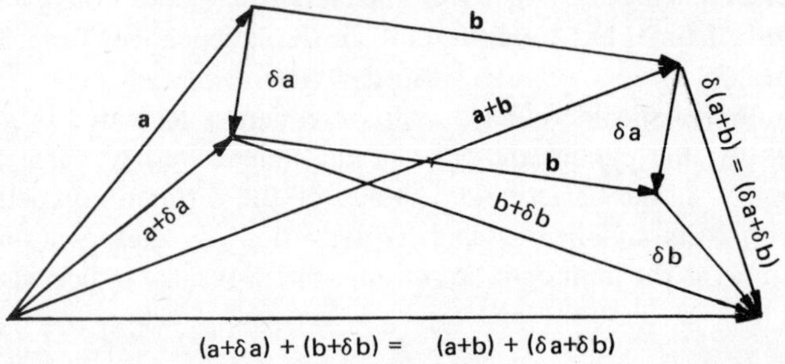

Fig. 2.21

$(\mathbf{b} + \delta\mathbf{b})$ to $(\mathbf{a} + \delta\mathbf{a})$ to form the final sum $(\mathbf{a} + \mathbf{b}) + \delta(\mathbf{a} + \mathbf{b})$, and it is obvious that the variation in the sum, $\delta(\mathbf{a} + \mathbf{b})$, is in fact equal to the sum of the several changes in **a** and **b**, $\delta\mathbf{a} + \delta\mathbf{b}$.

Thus
$$\frac{d(\mathbf{a} + \mathbf{b})}{ds} = \underset{\delta s \to 0}{\text{Limit}} \frac{\delta(\mathbf{a} + \mathbf{b})}{\delta s}$$

$$= \underset{\delta s \to 0}{\text{Limit}} \frac{\delta\mathbf{a} + \delta\mathbf{b}}{\delta s}$$

that is,
$$\frac{d(\mathbf{a} + \mathbf{b})}{ds} = \frac{d\mathbf{a}}{ds} + \frac{d\mathbf{b}}{ds} \tag{2.13}$$

The Differentiation of a Product

It can also be shown that products of vector functions can also be differentiated by the same rule as applies to products of scalars, though in the case of a vector product of vectors the sequence of the factors is significant because a vector product is not commutative. As an example, consider the differentiation of the scalar product $\mathbf{a} \cdot \mathbf{b}$, of two vector functions of scalar s.

By definition
$$\frac{d(\mathbf{a} \cdot \mathbf{b})}{ds} = \underset{\delta s \to 0}{\text{Limit}} \frac{(\mathbf{a} + \delta\mathbf{a}) \cdot (\mathbf{b} + \delta\mathbf{b}) - \mathbf{a} \cdot \mathbf{b}}{ds}$$

But a scalar product is distributive over a sum, and the second order term being ignored, it follows that

$$\frac{d(\mathbf{a}\cdot\mathbf{b})}{ds} = \underset{\delta s\to 0}{\text{Limit}} \frac{\mathbf{a}\cdot\delta\mathbf{b} + \mathbf{b}\cdot\delta\mathbf{a}}{\delta s}$$

that is

$$\frac{d(\mathbf{a}\cdot\mathbf{b})}{ds} = \mathbf{a}\cdot\frac{d\mathbf{b}}{ds} + \frac{d\mathbf{a}}{ds}\cdot\mathbf{b} \tag{2.14}$$

In fact, if u and v represent two scalar functions, whilst **a** and **b** represent two vector functions, all of the same scalar argument s, then a similar result will hold for any product of any two.

Thus

$$\begin{aligned}
\frac{d(u\,v)}{ds} &= u\,\frac{dv}{ds} + \frac{du}{ds}\,v \\
\frac{d(u\,\mathbf{a})}{ds} &= u\,\frac{da}{ds} + \frac{du}{ds}\,\mathbf{a} \\
\frac{d(\mathbf{a}\cdot\mathbf{b})}{ds} &= \mathbf{a}\cdot\frac{d\mathbf{b}}{ds} + \frac{d\mathbf{a}}{ds}\cdot\mathbf{b} \\
\frac{d(\mathbf{a}\times\mathbf{b})}{ds} &= \mathbf{a}\times\frac{d\mathbf{b}}{ds} + \frac{d\mathbf{a}}{ds}\times\mathbf{b}
\end{aligned} \tag{2.15}$$

The Integration of a Vector Function

The operation of integration is applicable only to a property that is smoothly distributed throughout a continuum such that a specific quantity is associated with each arbitrary part of the continuum, and the integral of such a quantity over any domain of the continuum is defined as the limiting sum of the quantities associated with all the elements that make up the domain. When the summation is performed over a specific domain the result is a specific quantity called a definite integral, but for a variable domain the result is a function of the boundary, and such is called an indefinite integral.

To apply the concept more specifically to a smoothly continuous vector function **a** over a specific range of its scalar argument from $s = s_1$ to $s = s_2$, let the range be divided into an arbitrary number of increments, and taking for each the product of the starting value **a** of the dependent variable with the increment δs in the independent variable, consider the sum of the products of all the increments. Symbolically, the sum of a finite set of quantities is indicated by Σ (sigma), and the range over which the sum is to be performed is indicated by writing the lower limit below the symbol, and the upper limit above. The required sum can therefore be indicated by $\Sigma_{S_1}^{S_2}(\mathbf{a}\,\delta s)$, though this would determine, not a specific quantity but one which varied with both the number of increments, and their disposition within the range. But suppose that the size of the increments is progressively reduced by increasing their number. Then the function being smoothly continuous, the sum would tend to a specific limiting value, and this is called the definite integral of **a** with respect to s over the given range $s_1 < s < s_2$.

Symbolically a limiting sum of indefinitely small increments is distinguished from a sum of a finite set by replacing Σ by the expanded letter s of the integral sign: $\int$, so the foregoing definite integral, and its derivation as a limiting sum are indicated as follows

$$\int_{S_1}^{S_2} \mathbf{a}\,\mathrm{d}s = \underset{\delta s \to 0}{\text{Limit}} \sum_{S_1}^{S_2} \mathbf{a}\,\delta s\ . \tag{2.16}$$

Given a function $\mathbf{a} = \mathbf{a}(s)$, consider the set of its integrals from a common lower bound $s = s_1$, to different upper bounds. The upper bound is then represented by the variable s, and since a unique value of the integral is associated with each value of the upper bound, the integrals of the set may be described as another function of s, and this is called an indefinite integral of the given function **a**.

Of course, each value of the lower limit would be associated with its own indefinite integral, but let s_1 and s_1' represent alternative values of the lower limit. Since s is a scalar, the range from s_1 to s is evidently equal to the sum of the ranges from s_1 to s_1' and from s_1' to s. The indefinite integrals from s_1 to s and from s_1' to s therefore differ only by the definite integral between the alternative lower limits from s_1' to s_1, and this is a specific quantity which is independent of s. The indefinite integrals of a given function are therefore recognised as a family of functions, each pair of which differ only by a constant, so if to any indefinite integral of the given function we add an arbitrary constant of integration, the result can be regarded as a general description of all the integrals of a given function, and it can be converted to a specific indefinite integral, simply by assigning the appropriate value to the arbitrary constant of integration.

Given an indefinite integral $\phi = \phi(s)$ of a function $\mathbf{a} = \mathbf{a}(s)$, the definite integral of **a** over a given range from $s = s_1$ to $s = s_2$ can, of course, be determined as the difference between the values of ϕ at the limits

Thus
$$\int_{S_1}^{S_2} \mathbf{a}\,\mathrm{d}s = \phi(s_2) - \phi(s_1) \tag{2.17}$$

Furthermore, since an integral is itself defined as a sum, it is obvious that the operation is distributive over a sum, so that if **a** and **b** are vector functions of scalar s

$$\int (\mathbf{a} + \mathbf{b})\, ds = \int \mathbf{a}\, ds + \int \mathbf{b}\, ds\ . \tag{2.18}$$

It is significant that there is no technique for the direct analytical determination of an indefinite integral, but it can be shown that integration is the inverse of differentiation. For if $\mathbf{f} = \mathbf{f}(s)$ is the derivative of $\mathbf{a} = \mathbf{a}(s)$,

that is, $$\frac{d\mathbf{a}}{ds} = \mathbf{f}$$

Then $$d\mathbf{a} = \mathbf{f}\, ds$$

and $$\int d\mathbf{a} = \int \mathbf{f}\, ds$$

or $$\mathbf{a} = \int \mathbf{f}\, ds + c\ .$$

Thus if **f** is the derivative of **a**, then **a** is an indefinite integral of **f**, so that we can identify the integral of a given function provided we can recognise the function of which the given function is the derivative.

2.8 SCALAR EQUIVALENTS

Just as a particular value of a vector is defined by specifying a rectangular set of components, so is the variation of a vector defined by specifying the variations of the set of components, and with fixed rectangular coordinates we choose these components in the fixed directions of coordinates. Thus whether a is a constant or a variable, we still write $\mathbf{a} = a_x.\mathbf{i} + a_y.\mathbf{j} + a_z.\mathbf{k}$, and we take it as understood that when **a** represents a constant, then a_x, a_y and a_z represent scalar constants, but when **b** represents a vector function, then a_x, a_y and a_z represent scalar functions of the same argument. However the unit vectors **i**, **j** and **k**, being fixed in both magnitude and direction, are wholly constant, and we recognise that when **a** represents a variable, each of its components like $a_x.\mathbf{i}$ represents a product of a scalar function like a_x with a constant like **i**. So let **a** represent a vector function of scalar s, and consider the application to the function of the operations of differentiation and integration, remembering, first that both operations are distributive over a sum, and second, that **i**, **j** and **k** are constants.

We have $\mathbf{a} = a_x.\mathbf{i} + a_y.\mathbf{j} + a_z.\mathbf{k}$

Therefore $$\frac{d\mathbf{a}}{ds} = \frac{d(a_x.\mathbf{i})}{ds} + \frac{d(a_y.\mathbf{j})}{ds} + \frac{d(a_z.\mathbf{k})}{ds}$$

or $$\frac{d\mathbf{a}}{ds} = \frac{da_x}{ds}.\mathbf{i} + \frac{da_y}{ds}.\mathbf{j} + \frac{da_z}{ds}.\mathbf{k} \quad (2.19)$$

Similarly $\int \mathbf{a}.ds = \int (a_x.\mathbf{i}).ds + \int (a_y.\mathbf{j}).ds + \int (a_z.\mathbf{k}).ds$

that is $$\int \mathbf{a}.ds = (\int a_x.ds).\mathbf{i} + (\int a_y.ds).\mathbf{j} + (\int a_z.ds).\mathbf{k} \quad (2.20)$$

We therefore conclude that when they are referred to an observer who is fixed in the rectangular coordinates in which a vector is defined, the operations of differentiation and integration can be applied to the vector simply by applying the operations to each of its scalar elements in turn.

2.9 VECTOR QUANTITIES

Although we chose to develop vector algebra only in application to a particular geometrical vector, we anticipated that, due account being taken of any additional effect due to location, the algebra would, in fact, prove to be relevant to a wide range of directed qunantities. But it is not necessarily appropriate to all such quantities, so the criterion of its applicability is clearly of some significance.

Perhaps this question can best be approached through the parallel question in the more familiar algebra of numbers. In effect this is built around a pair of operations whose definitions are arbitrary except that one precisely reverses the effect of the other, so that any sequence of the operations is necessarily reversible. This basic pair may be taken to be the operations of addition and subtraction, and all other operations of the algebra are then so defined that each is equivalent to a particular sequence of the basic pair. Thus a product is a repeated sum, a power is a repeated product, and so on, though the implied patterns of addition and subtraction may readily become so complex and extensive as to render their performance impracticable. The object of the algebra is therefore to avoid the necessity for performing complex sequences of addition and subtraction by devising the rules which govern the more complex operations, and the touchstone by which these rules are determined is, of course, that the direct performance of the operations must give the same result as if they were performed by the appropriate sequences of addition and subtraction. The essential feature of the algebra of numbers is therefore its unvarying consistency with the rule of numerical addition, and thus it is that the algebra is found to be appropriate to any situation in which two quantities of the same kind, acting together, have an effect which is equal to that of a single quantity equal to their numerical sum.

In just the same way, the algebra of vectors is built around the rule of vector addition, so in principle we determine whether or not a directed quantity may be treated as a vector quantity by performing an experiment to determine whether or not two of the quantities, acting together, have the same effect as a single quantity equal to their vector sum, though in practice it is often unnecessary to perform a special experiment, because the result is obvious from past experience.

In the main, the various vector quantities will be left to emerge incidently, case by case, but we here recognise one directed quantity that cannot be treated as a vector quantity, and one which can.

Finite Rotation

Referring to Fig. 2.22, consider a body which turns through a given angle about a given axis. By convention the positive sense along the axis is related

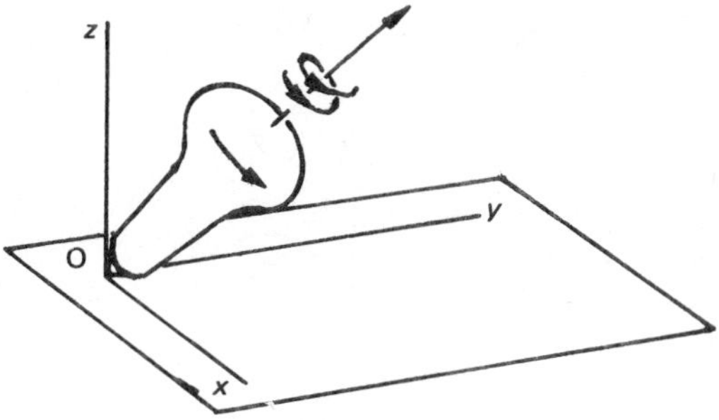

Fig. 2.22

to the sense of the rotation, as the sense of advance is related to the sense of rotation in a right-handed screw. The rotation can therefore be described as a directed quantity whose modulus is the angle turned through and whose direction is that of the axis. But referring to Fig. 2.23, suppose that the body indicated suffers one rotation of, say, $\pi/2$ about the x axis, and another of the same amount about the y axis. The vector sum of the two has a modulus of $\pi/\text{N}2$ about

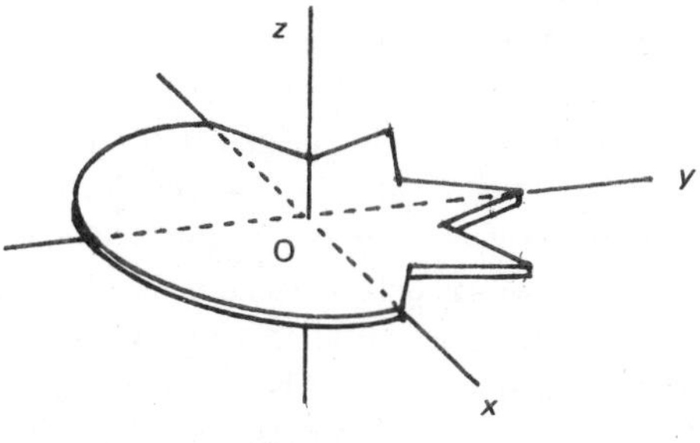

Fig. 2.23

an axis which bisects the angle between the x and y axes, and it is clear that this single rotation would not give the same result as the two rotations about the x and y axes. Indeed, the net effect of the two rotations depends on the order in which they are performed, so the two rotations are not even commutative, and we conclude that although a finite rotation can be described as a directed quantity, it cannot be treated as a vector quantity.

Vector Area

In many situations we are concerned with the area of a surface only as a scalar quantity, but sometimes the orientation of the surface is crucial, and then it is necessary to describe the area of the surface as a directed quantity.

Referring to Fig. 2.24, consider the inclination of the plane ABCD to the base plane of the frame of reference. An angle is necessarily measured between two lines, and different lines in the two planes contain different angles; it is therefore necessary to specify a particular pair, and in practice we choose a pair,

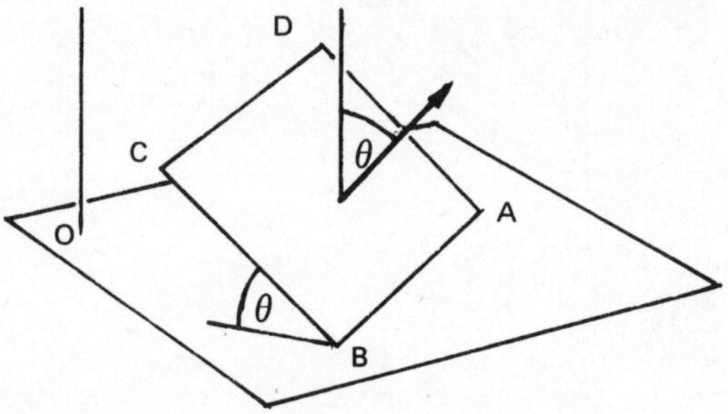

Fig. 2.24

both of which are at right-angles to the straight line or 'trace' in which the one plane (extended if necessary) intersects the other. However, it is then obvious that the angle θ between the planes is equal to the angle between their normals, and it is therefore more convenient to define the orientation of a plane by referring not to the angles which the plane itself makes with a set of reference planes, but to the angles which its normal makes with a set of reference axes.

The area of the plane surface is therefore described as a directed quantity whose direction is determined by its normal. It is true that we cannot usefully associate the area itself with a particular sense along its normal, but we are usually concerned with surfaces only as a boundary which is either truly closed, or else effectively closed by its being extended to infinity, and in this case we adopt a convention whereby the outward sense is taken as the positive. Otherwise it is necessary to define the positive sense along the normal arbitrarily. The area vector of a plane surface is therefore defined as one whose modulus is the superficial (scalar) area of the surface and whose direction is that of the normal to the surface. Where the surface forms part of a closed boundary the positive direction along the normal is taken as that which points outwards, but otherwise it is necessary to define the positive sense arbitrarily.

Vector Area by Vector Product

By definition, the vector product **a** X **b** determines a vector which is normal to the plane that contains both **a** and **b**, and whose modulus is equal to |**a**|.|**b**|. sinθ. But reference to Fig. 2.25 will show that if **a** and **b** are the vectors of the adjacent sides of either a parallelogram or a triangle, the modulus of **a** X **b** will be equal to the area of the parallelogram, and to twice the area of the triangle.

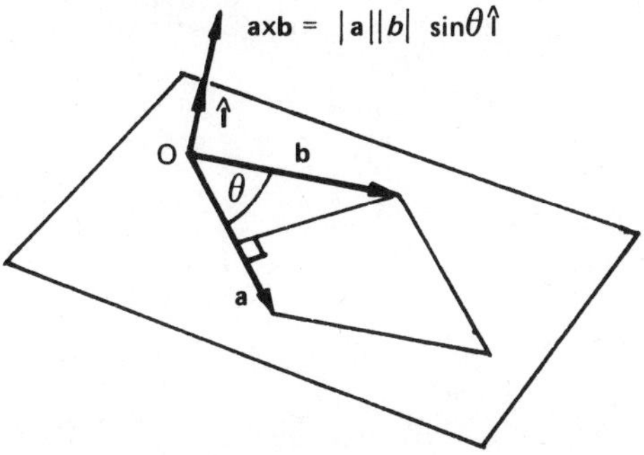

Fig. 2.25

We therefore conclude that if **a** and **b** are the vectors of two adjacent sides of either a parallelogram or a triangle, then depending upon which direction along the normal is reckoned to be positive, either **a** X **b** or **b** X **a** will determine the vector area of the parallelogram or twice the vector area of the triangle.

Projected Area

The projection of a point on a given plane is defined as the point in the plane in which it is intersected by the normal dropped from the point. Thus in Fig. 2.26, the points P, Q and R lie in the inclined plane, and their projections in the base plane are P′, Q′, and R′.

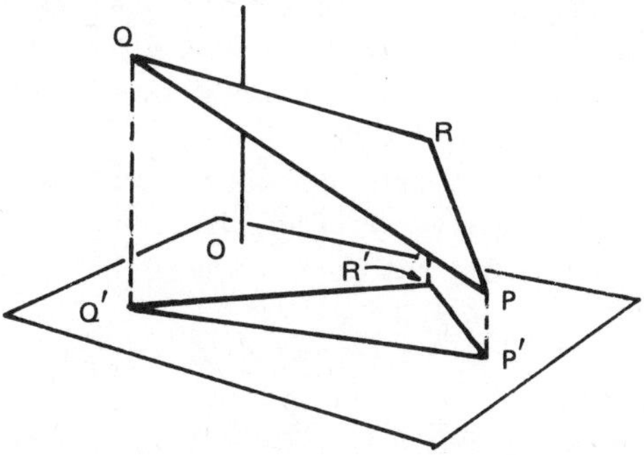

Fig. 2.26

Consider the ratio of the area of a projection to the area of the surface from which it derives. In Fig. 2.27, P′Q′R′ represents a projection of the triangle PQR, HT is the trace of PQR in the plane of projection, and PA and QB are lines in PQR which are drawn parallel and normal to HT respectively. Reflection will

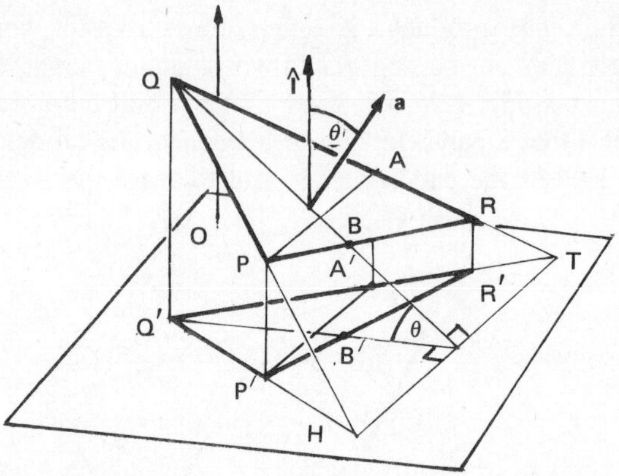

Fig. 2.27

show that the projection P′A′ will be parallel to the line PA itself, and it follows that any line in PQR which, like PA, is parallel to the trace HT will appear in the projection true size. On the other hand, any line in PQR which, like QB, is normal to the trace will be foreshortened by the cosine of the angle θ which PQR makes with the projection plane, and it follows that the area of the projection will be equal to the true area of the surface from which it derives times the cosine of the angle which the surface makes with the projection plane. But, since the modulus of the vector area of the surface is its superficial area, this may be equated to the scalar product of the area vector of the surface with the unit vector normal to the plane of projection, and we may therefore summarise as follows.

The area of the projection of any plane surface on any given plane is equal to the true area of the surface times the cosine of the angle which it makes with the plane of the projection. It may be determined as the scalar product of the vector area of the surface with the unit vector normal to the projection plane, and it may therefore be interpreted as the scalar component of the area vector resolved along the normal to the projection plane. In particular, we note that the scalar component of the area vector of any surface may be equated to the areas of its projections on the coordinate planes normal to the x, y and z axes in turn.

Example 2.1

The coordinates of two points A and B are given as A = (3, 0, −4) m, and B = (1, 2, −3) m.

(a) Determine the scalar elements of the radius vectors to A and to B, and of the vector from A to B, and verify the rule for the scalar equivalent of vector addition for this case.

(b) For the vector **AB**, determine its modulus, the unit vector $\hat{\mathbf{AB}}$, and its inclination to the y axis.

Solution

(a) In general, the scalar elements of a vector in rectangular coordinates can be determined by subtracting the coordinates of the initial point from the coordinates of the final point. However, a radius vector is one that is drawn from the origin of coordinates, and since the coordinates of its initial point are zero, the elements of a radius vector can be equated to the coordinates of the final point to which it is drawn.

Thus
$$\mathbf{OA} = \qquad = (3, 0, -4)\ \text{m}$$
$$\mathbf{OB} = \qquad = (1, 2, -3)\ \text{m}$$
$$\mathbf{AB} = (1, 2, -3) - (3, 0, -4) = (-2, 2, 1)\ \text{m}$$

By the rule of vector addition **OB** = **OA** + **AB**. But by the rule for the scalar equivalent, the sum of **OA** and **AB** can be determined by adding the corresponding pairs of elements.

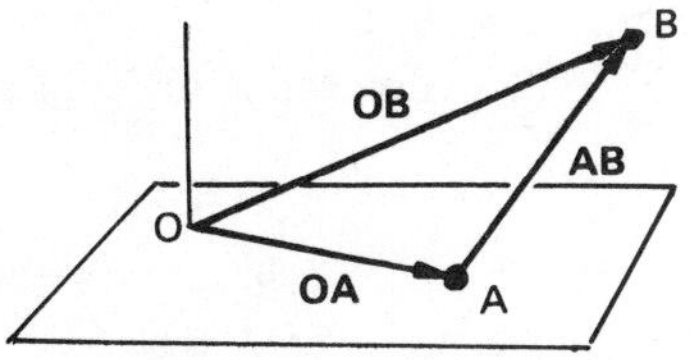

Thus
$$\mathbf{OA} + \mathbf{AB} = (3, 0, -4) + (-2, 2, 1)$$
$$= \{(3-2), (0+2), (-4+1)\}$$
$$= (1, 2, -3)\ \text{m}\ .$$

This sum is indeed equal to the value of **OB** above.

(b) By Pythagoras' theorem the modulus of a vector can be equated to the square root of the sum of the squares of its scalar components in any set of rectangular directions, and the scalar elements of **AB** are such a set. Furthermore, by the definition of multiplying by a number, any vector may be equated to its modulus times the unit vector in its own direction. The required unit vector $\hat{\mathbf{AB}}$ can therefore be equated to the inverse of the modulus $|\mathbf{AB}|$ times the vector **AB**.

Thus $$|\mathbf{AB}| = \sqrt{(-2)^2 + 2^2 + 1^2} \text{ m}$$

that is $$|\mathbf{AB}| = 3 \text{ m}$$

And $$\hat{\mathbf{AB}} = |\mathbf{AB}|^{-1}.\mathbf{AB}$$

$$= \frac{1}{3}(-2, 2, 1) \text{ m}$$

that is $$\hat{\mathbf{AB}} = \left(-\frac{2}{3}, \frac{2}{3}, \frac{1}{3}\right) \text{m} .$$

The scalar elements of a unit vector are equal to its direction cosines, i.e., to the cosines of the angles which it makes with the x, y, and z axes in turn. Therefore the angle which **AB** makes with the y axis is the angle whose cosine is equal to the j element of its unit vector $\hat{\mathbf{AB}}$.

$$\theta_y = \cos^{-1}\left(\frac{2}{3}\right)$$

Example 2.2

A point C lies on the straight line through the points A and B at a distance of 2 m from A on the side opposite to B. Given that the coordinates of A and B are A = (2, 1, −2) m; B = (5, 1, 2) m, determine the coordinates of C.

Solution

The coordinates of C can be equated to the scalar elements of the radius vector **OC**, and this can be equated to the sum of **OA** and **AC**. The scalar elements of **OA** can be equated to the given coordinates of its terminal point A. The vector **AC** can be equated to its given modulus times the unit vector $\hat{\mathbf{AC}}$, and since B, A and C are colinear in the sequence indicated in the diagram, the unit vectors $\hat{\mathbf{AC}}$ and $\hat{\mathbf{BA}}$, having the same unit modulus and the same direction, are equal.

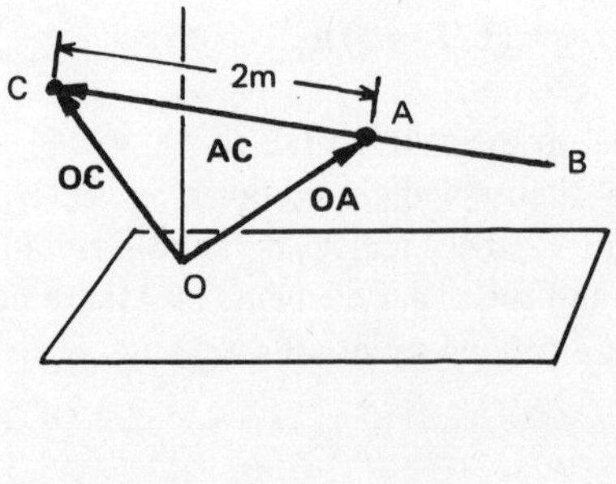

Thus $\quad \mathbf{OC} = \mathbf{OA} + \mathbf{AC}$

Or $\quad \mathbf{OC} = \mathbf{OA} + |\mathbf{AC}|.\hat{\mathbf{AC}}$

But $\quad \hat{\mathbf{AC}} = \hat{\mathbf{BA}}$

Therefore $\quad \mathbf{OC} = \mathbf{OA} + |\mathbf{AC}|.\mathbf{BA}$

Whence $\quad \mathbf{OC} = \mathbf{OA} + |\mathbf{AC}|.|\mathbf{BA}|^{-1}.\hat{\mathbf{BA}}$

But $\quad \mathbf{BA} = \{(2-5), (1-1), (-2-2)\}$

$$= (-3, 0, -4)\text{ m}$$

And $\quad |\mathbf{BA}| = \sqrt{(-3)^2 + 0^2 + (-4)^2} = 5\text{ m}$

Therefore $\quad \mathbf{OC} = (2, 1, -2) + (2/5)(-3, 0, -4)\text{ m}$

that is $\quad \mathbf{OC} = (2, 1, -2) + (-1{\cdot}2, 0, -1{\cdot}6)\text{ m}$

Or $\quad \mathbf{OC} = (0{\cdot}8, 1, -3.6)\text{ m}$.

Example 2.3

The coordinates of two points A and B are given as A = (1, −1, 1) m, B = (3, 0, −1) m. Determine the coordinates of the point D in which the straight line through A and B cuts the xy plane, and the point E in the line which is equidistant from the xy and xz planes.

Solution

The radius vector to a general point P on the line can readily be described in terms of λ, its distance in the sense $\vec{AB}$ from A. The value of λ for the point in which the line cuts the xy plane can then be determined by equating the k

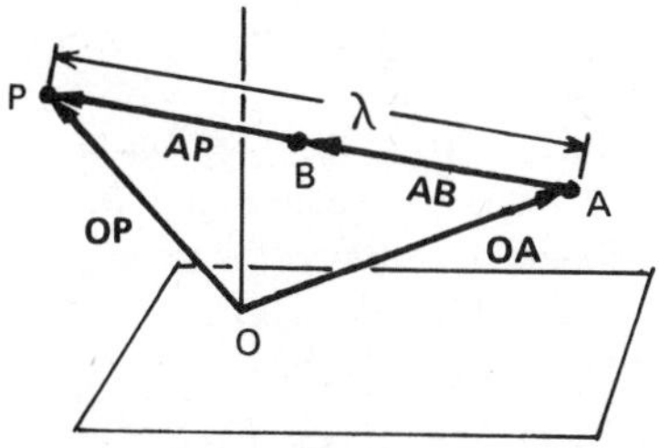

component of **OP** to zero, and the value for the point which is equidistant from the xy and xz planes can be determined by equating the k component of **OP** to its j component.

Thus $\quad \mathbf{OP} = \mathbf{OA} + \mathbf{AP} = \mathbf{OA} + |\mathbf{AP}|.\hat{\mathbf{AP}}$

Since **AP** and **AB** are colinear, $\hat{\mathbf{AP}} = \hat{\mathbf{AB}}$, so representing $|\mathbf{AP}|$ by λ:

$$\mathbf{OP} = \mathbf{OA} + \lambda.\hat{\mathbf{AB}}$$

$$= \mathbf{OA} + \lambda.|\mathbf{AB}|^{-1}.\mathbf{AB}$$

$$= (1,-1,1) + (\lambda/3)(2,1,-2)$$

$$\mathbf{OP} = \{(1+2\lambda/3), -(1-\lambda/3), (1-2\lambda/3)\} .$$

For the point D in which the line cuts the xy plane, the z component of OP is zero

that is $\quad (1-2\lambda_D/3) = 0$

$$\lambda_D = 3/2 .$$

Substituting accordingly

$$\mathbf{OD} = (2, -0{\cdot}5, 0) .$$

For the point E which is equidistant from the xy and xz planes the z component of OE is equal to its y component

that is $\quad (1-2\lambda_E/3) = -(1 - \lambda_E/3)$

$$\lambda_E = 2$$

Substituting accordingly

$$\mathbf{OE} = \frac{1}{3}(7,-1,-1) .$$

Example 2.4

Given that $\mathbf{a} = \mathbf{i} - 2\mathbf{j} + 3\mathbf{k}$; $b = 4\mathbf{i} + 2\mathbf{j} + 2\mathbf{k}$, and $\mathbf{a}\cdot\mathbf{c} = 18$, determine $\mathbf{a}\cdot\mathbf{b}$; $\mathbf{a}\cdot 3\mathbf{c}$; $\mathbf{a}\cdot(2\mathbf{b}-3\mathbf{c})$, and the angle between **a** and **b**. Determine the vector **c**, given that it has the same direction as the vector $(2\mathbf{i}-\mathbf{j}+2\mathbf{k})$.

Solution

$$\mathbf{a}\cdot\mathbf{b} = (1,-2,3)\cdot(4,2,2)$$

$$\mathbf{a}\cdot\mathbf{b} = 4 - 4 + 6 = 6$$

$$\mathbf{a}\cdot 3\mathbf{c} = 3(\mathbf{a}\cdot\mathbf{c}) = 54$$

$$\mathbf{a}\cdot(2\mathbf{b}-3\mathbf{c}) = 2(\mathbf{a}\cdot\mathbf{b}) - 3(\mathbf{a}\cdot\mathbf{c})$$

$$\mathbf{a}\cdot(2\mathbf{b}-3\mathbf{c}) = 12 - 54 = -42$$

$$\mathbf{a}\cdot\mathbf{b} = |\mathbf{a}|.|\mathbf{b}|.\cos\theta$$

$$\theta = \cos^{-1}\frac{\mathbf{a}\cdot\mathbf{b}}{|\mathbf{a}|.|\mathbf{b}|}$$

$$\theta = \cos^{-1}\frac{6}{\sqrt{(1+4+9).(16+4+4)}}$$

$$\theta = \cos^{-1}\sqrt{(3/28)}\ .$$

Since $\mathbf{c}$ is parallel to $(2\mathbf{i}-\mathbf{j}+2\mathbf{k})$, the two vectors have the same unit vector, so it can be represented by $(\lambda/3)(2\mathbf{i}-\mathbf{j}+2\mathbf{k})$, where λ is its modulus

$$\mathbf{a}\cdot\mathbf{c} = \mathbf{a}\cdot(\lambda/3)(2\mathbf{i}-\mathbf{j}+2\mathbf{k})$$

$$\lambda = \frac{3(\mathbf{a}\cdot\mathbf{c})}{(\mathbf{i}-2\mathbf{j}+3\mathbf{k})\cdot(2\mathbf{i}-\mathbf{j}+2\mathbf{k})}$$

$$= \frac{54}{(2+2+6)} = 5{\cdot}4$$

$$\mathbf{c} = 1{\cdot}8(2\mathbf{i}-\mathbf{j}+2\mathbf{k})$$

$$\mathbf{c} = 3{\cdot}6\mathbf{i} - 1{\cdot}8\mathbf{j} + 3{\cdot}6\mathbf{k}\ .$$

Example 2.5

By definition, the work done by a constant force is equal to the modulus of the force times the scalar component of its displacement resolved in the direction of the force. Determine the work done by the constant force $\mathbf{f} = 2\mathbf{i}+\mathbf{j}+3\mathbf{k}$ N in the displacement $\mathbf{r} = 2\mathbf{i}-2\mathbf{j}+\mathbf{k}$ m.

Solution

Let θ represent the angle between the force $\mathbf{f}$ and the displacement $\mathbf{r}$. Then the component of the displacement resolved in the direction of the force $= |\mathbf{r}|.\cos\theta$, and the work done is equal to $|\mathbf{f}|.|\mathbf{r}|.\cos\theta$. But this is equal to the scalar product of $\mathbf{f}$ with $\mathbf{r}$, so the definition implies that the work done by a constant force is equal to the scalar product of the force vector with the displacement vector. Thus for the force $\mathbf{f}$ in the displacement $\mathbf{r}$

$$\text{Work done} = \mathbf{f}\cdot\mathbf{r} = (2,1,3)\cdot(2,-2,1)\ \text{N m} = 4 - 2 + 3\ \text{N m} = 5\ \text{N m}\ .$$

Example 2.6

One end of a linear spring is attached to the point A whose coordinates are (1, 1, 4) m, whilst the other end is attached to a slider which runs on a straight track through B = (6, 1, 0) m, and C = (3, 7, 0) m. Determine the coordinates of the position of the slider for which the length of the spring is a minimum. Also determine the spring stiffness given that the force in the spring has a minimum value of 20 N, and a value when the slider is at B of 230 N.

Solution

Let P be a general point on the track at the parametric distance λ from A.

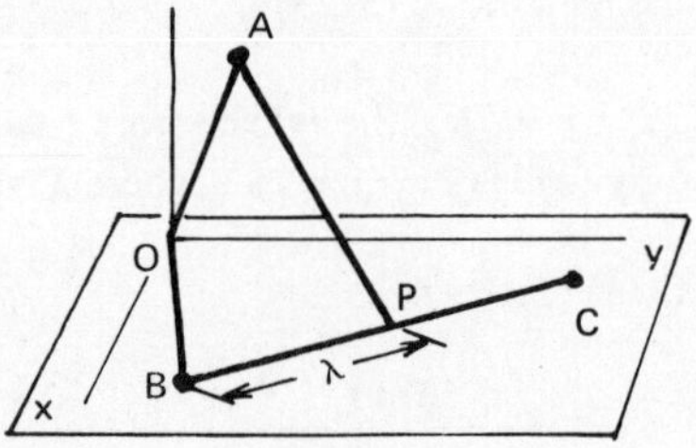

Then $$\mathbf{OA} = \mathbf{OB} + \mathbf{BP} + \mathbf{PA}$$

or $$\mathbf{PA} = \mathbf{OA} - \mathbf{OB} - \mathbf{BP}\ .$$

But **BP** and **BC** are colinear. Therefore $\mathbf{BP} = \lambda.\ \hat{\mathbf{BC}} = \lambda\ .\ |\mathbf{BC}|^{-1}\ .\ \mathbf{BC}$ and substituting accordingly

$$\mathbf{PA} = \mathbf{OA} - \mathbf{OB} - \lambda\ .\ |\mathbf{BC}|^{-1}\ .\ \mathbf{BC}$$

$$= (1, 1, 4) - (6, 1, 0) - \lambda(-3, 6, 0)$$

$$= \{(3\lambda - 5), -6\lambda, 4\}\ \text{m}\ .$$

The length of the spring will be a minimum when it is at right-angles to the track, i.e. when **PA** is normal to **BC**, in which case $\mathbf{PA} \cdot \mathbf{BC}$ is equal to zero.

$$O = \{(3\lambda - 5), -6\lambda, 4\} \cdot (-3, 6, 0)$$

$$O = -9\lambda + 15 - 36\lambda$$

$$\lambda = 1/3$$

Therefore
$$\mathbf{OP'} = 1/3(-3, 6, 0) = (-1, 2, 0)$$
$$\mathbf{P'A} = (-4, -2, 4)\ \text{m}$$
$$|\mathbf{P'A}| = 6\ \text{m}$$
$$\mathbf{BA} = (-5, 0, 4)\ \text{m}$$
$$|\mathbf{BA}| = \sqrt{(25 + 0 + 16)} = \sqrt{(41)}\ \text{m}\ .$$

Between P′ and B, the stretch in the spring is equal to $\sqrt{41}-6$ m, and the increase in the force in the spring is (230−20) N. Therefore necessary spring stiffness is given by

$$\text{Spring stiffness} = \frac{210\ \text{N}}{\sqrt{41}-6} = 520\ \text{N/m}$$

Example 2.7
The scalar elements of three vectors **a**, **b** and **c** are

$$\mathbf{a} = (2, -3, 4);\quad \mathbf{b} = (1, 3, -1);\quad \mathbf{c} = (3, 3, 1)$$

Determine (a) $\mathbf{a}\times\mathbf{b}$; (b) $\mathbf{a}\times\mathbf{c}$; (c) $\mathbf{c}\times\mathbf{a}$; and (d) $\mathbf{a}\times(\mathbf{b}-2\mathbf{c})$.

Solution

(a)
$$\mathbf{a}\times\mathbf{b} = \begin{vmatrix} \mathbf{i} & \mathbf{j} & \mathbf{k} \\ 2 & -3 & 4 \\ 1 & 3 & -1 \end{vmatrix} = (3-12)\mathbf{i} - (-2-4)\mathbf{j} + (6+3)\mathbf{k} = (-9, 6, 9)$$

(b)
$$\mathbf{a}\times\mathbf{c} \triangleq \begin{vmatrix} \mathbf{i} & \mathbf{j} & \mathbf{k} \\ 2 & -3 & 4 \\ 3 & 3 & 1 \end{vmatrix} = (-3-12)\mathbf{i} - (2-12)\mathbf{j} + (6+9)\mathbf{k} = (-15, 10, 15)$$

$$\mathbf{c}\times\mathbf{a} = -(\mathbf{a}\times\mathbf{c}) = (15, -10, -15)$$

(c)
$$\mathbf{a}\times(\mathbf{b}-2\mathbf{c}) = (\mathbf{a}\times\mathbf{b}) - 2(\mathbf{a}\times\mathbf{c})$$
$$= (-9, 6, 9) - 2(-15, 10, 15)$$
$$= (21, -14, -21)$$

Alternatively

$$(\mathbf{b}-2\mathbf{c}) \quad = (-5,-3.-3)$$

$$\mathbf{a}\times(\mathbf{b}-2\mathbf{c}) = \begin{vmatrix} \mathbf{i} & \mathbf{j} & \mathbf{k} \\ 2 & -3 & 4 \\ -5 & -3 & -3 \end{vmatrix}$$

$$= (21,-14,-21)$$

Example 2.8
Show that the vectors **a**, **b** and **c** of Example 2.7 are coplanar, and determine the angle which the plane makes with the xy plane.

Solution
If **a**, **b** and **c** are coplanar, the cross-product of any two will be at right-angles to the third. We can therefore show that the vectors are coplanar by showing that the dot product of any one with the vector product of the other two is zero. Thus

$$(\mathbf{a}\times\mathbf{b})\cdot\mathbf{c} = (-9,6,9)\cdot(3,3,1) = 0 \ .$$

Since $\mathbf{a}\times\mathbf{c}$ is normal to the plane, and the elements of a unit vector are equal to its direction cosines, it follows that the k component of the unit of $\mathbf{a}\times\mathbf{b}$ is the cosine of the angle which the plane makes with the xy plane.

Thus $\quad (\mathbf{a}\times\mathbf{b}) = |\mathbf{a}\times\mathbf{b}|^{-1}.(\mathbf{a}\times\mathbf{b})$

$$= (1/3\sqrt{22})(-9,6,9)$$

Therefore $\quad \theta_{xy} = \cos^{-1}(3/\sqrt{22}) \ .$

Example 2.8
The rectangular coordinates in metres of three points A, B and C are

$$A = (-2,1,3); \quad B = (-3,1,5); \quad C = (5,7,1)$$

For the triangle ABC determine: (a) its superficial area, (b) the angle it makes with the xz plane, (c) the areas of its projections on the coordinate planes, (d) the area of its projection on the plane normal to the line which makes an angle of 60° with the x axis and 45° with the y axis, and (e) the direction cosines of an auxiliary set of axes x', y' and z' which are drawn with x' and y' in the plane of the triangle, x' being parallel to AB in the same sense.

Solution
The vector area of the triangle can be determined by taking half the cross-product of any two sides.

Thus $\quad \mathbf{AB} = (-1, 0, 2); \quad \mathbf{AC} = (7, 6, -2)$

$$\frac{1}{2}(\mathbf{AB} \times \mathbf{AC}) = \frac{1}{2}\begin{vmatrix} \mathbf{i} & \mathbf{j} & \mathbf{k} \\ -1 & 0 & 2 \\ 7 & 6 & -2 \end{vmatrix} = \frac{1}{2}(-12, 12, -6) = 3(-2, 2. -1)$$

(a) The superficial area of the triangle is equal to the modulus of its vector area. Thus

$$\text{Area of triangle ABC} = 3\sqrt{(-2)^2 + 2^2 + 1^2}$$
$$= 9\ \text{m}^2$$

(b) The areas of the projections of a surface on the coordinate planes normal to x, y and z are equal to scalar elements of its vector area in turn. Thus

The areas of the projections of ABC on the yz, xz and xy planes are 6 m^2, 6 m^2 and 3 m^2 respectively.

(c) Dividing the vector area of ABC by its superficial area gives a unit vector normal to ABC. But the elements of a unit vector are equal to its direction cosines (i.e. to the cosines of the angles which it makes with the x, y and z axes in turn), and the angle which ABC makes with the xz plane is equal to the angle which its normal makes with the y axis. Thus if $\hat{\mathrm{l}}$ represents the unit vector normal to ABC: $\hat{\mathrm{l}} = (1/3)\,(-2, 2, -1)$, and the angle which ABC makes with the xz plane $= \cos^{-1}(2/3)$.

(d) the area of the projection of ABC on any plane is equal to the component of its area vector resolved in the direction normal to the projection plane.

Since the normal to the projection plane makes an angle of 60° with x and of 45° with y, and remembering that the modulus of a unit vector is unity, the unit vector $\hat{\mathrm{l}}'$, normal to the plane of projection is given by

$$\hat{\mathrm{l}}' = (\cos 60°, \cos 45°, \sqrt{1 - \cos^2 60° - \cos^2 45°})$$

that is $\quad \hat{\mathrm{l}}' = (1/2, 1/\sqrt{2}, 1/\sqrt{2})$

The area of the projection of ABC on the plane normal to $\hat{\mathbf{I}}'$ is therefore given by

$$\text{Area of projection} = |(1/2)(1,\sqrt{2},\sqrt{2})\cdot 3(-2,2,-1)|\,\mathrm{m}$$
$$= (3/2)(2-\sqrt{2})\,\mathrm{m}\ .$$

Let $\mathbf{i}'$, $\mathbf{j}'$ and $\mathbf{k}'$ be the unit vectors along the axes x', y' and z'. Then since $\mathbf{i}'$ and $\mathbf{j}'$ lie in the plane ABC, the unit vector z' is the unit vector $\hat{\mathbf{I}}$ normal to ABC.

that is $\quad \mathbf{k}' = \hat{\mathbf{I}} = (1/3)(-2,2,-1)$

Since x has the same direction as **AB**

$$\mathbf{i}' = \hat{\mathbf{AB}} = |\mathbf{AB}|^{-1}\,AB = (1/\sqrt{5})(-1,0,2)\ .$$

Finally since $\mathbf{i}'$, $\mathbf{j}'$ and $\mathbf{k}'$ constitute a right-handed rectangular set

$$\mathbf{j}' = \mathbf{k}'\times\mathbf{i}' = \frac{1}{3\sqrt{5}}\begin{vmatrix} \mathbf{i} & \mathbf{j} & \mathbf{k} \\ -2 & 2 & 1 \\ -1 & 0 & 2 \end{vmatrix} = (1/3\sqrt{5})(4,3,2)\ .$$

Example 2.10

The coordinates of point P in fixed rectangular coordinates vary with time t according to $P = \{(2 + t^3), 2t^2, (4 + 2t)\}$. Given that the velocity **v** and the acceleration **a** of the point in the given reference are the first and second derivatives (as seen by an observer fixed in the coordinates) of the radius vector to the point with respect to time, determine the velocity and acceleration at time $t = 10$.

Solution

In fixed rectangular coordinates, the scalar elements of the radius vector to a point are equal to its coordinates, and the elements of the derivative of a vector are equal to the derivatives of its several elements. Thus

$$\mathbf{r} = (2 + t^3)\mathbf{i} + 2t^2\mathbf{j} + (4 + 2t)\mathbf{k}$$

Whence $\quad \mathbf{v} = \dfrac{\mathrm{d}\mathbf{r}}{\mathrm{d}t} = 3t^2\,\mathbf{i} + 4t\,\mathbf{j} + 2\mathbf{k}$

and $\quad \mathbf{a} = \dfrac{\mathrm{d}^2\mathbf{r}}{\mathrm{d}t^2} = 6t\,\mathbf{i} + 4\,\mathbf{j}\ .$

At $t = 10$ $\quad \mathbf{v}_{(t=10)} = 300\,\mathbf{i} + 40\,\mathbf{j} + 2\mathbf{k}$

$$\mathbf{a}_{(t=10)} = 60\,\mathbf{i} + 4\,\mathbf{j}\ .$$

Example 2.11

The acceleration of a point in fixed rectangular coordinates varies with time t according to $\mathbf{a} = t^2\mathbf{i} + 2t\mathbf{j} + \mathbf{k}$. Given that the velocity and the radius vector **r** are equal to the first and second integrals of the acceleration with respect to time, determine the radius vector to the point at $t = 2$, given that time was reckoned from the instant of rest at the position whose coordinates are (1, 2, 3).

Solution

To the observer fixed in the rectangular coordinates, the elements of the integral of a vector are equal to the integrals of its several elements.

Given that $\mathbf{a} = t^2\mathbf{i} + 2t\mathbf{j} + \mathbf{k}$

Then $$\mathbf{v} = \int \mathbf{a}\,\mathrm{d}t + \mathbf{c}_1 = \frac{t^3}{3}.\mathbf{i} + t^2\mathbf{j} + t\mathbf{k} + \mathbf{c}_1$$

and $$\mathbf{r} = \int \mathbf{v}\,\mathrm{d}t + \mathbf{c}_2 = \frac{t^4}{12}.\mathbf{i} + \frac{t^3}{3}\mathbf{j} + \frac{t^2}{2}\mathbf{k} + \mathbf{c}_1 t + \mathbf{c}_2\ .$$

But when $t = 0$, then $\mathbf{v} = 0$. Therefore $\mathbf{c}_1 = 0$. Also when $t = 0$, then $\mathbf{r} = \mathbf{i} + 2\mathbf{j} + 3\mathbf{k}$, so $\mathbf{c}_2 = \mathbf{i} + 2\mathbf{j} + 3\mathbf{k}$.

Therefore $$\mathbf{r} = \left(1 + \frac{t^4}{12}\right)\mathbf{i} + \left(2 + \frac{t^3}{3}\right)\mathbf{j} + \left(3 + \frac{t^2}{2}\right)\mathbf{k}$$

and $$\mathbf{r}_{(t=2)} = \frac{7}{4}\mathbf{i} + \frac{14}{3}\mathbf{j} + 4\mathbf{k}\ .$$

Example 2.12

A point whose coordinates in fixed rectangular coordinates vary with time t according to $P = \{(2 + t), t^3, 2t^2\}$ is subjected to a force $\mathbf{f}$ which varies according to $\mathbf{f} = 2t\,\mathbf{i} + 4\,\mathbf{k}$. Given that the work done in the motion from an initial position $\mathbf{r} = \mathbf{r}_1$ to a final position $\mathbf{r} = \mathbf{r}_2$ is given by: Work done $= \int_{r_1}^{r_2} \mathbf{f}\cdot\mathrm{d}\mathbf{r}$, determine the work done during the interval of time from $t = 0$ to $t = 2$.

Solution

For vector functions as for scalar, the operations of differentiation and integration are defined only for functions of a single scalar argument, each with respect to its own argument. To determine the required integral it is therefore necessary to substitute for both $\mathbf{f}$ and $\mathrm{d}\mathbf{r}$ in terms of a single argument, remembering to adjust the limits of the integration in accord with the change of the independent variable. Thus, we are given that

$$\mathbf{f} = 2t\mathbf{i} + 4\mathbf{k}$$

and that $$\mathbf{r} = (2 + t)\mathbf{i} + t^3\mathbf{j} + 2t^2\mathbf{k}$$

Therefore $$\frac{\mathrm{d}\mathbf{r}}{\mathrm{d}t} = \mathbf{i} + 3t^2\mathbf{j} + 4t\mathbf{k}$$

or $$\mathrm{d}\mathbf{r} = (\mathbf{i} + 3t^2\mathbf{j} + 4t\mathbf{k})\,\mathrm{d}t$$

Thus $$\mathbf{f}\cdot\mathrm{d}\mathbf{r} = (2t + 16t)\,\mathrm{d}t = 18t\,\mathrm{d}t\ .$$

Therefore if t_1 and t_2 represent the times at position $\mathbf{r}_1$ and $\mathbf{r}_2$, we have

$$\text{Work done} = \int_{r_1}^{r_2} \mathbf{f} \cdot \mathrm{d}\mathbf{r} = \int_{t_1}^{t_2} 18t \, \mathrm{d}t$$

$$= \left[9t^2\right]_{t_1}^{t_2} = 9(t_2^2 - t_1^2) \; .$$

Therefore in the interval from $t = 0$ to $t = 2$

Work done = 36 units.

3

Point Kinematics

3.1 INTRODUCTION

Kinematics is the name given to the study of motion, without regard to the effects which cause a change of motion.

Here we consider only a limited range of cases which is restricted, first by considering the motions only of point particles and not of finite bodies, and second, by considering alternative observers only for the case in which there is no relative rotation between them. Throughout we assume that each point traces a path that appears in the frame of reference as a smoothly continuous curve, so we approach the study of kinematics by first considering their chief properties.

3.2 PROPERTIES OF SMOOTH CURVES

Given a curve that is drawn in a given frame of reference, let P be a fixed point on the curve, and let P′ be a second point on the curve at a variable position adjacent to P. In general, the slope of the chord PP′ would vary with the position of P′, but if P′ were moved towards P, in the limit, as the length of the arc PP′ tended to zero, the slope of the chord would commonly tend towards a specific limiting value, and where this is true for each point P, the curve is said to be **smoothly continuous** (Fig. 3.1).

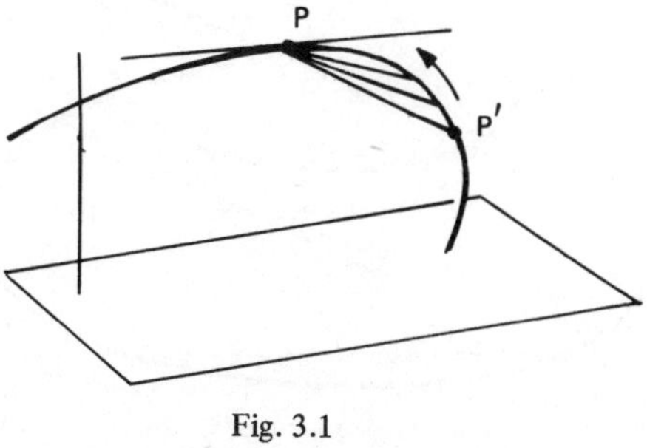

Fig. 3.1

Position

Let any convenient point on the path be chosen as datum, and let either direction along the path to be chosen as the positive. Then the position of a general point P on the path can be determined by its scalar arc distance s measured around the curve from the datum, as indicated in Fig. 3.2.

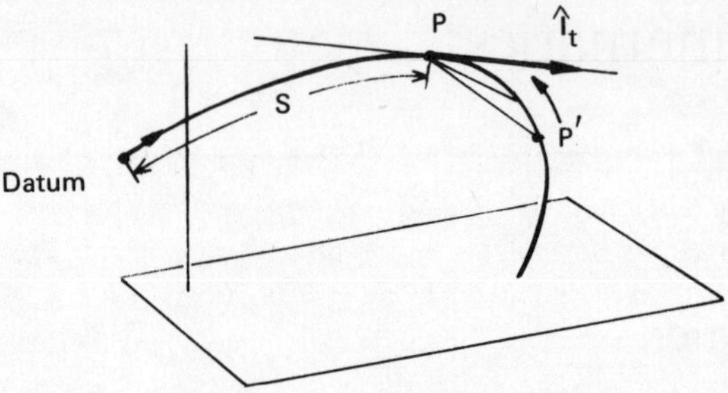

Fig. 3.2

The Tangent

The tangent to the path at P is defined as the straight line through P in the limiting direction of the chord from P to the adjacent point P′. The direction along the tangent is reckoned positive in the sense of s increasing, and it can conveniently be defined by the unit tangent vector $\hat{1}_t$.

It follows that a curve may be said to be smoothly continuous only when each point upon it has a unique tangent, and intuition suggests that in passing through any point, a particle which traces the curve would be moving instantaneously along the tangent to the curve at the point.

The Osculating Plane

Broadly, the osculating plane to a curve at any point is the plane that contains an indefinitely short arc of the curve at the point. More formally, consider the

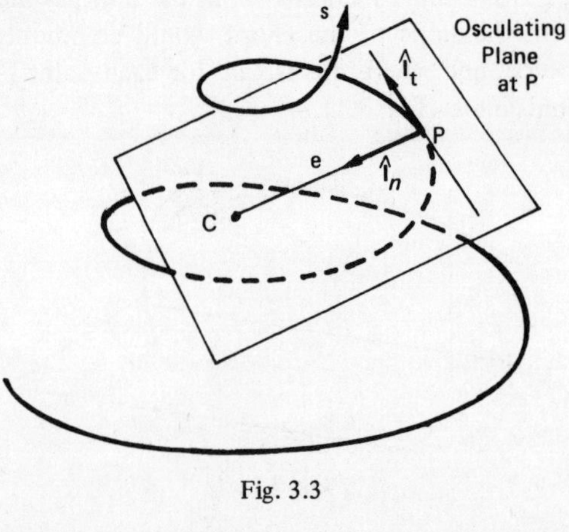

Fig. 3.3

plane which contains the tangent at a given point P and a variable adjacent point P′ which moves towards P. In general, the inclination of the plane would vary with the position P′, but the curve being smooth, in the limit, as P′ tended toward P, the plane would tend toward a specific limiting direction, and the limiting plane at P is called the osculating plane to the curve at the point.

Intuition suggests that in passing through any point, a particle would have a motion which was instantaneously confined to the osculating plane to the curve at the point, with no component in the direction normal to the plane.

The Centre and Radius of Curvature

Each indefinitely short arc of a curve is indistinguishable from an arc of a particular circle which lies in the osculating plane for the position, and the centre and radius of this circle are called the centre and radius of curvature of the curve at the position. A radius of curvature is indicated symbolically by ρ.

Intuition suggests that in passing through any point, a particle which moves along the curve can be considered to move instantaneously on a circular arc whose radius is equal to the radius of curvature of the curve at the point, and which lies in the osculating plane with its centre at the centre of curvature of the curve at the point.

The Normals

The principal normal to the curve at a given point P is the straight line through P which lies in the osculating plane at right-angles to the tangent, in the sense which points towards the centre of curvature for the point. The unit normal vector which defines the direction of the principal normal is indicated symbolically by $\hat{1}_n$, as in Fig. 3.3. The bi-normal at P is the straight line through P normal to the osculating plane. With the tangent and the principal normal, the bi-normal completes a set of mutually rectangular directions.

Curvature

Curvature measures the rate of change of direction (as defined by the tangent) with respect to distance along the curve. Thus, if δs represents the arc distance around the curve from a given point P to a variable, adjacent point P′, and if $\delta\theta$ is the angle between the tangents at P and at P′, the curvature at P is defined as

$$\text{Curvature} = \frac{d\theta}{ds} = \underset{\delta s \to 0}{\text{Limit}} \frac{\delta\theta}{\delta s}$$

Some Useful Relationships

It can be shown, first, that the curvature $d\theta/ds$ is equal to the inverse of the radius of curvature ρ, and second, that the derivative of the radius vector $\mathbf{r}$ to any point on the curve with respect to the distance along the curve s determines the unit tangent vector $\hat{\mathbf{I}}_t$. Furthermore, the derivative of $\hat{\mathbf{I}}_t$ with respect to s determines a vector in the direction of the unit normal vector $\hat{\mathbf{I}}_n$, whose modulus is equal to the curvature $\mathrm{d}\theta/\mathrm{d}s = 1/\rho$.

By definition, the angle subtended by an arc of a circle at its centre is equal to the ratio of the arc length to the radius. The ratio of the angle to the arc length can therefore be equated to the inverse of the radius. But referring to Fig. 3.4, the radius of the arc PP′ is the radius of curvature ρ of the curve at P,

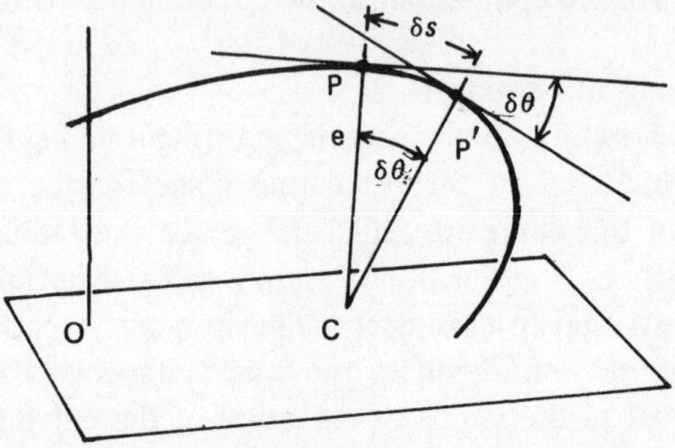

Fig. 3.4

and the angle subtended by PP′ at the centre of curvature is equal to the angle $\delta\theta$ which separates the tangents at P and P′. The limiting value of the ratio $\delta\theta/\delta s$ can therefore be equated to the inverse of the radius of curvature, and we may write

$$\text{Curvature} = \frac{\mathrm{d}\theta}{\mathrm{d}s} = \underset{\delta s\to 0}{\text{Limit}}\ \frac{\delta\theta}{\delta s} = \frac{1}{\rho} \tag{3.1}$$

Now consider the ratio $\delta\mathbf{r}/\delta s$. Referring to Fig. 3.5, we recognise that δs is the length of the arc PP′, whilst $\delta\mathbf{r}$ is the vector of its chord, so $\mathrm{d}\mathbf{r}/\mathrm{d}s$ can be recognised as a vector in the limiting direction of the chord $\delta\mathbf{r}$, whose modulus is equal to the limiting ratio of the length of the chord $|\delta\mathbf{r}|$, to the length of

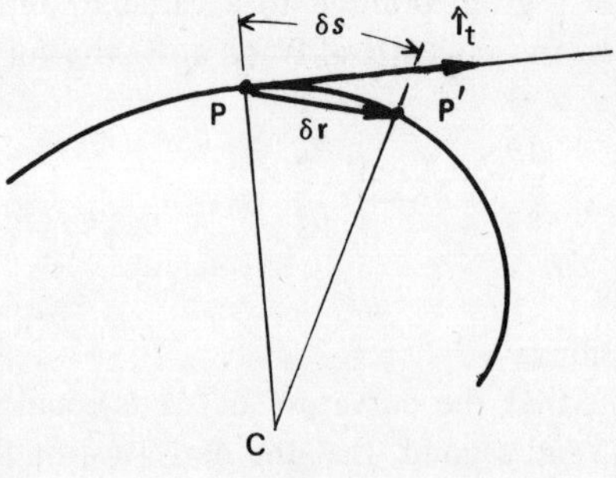

Fig. 3.5

the arc δs. But the former defines the direction of the tangent, and the latter is evidently unity, and we conclude that $\mathrm{d}\mathbf{r}/\mathrm{d}s$ determines a unit vector in the direction of the tangent.

that is
$$\frac{\mathrm{d}\mathbf{r}}{\mathrm{d}s} = \underset{\delta s \to 0}{\text{Limit}} \frac{\delta \mathbf{r}}{\delta s} = \hat{\mathbf{1}}_t \qquad (3.2)$$

The distance along the path to P and P′ being indicated by s and $s+\delta s$, the unit tangent vectors at P and P′ are indicated by $\hat{\mathbf{1}}_t$ and $\hat{\mathbf{1}}_t+\delta\hat{\mathbf{1}}_t$, whilst the angle between them is indicated by $\delta\theta$, and we now consider the derivative $\mathrm{d}\hat{\mathbf{1}}_t/\mathrm{d}s$, the limiting value of the ratio $\delta\hat{\mathbf{1}}_t/\delta s$. Now

$$\frac{\mathrm{d}\hat{\mathbf{1}}_t}{\mathrm{d}s} = \frac{\mathrm{d}\hat{\mathbf{1}}_t}{\mathrm{d}\theta} \cdot \frac{\mathrm{d}\theta}{\mathrm{d}s}.$$

and referring to Fig. 3.6 it is clear that in the limit, as δs and $\delta\theta$ tend to zero, the direction of $\delta\hat{\mathbf{1}}_t$ tends to the direction of $\hat{\mathbf{1}}_n$ at right-angles to $\hat{\mathbf{1}}_t$ itself. Furthermore, since the length of a unit vector is unity, the limiting value of the ratio

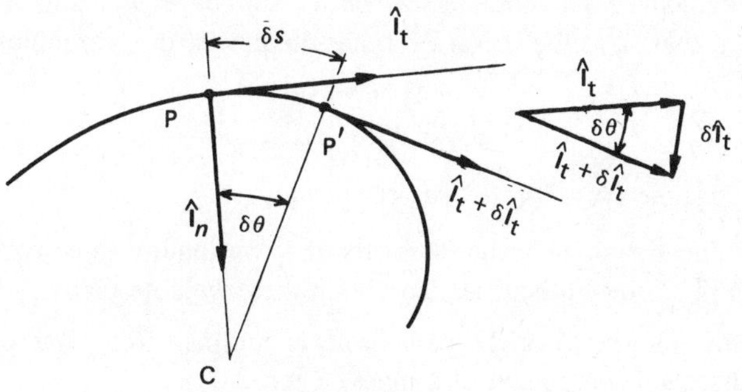

Fig. 3.6

$|\delta\hat{\mathbf{1}}_t|/\delta\theta$ is unity. The derivative $\mathrm{d}\hat{\mathbf{1}}_t/\mathrm{d}\theta$ therefore determines the unit vector $\hat{\mathbf{1}}_n$, and since we have already seen that the curvature $\mathrm{d}\theta/\mathrm{d}s$ is equal to the inverse of the radius of the curvature, we may substitute accordingly to give

$$\frac{\mathrm{d}\hat{\mathbf{1}}_t}{\mathrm{d}s} = \frac{\mathrm{d}^2\mathbf{r}}{\mathrm{d}s^2} = \rho^{-1}.\hat{\mathbf{1}}_n \qquad (3.3)$$

Example 3.1

A particle moves in a plane parallel to the yz plane of a given reference. For a point P = (2, 2, 4) on the path of the motion, the centre of curvature is at C = (2, −2, 1). For the path at P, determine the radius of curvature ρ, the unit normal vector $\hat{\mathbf{1}}_n$, and the unit tangent vector in the sense of z increasing $\hat{\mathbf{1}}_t$.

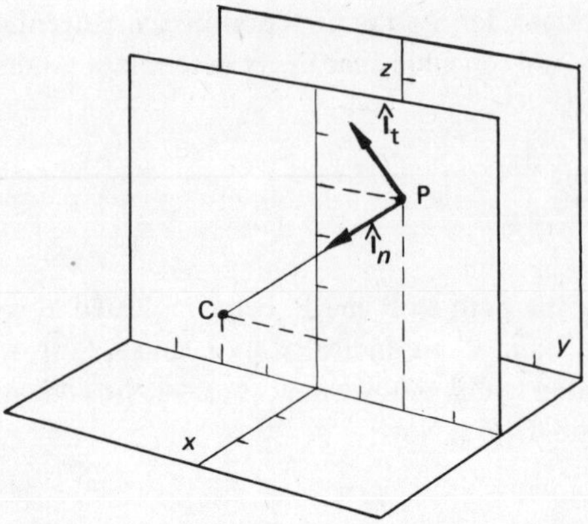

Solution

We know that the unit normal vector at any point on the path is directed along the straight line which joins the point to the centre of curvature for the point, and that the radius of curvature is equal to the length of this line. Therefore, given the coordinates of P and C, the vector **PC** can be written at sight, when ρ can be equated to its modulus, whilst $\hat{1}_n$ can be equated to the coincident unit vector, that is, to the vector **PC** times the inverse of its modulus. Thus

$$\mathbf{PC} = -4\mathbf{j} + -3\mathbf{k}$$

and $$\rho = |\mathbf{PC}| = (4^2 + 3^2)^{1/2} = 5$$

and $$\hat{1}_n = |\mathbf{PC}|^{-1}\mathbf{PC} = -0{\cdot}8\mathbf{j} - 0{\cdot}6\mathbf{k}\ .$$

To determine the three scalar elements of $\hat{1}_t$ we require three equations which relate them to known quantities. To obtain these we note that

1. Since the tangent, like the path itself, is confined to a plane parallel to the *yz* plane, the **i** component of $\hat{1}_t$ must be zero.
2. Since $\hat{1}_t$ is normal to $\hat{1}_n$, the scalar product of the two must be equal to zero.
3. The sum of the squares of the elements of any unit vector must be equal to unity.

Noting 1 above, let $\hat{1}_t = b\mathbf{j} + c\mathbf{k}$. Then from 2 above

$$-(0{\cdot}8b + 0{\cdot}6c) = 0 \tag{i}$$

And from 3 above

$$b^2 + c^2 = 1\ . \tag{ii}$$

Between (i) and (ii) $b = \pm\,0{\cdot}6$ and $c = \mp\,0{\cdot}8$.

Finally, the required sense for $\hat{1}_t$ is that for which z increases, that is, that for which the **k** component is positive. Therefore

$$\hat{1}_t = -0{\cdot}6\mathbf{j} + 0{\cdot}8\mathbf{k}\ .$$

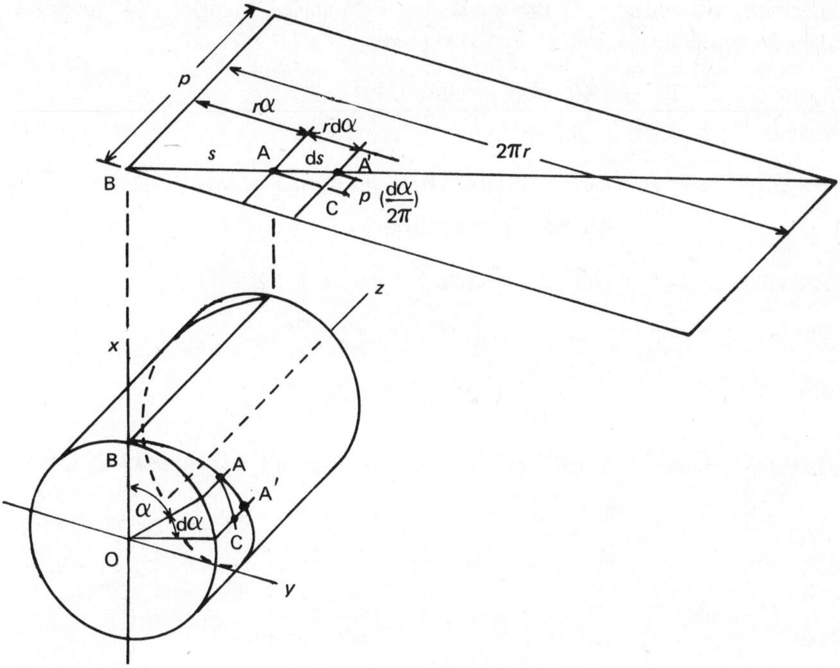

Example 3.2
The figure illustrates a uniform helix of pitch p drawn on a cylinder of radius r whose axis coincides with the z axis of rectangular coordinates. A general point A on the helix being identified by the angle α, determine the unit tangent vector $\hat{\mathbf{I}}_t$, the radius of curvature ρ, and the unit normal $\hat{\mathbf{I}}_n$ as functions of α.

Note In a uniform helix the displacement parallel to the axis is proportional to the angle α. In the plane development of the cylindrical surface the helix therefore appears as a straight line, as indicated.

Solution
In this case the position of the Centre of Curvature is not apparent. Nor can the radius vector $\mathbf{r}$ to the point A be readily described as an explicit function of the distance along the path s. However, the relationship between $\mathbf{r}$ and s can readily be described in parametric form by describing each as a function of α, when d/ds can be equated to $(d/d\alpha)/(ds/d\alpha)$.

From the plane development of the surface of the cylinder, it is obvious that, by Pythagoras' Theorem

$$s = \sqrt{((r\alpha)^2 + (p\alpha/2\pi)^2)}$$
$$= r\alpha\sqrt{(1 + (p/2\pi r)^2)}$$

Therefore $\quad ds/d\alpha = r\sqrt{(1 + (p/2\pi r)^2)}$.

Alternatively, considering the elemental, right-angled triangle ACA′ drawn to the adjacent points A and A′, by Pythagoras

$$ds = \sqrt{((r\,.\,d\alpha)^2 + (p d\alpha/2\pi)^2)}$$

whence $\quad ds/d\alpha = r\sqrt{(1 + (p/2\pi r)^2)}$, as before.

The coordinates of A can be described in terms of α as follows

$$A = \{r\cos\alpha,\quad r\sin\alpha,\quad (p\alpha/2\pi)\}\ .$$

Therefore $\quad \mathbf{r} = r\{\cos\alpha\,\mathbf{i} + \sin\alpha\,\mathbf{j} + (p/2\pi r)\alpha\,\mathbf{k}\}$

and $\quad d\mathbf{r}/d\alpha = \mathbf{r}\{-\sin\alpha\,\mathbf{i} + \cos\alpha\,\mathbf{j} + (p/2\pi r)\mathbf{k}\}$

But $\quad \hat{\mathbf{I}}_t = \dfrac{d\mathbf{r}}{ds} = \dfrac{d\mathbf{r}}{d\alpha}\Big/\dfrac{ds}{d\alpha}$

that is $\quad \hat{\mathbf{I}}_t = \{1 + (p/2\pi r)^2\}^{-\frac{1}{2}}\{-\sin\alpha\,\mathbf{i} + \cos\alpha\,\mathbf{j} + (p/2\pi r)\mathbf{k}\}\ .$

Also $\quad \dfrac{1}{\rho}\hat{\mathbf{I}}_n = \dfrac{d\hat{\mathbf{I}}_t}{ds} = \dfrac{d\hat{\mathbf{I}}_t}{d\alpha}\ \dfrac{ds}{d\alpha}$

that is $\quad \dfrac{1}{\rho}\hat{\mathbf{I}}_n = r\left\{1 + \left(\dfrac{p}{2\pi r}\right)^2\right\}^{-1}(-\cos\alpha\,\mathbf{i} - \sin\alpha\,\mathbf{j})\ .$

Since $\cos^2\theta + \sin^2\theta = 1$, the vector $(-\cos\alpha\,\mathbf{i} - \sin\alpha\,\mathbf{j})$ is evidently a unit vector, whence

$$\rho = r\left\{1 + \left(\frac{p}{2\pi r}\right)^2\right\}\ .$$

$$\hat{\mathbf{I}}_n = -(\cos\alpha\,\mathbf{i} + \sin\alpha\,\mathbf{j})\ .$$

3.3 MOTION OF A POINT ALONG A GIVEN FIXED PATH

Consider the motion of a point particle along a path which is defined in a given frame of reference as a smoothly continuous curve.

Position

The position of the particle at any instant can be defined by its scalar arc-distance s along the path, and this can be regarded as a function of time.

Speed

The speed v of the particle at the instant of passing through any point on the path is defined as the instantaneous time-rate of change of the distance along the path. Thus if s is the distance along the path of the position P at time t, and if $s+\delta s$ is the distance along the path of the position P′ at time $t+\delta t$, the distance travelled in the interval of time δt is represented by δs, and

by definition

$$v = \frac{ds}{dt} = \underset{\delta s \to 0}{\text{Limit}} \frac{\delta s}{\delta t} \tag{3.4}$$

Acceleration-along-the-path

The acceleration-along-the-path is defined as the instantaneous time-rate of change of speed. Thus if v represents the speed of the particle in passing through P at time t, the speed in passing through P′ at time $t+\delta t$ is represented by $v+\delta v$, and

by definition

$$a = \frac{dv}{dt} = \underset{\delta t \to 0}{\text{Limit}} \frac{\delta v}{\delta t} \tag{3.5}$$

Since both time and the distance along the path are scalars, the speed and the acceleration-along-the-path are also scalars.

Derived Relationships

If the motion of the particle is determined by describing the distance s as a function of time t, the definitions provide for the direct determination of the speed v and the acceleration-along-the-path a simply by repeated differentiation But with certain reservations, the motion can as well be determined by describing any of the quantities: s, v, t and a as a function of any other, and since the standard differentials and integrals are applicable only to the differentiation of a function with respect to its own argument, it is then necessary to manipulate the definitive relationships into different forms.

In fact, the four quantities t, s, v and a can be combined in groups of three in four ways: (v, s and t), (a, v and t), (a, s and t), and (a, v and s). Relationships between the first two sets are provided (in differential form) directly by the definitions thus: $v = ds/dt$ (i); and $s = dv/dt$ (ii).

The substitution of (i) in (ii) gives a relationship between a, s and t as follows: $a = d^2s/dt$.

A relationship between the fourth set may be determined as follows

By definition

$$a = \frac{dv}{dt} = \frac{dv}{ds} \cdot \frac{ds}{dt} .$$

But ds/dt is equal to v by definition, whence

$$a = v \cdot \frac{dv}{ds} = \frac{1}{2} \frac{d(v^2)}{ds} .$$

Each of the differential relationships can also be written in integral form. For example, by definition, the speed is given by $v = \mathrm{d}s/\mathrm{d}t$; but suppose that v is given as a function of s and we require to determine t, or as a function of t and we require to determine s. The definitive equation can then be rewritten as follows.

By definition

$$v = \mathrm{d}s/\mathrm{d}t\ .$$

Therefore $\mathrm{d}s = v.\mathrm{d}t$; or $\mathrm{d}t = (1/v).\mathrm{d}s$.

Either of these equations can then be integrated between limits to give, in the one case the distance travelled in a given interval of time, or in the other, the time taken to travel over a given section of the path. Thus if s_1 and s_2 represent the distances along the path at times t_1 and t_2, the foregoing equations can be integrated between these limits to give

$$\int_{s_1}^{s_2} \mathrm{d}s = \int_{t_1}^{t_2} v.\mathrm{d}t; \quad \text{or} \quad \int_{t_1}^{t_2} \mathrm{d}t = \int_{s_1}^{s_2} v^{-1}.\mathrm{d}s$$

or $$(s_2 - s_1) = \int_{t_1}^{t_2} v.\mathrm{d}t; \quad \text{or} \quad (t_2 - t_1) = \int_{s_1}^{s_2} v^{-1}.\mathrm{d}s\ .$$

Alternatively, the indefinite integral of one side of the equation can be equated to the indefinite integral of the other plus a constant integration. Thus

$$s = \int v.\mathrm{d}t + c; \quad \text{or} \quad t = \int v^{-1}.\mathrm{d}s + c\ .$$

Then provided we are given the value of s at one time t, this information can be substituted in either of the equations to determine the appropriate value of c, and the result will then determine, in the one case s as a function of t, or in the other, t as a function of s.

To summarise, the speed is defined as the derivative of the distance along the path, whilst the acceleration-along-the-path is defined as a derivative of the speed, both with respect to time. But taken together, these definitions imply that the acceleration can alternatively be equated either to the second derivative of the distance with respect to time, or to the speed times its derivative with respect to distance, or to half the derivative of the square of the speed with respect to distance. Furthermore, each of these equations can be rewritten in equivalent integral forms, and the complete set of relationships are indicated in Fig. 3.7.

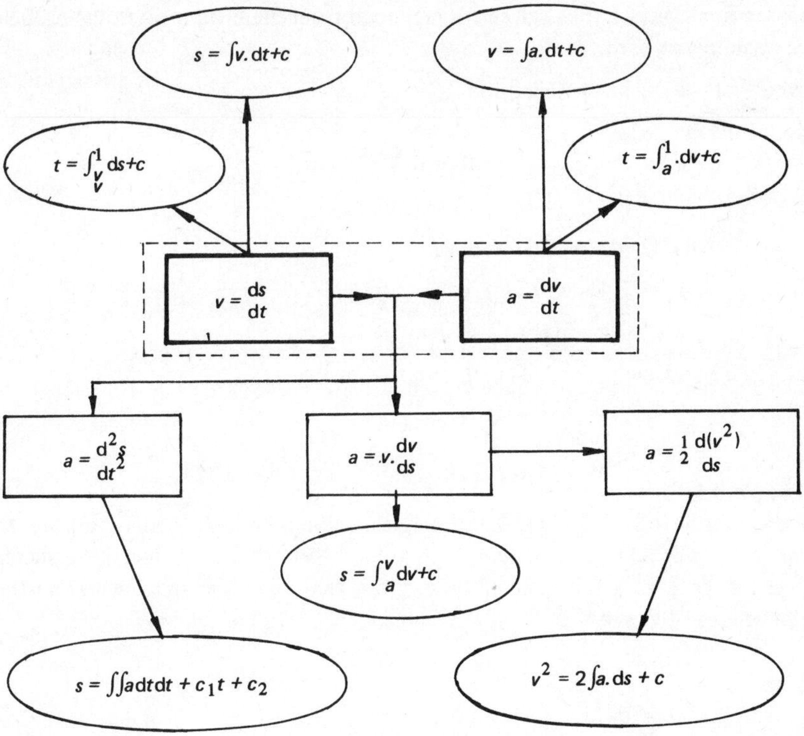

Fig. 3.7

Example 3.3
In the motion of a particle along a fixed path, the distance along the path s metres, varies with time t secs. according to $s = 4 + 2t^2 + t^3$. Determine the speed and the acceleration-along-the-path at time $t = 4$ seconds.

Solution

Given that $s = 4 + 2t^2 + t^3$

By definition

$$v = ds/dt = 4t + 3t^2$$

and $a = dv/dt = 4 + 6t$

Therefore $v_{t=4} = 4.4 + 3.4^2 = 64$ m/s

and $a_{t=4} = 4 + 6.4 = 28$ m/s^2 .

Example 3.4
The speed v m/s of a moving particle varies with time t seconds, thus: $v = 4t + 0.1t^2$. Determine the acceleration-along-the-path and the distance along the

path as functions of time, given t is reckoned from the instant when the particle is 2 m along the path.

Given that $v = 4t + 0{\cdot}3t^2$ m/s

By definition

$$a = \mathrm{d}v/\mathrm{d}t = 4 + 0{\cdot}6t \text{ m/s}^2$$

Also $v = \mathrm{d}s/\mathrm{d}t = 4t + 0{\cdot}3t^2$ m/s

$$\mathrm{d}s = (4t + 0{\cdot}3t^2)\,.\,\mathrm{d}t \text{ m}$$

$$s = \int(4t + 0{\cdot}3t^2)\,.\,\mathrm{d}t + c \text{ m}$$

$$s = (2t^2 + 0{\cdot}1t^3) + c \text{ m}$$

But when $t = 0$, then $s = 2$. Therefore $c = 2$, giving

$$s = 2 + 2t^2 + 0{\cdot}1t^3 \text{ m}\ .$$

Example 3.5

In the motion of a particle the acceleration-along-the-path a m/s^2 varies with time t seconds according to $a = 1 + 0{\cdot}6t + 0{\cdot}12t^2$. Determine the distance along the path as a function of time, given that time was reckoned from the position of rest at $s = 8$ m.

Solution

$$a = \mathrm{d}v/\mathrm{d}t$$

therefore $\mathrm{d}v = a\,.\,\mathrm{d}t$.

Integrating $v = \int a\,.\,\mathrm{d}t + c_1$

that is $v = \int(1 + 0{\cdot}6t + 0{\cdot}12t^2)\,.\,\mathrm{d}t + c_1$

or $v = t + 0{\cdot}3t^2 + 0{\cdot}04t^3 + c_1$. (a)

Also $v = \mathrm{d}s/\mathrm{d}t$.

Therefore $s = \int v\,.\,\mathrm{d}t + c_2$

that is $s = 0{\cdot}5t^2 + 0{\cdot}1t^3 + 0{\cdot}01t^4 + c_1t + c_2$. (b)

But when $t = 0$, then $v = 0$ and $s = 8$, and substituting accordingly in (a) and (b) respectively gives: $c_1 = 0$, and $c_2 = 8$. Therefore

$$v = t + 0{\cdot}3t^2 + 0{\cdot}04t^3 \text{ m/s}$$

and $s = 8 + 0{\cdot}5t^2 + 0{\cdot}1t^3 + 0{\cdot}01t^4$ m .

Example 3.6

A particle moves so that its speed v m/s varies with distance s m from rest according to $v = \sqrt{(16 - 9s^2)}$. Determine the scalar acceleration and the time from rest, both as functions of distance.

We have that

$$a = \frac{1}{2} \cdot \frac{d(v^2)}{ds} = \frac{1}{2} \cdot \frac{d(16 - 9s^2)}{ds} ,$$

that is $a = -9s \text{ m/s}^2$

Alternatively

$$a = v \cdot \frac{dv}{ds} = (16 - 3s^2)^{\frac{1}{2}} . \frac{d(16 - 3s^2)^{\frac{1}{2}}}{ds}$$

Whence $a = -9s \text{ m/s}^2$

By definition

$$v = ds/dt .$$

Therefore $t = \int v^{-1} . ds + c$

that is $$t = \frac{ds}{\sqrt{(16-9s^2)}} + c ,$$

giving $$t = \frac{1}{3} . \sin^{-1}\left(\frac{3s}{4}\right) + c .$$

Substituting $t = 0$ when $s = 0$, gives $c = -n\pi/3$; (n = integer.

Therefore $$t = \frac{1}{3}\left\{\sin^{-1}\left(\frac{3s}{4}\right) - n\pi\right\} s .$$

Example 3.7

In the motion of a particle the accleration-along-the-path a varies with the speed v according to $a = v(6-v)$. Determine v as a function of distance s, given that $v = 4$ when $s = 0$.

Solution

Given that $a = v . (6 - v)$

and that $a = v . dv/ds$,

then $dv/ds = 6 - v$

Therefore $$ds = \frac{dv}{6 - v}$$

Integrating $s = -\log(6 - v) + \log_e c$

or $$e^s = \frac{c}{6 - v} .$$

Substituting $v = 4$ when $s = 0$ gives $c = 2$

$$6 - v = 2e^{-s}$$
$$v = 6 - 2e^{-s}$$

3.4 MOTION OF A POINT AS SEEN BY A SINGLE OBSERVER

We have seen that when a point moves along a known path, its position at any instant can be determined by its scalar distance along the path, and we idenfity two scalar features of the motion: the speed, which is defined as the instantaneous time-rate of change of distance along the path, and the acceleration-along-the-path, which is defined as the instantaneous time-rate of change of speed. But more generally, when the path of the motion is not independently defined, the scalar approach gives way to a vectorial treatment. In brief, the position of the point of interest relative to a given datum is then defined by the vector drawn between them, and we identify two vectorial features of the motion: the velocity, which is defined as the instantaneous time-rate of change of position, and the acceleration, which is defined as the instantaneous time-rate of velocity. The treatment of the general case will therefore be closely similar to that accorded to the special case, but whereas we were then concerned with scalar derivatives of scalars, we are now concerned with vectorial derivatives of vectors. It is therefore particularly significant that whereas the derivative of a scalar is independent of the observer, the derivative of a vector may have different values for different observers, depending on their relative rotation. In particular, it follows that the concepts of velocity and acceleration necessarily imply, not only a particular datum to which the position of the point of interest is referred, but also a particular observer to whom the operation of differentiation is referred.

In general, the datum may be chosen as any point whose motion is known, but particular interest attaches to the special case in which the datum is fixed relative to the observer, when the observer of the motion effectively constitutes its datum also.

As to the observer, he is effectively treated as a rigid body whose essential features are his position and his orientation. However these can more conveniently be defined by referring, not to the observer himself, but to a frame of reference that is fixed within him, so each observer is thus provided, and we distinguish between different observers by referring to the origins of their frames of reference.

From the need thus to distinguish between different observers, it follows that the dot convention that proves so convenient for indicating the operation of differentiating a scalar with respect to time, is not generally suitable for use with vectors. Instead, in application to vectors, a single differentiation will be indicated by the symbol D, whilst a double differentiation will be indicated by D^2, and we shall then be able to indicate the intended observer by nominating the origin of his frame of reference in a suffix. We note, however, that D^2 indicates the second derivative d^2/dt^2, and not the square of the first derivative $(d/dt)^2$.

Motion Relative to an Arbitrary Datum

Irrespective of the observer, and even though the datum is itself in independent motion, the position of a given point P (Fig. 3.8) relative to a given datum A is defined by the vector $\mathbf{r}_{PA}$ that is drawn to P from the datum A.

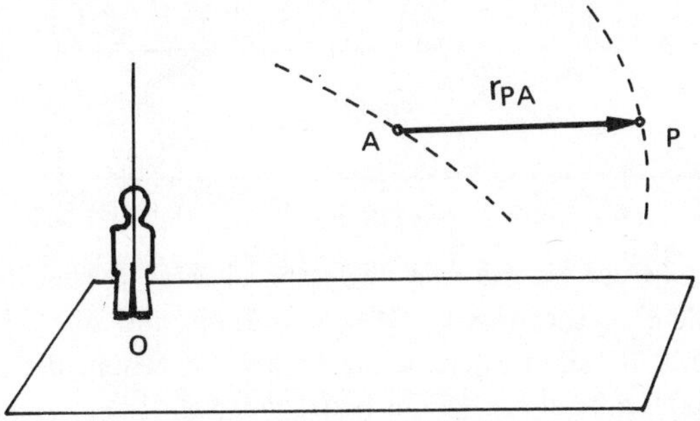

Fig. 3.8

Taking it as understood that we refer throughout to rates of change, velocities and accelerations as seen by a single observer O, the velocity of P relative to A is then defined as the instantaneous time-rate of change of the position of P relative to A, whilst the acceleration of P relative to A is defined as the instantaneous time-rate of change of the velocity of P relative to A.

Symbolically, a velocity is represented by $\mathbf{v}$, and an acceleration by $\mathbf{a}$, each being qualified by a first suffix which identifies the observer. This is followed by a pair of suffices which, as in the case of the position vector, identify first the point of interest, and then the datum point. Thus, for the motion as seen by an observer O, of the point P relative to an arbitrary datum A we may write

By definition

$$\mathbf{v}_{O/PA} = D_O \mathbf{r}_{PA} \tag{3.6a}$$

and

$$\mathbf{a}_{O/PA} = D_O \mathbf{v}_{O/PA} = D_O^2 \mathbf{r}_{PA} \tag{3.6b}$$

Motion Relative to the Observer Himself

The position of the point P relative to the observer himself can be determined by the vector drawn from any given point that is fixed relative to the observer, and it follows that for a given observer, a given point has a unique motion relative to all datum points that are fixed in the observer himself. Thus, referring to Fig. 3.9, let A and B represent alternative datum points, each of which is fixed in the observer's own frame or reference. Then whereas the position of P relative to A is defined by the vector $\mathbf{r}_{PA}$, its position relative to B is defined

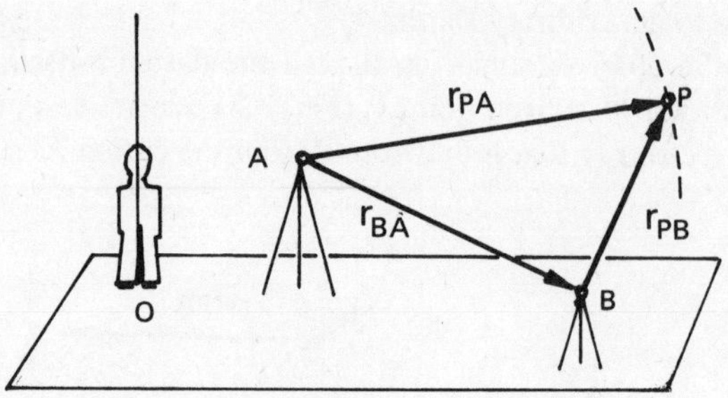

Fig. 3.9

by the vector $\mathbf{r}_{PB}$; yet by the rule of vector addition these differ only by the vector $\mathbf{r}_{BA}$, which, to the observer O is a constant, and since the derivative of a constant is zero, it follows that to the observer in whom the points A and B are fixed, the derivatives of $\mathbf{r}_{PB}$ will be equal to those of $\mathbf{r}_{PA}$.

To indicate such a motion in which the observer O himself also constitutes the datum we shall refer simply to the motion of P relative to the observer O, or equivalently, to the motion of P in the frame of reference O, taking it as understood that the use of either expression implies that the datum may be chosen as any point that is fixed relative to the nominated observer of the motion.

A Useful Relationship

Referring to Fig. 3.10, it is obvious that by the rule of vector addition

$$\mathbf{r}_{PA} = \mathbf{r}_{PO} - \mathbf{r}_{AO}$$

therefore $$D_O\mathbf{r}_{PA} = D_O\mathbf{r}_{PO} - D_O\mathbf{r}_{AO}$$

whence $$\mathbf{v}_{O/PA} = \mathbf{v}_{O/PO} - \mathbf{v}_{O/AO} \tag{3.7a}$$

and $$\mathbf{a}_{O/PA} = \mathbf{a}_{O/PO} - \mathbf{a}_{O/AO} \tag{3.7b}$$

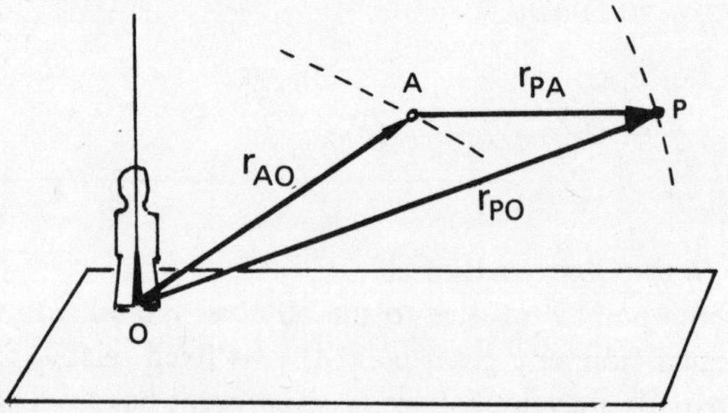

Fig. 3.10

We therefore conclude that the motion as seen by the observer O of a point P relative to an arbitrary datum A can be determined as the difference between the motions of the point P and the datum A, both relative to the observer O.

3.5 ALTERNATIVE OBSERVERS IN SIMPLE, RELATIVE TRANSLATION

The derivative of a given vector depends on the observer only in respect of his rotation, and not of his translation, and it follows that although the various velocities and accelerations in the foregoing relationships were all attributed to the one observer O, any or all of them are equally attributable to any observer,

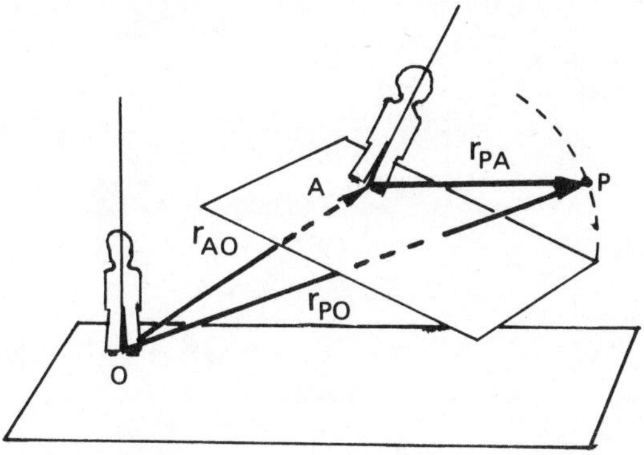

Fig. 3.11

provided only that he does not rotate relative to the observer O. So referring to Fig. 3.11, let A now represent the origin of the frame of reference of a second observer, who, relative to observer O, is in simple translation without rotation. In the terms on the left of the foregoing relationships we may then substitute the observer A for the observer O to write

$$\mathbf{v}_{A/PA} = \mathbf{v}_{O/PO} - \mathbf{v}_{O/AO} \tag{3.8a}$$

and

$$\mathbf{a}_{A/PA} = \mathbf{a}_{O/PO} - \mathbf{a}_{O/AO} \tag{3.8b}$$

In each term, the datum (as indicated by the last suffix) now falls at the origin of the observer's frame of reference, (as indicated by the first suffix). We therefore conclude that

The motion of P relative to an observer A who does not rotate relative to an observer O	=	The motion of P relative to observer O	−	The motion of A relative to observer O

The significance of this relationship is that it allows us to infer from observations made from one frame of reference, what the motion of a given point would be when viewed from another, provided only that there is no real relative rotation between the references. To illustrate the point, Fig. 3.12 shows a shore-based observer O who studies the motion of two ships. One ship A steers a path which appears in the frame of reference O as a straight line, whilst the other B steers a path which appears as a curve. This implies that whereas an observer fixed in ship A retains a fixed orientation in the frame of reference O, an observer fixed in ship B would not, so that whereas the shore-based observations on

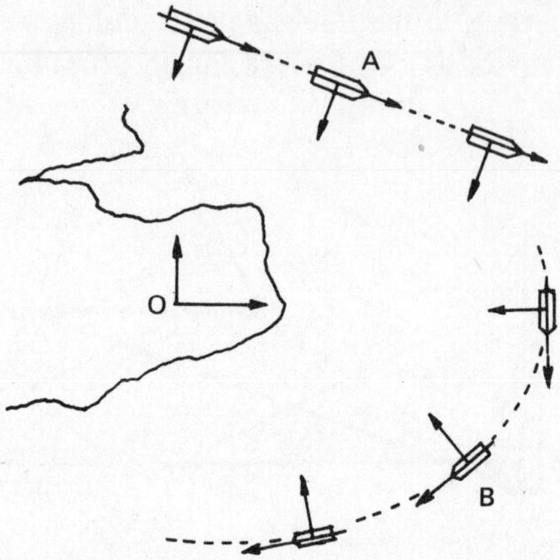

Fig. 3.12

A and B could be used to infer the motion of B relative to observer A, they could not be used to infer the motion of A relative to observer B. We note, however, that the impediment arises, not directly from the curvature of the path of ship B, but from the rotation that results from the steering of the ship. For example, suppose that B′ represents a second observer in ship B, but one who is fixed, not in the ship itself, but in a steerable platform that is mounted within it. If the platform is steered so that it retains a fixed heading, the shore-based observations could be used to infer the motion of A relative to the observer B′, irrespective of the path of the ship in which his platform is mounted.

3.6 GENERAL POINT MOTION IN FIXED RECTANGULAR COORDINATES

In applying the foregoing conclusions we are free to adopt the system of coordinates that best accords with the general form of the motion, but rectangular

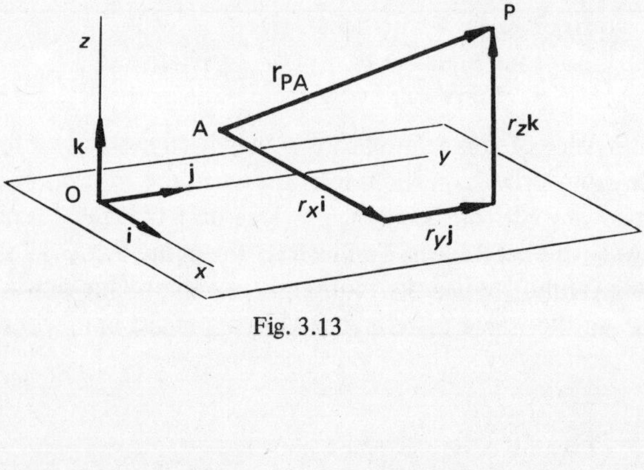

Fig. 3.13

coordinates that are fixed in the observer's own frame of reference have a particular convenience, because the unit vectors **i**, **j** and **k** which then define the fixed directions in which all vectors are resolved are, to the observer constants, and we have argued that for him, the operations of differentiation and integration can be applied to a vector simply by applying the operations to each of the scalar elements of the vector in turn. Thus let $\mathbf{r}_{PA}$ represent the vector function which describes the variation in the position of a given point P relative to a given datum A, and let r_x, r_y and r_z represent the scalar functions which describe the variations of its scalar components in the directions of rectangular coordinates fixed in the observer O.

Then $\mathbf{r}_{PA} = r_x\,.\mathbf{i} + r_y\,.\mathbf{j} + r_z\,.\mathbf{k}$

Whence $\mathbf{v}_{O/PA} = D_O\mathbf{r}_{PA} = \dot{r}_x\,.\mathbf{i} + \dot{r}_y\,.\mathbf{j} + \dot{r}_z\,.\mathbf{k}$ (3.9a)

and $\mathbf{a}_{O/PO} = D_O^2\mathbf{r}_{PA}\ \ddot{r}_x\,.\mathbf{i} + \ddot{r}_y\,.\mathbf{j} + \ddot{r}_z\,.\mathbf{k}$ (3.9b)

Thus, in rectangular coordinates that are fixed in the observer of a motion, the elements of the velocity and acceleration vectors of the motion can be determined as the first and second derivatives of the scalar elements of the appropriate position vector.

Example 3.8

In rectangular coordinates Oxyz, the coordinates of two points P and A vary with time t according to

$$P = \{(4 + 2t), 4t^2, 2t^3\}; \quad \text{and} \quad A = \{t^2, (3 + t^2), 2t\} .$$

Determine the velocities and accelerations of P and A relative to the observer O, and the velocity, as seen by the observer O, of P relative to A.

Solution

$\mathbf{r}_{PO} = (4 + 2t)\,.\mathbf{i} + 4t^2.\mathbf{j} + 2t^3.\mathbf{k}$.

Therefore $\mathbf{v}_{O/PO} = D_O\mathbf{r}_{PO} = 2\,.\mathbf{i} + 8t\,.\mathbf{j} + 6t^2.\mathbf{k}$,

and $\mathbf{a}_{O/PO} = D_O^2\mathbf{r}_{PO} = 8\,.\mathbf{j} + 12t\,.\mathbf{k}$

Also $\mathbf{r}_{AO} = t^2\mathbf{i} + (3 + t^2)\,.\mathbf{j} + 2t\,.\mathbf{k}$

$\mathbf{v}_{O/AO} = D_O\mathbf{r}_{AO} = 2t\,.\mathbf{i} + 2t\,.\mathbf{j} + 2\,.\mathbf{k}$

$\mathbf{a}_{O/AO} = D_O^2\mathbf{r}_{PO} = 2\,.\mathbf{i} + 2\,.\mathbf{j}$.

The motion, as seen by observer O, of P relative to A can now be determined either by differenting the vector $\mathbf{r}_{PA}$, or as the difference between the motions of P and A relative to the observer O. Thus, determining the elements of $\mathbf{r}_{PA}$ as the difference between the coordinates of the final point P and the initial point A, we have

$$\mathbf{r}_{PA} = (4 + 2t - t^2).\mathbf{i} + 3(t^2 - 1).\mathbf{j} + 2(t^3 - t).\mathbf{k}$$

whence $\mathbf{v}_{O/PA} = D_O\mathbf{r}_{PA} = 2(1 - t).\mathbf{i} + 6t.\mathbf{j} + 2(3^2 - 1).\mathbf{k}$

and $\mathbf{a}_{O/PA} = D_O^2\mathbf{r}_{PA} = -2.\mathbf{i} + 6.\mathbf{j} + 12t.\mathbf{k}$.

Alternatively the same results can be obtained by writing

$$\mathbf{v}_{O/PA} = \mathbf{v}_{O/PO} - \mathbf{v}_{O/AO}$$

that is $\mathbf{v}_{O/PA} = (2.\mathbf{i} + 8t.\mathbf{j} + 6t^2\mathbf{k}) - (2t.\mathbf{i} + 2t.\mathbf{j} + 2.\mathbf{k})$

or $\mathbf{v}_{O/PA} = 2(1 - t).\mathbf{i} + 6t.\mathbf{j} + 2(3t^2 - 1).\mathbf{k}$.

Similarly $\mathbf{a}_{O/PA} = \mathbf{a}_{O/PO} - \mathbf{a}_{O/AO}$

that is $\mathbf{a}_{O/PA} = -2.\mathbf{i} + 6.\mathbf{j} + 12t.\mathbf{k}$.

Example 3.9

The illustration shows a slider P which runs along a horizontal arm B, which itself both turns about the fixed vertical post C, and slides vertically along it.

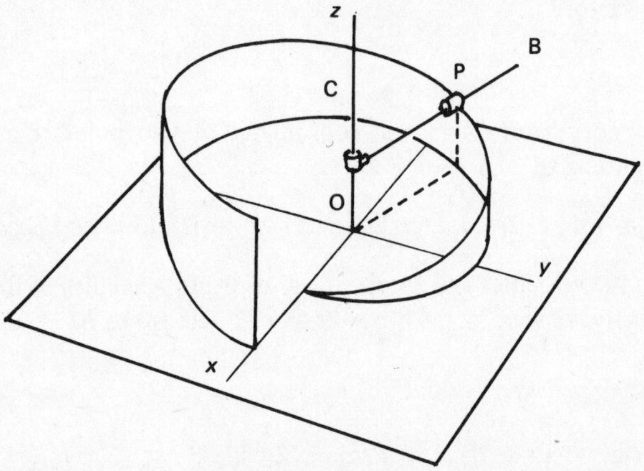

Initially the arm lies along the x axis with the slider at a distance of 2 m from the axis of the vertical post C, from which position the slider runs outward along the arm at a constant speed of 4 m/s, whilst the arm rotates about the post at a constant speed of 5 rev/s, and slides vertically at a constant speed of 6 m/s.

Determine the velocity and acceleration of the slider in the frame of reference Oxyz (i.e., relative to an observer O who is fixed in Oxyz), at time $t = 10$ s.

Solution

From the given information the coordinates of the slider can be described as a function of time t, and since these are then the scalar elements of the radius vector which describes the position of the slider relative to a point (the origin O) that is fixed in the observer O, the scalar elements of the velocity and acceleration of the slider relative to the observer O can be determined by the successive differentiation of the coordinates of P.

Thus in time t seconds, the arm rotates about the post through the angle $\phi = \omega t = 10\,\pi\,t$, whilst the slider moves outwards along the arm to the radius $r = 2 + 4t$ m, and the arm moves vertically a distance of 6t from the xy plane. Therefore, equating the scalar elements of the vector $\mathbf{r}_{PO}$ to the coordinates of its final point P we have

$$\mathbf{r}_{PO} = (2+4t)\cos 10\pi t\,.\mathbf{i} + (2+4t)\sin 10\pi t\,.\mathbf{j} + 6t\,.\mathbf{k}$$

therefore $$\mathbf{v}_{O/PO} = \mathrm{D_O}\mathbf{r}_{PO} = \{4\cos 10\pi t - (2+4t)10\pi\sin 10\pi t\}..\mathbf{i} + \{4\sin 10\pi t + (2+4t)10\pi\cos 10\pi t\}.\mathbf{j} + 6.\mathbf{k}$$

And $$\mathbf{a}_{O/PO} = \mathrm{D_O^2}\mathbf{r}_{PO} = \{-80\pi\sin 10\pi t - (2+4t)100\pi^2\cos 10\pi t\}.\mathbf{i} + \{80\pi\cos 10\pi t - (2+4t)100\pi^2\sin 10\pi t\}.\mathbf{j}\;.$$

The velocity and acceleration at time $t = 10\,s$ can now be determined by substituting accordingly to give

$$\mathbf{v}_{O/PO} = 4.\mathbf{i} + 420\pi.\mathbf{j} + 6.\mathbf{k}$$
$$\mathbf{a}_{O/PO} = -420\pi^2\mathbf{i} + 80\pi.\mathbf{j}\;.$$

Note

You are invited to make use of your present knowledge of velocity and acceleration in circular motion to interpret the given data directly in terms of their radial and circumferential components, and to describe these directly in terms of their scalar components in the directions of the given coordinates.

3.7 GENERAL POINT MOTION IN CYLINDRICAL COORDINATES

In the preceding section we were able to differentiate a vector simply by applying the operation to each of the scalar element of the vector in turn. However, this was so only because the unit vector **i, j** and **k** which there defined the directions of resolution could be treated as constants, and this in turn depended upon the use of rectangular coordinates which were drawn in the observer's own frame of reference. For with each system of coordinates it is appropriate to choose the directions of resolution to agree with the directions of the coordinates themselves, and since it is only the Cartesian system that makes use solely of linear coordinates whose directions are fixed in the frame of reference in which they are drawn, it follows that the simple rule of differentiation is valid only in rectangular coordinates drawn in a frame of reference which does not rotate relative to the observer. With any coordinates other than rectangular, and even with these when they are drawn in a frame of reference which rotates, it is necessary to take account not only of the variations in the scalar elements of the vector, but also in the unit vectors which define their directions, and with cylindrical coordinates this is effectively achieved by describing the derivatives of vectors as sums of rectangular components of which one is colinear with the vector itself and depends solely on the variation of its modulus, whilst the other is normal to the vector itself and depends solely on the variation in its direction.

Derivatives of Vectors in Terms of their Colinear and Normal Components

Thus in Fig. 3.14, **r** and **r** + d**r** represent the values adopted by a vector function when its scalar argument suffers an infintesimal variation from s to $s + ds$. The increment d**r** is determined by the rule of vector addition as indicated, and in

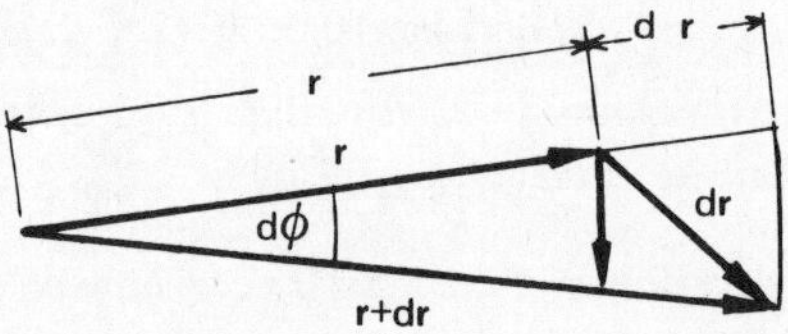

Fig. 3.14

the limit, it can evidently be described as a sum of two rectangular components of which one is colinear with **r** itself and has a modulus equal to the variation $d|\mathbf{r}|$ in its modulus, whilst the other is normal to **r** itself and has a modulus equal to the modulus of **r** times the angle $d\phi$ which specifies the variation in its direction. It then follows that the derivative of **r** may similarly be described as a sum of two rectangular components, of which one is colinear with **r** itself and has a magnitude equal to the derivative $d|\mathbf{r}|/ds$ of its modulus, whilst the other is normal to **r** and has a magnitude equal to the product of its modulus $|\mathbf{r}|$ with the derivative $d\phi/ds$ which determines the rate of change of its direction.

More particularly, when the argument of **r** is time t, the derivative of **r** may be indicated by the operator D, whilst the derivatives of the scalars may be indicated in the dot convention, and if $\hat{\mathbf{1}}_r$ represents the unit vector which coincides with **r** whilst $\hat{\mathbf{1}}_\phi$ represents the unit vector in the appropriate sense normal to **r**, the derivative of **r** can be described vectorially as follows

$$D_o\mathbf{r} = |\dot{\mathbf{r}}|\,\hat{\mathbf{1}}_r + \mathbf{r}\,\dot{\phi}\,\hat{\mathbf{1}}_\phi \ .$$

We note that the rate of change of direction $\dot{\phi}$ is not generally the same as the angular speed of the vector. Thus, Fig. 3.15(a) shows a vector which varies around the surface of a cone, and it is clear that the angular speed of the vector is then given by $\dot{\theta}$ rather than by $\dot{\phi}$. However, we are now effectively concerned only with vectors which are confined to a plane, (Fig. 3.15(b), and since the distinction then vanishes, for this case we may summarise as follows.

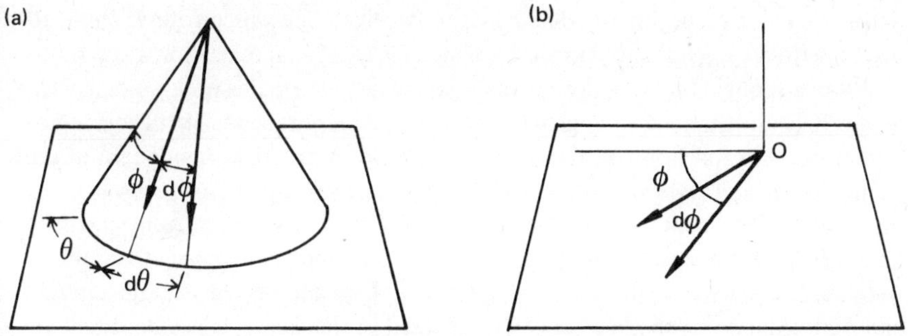

Fig. 3.15

When a vector function of time is confined to a plane, its derivative is confined to the same plane, and can be described as a sum of two rectangular components, of which one is colinear with the vector itself and has a modulus equal to the derivative of its modulus. The other, which is normal to the vector itself depends solely on the rotation of the vector and has a magnitude equal to the product of its modulus with its angular speed. In particular, for a vector of constant modulus, the colinear component of its derivative is zero, and when the constant value of the modulus is unity, the magnitude of the remaining normal component of the derivative is equal to the angular speed of the vector.

Cylindrical Coordinates, and the unit vectors ($\hat{1}_r, \hat{1}_\phi, \hat{1}_z$)

As indicated in Fig. 3.16(a), the cylindrical (r, ϕ, z) system employs only two linear coordinates, of which z is measured in the fixed direction normal to the datum plane, whereas the r coordinate is measured in the variable direction

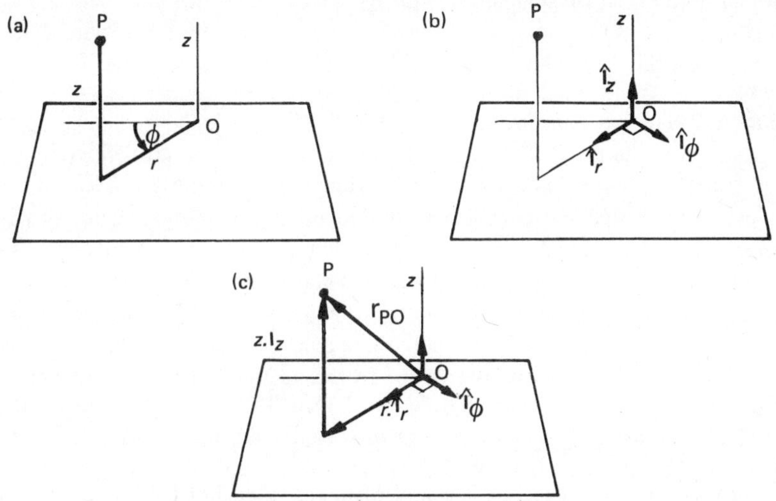

Fig. 3.16

defined by the ϕ coordinate. The r axis is therefore one which rotates about the z axis, with an angular speed given by $\dot{\phi}$, and an angular acceleration given by $\ddot{\phi}$.

But although the direction of the r axis varies, it remains in the datum plane where it is always at right-angles to the z axis. An appropriate set of rectangular directions of resolution can therefore be defined by the right-handed set of unit vector $(\hat{1}_r, \hat{1}_\phi, \hat{1}_z)$ indicated in Fig. 3.16(b), where $\hat{1}_r$ and $\hat{1}_z$ are chosen in the directions of the r and z axes. Then for a right-handed set, $\hat{1}_\phi$ must be chosen to lie, with $\hat{1}_r$, in the datum plane, but in a direction which is advanced from that of $\hat{1}_r$ by $\pi/2$. Relative to the frame of reference, $\hat{1}_z$ is then fixed in both modulus and direction, so to an observer O who is fixed in the frame, $\hat{1}_z$ is a constant, and its derivative is zero. However, since $\hat{1}_r$ and $\hat{1}_\phi$ are fixed to the rotating r axis, to the observer O they are variables, and their derivatives have a common modulus equal to their common angular speed $\dot{\phi}$. Furthermore, since $\hat{1}_r$ and $\hat{1}_\phi$ are rectangular vectors of constant modulus, the derivative of each is parallel to the

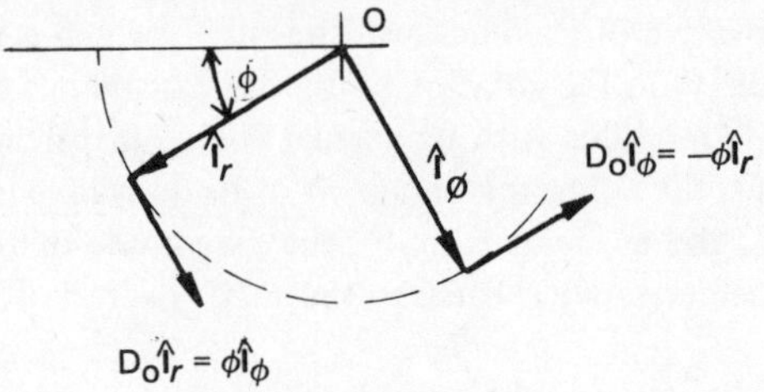

Fig. 3.17

other. But Fig. 3.17 shows that whereas the direction of the derivative of $\hat{1}_r$ is the same as that of $\hat{1}_\phi$, the direction of the derivative of $\hat{1}_\phi$ is opposed to that of $\hat{1}_r$, and we may therefore conclude as follows.

$$D_o\hat{1}_r = \dot{\phi}\,\hat{1}_\phi;\ \ D_o\hat{1}_\phi = -\dot{\phi}\,\hat{1}_r;\ \ D_o\hat{1}_z = 0 \tag{3.10}$$

Motion in Cylindrical Coordinates

The position of the point P in a given frame of reference may be defined by the radius vector $\mathbf{r}_{po}$ which is drawn to P from the origin O, and Fig. 3.14(c) shows, that given the cylindrical coordinates of the point, this radius vector can be described as a sum of only two rectangular components, of which one has a modulus equal to the r coordinate and lies in the variable direction defined by $\hat{1}_r$, whilst the other has a modulus equal to the z coordinate, and lies in the fixed direction defined by $\hat{1}_z$. Vectorially these components are therefore defined by $r\,\hat{1}_r$ and $z\,\hat{1}_z$, and equating the vector to the vector sum of its components, we have

$$\mathbf{r}_{po} = r\,\hat{1}_r + z\,\hat{1}_z\ .$$

of course, if P moves, then r, ϕ and z are variables, but to the observer O, so also are $\hat{1}_r$ and $\hat{1}_\phi$, and only $\hat{1}_z$ is a constant. So treating the factors of the various products accordingly, and substituting from (3.10), the observer O determines the velocity and acceleration of P by the successive differentiation of the radius vector, as follows

$$\mathbf{v}_{o/po} = \mathrm{D_o}\mathbf{r}_{po} = \mathrm{D_o}(r\,\hat{\mathbf{1}}_r) + \mathrm{D_o}(z\,\hat{\mathbf{1}}_z)$$

$$= (\dot{r}\,\hat{\mathbf{1}}_r + r\,\mathrm{D_o}\hat{\mathbf{1}}_r) + \dot{z}\,\hat{\mathbf{1}}_z$$

that is

$$\mathbf{v}_{o/po} = \dot{r}\,\hat{\mathbf{1}}_r + r\dot{\phi}\,\hat{\mathbf{1}}_\phi + \dot{z}\,\hat{\mathbf{1}}_z \tag{3.11a}$$

$$\mathbf{a}_{o/po} = \mathrm{D_o}\mathbf{v}_{o/po} = \mathrm{D_o}(\dot{r}\,\hat{\mathbf{1}}_r) + \mathrm{D_o}(r\dot{\phi}\,\hat{\mathbf{1}}_\phi) + \mathrm{D_o}(\dot{z}\,\hat{\mathbf{1}}_z)$$

$$= (\ddot{r}\,\hat{\mathbf{1}}_r + \dot{r}\,\mathrm{D_o}\hat{\mathbf{1}}_r) + \{(r\ddot{\phi} + \dot{r}\dot{\phi})\,\hat{\mathbf{1}}_\phi + r\dot{\phi}\,\mathrm{D_o}\hat{\mathbf{1}}_\phi\} + \ddot{z}\,\hat{\mathbf{1}}_z$$

$$(\ddot{r}\,\hat{\mathbf{1}}_r + \dot{r}\dot{\phi}\,\hat{\mathbf{1}}_\phi) + \{(r\ddot{\phi} + \dot{r}\dot{\phi})\,\hat{\mathbf{1}}_\phi - r\dot{\phi}^2\hat{\mathbf{1}}_r + \ddot{z}\,\hat{\mathbf{1}}_z$$

that is

$$a_{o/po} = (\ddot{r} - r\dot{\phi}^2)\,\hat{\mathbf{1}}_r + (r\ddot{\phi} + 2\,\dot{r}\dot{\phi})\,\hat{\mathbf{1}}_\phi + \ddot{z}\,\hat{\mathbf{1}}_z \tag{3.11b}$$

Tracing the origins of the various terms, first consider the effects in the z direction. Reflection will show that the z axis of the cylindrical system is identical with the z axis of the rectangular system, and $\hat{\mathbf{1}}_z$ is identical with **k**. In particular, $\hat{\mathbf{1}}_z$, like **k**, is constant in the frame of reference in which the coordinates are drawn, and it follows that, with cylindrical coordinates as with rectangular, for an observer who is fixed in the frame of reference, the z elements of the velocity and acceleration vectors may be determined simply as the first and second derivatives of the z elements of the radius vector.

The independent effects in the z direction being taken for granted, the remaining effects may be reviewed in the special case of a motion which is confined to the (r, ϕ) plane, where the modulus of the radius vector to a point is equal to its r coordinate. Then the velocity of the point being determined as the derivative of the radius vector, it may be described as the sum of two rectangular components, of which one is colinear with the radial direction of the radius vector itself and has a modulus equal to the derivative $\dot{r}$ of its modulus. The other lies in the circumferential direction normal to the radius vector itself, and has a modulus equal to $r\dot{\phi}$, the product of the modulus of the radius vector with its angular speed (Fig. 3.18a).

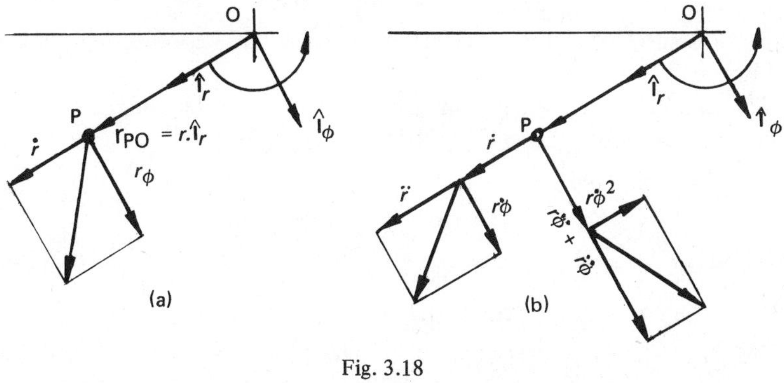

Fig. 3.18

The acceleration is then defined as the derivative of the velocity. However, this may be equated to the sum of the derivatives of the radial and circumferential components of the velocity, and each component of the acceleration can therefore be described as a sum of two parts, of which one derives from the variation in the modulus of the colinear component of the velocity, whilst the other derives from the rotation of the normal component of the velocity. Thus, the radial component of the velocity has a modulus $\dot{r}$, whose derivative is $\ddot{r}$, whilst the circumferential component has a modulus equal to the product $r\dot{\phi}$ whose derivative is equal to $(r\ddot{\phi} + \dot{r}\dot{\phi})$. Furthermore, they both rotate with a common angular speed $\dot{\phi}$, and we recognise that in the radial component of the acceleration, the term $\ddot{r}$ derives from the variation in the modulus of the colinear, (radial), component of the velocity, whilst the term $-r\dot{\phi}^2$ derives from the rotation of the rectangular (circumferential) component of the velocity, and the latter is called the centripetal component of acceleration. Similarly, to the circumferential component of the acceleration, the variation in the modulus of the colinear (circumferential) component of the velocity contributes an amount equal to $(r\ddot{\phi} + \dot{r}\dot{\phi})$. To this, the rotation of the rectangular (radial) component of the velocity adds a further $\dot{r}\dot{\phi}$, so the total circumferential component is equal to $(r\ddot{\phi} + 2\dot{r}\dot{\phi})$, and the component $2\dot{r}\dot{\phi}\,\hat{1}_\phi$ is called the Coriolis component of acceleration. Thus whereas the centripetal acceleration is radial and derives solely from the rotation of the circumferential component of velocity, the Coriolis acceleration is circumferential and arises as two equal parts, of which only one derives from the rotation of the radial component of the velocity. The other, together with the component $r\ddot{\phi}\,\hat{1}_\phi$, arises from the variation in the modulus of the circumferential component of the velocity.

Example 3.10

Repeat Example 3.9 in cylindrical coordinates.

Solution

In the specification of the problem we are given

$$\dot{r} = 4\text{ m/s};\quad \dot{\phi} = 10\pi\text{ rad/s};\quad \dot{z} = 6\text{ m/s}$$

$$\ddot{r} = 0 \qquad \ddot{\phi} = 0 \qquad \ddot{z} = 0\,.$$

Furthermore, it can be inferred from the data that the r coordinate at time t is given by $r = r_{t=0} + \dot{r}.t = 2 + 4t$.

The cylindrical components of the velocity and acceleration can therefore be written, at sight, as follows

$$\begin{aligned}
\mathbf{v} &= \dot{r}.\hat{1}_r + r\dot{\phi}.\hat{1}_\phi + \dot{z}.\hat{1}_z \\
\mathbf{v} &= 4.\hat{1}_r + (2+4t).10\pi.\hat{1}_\phi + 6.\hat{1}_z\text{ m/s} \\
\mathbf{v}_{t=10} &= 4.\hat{1}_r + 420\pi.\hat{1}_\phi + 6.\hat{1}_z\text{ m/s} \\
\mathbf{a} &= (\ddot{r}-r\dot{\phi}^2).\hat{1}_r + (r\ddot{\phi}+2\dot{r}\dot{\phi}).\hat{1}_\phi + \ddot{z}.\hat{1}_z \\
\mathbf{a} &= -(2+4t).100\pi^2\hat{1}_r + 80\pi.\hat{1}_\phi \\
\mathbf{a}_{t=10} &= -4200\pi^2\hat{1}_r + 80\pi.\hat{1}_\phi\,.
\end{aligned}$$

Example 3.11

The diagram shows a horizontal turntable A which rotates about a vertical axis at an angular speed of 180 rev/min in the sense indicated. It carries equipment powered through a chain drive which stands in a vertical plane which contains the axis of rotation. The drive wheel B of radius 0·2 m turns at 3000 rev/min in the sense indicated.

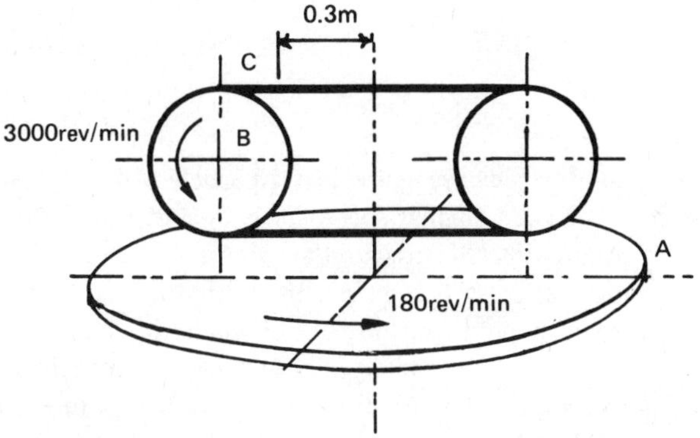

Determine the acceleration relative to the ground of the link C which is at a distance of 0·3 m from the axis of the turntable.

Solution

By the specification of the problem, in cylindrical coordinates fixed in the ground, we are given that

$$r = 0{\cdot}3\ \text{m};\ \dot{r} = 0{\cdot}2.100\pi;\ \ddot{r} = 0$$

$$\dot{\phi} = 6\pi\ \text{rad/s};\ \ddot{\phi} = 0;\ \dot{z} = \ddot{z} = 0$$

therefore $\mathbf{a} = (\ddot{r} - r\dot{\phi}^2).\hat{1}_r + (r\ddot{\phi} + 2\dot{r}\dot{\phi}).\hat{1}_\phi + \ddot{z}.\hat{1}_z$

$$\mathbf{a} = -0{\cdot}3(6\pi)^2.\hat{1}_r + 2.20\pi.6k.\hat{1}_\phi$$

$$\mathbf{a} = \pi^2(-10{\cdot}8\,.\hat{1}_r + 240\,.\hat{1}_\phi)\ .$$

3.8 MOTION IN TANGENTIAL AND NORMAL COMPONENTS

The velocity and acceleration of a point can also be described in terms of sets of components whose directions are chosen to agree, not with the directions of the coordinates, but with the cardinal directions of the path of the motion.

In fact it is intuitively obvious that at each point on the path, the velocity vector will be directed along the tangent to the path at the point, and will have a modulus equal to the speed v along the path. Thus intuition suggests that velocity vector can be described as a single component $\mathbf{v} = v\,.\hat{1}_t$, where v represents the speed of the point, and $\hat{1}_t$ is the unit tangent vector to the path at the point, as shown in Fig. 3.19(a).

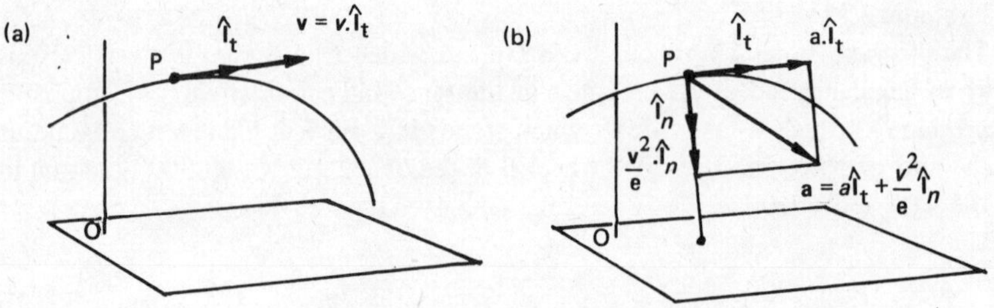

Fig. 3.19

As to the acceleration, we recognise that a particle which runs with speed v on a circle of radius r has an angular speed given by $\dot{\phi} = v/r$. It therefore has a centripetal acceleration directed toward the centre of magnitude $r\dot{\phi}^2 = v^2/r$, and intuition therefore suggest that in passing with speed v, and acceleration-along-the-path a through a point on the path at which the radius of curvature is ρ, a particle would have one component of acceleration of magnitude a directed along the tangent to the path, and another of magnitude v^2/ρ directed along the unit normal, toward the Centre of Curvature. Intuition therefore suggests that the acceleration can be described as the sum of its tangential component $a\,.\hat{\mathbf{1}}_t$, and its normal component, $(v^2/\rho)\hat{\mathbf{1}}_n$, as indicated in Fig. 3.19(b).

These expectations can be confirmed algebraically as follows. By definition

$$\mathbf{v} = \frac{d\mathbf{r}}{dt} = \frac{d\mathbf{r}}{ds} \cdot \frac{ds}{dt} .$$

But ds/dt is, by definition, equal to the speed v, and we have seen that $d\mathbf{r}/ds$ can be equated to the unit tangent vector $\hat{\mathbf{1}}_t$.

Therefore $\quad \mathbf{v} = v\,.\hat{\mathbf{1}}_t \qquad$ (3.12a)

Furthermore, by definition

$$\mathbf{a} = \frac{d\mathbf{v}}{dt} = \frac{d}{dt}(v\,.\hat{\mathbf{1}}_t) .$$

Therefore $\quad \mathbf{a} = \dot{v}\,.\hat{\mathbf{1}}_t + v\,.\dfrac{d\hat{\mathbf{1}}_t}{dt}$

$$= \dot{v}\,.\hat{\mathbf{1}}_t + v\,.\frac{d\hat{\mathbf{1}}_t}{ds} \cdot \frac{ds}{dt} .$$

But, by definition, $\dot{v} = a$, and $\dot{s} = v$. Furthermore, we have seen that $d\hat{\mathbf{1}}_t/ds$ can be equated to $\rho^{-1}\hat{\mathbf{1}}_n$.

Therefore $\quad \mathbf{a} = a\,\hat{\mathbf{1}}_t + \dfrac{v^2}{\rho} \cdot \hat{\mathbf{1}}_n \qquad$ (3.12b)

Example 3.12

A particle moves on a plane circular path of radius 5 m, centred on the origin of rectangular coordinates. The normal to the plane lies in the xy plane at an angle of 30° to the y axis as indicated in the diagram. At the point P where the path cuts the z axis, the particle moves in the sense indicated with a speed of 10 m/s, and an acceleration-along-the-path of 20 m/s².

Determine the velocity and acceleration at P in terms of their tangential and normal components.

Solution

The unit normal vector $\hat{\mathbf{1}}_n$ is directed toward the centre of curvature , O. Therefor for the point P it lies in the negative direction of the z axis, and is described vectorially by $\hat{\mathbf{1}}_n = -\mathbf{k}$.

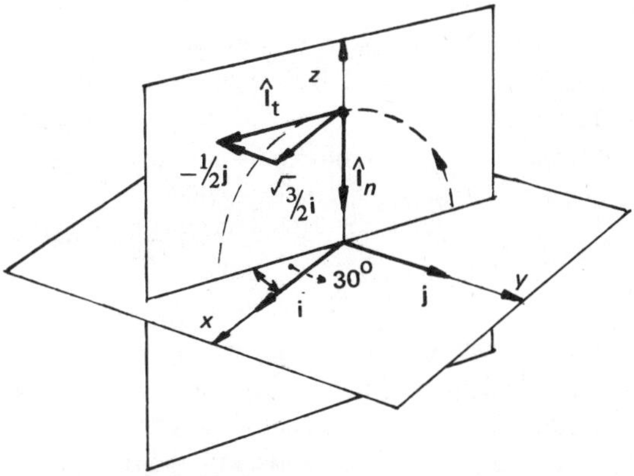

The unit tangent vector lies in the plane of the motion, at right-angles to the z axis, and its components in the x and y directions are $\sqrt{3}/2$ and $-1/2$ respectively. Therefore $\hat{\mathbf{1}}_t = (\sqrt{3}.\mathbf{i} - \mathbf{j})/2$. Thus, given $v = 10$ m/s, and $a = 20$ m/s², we have

$$\mathbf{v} = v\,\hat{\mathbf{1}}_t$$

or $$\mathbf{v} = 5(\sqrt{3}\mathbf{i} - \mathbf{j})\text{ m/s}$$

and $$\mathbf{a} = a.\hat{\mathbf{1}}_t + \frac{v^2}{\rho}\cdot\hat{\mathbf{1}}_n$$

or $$= 10(\sqrt{3}\mathbf{i} - \mathbf{j}) + (10^2/5)(-\mathbf{k})$$

whence $$\mathbf{a} = 10(\sqrt{3}.\mathbf{i} - \mathbf{j} - 2\mathbf{k})\ .$$

4

Particle Dynamics

4.1 THE NEWTONIAN INTERPRETATION

Dynamics is concerned with changes of motion and their relationship to the effects from which they arise, and here we consider the wide range of common situations that yield to the classical Newtonian interpretation. In this, changes of motion are conceived to result from forces of interactions in which all bodies are considered to engage, but whilst it quantifies the effects of the forces on the bodies themselves, the approach offers no explanation of the manner in which the forces are transmitted between them. Rather, it accepts the concept of action at a distance in which both the forces of any interaction are considered to be instantaneously effective on both the bodies simultaneously.

When a force is applied to a body its effects are seen not only in the translation of the body, but in its rotation and deformation also, and this is reflected in complex relative motions between the parts of the body. As a result, a law which attempted directly to describe the effects of forces on real bodies would be somewhat complex, so instead a body is conceived as an assembly of particles, each of which is geometrically equivalent only to a point. The possibility of relative motions between the parts of a particle is then precluded, and postulating laws of motion as a description of the effect of a force on the translation of one such particle, we treat a body as a set of particles that are assembled in an appropriate manner.

For example, the particles of a body are sometimes treated as discrete entities in independent motion (as in the kinetic theory of gases) and sometimes as the infinitesimal elements of a coherent continuum, but in either case a Newtonian particle is conceived to be so small that rotational effects, which depend on the fourth power of the dimensions, can be ignored compared with the translation effects, which depend only on the third power. However, such a particle is then necessarily devoid of a structural form which is capable of modification. It is therefore essentially different from, say, an atom, in that it is incapable of those microscopic motions between its parts that are essential to the concepts of thermodynamic state and internal energy, but which, being periodic, are not reflected in the apparent macroscopic motion of the assembly as a whole. The Newtonian particle is therefore incapable of reflecting the effects of a change in its internal energy or thermodynamic state, and it follows that the conclusions of the Newtonian interpretation will need to be modified (in effect into the First Law of Thermodynamics) before they can be applied to situations in which such effects are significant.

Apart from position and its derivatives, the assumptions effectively invest a Newtonian particle with but a single property called inertia. This is attributed to all bodies as a property by which they resist any change in their motions, and the translational inertia of a particle is called its mass. Since the inertia of a particle is found to be equally effective in all directions, mass is necessarily treated as a scalar quantity, and the mass of a particle which forms a part of a coherent, finite body is assumed to be constant. This is contrary to the assumptions of theoretical physics where the mass of a given particle is treated as a function of its speed, but the variation of mass with speed is significant only at speeds which are commensurate with the speed of light, and these are feasible only in the independent motions of 'impermanent', sub-atomic particles, and not in the macroscopic motions of even the smallest of 'permanent', tangible bodies.

The product of the mass m of a particle with its velocity $\mathbf{v}$ is called the momentum of the particle, and mass being a scalar, momentum is recognised as a vector quantity whose direction is that of the velocity, and whose modulus is equal to the product of the mass of the particle with its speed.

But the momentum of a given particle at a given instant is not a unique quantity because the velocity of a given motion may have different values in different frames of reference. It follows that the observations of Newtonian Laws must be referred not to frames of reference in general, but only to the frames of a particular equivalent set. These are known as inertial frames, and as a result of experiment, it is generally conceded that a frame of reference that is based on a set of fixed stars is truly inertial. However, it can be shown that any frame of reference that is in simple uniform translation relative to an inertial frame is itself an inertial frame, so any frame that neither accelerates nor rotates relative to the fixed stars is treated as truly inertial.

Evidently, a frame of reference fixed in the Earth is therefore not truly inertial. However, in the motions of the Earth its daily rotation is dominant, and even this gives rise to little discrepancy. Therefore, for the great majority of purposes in engineering mechanics an Earth frame is, in fact, treated as inertial, and in the absence of any indication to the contrary, we shall treat it as such throughout this text. The rare exceptions occur only in events of very long range in distance or time; but even in these it is never necessary to take account of any aspect of the earth's motion other than its daily rotation, and for this we can adopt a frame of reference which conveniently moves with the centre of the earth, but with respect to which the earth simply rotates once per day.

4.2 NEWTON'S LAWS OF POINT MOTION

On the foregoing basis, and all observations being referred to an inertial frame, Newton's Laws of Motion for a permanent, point particle of fixed identity and thermodynamic state can be paraphrased in the modern idiom as follows

1. The momentum of a point particle can be changed only by the force of an interaction with the surroundings.
2. The force of an interaction is proportional to the instantaneous time-rate of change of momentum that it causes in the particle on which it acts.
3. Every action of the surroundings on a particle is associated with an equal but opposite reaction of the particle on the surroundings.

Units

In the Système International, the dimensions of length, mass and time are included in the basic set, so the unit metre of length, the unit kilogramme of mass and the unit second of time are all determined by agreed, arbitrary prototypes. But the unit of force is treated as derived, and the unit Newton is so defined as to give a unit constant of proportionality in the Second Law. In this system, the force of an interaction can therefore be equated to the resulting time-rate of change of momentum, and it follows that the unit Newton can be equated, in basic units, to 1 kg m/s^2. In what follows all equations are expressed in SI units.

The Vector Equation of Point Motion

Experiment shows that in the context of particle dynamics, the forces of a concurrent set combine as vectors. Therefore, for a particle of mass m, position $\mathbf{r}$, velocity $\mathbf{v}$ and acceleration $\mathbf{a}$, subjected to a set of forces whose vector sum is represented by $\Sigma\mathbf{f}$, Newton's Second Law of Motion can be expressed by a vector equation, as follows

$$\Sigma\mathbf{f} = m.\mathbf{a} = m\cdot\frac{\mathrm{d}\mathbf{v}}{\mathrm{d}t} = m\cdot\frac{\mathrm{d}^2\mathbf{r}}{\mathrm{d}t^2} \tag{4.1}$$

The Lumped Parameter Equation

The foregoing equation refers directly only to a point particle in independent motion, and it requires a measure of development before it can be applied generally to a particle which forms part of a finite body. But in every such body there is one particular point which is amenable to the law as formulated, and this point, which we shall later identify formally as the Centre of Mass of the body, we may, for the present, refer to as its Centre of Inertia.

The experimental evidence for the existence of this point is indicated diagrammatically in Fig. 4.1. This shows that when a given force is applied to a given body on a given line of action, the effect on different points such as A, B and C generally varies from point to point. Equally, if a given force is applied to a given body on different lines of action, the effect on a given point such as

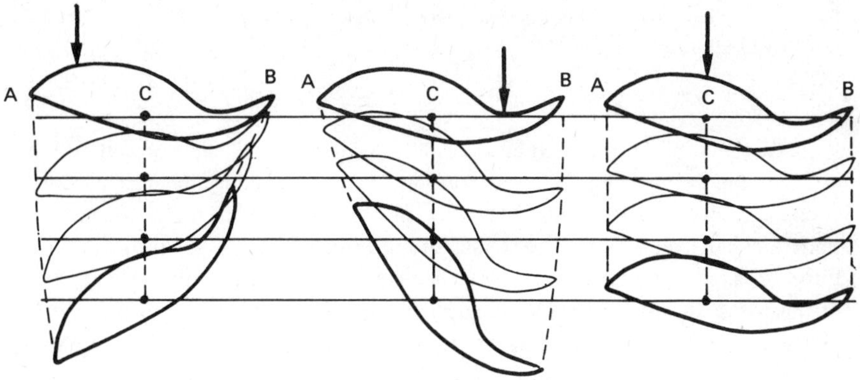

Fig. 4.1

A or B generally varies with the line of action. Yet in every case there is one point C on which the effect of a given force is independent of its line of action, and if any force is applied on a line which passes through C, it has the same effect on every point in the body.

Since the effect of any force on C is independent of its line of action, it follows that in studying the motion of this particular point, we may ignore the lines of action of the forces that act on the body, and treat them as though they were concurrent at C. But then every particle of the body would suffer the same effect, so the motion of the Centre of Inertia can be studied by treating the distributed system of mass and forces as though it were shrunk to a point at the Centre. The resulting equation is called the Lumped Parameter Equation for the Motion of the Centre of Inertia, and if $\Sigma\mathbf{f}$ represents the vector sum of the forces which act on a body of total mass Σm, whilst $\mathbf{r}_c$, $\mathbf{v}_c$ and $\mathbf{a}_c$ represent the position, velocity and acceleration of its Centre of Inertia, the equation may be represented as

$$\Sigma\mathbf{f} = (\Sigma m).\mathbf{a}_c = (\Sigma m)\cdot\frac{d\mathbf{v}_c}{dt} = (\Sigma m)\cdot\frac{d^2\mathbf{r}_c}{dt^2} \qquad (4.2)$$

Throughout this text we shall take it as understood that where we refer simply to the motion of a body, we intend the motion of its Centre of Inertia.

Scalar Equations of Point Motion

The single vector equation of point motion is, of course, equivalent to a set of three scalar equations. For example, if the resultant of the forces on a body, and the acceleration of its Centre of Inertia are both described in terms of their components in the same rectangular coordinates we have

$$\Sigma\mathbf{f} = (\Sigma m).\mathbf{a}_c$$

or

$$\Sigma f_x.\mathbf{i} + \Sigma f_y.\mathbf{j} + \Sigma f_z.\mathbf{k} = \Sigma m.a_x.\mathbf{i} + \Sigma m.a_y.\mathbf{j} + \Sigma m.a_z.\mathbf{k}\ .$$

But two vectors are equal only when the two components for each direction of resolution are equal. Therefore

$$\Sigma f_x = \Sigma m.a_x; \text{ and } \Sigma f_y = \Sigma m.a_y; \text{ and } \Sigma f_z = \Sigma m.a_z \qquad (4.3)$$

It follows that for any direction of resolution, the sum of the resolved components of the forces is equal to the total mass times the component of the acceleration of the Centre of Inertia resolved in the same direction. The scalar equations are particularly convenient in a situation in which a body moves under a force of fixed direction, because irrespective of any variation in the magnitude of the force, it will produce no change of speed in any direction at right-angles to its own line. It is then necessary to consider only the single scalar equation for the motion in the fixed direction of the force.

Kinematic Constraints

Given the resultant of the forces on a body, the lumped parameter equation determines the acceleration of its Centre, and given the acceleration of its Centre, the Lumped Parameter Equation determines the resultant of the forces; but in practice we commonly require to determine the effect of given forces when the body is also subjected to unknown forces which impose on the motion of the body known restraints. The equation(s) of motion for the body will then contain the unknown forces, and it will be necessary to solve it simultaneously with equations which formally define the proscription which the unknown forces place on the motion. These latter equations are known as equations of kinematic constraint, and the unknown forces which impose them are known as forces of constraint.

Contact Forces and Friction

In practice, kinematic constraints are imposed on a body by its contact with guiding surfaces, and the necessary forces are generated as contact stresses which are exchanged between the two contacting surfaces at their mutual interface.

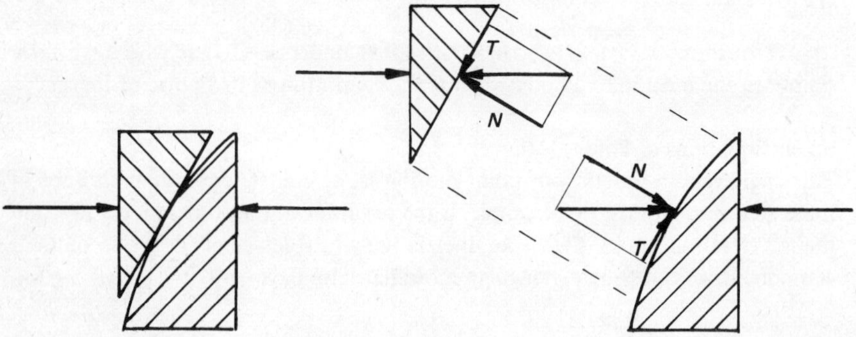

Fig. 4.2

The aggregate force that is thus engendered on the body at the position of contact is called a force of kinematic constraint, and contact forces of this kind are described as a pair of rectangular components, of which one is normal to the mutual interface of contact, whilst the other lies along it. Evidently, the maintenance of the contact requires that the normal components of any pair of contact forces be compressive. As to the tangential components, these, by the third law of motion, must be oppositely directed, and they therefore tend to cause relative slide (or shear) between the two surfaces along the mutual interface. In a contact between two solid bodies, this tendency is resisted by a phenomenon known as friction. But the frictional capacity of a given pair of surfaces is limited, and for most practical purposes the rule of this limitation can be expressed as follows:

> 'Relative slide (or shear) occurs along the mutual interface of contact between two solid bodies when the tangential components of the contact forces are a specific proportion of the normal components. This proportion is a characteristic property of the particular pair of surfaces, and it is called their coefficient of friction.'

A coefficient of friction is commonly represented by the symbol μ, and if N and T represent the magnitudes of the normal and tangential components of the force of contact of a pair of solid surfaces whose coefficient of friction is μ, then relative slip will occur when $T = \mu . N$, and this proportion will be maintained for so long as the slip continues.

For a frictionless contact, μ us zero, and the contact forces are normal to the mutual interface, even during slip.

Gravitational Forces

With the Laws of Motion, Newton also postulated the Law of Gravitational Attraction. This states that two point masses interact with forces of gravitational attraction which act along the straight line which joins the masses, and whose magnitudes are proportional to the product of the masses, and inversely proportional to the square of the distance between them. The constant of proportionality G is a universal constant.

Let the gravitational force exerted on a mass m_A at A by a mass m_B at B, be represented by $\mathbf{f}_{AB}$. The position of m_A relative to m_B is defined by the vector $\mathbf{r}_{AB}$, and since $\mathbf{f}_{AB}$ is directed from A to B, its direction is defined by the unit vector $-\hat{\mathbf{r}}_{AB}$. Furthermore, the distance of m_A from m_B is equal to the

Fig. 4.3

modulus of $\mathbf{r}_{AB}$, and the Law of Gravitational Attraction can therefore be extpressed as

$$\mathbf{f}_{AB} = -G \cdot \frac{m_A m_B}{|\mathbf{r}_{AB}|^2} \cdot \hat{\mathbf{r}}_{AB} \tag{4.4}$$

The Earth approximates to a sphere in which the density (that is, the mass per unit volume) is a function of the distance from the centre, in which case it can be shown that, at positions outside its boundary, the gravitational effect of the sphere is equal to that of its total mass lumped at the centre. The gravitational force of the Earth on any body at or above its surface is therefore directed radially toward its centre, and its magnitude varies with the square of the distance from the centre. But, within a span of a few miles, either horizontally or vertically, the effect of the variation of direction and distance is, for most purposes in terrestrial mechanics, negligible, and the Earth's gravitational 'field' in the vicinity of its surface is commonly assumed to be uniform in both magnitude and direction.

The gravitational force that the Earth exerts on a body is called the weight of the body, and at any given position, the weights of different bodies are proportional to their masses. It therefore follows from the Second Law of Motion, that at any given position, the resulting gravitational accelerations of different bodies are equal, irrespective of their masses, and the gravitational acceleration g at any position on the surface of the Earth is approximately equal to $9{\cdot}8\ \mathrm{m/s^2}$.

Example 4.1

In an inertial frame of reference, the rectangular coordinates of a particle of mass m kg vary with time t seconds according to

$$(x, y, z) = \{(1+t^2), (2t^2+t^3), 4t^3\}\ \mathrm{m} \quad .$$

Determine the force on the particle as a function of time.

Solution

The position vector of the particle in the given coordinates is defined by its radius vector $\mathbf{r}$, and since the initial point of a radius vector is chosen at the origin, it follows that the coordinates of the particle are the elements of its radius vector.

Thus $\quad \mathbf{r} = \{(1+t^2).\mathbf{i} + (2t^2+t^3).\mathbf{j} + 4t^3.\mathbf{k}\}\ \mathrm{m}$

and $\quad \mathbf{v} = \mathrm{d}\mathbf{r}/\mathrm{d}t = \{2t.\mathbf{i} + (4t+3t^2).\mathbf{j} + 12t^2.\mathbf{k}\}\ \mathrm{m/s}$

and $\quad \mathbf{a} = \mathrm{d}\mathbf{v}/\mathrm{d}t = \{2.\mathbf{i} + (4+6t).\mathbf{j} + 24t.\mathbf{k}\}\ \mathrm{m/s^2}$

By the 2nd Law

$$\mathbf{f} = m.\mathbf{a}$$

$$\mathbf{f} = 2m\{\mathbf{i} + (2+3t).\mathbf{j} + 12t.\mathbf{k}\}\ \mathrm{N}$$

Example 4.2

A linear actuator is fitted around a rotating arm to move a slider of mass 0·5 kg along it.

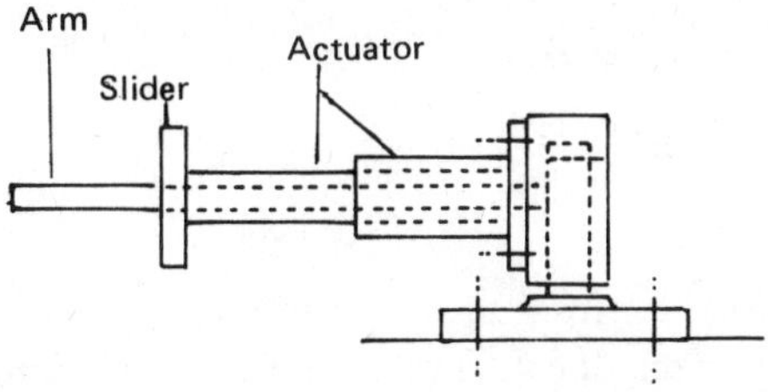

(a) Treating the slider as a particle, determine, in terms of its components along and normal to the arm, the force required to accelerate the slider when it moved inwards through the position at a radius of 0·4 m at a constant speed relative to the arm of 40 m/s, whilst the arm rotates at a constant angular speed of 60 rev/min.

(b) Ignoring the gravitational force, determine the radial force required of the actuator to perform the manoeuvre described in (a), if the coefficient of friction between the slider and the arm is 0·3.

Solution

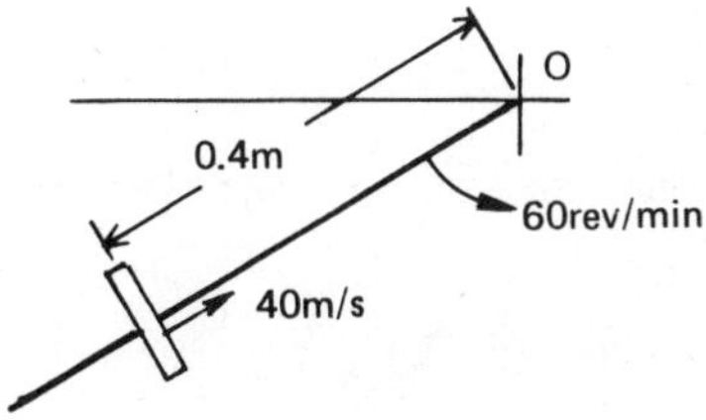

$$\mathbf{a} = (\ddot{r} - r\dot{\phi}^2)\,.\,\hat{\mathbf{e}}_r + (r\ddot{\phi} + 2\dot{r}\dot{\phi})\,.\,\hat{\mathbf{e}}_\phi$$

But $\ddot{r} = \ddot{\phi} = 0;\ r = 0{\cdot}4\text{ m};\ \dot{r} = -40\text{ m/s};\ \dot{\phi} = 2\pi\text{ rad/s}$

Therefore $\mathbf{a} = -0{\cdot}4.(2\pi)^2\,.\,\hat{\mathbf{e}}_r + 2.(-40)\,.\,2\pi\,.\,\hat{\mathbf{e}}_\phi\text{ m/s}^2$

that is $\mathbf{a} = -1{\cdot}6\,.\,\pi^2\,.\,\hat{\mathbf{e}}_r - 160\,.\,\pi\,.\,\hat{\mathbf{e}}_\phi\text{ m/s}^2\ .$

By the Second Law of Motion, the required force is therefore

$$\mathbf{f} = m.\mathbf{a} = -0{\cdot}8\pi^2.\hat{\mathbf{e}}_r - 80\pi.\hat{\mathbf{e}}_\phi\text{ N}\ .$$

The first component derives from the centripetal effect, and the second from the Coriolis effect.

(b) The actuator must provide the force required for the radial component of acceleration of the slider, after overcoming the frictional force between the slider and the arm. The normal component of the contact force is the circumferential component of the acceleration force required by the slider. Total, inward radial force of the actuator is therefore

$$\mathbf{f} = (0{\cdot}8.\pi^2 + 0{\cdot}3.80.\pi)\,\mathrm{N}$$
$$= (7{\cdot}9 + 75{\cdot}4)\,\mathrm{N}$$
$$\mathbf{f} = 83{\cdot}3\,\mathrm{N}\ .$$

Example 4.3

A projectile is discharged from the origin of a rectangular Earth frame, with a speed of 1470 m/s in the direction of the unit vector $(2\mathbf{i} + 2\mathbf{j} + \mathbf{k})/3$. Ignoring air resistance, determine the coordinates of the point in which the projectile falls back to the xy plane.

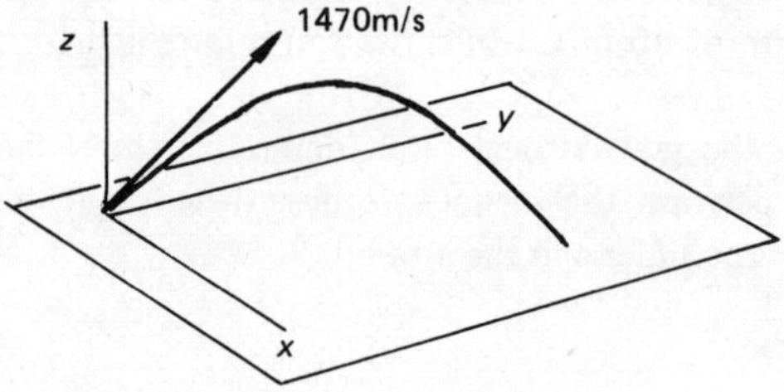

Solution

The constant gravitational force on the projectile is $-mg.\mathbf{k}$, and, by the Second Law of Motion

$$-mg.\mathbf{k} = m\,.\,(\mathrm{d}\mathbf{v}/\mathrm{d}t)\ ,$$

therefore $\quad \mathrm{d}\mathbf{v} = -g.\mathrm{d}t.\mathbf{k}\ .$

Integrating $\mathrm{d}\mathbf{r}/\mathrm{d}t = \mathbf{v} = -gt.\mathbf{k} + \mathbf{c}_1\ .$

Integrating again

$$\mathbf{r} = -\frac{g}{2}t^2\mathbf{k} + \mathbf{c}_1.t + \mathbf{c}_2\ .$$

But when $t = 0$, then $\mathbf{r} = (0, 0, 0)$ and $\mathbf{v} = 490(2\mathbf{i} + 2\mathbf{j} + \mathbf{k})$, and substituting accordingly gives

$$\mathbf{c}_1 = 490(2\mathbf{i} + 2\mathbf{j} + \mathbf{k});\quad \mathbf{c}_2 = (0, 0, 0)\ ,$$

Therefore $\quad \mathbf{v} = 490(2\mathbf{i} + 2\mathbf{j} + \mathbf{k}) - gt.\mathbf{k}$

$$\mathbf{r} = 490t(2\mathbf{i} + 2\mathbf{j} + \mathbf{k}) - \frac{g}{2}t^2\mathbf{k}\ .$$

To determine the value of time $t_{z=0}$ at which the projectile falls back to the xy plane we equate the **k** component of **r** to zero. Thus

$$490t_{z=0} - \frac{g}{2}t_{z=0}^2 = 0; \quad \text{whence } t_{z=0} = 980/g = 100 \text{ s} ,$$

therefore $\mathbf{r}_{z=0} = (49\times10^3)(2\mathbf{i} + 2\mathbf{j} + \mathbf{k}) - 49.10^3.\mathbf{k}$

that is $\mathbf{r}_{z=0} = 98.10^3(\mathbf{i} + \mathbf{j})\, m$.

Example 4.4

If one of a pair of parallel plates at separation d is charged to a voltage V above the other, then a charge q at any position between the plates would experience an electrostatic force $q\mathbf{E}$, where **E** is a vector of magnitude V/d, normal to the plates, in the sense which is directed away from the plate of higher voltage, and toward the plate at lower voltage. Outside the space between the plates, the force is negligible. An electron has a mass represented by m, and bears a negative charge $-e$.

(a) An electron accelerator comprises a pair of parallel plates at separation d, charged to a voltage difference V. Electrons emitted at negligible speed from the plate of lower voltage are accelerated between the plates to emerge from a central hole in the plate of higher voltage. Show that the stream of emerging electrons has a characteristic speed, $v = \sqrt{(2eV/d)}$.

(b) Electrons discharged from the accelerator in (a) enter an electrostatic deflector as shown. This comprises a second pair of parallel plates of length l and separation b, at right-angles to the accelerator, and charged to a voltage difference U. Determine the displacement δ, and the inclination θ of the emerging beam.

Solution

(a) Since the field **E** is uniform, and the initial velocities of the electrons are negligible, the motions of the electrons are rectilinear, and it is therefore sufficient to consider only the scalar motions of electrons in a direction parallel to the field, normal to the plates.

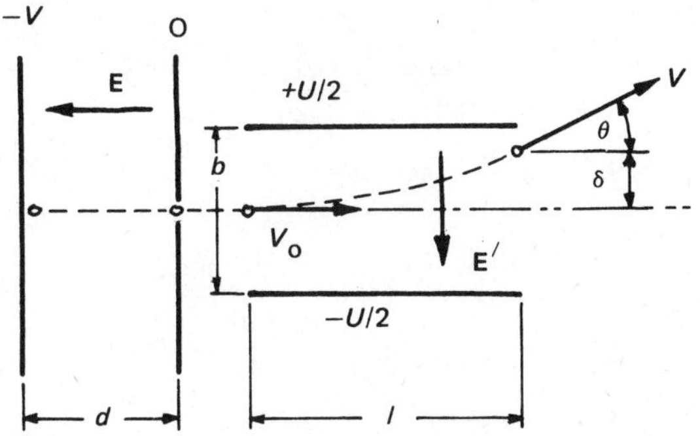

Noting that the charge of an electron is negative, we choose positive x in the negative of the field direction. Furthermore, since we require to determine the speed v at a known distance x, we represent the acceleration as half the derivative of v^2 with respect to x. Thus

By electrical theory

$$f_x = -(-e).|\mathbf{E}| = eV/d \tag{1}$$

By the 2nd Law of Motion

$$f_x = m.a_x = \frac{m}{2}\cdot\frac{\mathrm{d}(v_x^2)}{\mathrm{d}x} \tag{2}$$

Between (1) and (2)

$$\frac{e.V}{d} = \frac{m}{2}\cdot\frac{\mathrm{d}(v_x^2)}{\mathrm{d}x} ,$$

therefore $\mathrm{d}(v_x^2) = (2eV/md)\,.\,\mathrm{d}x$

Integrating $v_x^2 = (2eV/md).x + c_1$

But when $x = 0$, then $v_x = 0$. Therefore $c_1 = 0$, and at $x = d$

$$v_{x=d} = \sqrt{(2eV/m)} .$$

(b) Since the field of the deflector, $\mathbf{E}' = -(U/b).k$, is uniformly parallel to y, the x component of the electron velocity remains constant at $v_x = v_0 = \sqrt{(2eV/m)}$, and the time for the transit of an electron through the field is evidently given by $t_{x=l} = l/v_0$. In this case it is therefore convenient to interpret the acceleration as the first derivative of speed v with respect to time t.

Thus for the scalar motion parallel to y

$$f_y = -(-e).U/b = m\,.\,(\mathrm{d}v_y/\mathrm{d}t) ,$$

therefore $\mathrm{d}v_y = (eU/bm).\mathrm{d}t$.

Integrating $v_y = (eU/bm).t + c_1$.

Integrating again

$$y = (eU/2bm).t^2 + c_1t + c_2 .$$

But when $t = 0$, then $y = 0$ and $v_y = 0$. Therefore $c_1 = c_2 = 0$, and we determine y and v_y at $x = l$ by substituting $t_{x=l} = l/v_0$.

Thus at $x = l;\ (v_y)_{x=l} = (eU/bm).l/v_0$,

therefore $\theta = \tan^{-1}\dfrac{(v_y)_{x=l}}{v_0} = \tan^{-1}(eUl/bmv_0^2)$,

and substituting $v_0 = \sqrt{(2eV/m)}$ gives

$$\theta = \tan^{-1}(Ul/2Vb) .$$

Also $$\delta = y_{x=l} = \frac{eU}{2bm} \cdot \frac{l^2 m}{2eV},$$

that is $$\delta = \frac{l^2}{4b} \cdot \frac{U}{V}.$$

Example 4.5

A rigid cylinder slips as it rolls down a plane which is inclined at the angle θ to the horizontal. If the coefficient of friction between the cylinder and the plane is represented by μ, and gravitational acceleration by g, determine the scalar acceleration of the centre of the cylinder in terms of θ, μ and g.

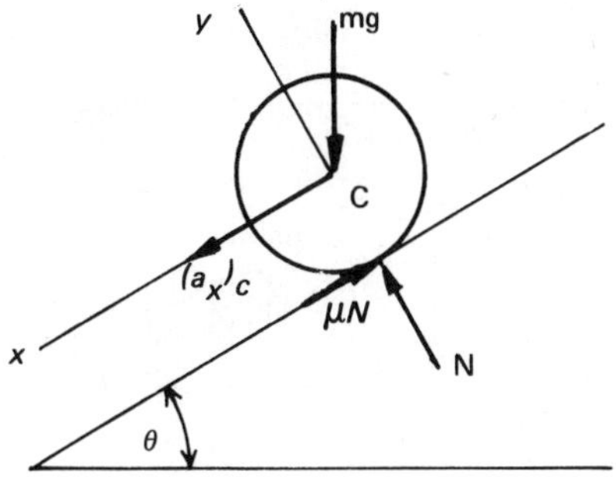

Solution

Let N represent the normal component of the force of contact between the cylinder and the plane. Then, since slip occurs, the tangential component is μN, and it acts upwards to oppose the motion. Choosing x and y axes as shown, the Lumped Parameter Equation gives

$$(mg.\sin\theta - \mu N).\mathbf{i} + (N - mg.\cos\theta).\mathbf{j} = m.\mathbf{a}_c. \qquad (1)$$

The cylinder moves under a kinematic constraint, such that

$$(a_y)_c = 0\ , \qquad (2)$$

therefore, equating the y component of $\mathbf{a}_c$ to zero, we have

$$N - mg.\cos\theta = 0$$

whence $$N = mg.\cos\theta$$

and substituting accordingly in (1) gives

$$mg(\sin\theta - \mu.\cos\theta)\mathbf{i} = m.\mathbf{a}_c\ ,$$

from which the x component of $\mathbf{a}_c$ is given by

$$(a_x)_c = g(\sin\theta - \mu.\cos\theta)\ .$$

Example 4.6

A smooth, frictionless track comprises a straight length AB which blends to a circular arc of radius b. It stands in a vertical plane in a uniform gravitational field $-g.\mathbf{j}$. A point mass is released from rest at A. Show that

(a) The speed of the particle at P is $v_P^2 = 2g(h - b\sin\theta)$.

(b) If $h < 3b/2$, the particle will fall away from the track at C, where $\theta_C = \sin^{-1}(2h/3b)$.

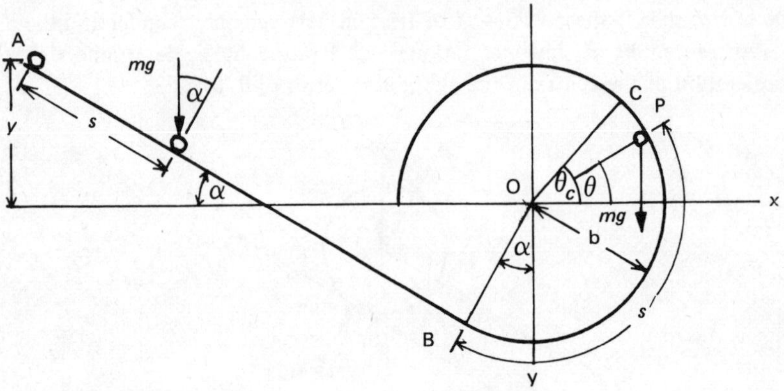

Solution

(a) Since the track is frictionless, the contact force is normal to the track at every position. It is therefore convenient to consider the scalar motion along the path, because this is independent of the unknown contact force and depends on the resolved component of the gravity field only. Furthermore, since we require to determine the speed v at known distances along the path s, it is convenient to identify the acceleration-along-the-path as half the derivative of v^2 with respect to distance s.

For Straight Track AB the 2nd Law of Motion for components resolved along the track is

$$mg.\sin\alpha = \frac{m}{2}\cdot\frac{dv^2}{ds}$$

$$d(v^2) = 2g.\sin\alpha.ds$$

$$v^2 = 2g.\sin\alpha.s + c\ .$$

But when $s = 0$, then $v = 0$. Therefore $c = 0$, and the speed at B is determined by substituting $s = s_B = (h.\sec\alpha + b.\cot\alpha)$.

Thus $$v_B^2 = 2g.(h + b.\cos\alpha) \qquad (1)$$

For the Curved Track BD. Representing distances around the curved track from B by s', the equation for the motion resolved along the track at P is

$$-mg.\cos\theta = \frac{m}{2} \cdot \frac{d(v_P^2)}{ds'}$$

or
$$d(v_P^2) = -2g.\cos\theta.ds' .$$

But:
$$s' = b(\alpha + \pi/2 + \theta) \quad \alpha = \text{Const.},$$

therefore, $ds' = b.d\theta$, and substituting accordingly gives

$$d(v_P^2) = -2gb.\cos\theta.d\theta .$$

Integrating
$$v_P^2 = -2gb.\sin\theta + c_1 .$$

But when $\theta = -(\pi/2 + \alpha)$, $v = v_B$. So $c_1 = v_B^2 - 2gb.\sin(\pi/2 + \alpha)$,

therefore:
$$v_P^2 = v_B^2 - 2gb\{\sin\theta - \sin(\pi/2 + \alpha)\}$$
$$= v_B^2 - 2gb\{\sin\theta + \cos\alpha\} ,$$

and substituting for v_B from (1) gives

$$v_P^2 = 2g(h - b.\sin\theta) . \tag{2}$$

(b) Contact ceases when the contract force falls to zero, and at this stage the radial (centripetal) acceleration of the particle must derive solely from, and be equal to, the component of the gravitational acceleration resolved along the radius.

that is
$$v_c^2/b = g.\sin\theta_c .$$

But from (2):
$$v_c^2 = 2g(h - b.\sin\theta_c) ,$$

therefore:
$$bg.\sin\theta_c = 2g(h - b.\sin\theta_c)$$

that is
$$3b.\sin\theta_c = 2h$$
$$\theta_c = \sin^{-1}(2h/3b) .$$

Since a sine cannot have a value greater than unity, there will be a real solution for θ_c only if $2h < 3b$, or $h < 3b/2$. It follows that there will be a point at which the particle falls away from the track, only if $h < 3b/2$.

Example 4.7

For a sphere of diameter d, in motion with speed v relative to a viscous fluid of kinematic viscosity ν and density ρ, the drag force opposing the motion has a magnitude D given by: $D = 9dv\rho\nu$.

A sphere of density ρ' is allowed to fall from rest, under gravity, through the stationary fluid.

(a) Determine the speed of fall and the distance from rest as functions of time.

(b) The speed tends towards a limiting terminal value. Given that:

$$d = 0{\cdot}02\text{ m};\quad \nu = 1\times10^{-4}\text{ m}^2/\text{s};\quad \rho = 0{\cdot}9\times10^3\text{ kg/m}^3;$$

$$\rho' = 2\times10^3\text{ kg/m}^3;$$

and $$g = 9{\cdot}85\text{ m/s}^2,$$

determine: the terminal speed, and the time and distance from rest to achieve 95% of the terminal speed.

Solution

(a) By Newton's 2nd Law of Motion for the vertical direction:

$$f = mg - D = m\,.\,(\mathrm{d}v/\mathrm{d}t)$$

$$\mathrm{d}v/\mathrm{d}t = g - (D/m) = g - Av$$

where $$A = \frac{9\mathrm{d}\nu}{(1/6)\pi \mathrm{d}^3}\,\frac{\rho}{\rho'} = \frac{54\nu}{\pi d^2}\,\frac{\rho}{\rho'} = \text{Const.},$$

therefore $$\mathrm{d}v/(g-Av) = \mathrm{d}t\;.$$

Integrating: $$-\frac{1}{A}\log_e(g-Av) = t + c_1\;.$$

But when $t = 0, v = 0$. Therefore $c_1 = -(1/A)\log_e g$,

that is $$At = \log_e \frac{g}{g-Av}$$

when $$v = \frac{g}{A}(1-e^{-At}) \tag{1}$$

that is $$\mathrm{d}s = \frac{g}{A}(1-e^{-At})\,\mathrm{d}t\;.$$

Integrating: $$s = \frac{g}{A}\left(t + \frac{1}{A}\mathrm{e}^{-At}\right) + c_2\;.$$

But when $t = 0, s = 0$. Therefore $c_2 = -g/A^2$

$$s = \frac{g}{A^2}(At + \mathrm{e}^{-At} - 1)\;. \tag{2}$$

(b) At the limiting velocity, v_t, the magnitudes of the drag and gravitational force balance.

that is $$mg = D$$

or $$g = D/m = Av_t$$

therefore $$v_t = g/A = \frac{\pi d^2}{54\nu}\cdot\frac{\rho'}{\rho}\cdot g$$

$$= \frac{\pi . 4\times 10^{-4}}{54\times 10^{-4}}\cdot\frac{2\times 10^3}{0{\cdot}9\times 10^3}\cdot 9{\cdot}85$$

$$v_t = 5{\cdot}094 \text{ m/s} .$$

Eqn. (1) may be written $v = v_t(1-e^{-At})$

When $v = 0{\cdot}95 v_t$

$$0.95 = 1 - e^{-At}$$

$$At = -\log_e 0{\cdot}05$$

$$At = 3$$

$$t = \frac{3.\pi d^2}{54\nu}\frac{\rho'}{\rho}$$

$$t = \frac{3.\pi.0{\cdot}02^2}{54.1\times 10^{-4}}\cdot\frac{2}{0{\cdot}9}$$

$$t = 1.55 \text{ s.}$$

From (2) $$s = \frac{gt^2}{(At)^2}(At + e^{-At} - 1)$$

$$= \frac{9{\cdot}85 \times 1{\cdot}55^2}{9}(3 + e^{-3} - 1)$$

$$s = 5{\cdot}39 \text{ m} .$$

4.3 PLANETARY ORBITS

Being relevant both to space travel and to the placement of artificial satellites, planetary orbits are not devoid of current interest, but historically this topic was crucial, for it was from Kepler's observations on planetary motion that Newton laid the foundations of classical dynamics by formulating the laws both of motion and of gravitational attraction.

In brief, Kepler's observations referred solely to the kinematics of planetary motion, and he asserted, first, that each planet moves on a fixed elliptical orbit in a plane which contains the Sun, such that the areal speed of the planet about the Sun is constant, and second, that in the orbits of different planets, the squares of their periods vary as the cubes of their semi-major axes.

Apart from the massive development of the associated mathematics, the essence of Newton's contribution was to recognise that if a mass moved under a force centred on the Sun according to what we know as the second law of motion, the moment of its momentum about the Sun would be constant. This in turn implied that its areal speed about the Sun would be constant, and, when coupled with an inverse square law of gravitational attraction, leads to motions that are entirely consistent with Kepler's observations.

Here we briefly trace the essential steps of Newton's argument, but first it is necessary to consider the chief properties of the conic sections, and to introduce the concept of areal speed.

Conic Sections

The conic sections are so called because they can be defined as the families of curves in which cones are intersected by planes. But although they can be thus defined as a single set, they can be distinguished in three sub-sets, according as the slope of the plane is less than, equal to, or greater than that of the generators of the cone, as exemplified by the sections 1, 2, and 3 in Fig. 4.4. For sections like 1 which cut only one half of the double cone the section takes the form of

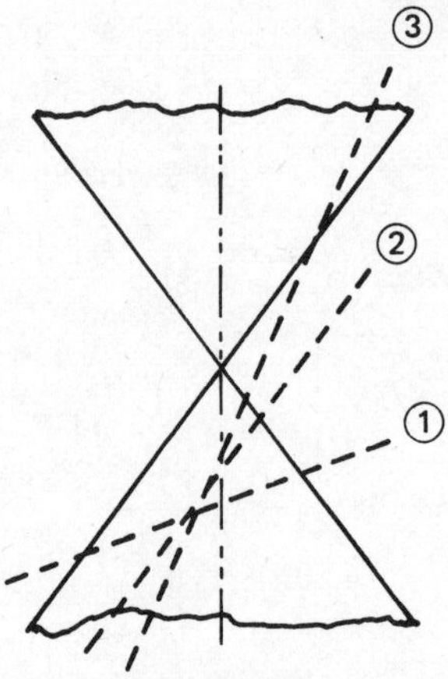

Fig. 4.4

an ellipse, whilst sections like 3 take the form of a pair of hyperbolae. Between these two domains lies the special case in which the plane is parallel to a generator of the cone, in which case the section is a parabola.

But for analytical purposes, a conic can more conveniently be defined as the locus of a point which moves in the plane which contains a given straight line called the directrix and a given point called the focus, so that its distance from the focus is a fixed proportion of its perpendicular distance from the directrix, and this ratio is called the eccentricity e of the conic. When e is less than unity the conic is an ellipse; when e is greater than unity the conic is a pair of hyperbolae, and between these two domains lies the special case in which e is equal to unity, when the conic is a parabola. A typical set of conics is shown in Fig. 4.5, and we note that the straight line through the focus at right-angles to the directrix is an axis of symmetry called the principal axis.

Referring to Fig. 4.6, consider the description of a conic in polar (r, ϕ) coordinates whose origin is chosen at the focus, and whose datum direction is chosen along the principal axis. For any point on the conic, its distance from

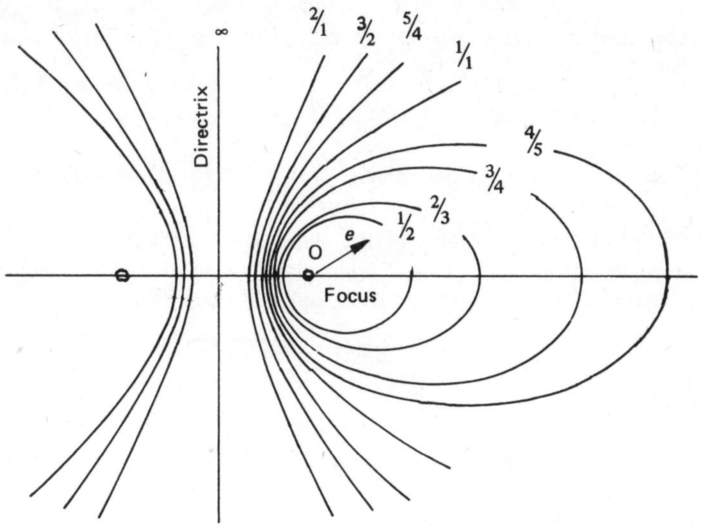

Fig. 4.5

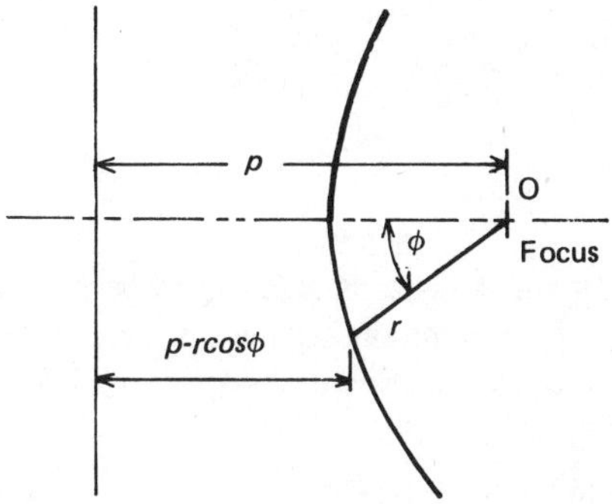

Fig. 4.6

the focus is then equal to its r coordinate. Furthermore, it is obvious from elementary trigonometry that, the distance of the focus from the directrix being represented by p, the perpendicular distance of the point on the conic from the directrix can be equated to $(p - r.\cos\phi)$.

From the definition of eccentricity we can therefore write:

$$e = \frac{\text{Distance from focus}}{\text{Distance from directrix}} = \frac{r}{p - r.\cos\phi}$$

whence $\quad e(p - r.\cos\phi) = r$

or $\quad 1 + e.\cos\phi = ep.r^{-1} \qquad (4.5)$

This is the standard equation of any conic in polar coordinates whose origin lies at the focus and whose datum direction lies along the principal axis, and it has the form

$$1 + e.\cos\phi = \mu.r^{-1} \tag{4.5a}$$

where e and μ are a pair of parameters, of which μ (or p) determines the scale of the conic, whilst e alone determines its form.

In elliptical orbits, the area of the ellipse is of particular importance, and this can be determined in terms of e and p (or of e and μ) by noting that the segment that lies between the point P = (r, ϕ), and the adjacent point P′ = $(r+\mathrm{d}r,$

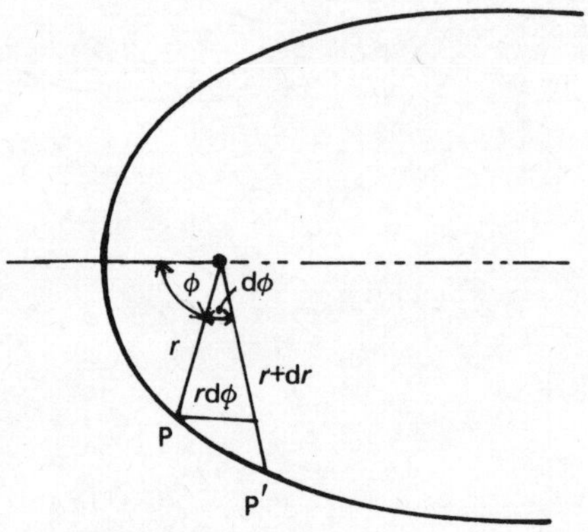

Fig. 4.7

$\phi+\mathrm{d}\phi)$ approximates to a triangle of base r, and perpendicular height $r.\mathrm{d}\phi$, as indicated in Fig. 4.7. The area of the segment is therefore equal to $r^2.\mathrm{d}\phi/2$, and integrating from $\phi = 0$ to $\phi = 2\pi$ gives

$$\text{Area of ellipse} = \int_0^{2\pi} (r^2/2).\mathrm{d}\phi = \int_0^{\pi} r^2.\mathrm{d}\phi \ .$$

But $\quad r = ep/(1 + e.\cos\phi)$,

therefore $$\text{Area} = (ep)^2 \int_0^{\pi} \frac{\mathrm{d}\phi}{(1 + e.\cos\phi)^2} ,$$

that is $$\text{Area} = \frac{(ep)^2}{1 - e^2}\left[\frac{-e.\sin\phi}{1 + e.\cos\phi} + \int \frac{\mathrm{d}\phi}{1 + e.\cos\phi}\right]_0^{\pi}$$

$$= \frac{2(ep)^2}{(1 - e^2)^{3/2}}\left[\tan^{-1}\frac{\sqrt{1 - e^2}.\tan\phi/2}{(1 - e)}\right]_0^{\pi} ,$$

whence: $\quad \text{Area} = \pi(ep)^2/(1 - e^2)^{3/2} = \pi.\mu^2/(1 - e^2)^{3/2}$.

Areal Speed

If either or both of two points P and Q are in motion, the straight line drawn between them sweeps out a surface, and the instantaneous time-rate of change of the area of the surface is called the areal speed of either of the points about the other. Thus, referring to Fig. 4.8, let P and Q represent the positions of the

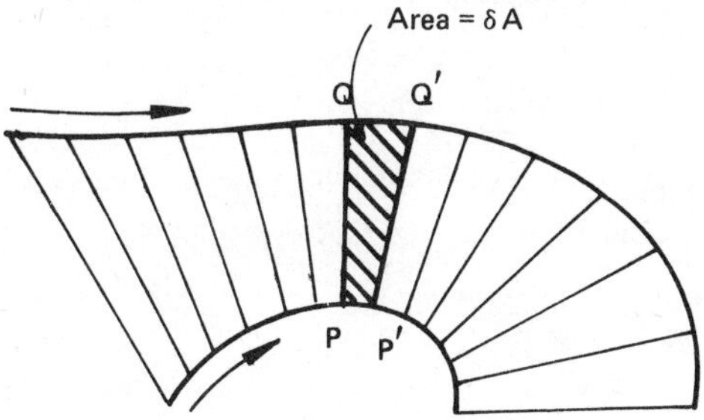

Fig. 4.8

points at time t, and let P′ and Q′ represent their positions at time $(t + \delta t)$. Then if δA represents the superficial (scalar) area of the element QPP′Q′ of the swept surface, the areal speed of either of the points about the other is defined as follows:

$$\text{Areal Speed} = \frac{dA}{dt} = \underset{\delta t \to 0}{\text{Limit}} \frac{\delta A}{\delta t} .$$

Consider the areal speed about the origin, of a point P which is confined to the (r, ϕ) plane of polar coordinates as indicated in Fig. 4.9. If $P = (r, \phi)$ represents the position of the point at time t, and if $P' = (r+\delta r, \phi+\delta\phi)$ represents its position at time $t + \delta t$, the surface swept out by the line OP in the interval of time δt approximates to a triangle of base r and perpendicular height $r.\delta\phi$. The area swept in time δt can therefore be equated to $\frac{1}{2}r^2\delta\phi$, and since the limit of $\delta\phi/\delta t$ is indicated by $\dot{\phi}$, it follows that the instantaneous time-rate of change

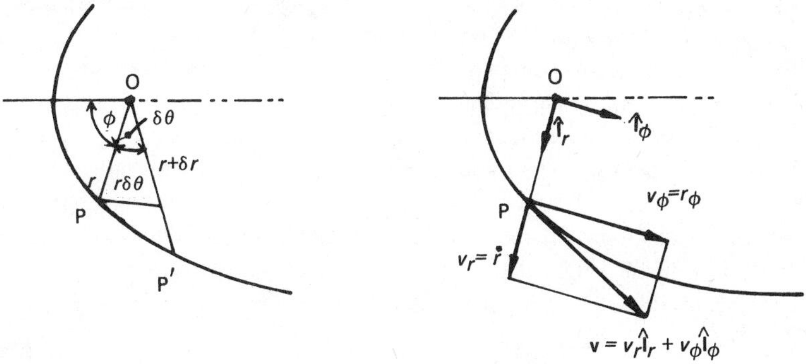

Fig. 4.9

of the area can be equated to $\frac{1}{2} r^2\dot{\phi}$. However, if the velocity of P is given in terms of its cylindrical components, $\mathbf{v} = v_r\hat{\mathbf{l}}_r + v_\phi\hat{\mathbf{l}}_\phi$, the circumferential component v_ϕ can be substituted for the product $r\dot{\phi}$. Therefore if $(r, \phi, 0)$ represent the coordinates of a point which moves in the $r\phi$ plane, and if v_ϕ represents the coefficient of the circumferential component of its velocity, the areal speed of the point about the origin is given by:

$$\text{Areal speed} = \frac{1}{2} r^2\dot{\phi} = \frac{1}{2} r\, v_\phi \qquad (4.5)$$

Motion in a Central Field

By definition, a central field of force is one in which the force is a function of position, such that it is always directed along the line that joins the position to a fixed Centre, and has a magnitude that is a function of the distance from the Centre. Thus if **r** is the vector which defines the position of any point relative to the Centre, the distance from the Centre is given by its modulus $|\mathbf{r}|$, and a central field can be described by $f(|\mathbf{r}|).\mathbf{r}$, where $f(|\mathbf{r}|)$ represents a scalar function of the distance from the Centre $|\mathbf{r}|$.

Consider the motion of a point mass which responds to such a field according to the Second Law of Motion.

By the 2nd Law:

$$f(|\mathbf{r}|).\mathbf{r} = m.\mathrm{d}\mathbf{v}/\mathrm{d}t$$

Taking the cross-product of both sides of the equation with **r**, we note that $\mathbf{r} \times \mathbf{r} = 0$.

Therefore: $0 = \mathbf{r} \times m(\mathrm{d}\mathbf{v}/\mathrm{d}t)$.

But the expression on the right can be equated to $\mathrm{d}(\mathbf{r} \times m\mathbf{v})/\mathrm{d}t$, because $\mathrm{d}\mathbf{r}/\mathrm{d}t = \mathbf{v}$, and $\mathbf{v} \times m.\mathbf{v}$ is necessarily zero.

Therefore: $0 = \mathrm{d}(r \times m\mathbf{v})/\mathrm{d}t$

or $\mathbf{r} \times m\mathbf{v} = \text{Constant}$.

Now $m\mathbf{v}$ is the momentum of the mass m, and the quantity $\mathbf{r} \times m\mathbf{v}$ is called the moment of momentum (or the angular momentum) of the mass about the centre of the field, and Newton recognised that if a point mass moved in a central field according to the Second Law of Motion, the moment of its momentum about the Centre would be constant. However, the mass being a constant, we could also say that $\mathbf{r} \times \mathbf{v}$ would be constant, when it follows from the definition of a cross-product that both **r** and **v** must be confined to a plane which is normal to the fixed direction of their product. Taking polar coordinates in this plane of the motion, with the origin at the centre of the field, the vectors **r** and **v** can now be described in terms of their components in these coordinates to write:

$$\mathbf{r} \times \mathbf{v} = r.\hat{\mathbf{l}}_r \times (\dot{r}.\hat{\mathbf{l}}_r + r\dot{\phi}.\hat{\mathbf{l}}_\phi)$$

whence $\mathbf{r} \times \mathbf{v} = r^2\dot{\phi}.\hat{\mathbf{l}}_z$.

Thus the modulus of $\mathbf{r} \times \mathbf{v}$ is now seen to be equal to the areal speed of the mass about the centre of the field, and Newton was therefore able to recognise that a point mass which moved in a central field according to the Second Law of Motion would move in a plane which contained the Centre with a constant areal speed, as observed in the motions of the planets about the Sun.

Motion in a Gravitational Field

Newton's Law of Gravitation postulates that any two masses interact with forces of mutual attraction which are such that the force on each is directed toward the other, and has a magnitude which is proportional both to each of the forces, and to the inverse of the square of the distance between them. A gravitational field of any mass is thus postulated as one that is centred on the mass itself, and we now consider the motion of mass m_p which responds according to the Second Law of Motion to the sole influence of a second mass m_o, which, for the present, is assumed to be fixed in inertial space.

We have already seen that such a motion is necessarily confined to a plane which contains the fixed Centre, and referring to Fig. 4.10, we analyse the motion in polar coordinates whose (r, ϕ) plane lies in the plane of the motion.

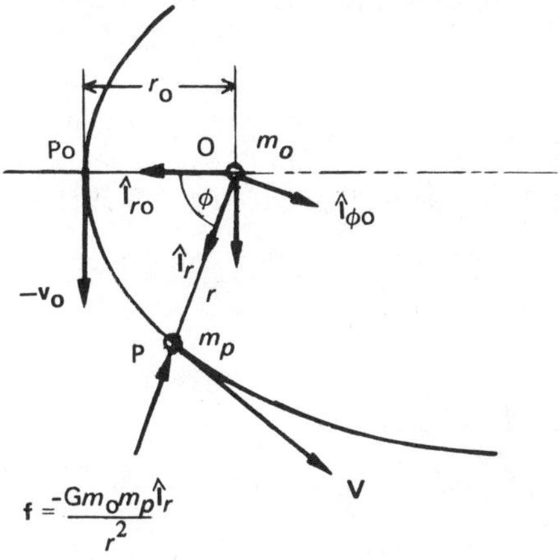

Fig. 4.10

The origin is chosen at the fixed position of m_o, and assuming such a point to exist, the datum of ϕ is chosen to pass through the point P_o on the path at which the velocity $\mathbf{v}_o$, (and therefore the tangent to the path), is normal to the r coordinate r_o, and parallel to the unit circumferential vector $\hat{\mathbf{1}}_{\phi o}$. For a general point on the path $P = (r, \phi)$, the direction from P to O is then defined by $-\hat{\mathbf{1}}_r$, whilst the distance of P from O is defined by its r coordinate, and by the simultaneous application to the mass m_p of Newton's Laws of Gravitation and of Motion we have:

$$f = -\frac{Gm_p m_o}{r^2} \cdot \hat{\mathbf{1}}_r = m_p \cdot (\mathrm{d}\mathbf{v}/\mathrm{d}t)$$

or: $$-\hat{\mathbf{1}}_r = \frac{r^2}{Gm_o} \cdot \frac{\mathrm{d}\mathbf{v}}{\mathrm{d}t}$$

where **v** represents the velocity of m_p in the inertial reference.

Now G and m_o are constants, and we have seen, firstly, that in a motion in a central field $\dot{\phi}r^2$ also is constant, and secondly, that $-\dot{\phi}\hat{\mathbf{1}}_r = \mathrm{d}\hat{\mathbf{1}}_\phi/\mathrm{d}t$. The foregoing equation can therefore be reduced to a readily integrable form by multiplying through by $\dot{\phi}$, and substituting $-\dot{\phi}\hat{\mathbf{1}}_r = \mathrm{d}\hat{\mathbf{1}}_\phi/\mathrm{d}t$.

Thus: $$\frac{\mathrm{d}\hat{\mathbf{1}}_\phi}{\mathrm{d}t} = \frac{\dot{\phi}r^2}{Gm_o} \frac{\mathrm{d}\mathbf{v}}{\mathrm{d}t} .$$

whence: $$\mathrm{d}\hat{\mathbf{1}}_\phi = \frac{\dot{\phi}r^2}{Gm_o} \mathrm{d}\mathbf{v}$$

And: $$\hat{\mathbf{1}}_\phi + \mathbf{c} = \frac{\dot{\phi}r^2}{Gm_o} \mathbf{v} . \tag{4.7}$$

Here **c** is the constant vector of integration which can be determined by substituting the value of $\dot{\phi}r^2/Gm_o$ for the motion, together with the velocity **v** for any known point on the path. Thus **c** is determined as the difference of two vectors, of which one has the direction of the velocity for any point, whilst the other is the unit vector $\hat{\mathbf{1}}_\phi$ for the same point. But P_o was so chosen that $\mathbf{v}_o$ has the same direction as $\hat{\mathbf{1}}_{\phi o}$, and it follows that **c** also must lie in that direction, as indicated in Fig. 4.11.

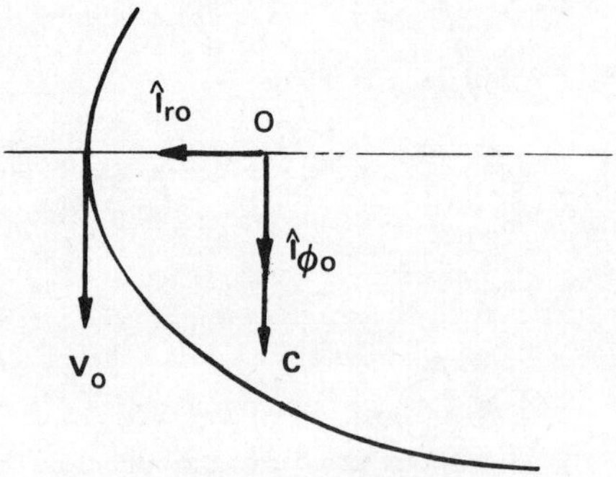

Fig. 4.11

Describing **v** in terms of its cylindrical component we then have:

$$\hat{\mathbf{1}}_\phi + \mathbf{c} = \frac{\dot{\phi}r^2}{Gm_o} \cdot (\dot{r}.\hat{\mathbf{1}}_r + r\dot{\phi}.\hat{\mathbf{1}}_\phi) ,$$

But $\hat{1}_r \times \hat{1}_r$ is necessarily zero, so the first term on the right can be eliminated by taking the cross-product of the equation with $\hat{1}_r$, and at the same time, we substitute $\hat{1}_r \times \hat{1}_\phi = \hat{1}_z$.

Thus: $$\hat{1}_z + \hat{1}_r \times \mathbf{c} = \frac{(\dot{\phi}r^2)^2}{Gm_o r} \cdot \hat{1}_z \ .$$

Now $\hat{1}_r$ and **c** both lie in the (r, ϕ) plane, so their cross-product lies in the direction defined by $\hat{1}_z$. Furthermore, Fig. 4.12 shows that for the general point P,

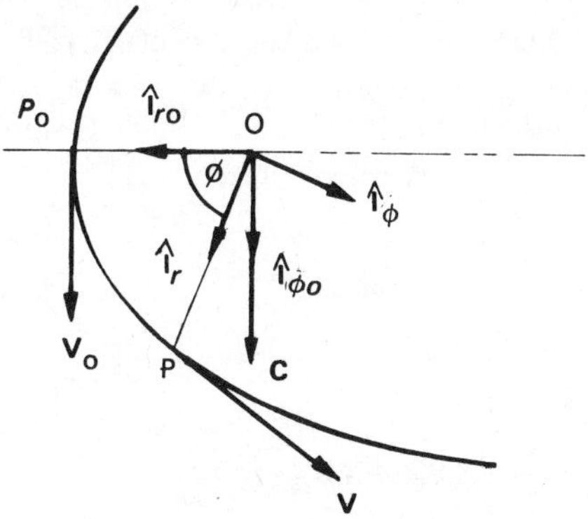

Fig. 4.12

the angle between $\hat{1}_r$ and **c** is equal to $(\pi/2 - \phi)$. Therefore by the definition of a cross-product:

$$\hat{1}_r \times \mathbf{c} = |\mathbf{c}|.\sin\left(\frac{\pi}{2} - \phi\right).\hat{1}_z$$

or: $$\hat{1}_r \times \mathbf{c} = |\mathbf{c}|.\cos\phi.\hat{1}_z \ .$$

Substituting accordingly, and dividing through by $\hat{1}_z$ then gives:

$$1 + |\mathbf{c}|.\cos\phi = \frac{(\dot{\phi}r^2)^2}{Gm_o} \cdot \frac{1}{r} \tag{4.8}$$

This is the equation of the path of m_p in the polar coordinates specified, and it can be recognised as the standard equation of the conic sections in which the modulus $|\mathbf{c}|$ determines the eccentricity, whilst $\dot{\phi}r^2/Gm_o$ determines the product of the eccentricity e with the distance of the focus from the directrix p, and thus determines the scale of the figure.

To use this result to determine the path of a particular motion we require to know only Gm_o, and the velocity **v** at any known point on the path. For in a motion that is deremined by any central field, $\dot{\phi}r^2$ has a common value at all points on the path, and since $\dot{\phi}r$ is the circumferential component of the velocity, the value of $\dot{\phi}r^2$ can be determined as the product of the circumferential component of the given velocity and the r coordinate of the position to which it applies. As to the constant of integration **c**, this can be determined simply by substituting the given data in eqn. (4.7), that is:

$$\mathbf{c} = \frac{\dot{\phi} r^2}{G m_o} \mathbf{v} - \hat{\mathbf{1}}_\phi \; .$$

It is now apparent that in choosing the datum to pass through the point on the path at which the tangent was normal to the r coordinate we effectively chose it along the principal axis of the resulting conic section, and we can now recognise the constant **c** as one whose direction determines the direction of the directrix, whilst its modulus determines the eccentricity. The determination of a given path is therefore simplest when we are given the velocity $\mathbf{v}_o$ and the r coordinate r_o for the vertex P_o, for in this case, $\mathbf{v}_o$ is wholly circumferential, and if v_o represents its modular speed we obtain:

$$\mathbf{v}_o = v_o \hat{\mathbf{1}}_{\phi o}$$

and

$$\dot{\phi} r^2 = v_o r_o$$

and

$$\mathbf{c} = \left[\frac{v_o^2 r_o}{G m_o} - 1\right] \cdot \hat{\mathbf{1}}_{\phi o} \; .$$

The eccentricity of the path is therefore given by:

$$e = |\mathbf{c}| = \frac{v_o^2 r_o}{G m_o} - 1 \; ,$$

and we conclude as follows:

(a) When $v_o^2 < 2Gm_o/r_o$, then $e < 1$, and the mass m_B is captive on an elliptical orbit.

(b) When $v_o^2 > 2Gm_o/r_o$, then $e > 1$, and the mass m_B is on a hyperbolic escape path.

(c) When $v_o^2 = 2Gm_o/r_o$, then $e = 1$, and this determines the minimum velocity for escape from the position on a parabolic path.

(d) A circle is the limiting case of an ellipse in which the two focuses coalesce and the eccentricity is zero. In this case $v_o^2 = Gm_o/r_o$, so that the speed of a circular orbit is $\sqrt{2}$ times the speed required for escape from the orbit.

Periods of Elliptical Orbits

Since the areal speed of any 'central' motion is constant it follows that the period of an elliptical orbit can be determined by dividing the area enclosed by the orbit by its areal speed. Thus, representing the period by T we have:

$$T^2 = \left(\frac{2\text{ Area}}{\dot{\phi} r^2}\right)^2 = \frac{4\pi^2 (ep)^4}{(1-e^2)^3} \cdot \frac{1}{(\dot{\phi} r^2)^2} \; ,$$

that is $$T^2 = \frac{4\pi^2(ep)}{(\dot{\phi}r^2)^2}\left[\frac{ep}{1-e^2}\right]^3 .$$

But in the standard equation of the conics, *ep* is the coefficient of $(1/r)$, which for a gravitational orbit is equal to the constant $(\dot{\phi}r^2)^2/Gm_o$, and substituting accordingly gives:

$$T^2 = \frac{4\pi^2}{Gm_o}\left(\frac{ep}{1-e^2}\right)^3 .$$

Now the major axis of an ellipse can evidently be equated to the differences of the r coordinates at $\phi = 0$ and $\phi = \pi$, and by substituting accordingly in the standard equation it can be shown that $ep/(1-e^2)$ is equal to the semi-major axis a.

Thus $$1 + e.\cos\phi = ep\frac{1}{r} ,$$

therefore $$r = ep/(1 - e.\cos\phi) .$$

Now $$2a = r_{\phi=o} - r_{\phi=\pi} ,$$

that is $$2a = \frac{ep}{1-e} + ep$$

or $$a = ep/(1-e^2) ,$$

therefore $$T^2 = \frac{4\pi^2}{Gm_o}\cdot a^3 .$$

Thus the square of the periods is proportional to the cube of the semi-major axes, as in Kepler's observation of the planets.

The Two Body Problem

So far we have only considered the case in which the mass m_o was fixed in inertial space, and the results were described in an inertial frame of reference. However, greater interest attaches to a case, like that of a rocket orbiting the Earth, in which each of the masses moves under the sole influence of the other. Furthermore there is some convenience in a result which, instead of describing the motions of each of the masses in an inertial frame, describes the motion of one in a frame of reference which moves with the other. So, referring to Fig. 4.13, now let m_p and m_o represent a pair of masses, each of which moves under the sole influence of the other, and identifying I as an inertial frame of reference, consider the motion of m_p in a frame of reference O whose origin moves with m_o, but without rotation in the inertial frame. This is not an inertial frame, but in the absence of relative rotation, the acceleration of P in the frame O can be equated to the difference in the accelerations of P and O in the inertial frame.

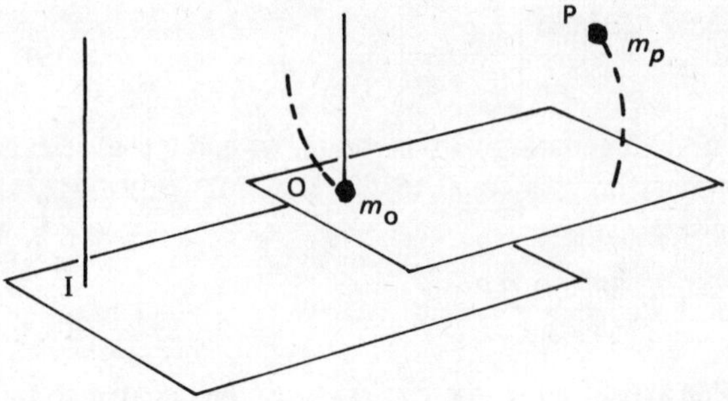

Fig. 4.13

Thus: $\mathbf{a}_{O/PO} = \mathbf{a}_{I/PI} - \mathbf{a}_{I/OI}$.

But by the simultaneous applications of the laws of gravitation and motion to the motion of m_p in the inertial frame, we have:

$$\mathbf{f}_{PO} = \frac{-Gm_p m_o}{|\mathbf{r}_{PO}|^2} \cdot \hat{\mathbf{r}}_{PO} = m_p . \mathbf{a}_{I/PI}$$

whence $$\mathbf{a}_{I/PI} = \frac{-Gm_p}{|\mathbf{r}_{PO}|^2} \cdot \hat{\mathbf{r}}_{PO} \ .$$

Similarly, for the motion of m_o in the inertial frame:

$$\mathbf{a}_{I/OI} = \frac{Gm_o}{|\mathbf{r}_{PO}|^2} \cdot \hat{\mathbf{r}}_{PO} \ ,$$

therefore $$\mathbf{a}_{O/PO} = \frac{-G(m_p + m_o)}{|\mathbf{r}_{PO}|^2} \cdot \hat{\mathbf{r}}_{PO} \ .$$

Now we have seen that when m_o is fixed in the inertial frame, then O also is inertial, and the controlling equation for the motion of P in this inertial frame is:

$$\mathbf{a}_{O/PO} = \frac{-Gm_o}{|\mathbf{r}_{PO}|^2} \cdot \hat{\mathbf{r}}_{PO} \ .$$

Evidently, these two equations are identical except that where the latter contains the constant factor m_o, the former contains the factor $(m_p + m_o)$. It follows that the previous results for the case in which m_o is fixed can be adapted to serve the case when both masses move simply by substituting $(m_p + m_o)$ for m_o, but the result must be interpreted as the motion of m_p in a frame of reference which moves with m_o, but without rotation in inertial space.

Consider, for example, the case of a rocket which orbits the Earth. The mass of the rocket is presumably negligible compared with that of the Earth and would invariably be ignored, but the result of applying the previous equations must still be interpreted as the motion of the rocket in a frame of reference in which the centre of the Earth occupies a fixed position at the focus of the path, but with respect to which the Earth rotates upon its N/S axis once per day.

Example 4.8

At burnout a rocket moves parallel to the surface of the Earth beneath at a speed of 8 km/s and an altitude of 500 km. Treating the Earth as a sphere of 6400 km radius at whose surface the gravitational acceleration is 9·85 m/s^2,

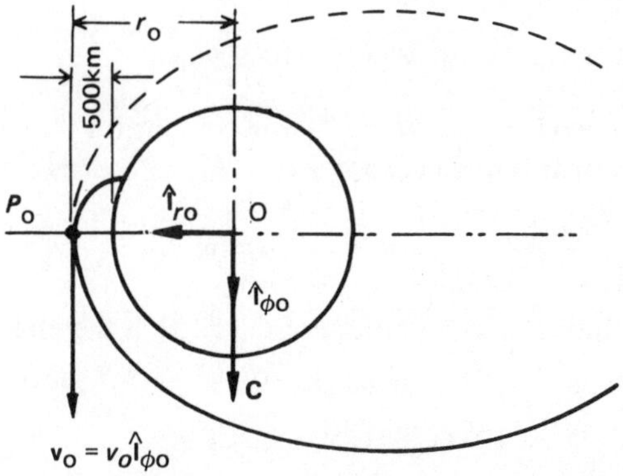

and treating the mass of the rocket as negligible compared with that of the Earth, determine the maximum altitude and the period of the orbit.

To determine Gm_o. Given the gravitational acceleration at the surface of the Earth, Gm_o can be determined by the simultaneous application of the laws of gravitational attraction and motion to a mass m at the Earth's surface.

Thus:
$$\frac{Gmm_o}{R^2} = mg \ ,$$

whence $\quad Gm_o = gR^2 = 9{\cdot}85(6{,}400 \times 10^3)^2$

that is $\quad Gm_o = 403{\cdot}5 \times 10^{12}$ kg m/s^2 .

To determine **c**. At burnout the motion is at right-angles to the r coordinate. This can therefore be recognised as the vertex P_o at which the velocity is equal to the speed v_o times the unit circumferential vector $\hat{1}_{\phi o}$. Thus by substituting the given conditions at burnout equation 4.7.

$$\underset{\sim}{c} = \frac{\dot{\phi}r^2}{Gm_o}\cdot\mathbf{v}_o - \hat{1}_{\phi o}$$

$$= \frac{55{\cdot}2 \times 10^9}{403{\cdot}5 \times 10^{12}}(8 \times 10^3.\hat{1}_{\phi o}) - \hat{1}_{\phi o}$$

$$= 1{\cdot}095\,\hat{1}_{\phi o} - \hat{1}_{\phi o}$$

$$\mathbf{c} = 0{\cdot}095\,\hat{1}_{\phi o} \ .$$

To determine the path. Substituting accordingly in equation 4.8 determines the path of the rocket in a frame of reference in which the position of the Earth is fixed at the focal origin, but with respect to which the Earth rotates on its N/S axis once per day.

Thus: $$1 + 0{\cdot}095 \cos\phi = \frac{55{\cdot}2^2 \times 10^{18}}{403{\cdot}5 \times 10^{12}} \cdot \frac{1}{r}$$

or: $$1 + 0{\cdot}095 \cos\phi = 7{\cdot}55 \times 10^6 \cdot \frac{1}{r} .$$

To determine the maximum altitude. The maximum (or minimum) value of r occurs when $\phi = \pi$, that is when $\cos\phi = -1$.

Thus: $$\hat{r} = \frac{7{\cdot}55 \times 10^6}{(1 - 0{\cdot}095)} = 8{\cdot}35 \times 10^6 \,\text{m} .$$

The maximum altitude above the surface of the Earth is therefore given by:

$$\hat{h} = 8{\cdot}35 \times 10^6 - 6400 \times 10^3$$

$$\hat{h} = 1{\cdot}95 \times 10^6 \,\text{m} \equiv 1950 \,\text{km} .$$

To determine the area enclosed by the orbit. In terms of the eccentricity e, and the distance p of the focus from the directrix, the area of an ellipse is given by:

$$\text{Area} = \frac{\pi(ep)^2}{(1 - e^2)^{3/2}} .$$

The values of e and (ep) can be read directly from the standard equation where e appears as the coefficient of $\cos\phi$, whilst (ep) appears as the coefficient of $\frac{1}{r}$.

$$\text{Area} = \frac{\pi(7{\cdot}55 \times 10^6)^2}{(1 - 0{\cdot}095^2)^{3/2}}$$

$$\text{Area} = 182 \times 10^{12} \,\text{m}^2 .$$

To determine the period of the orbit. The period of the orbit is equal to its area divided by the areal speed, and the areal speed is equal to $\dot{\phi}r^2/2$.

$$T = \frac{2 \text{ Area}}{\dot{\phi}r^2} = \frac{2 \times 182 \times 10^{12}}{55{\cdot}2 \times 10^9} \,\text{s}$$

$$T = 6594 \,\text{s} \equiv 1 \,\text{hr}\; 49 \,\text{m}\; 54 \,\text{s} .$$

Example 4.9

At burnout, a rocket has an altitude of 600 km, and speed v_i of 7,920 m/s at the angle $\theta_i = 5°$ to the circumferential, as indicated. Treating the earth as a sphere of 6400 km radius, and given that gravitational acceleration at its surface

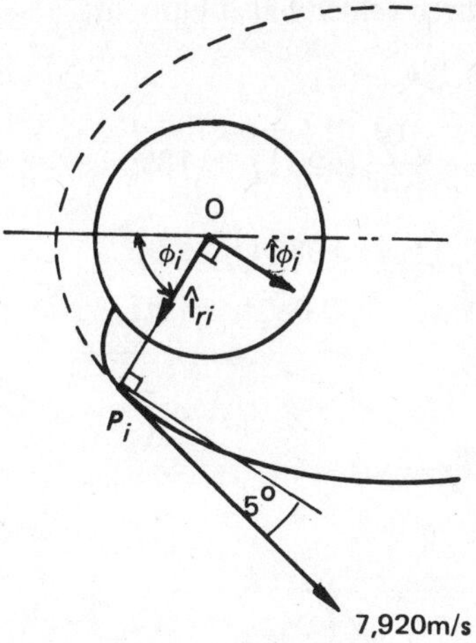

is 9·85 km/s², determine: the coordinate of the point of burnout ϕ_i, the minimum and maximum altitudes of the orbit, and the period of the orbit.

Solution

At the vertex of an orbit the velocity lies in the circumferential direction. In the present case, therefore, the vertex does not occur at the point of burnout, and it is necessary to determine the angle θ_i at which it occurs. Otherwise the calculations proceed as in the previous example.

From the given initial conditions at burnout:

$$\mathbf{v}_i = 7920(\sin 5° . \hat{\mathbf{1}}_{ri} + \cos 5° . \hat{\mathbf{1}}_{\phi i}) \text{ m/s};$$

that is $$\mathbf{v}_i = 690\, \hat{\mathbf{1}}_{ri} + 7890\, \hat{\mathbf{1}}_{\phi i} \text{ m/s } .;$$

To determine $\dot{\phi}r^2$ we take the product of $r_{\phi i}$ with the circumferential component of $\mathbf{v}_i$.

Thus: $$\dot{\phi}r^2 = (7 \times 10^6)(7{\cdot}89 \times 10^3) \text{ m}^2/\text{s}$$

that is $$\dot{\phi}r^2 = 55{\cdot}23 \times 10^9 \text{ m}^2/\text{s} .$$

To determine Gm_o from the given gravitational acceleration at the Earth's surface by the simultaneous application of the laws of gravitation and motion to a mass m:

$$Gmm_o/R^2 = m.g$$

$$Gm_o = gR^2 = 9{\cdot}85(6{\cdot}4 \times 10^6)^2 = 403{\cdot}5 \times 10^{12}$$

To determine **c** from the given velocity v_i at burnout:

$$\mathbf{c} = (\dot{\phi}r^2/Gm_o)\,\mathbf{v}_i - \hat{\mathbf{l}}_{\phi i}$$

$$\mathbf{c} = \frac{55{\cdot}23 \times 10^9}{403{\cdot}5 \times 10^{12}}(690\,\hat{\mathbf{l}}_{ri} + 7890\,\hat{\mathbf{l}}_{\phi i}) - \hat{\mathbf{l}}_{\phi i}$$

$$\mathbf{c} = 0{\cdot}09445\,\hat{\mathbf{l}}_{ri} + 0{\cdot}080\,\hat{\mathbf{l}}_{\phi i}$$

$$|\mathbf{c}| = \sqrt{(0{\cdot}09445^2 + 0{\cdot}080^2)} = 0{\cdot}1238\ .$$

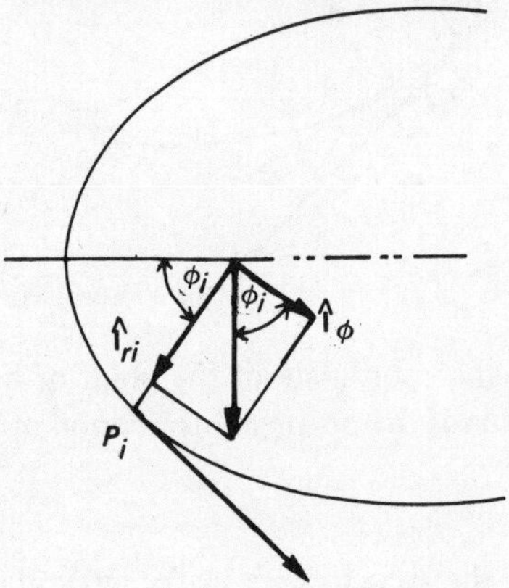

To determine ϕ_i we note that $\tan\phi_i$ is equal to the ratio of the $\hat{\mathbf{l}}_{ri}$ and $\hat{\mathbf{l}}_{\phi i}$ components of **c**.

Thus: $$\phi_i = \tan^{-1}\frac{0{\cdot}09445}{0{\cdot}080}$$

$$\phi_i = 0{\cdot}868 \text{ rad} \equiv 49{\cdot}74^\circ\ .$$

To determine the orbit we substitute $|\mathbf{c}|$, $\dot{\phi}r^2$ and Gm_o in:

$$1 + |\mathbf{c}|\cos\phi = \{(\dot{\phi}r^2)^2/Gm_o\}\,r^{-1}\ .$$

Thus. $$1 + 0{\cdot}1238\cos\phi = \frac{(55{\cdot}23.10^9)^2}{403{\cdot}5.10^{12}}\cdot r^{-1}$$

$$1 + 0{\cdot}1238\cos\phi = 7{\cdot}560.10^6\,r^{-1}\ .$$

Maximum and minimum altitudes occur at $\phi = 0$ and $\phi = \pi$:

Thus: $$\hat{r} = 7{\cdot}560/0{\cdot}8762 = 8{\cdot}628.10^6 \text{ m}$$

and $$\check{r} = 7{\cdot}560/1{\cdot}1238 = 6{\cdot}727.10^6 \text{ m}$$

Max. Alt = 8628 − 6400 = 2228 km

Max. Alt = 6727 − 6400 = 327 km .

To determine the period of the orbit we divide its area by its areal speed:

$$\text{Area} = \frac{\pi(ep)^2}{(1 - e^2)^{3/2}} = \frac{\pi(\dot{\phi}r^2/Gm_o)^2}{(1 - |\mathbf{c}|^2)^{3/2}}$$

$$T = \frac{2.\text{Area}}{\dot{\phi}r^2} = \sqrt{\left(\frac{4\pi^2(\dot{\phi}r^2)^2}{(Gm_o)^4\,(1-|\mathbf{c}|^2)^3}\right)}$$

$$T = \frac{4\pi^2 55{\cdot}23 \times 10^9}{(403{\cdot}5.10^{12})^4 \times (1-0{\cdot}1238^2)^3}$$

$$T = 6654 \text{ s} \equiv 1 \text{ hr } 50 \text{ m } 54 \text{ s} \;.$$

4.4 IMPULSE

The Second Law of Motion relates the force on a particle to its acceleration, and so far we have used this equation only by substituting a given force, and then integrating the resulting equation in order to determine the velocity and the displacement. But by integrating the law of motion itself (in one case with respect to time, and in another with respect to displacement) we can arrive at a pair of corollaries of the Second Law of Motion that directly relate integrals of the force to the velocity and the speed, and these are commonly of particular convenience.

Thus, if $\mathbf{f}$ represents the resultant force on a particle of mass m, and $\mathbf{v}$ its velocity, by the Second Law of Motion:

$$\mathbf{f} = m\frac{d\mathbf{v}}{dt} \;,$$

or
$$\mathbf{f}\, dt = m\, d\mathbf{v} \;,$$

and if $\mathbf{v}_1$ and $\mathbf{v}_2$ represent the velocities at times t_1 and t_2, this equation can be integrated between limits to give

$$\int_{t_1}^{t_2} \mathbf{f}\, dt = m\int_{\mathbf{v}_1}^{\mathbf{v}_2} d\mathbf{v}$$

or
$$\int_{t_1}^{t_2} \mathbf{f}\, dt = m(\mathbf{v}_2 - \mathbf{v}_1) \tag{4.9}$$

The term on the right can be recognised as the net change of momentum in the interval of time from $t = t_1$ to $t = t_2$, whilst the integral on the left is called the impulse of the force in the interval, so a corollary of the Second Law can be stated as follows:

The impulse of a force in a given interval of time = The net change of momentum it produces in the interval.

This relationship is evidently convenient where the force is a known function of time, and we require to determine the net change of velocity over a given interval of time. But it is also useful for determining the time-average value of a force that produces a known net change of velocity in a given interval of time – particularly in collisive encounters. For if a force **f** is a function of time t, the time-average of **f** over a given interval of time from $t = t_1$ to $t = t_2$, is, by definition, the constant value $\bar{\mathbf{f}}$ whose integral over the interval is equal to the integral of the actual force **f**. Thus:

By definition $$\int_{t_1}^{t_2} \bar{\mathbf{f}}.\mathrm{d}t = \int_{t_1}^{t_2} \mathbf{f}.\mathrm{d}t \qquad \bar{\mathbf{f}} = \text{const.}$$

or $$\bar{\mathbf{f}}.(t_2 - t_1) = \int_{t_1}^{t_2} \mathbf{f}.\mathrm{d}t\ ,$$

Therefore: $$\bar{\mathbf{f}} = \frac{\int_{t_1}^{t_2} \mathbf{f}.\mathrm{d}t}{(t_2 - t_1)} = \frac{m(\mathbf{v}_2 - \mathbf{v}_1)}{(t_2 - t_1)} \tag{4.9}$$

We therefore recognise that the time average value of a force over a given interval of time can be equated to the resulting net change of momentum per unit time.

Example 4.10

The magnitudes of the components of a force **f** vary with time t seconds according to

$$(f_x, f_y, f_z) = [\{2 + \pi\sin(\pi t/8)\}, \pi\cos(\pi t/8), t - 0{\cdot}3t^2]\ .$$

(a) Determine the impulse of the force in the interval $0 < t < 2$.

(b) If the force acts on a mass of 4 kg, and its velocity at time $t = 0$ is $(\mathbf{i} + 2\mathbf{j} + 3\mathbf{k})$ m/s, determine the velocity at time $t = 2$ s.

Solution

(a) By definition: Impulse $= \int_0^2 \mathbf{f}.\mathrm{d}t$

$$= \int_0^2 [\{2 + \pi\sin(\pi t/8)\}\mathbf{i} + \pi\cos(\pi t/8)\mathbf{j} + (t - 0{\cdot}3t^2)\mathbf{k}]\ .\mathrm{d}t$$

$$= \Big[\{2t - 8\cos(\pi t/8)\}\mathbf{i} + 8\sin(\pi t/8)\mathbf{j} + (0{\cdot}5t^2 - 0{\cdot}1t^3)\mathbf{k}\Big]_0^2$$

$$= \{(4 - 4\sqrt{2}).\mathbf{i} + 4\sqrt{2}.\mathbf{j} + (2 - 0{\cdot}8)\mathbf{k}\} - \{-8.\mathbf{i}\}$$

$$\text{Impulse} = 4(3 - \sqrt{2})\mathbf{i} + 4\sqrt{2}.\mathbf{j} + 1.2.\mathbf{k}$$

(b) The change of momentum in a given interval of time is equal to the impulse of the force in the interval. Therefore:

$$4\{\mathbf{v}_2 - (\mathbf{i} + 2\mathbf{j} + 3\mathbf{k})\} = 4(3 - \sqrt{2})\mathbf{i} + 4\sqrt{2}.\mathbf{j} + 1{\cdot}2.\mathbf{k}$$

therefore: $$4\mathbf{v}_2 = \{4(3 - \sqrt{2}).\mathbf{i} + 4\sqrt{2}.\mathbf{j} + 1{\cdot}2.\mathbf{k}\} + 4(\mathbf{i} + 2\mathbf{j} + 3\mathbf{k})$$

$$\mathbf{v}_2 = (4 - \sqrt{2})\mathbf{i} + (2 + \sqrt{2}).\mathbf{j} + 3{\cdot}3.\mathbf{k}\ \mathrm{m/s}\ .$$

Example 4.11

A point mass of 2 kg approaches a wall at an angle of 45° and rebounds at an angle of 30° as shown. The approach speed is 80 m/s, and the rebound speed is 60 m/s. If the mass is in contact with the wall for 10^{-3} seconds, determine the average force on the wall during contact.

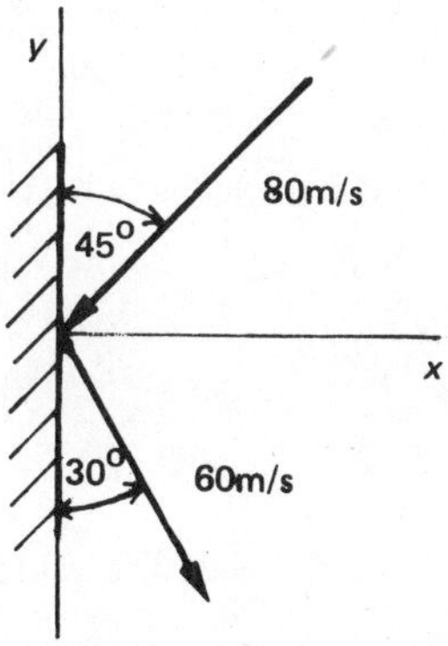

The time-average of the force is equal to the change of momentum per unit time:

$$\bar{\mathbf{f}} = m(\mathbf{v}_2 - \mathbf{v}_1)/(t_2 - t_1)$$
$$= 2\{(30.\mathbf{i} - 30\sqrt{3}.\mathbf{j}) - (-40\sqrt{2}.\mathbf{i} - 40\sqrt{2}.\hat{\mathbf{j}})\}/10^{-3}$$
$$= 20 \times 10^3\{(3 + 4\sqrt{2}).\mathbf{i} + (4\sqrt{2} - 3\sqrt{3}).\hat{\mathbf{j}}\}\ \mathrm{N}\ .$$

4.5 WORK AND KINETIC ENERGY

Under a given resultant force **f**, a given mass m would trace a particular path, and Fig. 4.14 shows an increment of displacement $\delta\mathbf{r}$ that the particle makes between a general point P on the path and an adjacent point P′. In general, this increment would not coincide with the path itself, but in the limit, as $\delta\mathbf{r}$ tends to zero, the two would tend to coincidence, so that even though a finite displacement is independent of the path by which the change of position is effected, a path can nevertheless be considered to comprise a succession of particular infinitesimal increments of displacement d**r**. So let **f** represent the resultant force that acts on the particle in the increment of displacement d**r**.

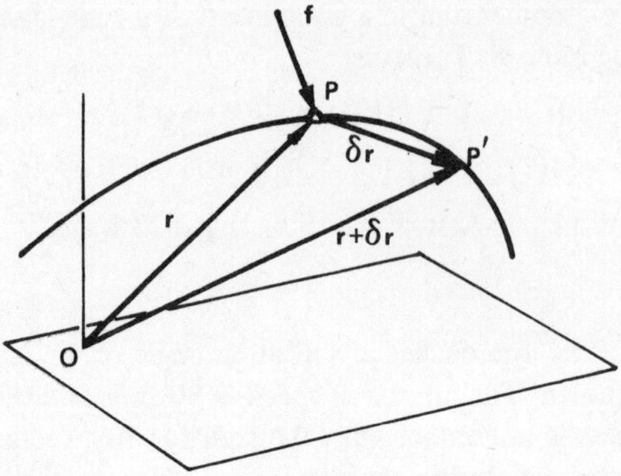

Fig. 4.14

By the 2nd Law:

$$\mathbf{f} = m\,(\mathrm{d}\mathbf{v}/\mathrm{d}t)\ ,$$

and taking the scalar product of each side with $\mathrm{d}\mathbf{r}$ gives:

$$\mathbf{f}\cdot\mathrm{d}\mathbf{r} = m\,(\mathrm{d}\mathbf{v}/\mathrm{d}t)\cdot\mathrm{d}\mathbf{r} = m\,\mathrm{d}\mathbf{v}\cdot(\mathrm{d}\mathbf{r}/\mathrm{d}t)$$

But $\mathrm{d}\mathbf{r}/\mathrm{d}t$ is, by definition, equal to $\mathbf{v}$.

Therefore: $\mathbf{f}\cdot\mathrm{d}\mathbf{r} = m\,\mathbf{v}\cdot\mathrm{d}\mathbf{v} = \dfrac{m}{2}\,\mathrm{d}(\mathbf{v}\cdot\mathbf{v})$

But $\mathbf{v}\cdot\mathbf{v} = |\mathbf{v}|^2$, and the modulus of the velocity $\mathbf{v}$ is the speed v, and we may therefore write:

$$\mathbf{f}\cdot\mathrm{d}\mathbf{r} = \frac{m}{2}\,d(v^2)\ .$$

Finally, if v_1 and v_2 represent the speeds of the particle at the positions $\mathbf{r} = \mathbf{r}_1$ and $\mathbf{r} = r_2$, we may integrate between limits to give:

$$\int_{\mathbf{r}_1}^{\mathbf{r}_2} \mathbf{f}\cdot\mathrm{d}\mathbf{r} = \frac{m}{2}\int_{v_1}^{v_2} \mathrm{d}(v^2) = \frac{m}{2}\,(v_2^2 - v_1^2) \qquad (4.10)$$

Work

The integral on the left is identified as the work done on the particle in the displacement of the force from $\mathbf{r}_1$ to $\mathbf{r}_2$ along the path specified, and we note that if $\mathbf{f}$ were constant, the value of the integral would be $\mathbf{f}\cdot(\mathbf{r}_2 - \mathbf{r}_1)$. But $(\mathbf{r}_2 - \mathbf{r}_1)$ can be recognised as the net displacement, so recognising that a constant force is a special case in which the work done in a given displacement is independent of the path by which the net change of position is achieved, we specify as follows:

"The work done by a constant force in any given displacement, or the increment of work done by a variable force in an infinitesimal increment of displacement, is, by definition, equal to the scalar product of the force and the displacement; and it follows from the definition of a scalar product that this may be equated to the magnitude of the force times the component of the displacement resolved in the direction of the force, or vice versa.

The work done by a variable force in a finite displacement is generally defined as a path integral which sums the increments of work in the increments of displacement which make up the particular path of the motion, as follows:

$$\text{Work done} = \int_{\mathbf{r}_1}^{\mathbf{r}_2} \mathbf{f} \cdot \mathbf{dr} \ .$$

Kinetic Energy

The quantity on the right of the previous equation, {that is, $(m/2)\,(v_2^2 - v_1^2)$}, is identified as the change in the kinetic energy of the particle, where the kinetic energy is defined as the product of half the mass with the square of the speed:

$$\text{Kinetic energy} = \frac{m}{2} \cdot v^2 \ .$$

Work v Kinetic Energy

The previous equation can therefore be expressed as a corollary of the Second Law of Motion as follows:

> "When a particle moves under a force, the kinetic energy of the particle is varied by an amount equal to the work done upon it."

Example 4.12

The figure shows the path of a particle which moves from A to B along a cylindrical helix of radius a, and uniform pitch. The particle supports a force f which

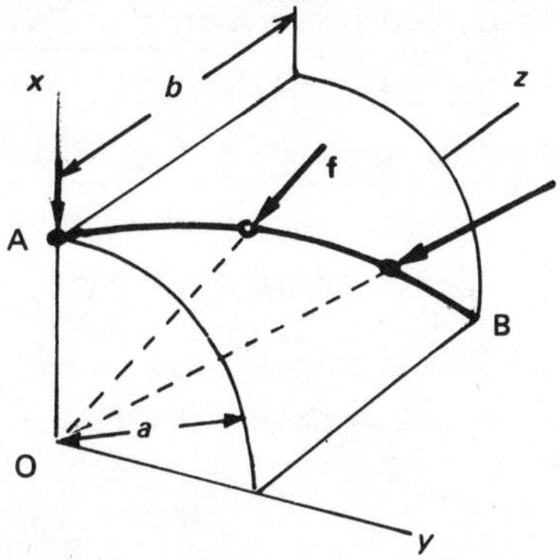

is directed toward the origin, and whose magnitude varies from F at A, inversely as the cube of the distance from O. Determine the work done on the particle.

Solution

To determine $\int \mathbf{f} \cdot d\mathbf{r}$, it is necessary to describe f and dr (or r) as functions of a single scalar argument. But the magnitude and direction of f are both described in terms of the magnitude and direction of the radius vector r, and since the elements of a radius vector are equal to the coordinates of its final point, the

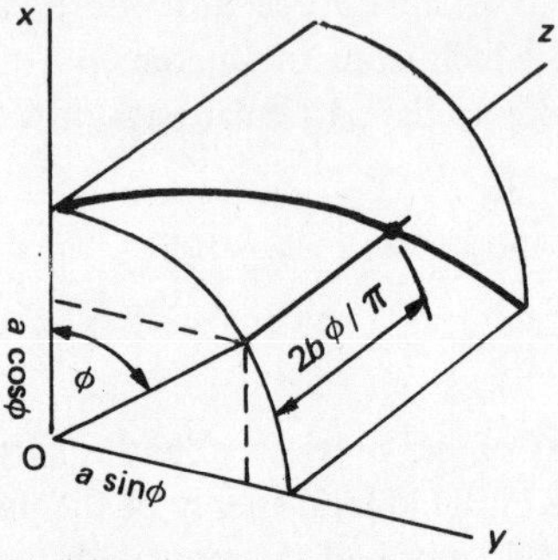

first essential step is simply that of describing the coordinates of a point P on the path as a function of a single scalar argument. Any of the given coordinates could well serve as the independent variable, but a helix of uniform pitch is one in which the advance parallel to the axis is proportional to the angle turned about the axis, so we choose as the independent variable, the angle ϕ indicated in the diagram.

$$\mathbf{r} = \{a.\cos\phi.\mathbf{i} + a.\sin\phi.\mathbf{j} + (2b\phi/\pi).\mathbf{k}\}$$

$$|\mathbf{r}| = \{a^2 + (2b\phi/\pi)^2\}^{\frac{1}{2}}$$

$$d\mathbf{r} = \{-a.\sin\phi.\mathbf{i} + a.\cos\phi.\mathbf{j} + (2b/\pi).\mathbf{k}\}.d\phi$$

$$\mathbf{f} = -F\left(\frac{a}{|\mathbf{r}|}\right)^3.\hat{\mathbf{r}} = \frac{-F.a^3}{|\mathbf{r}|^4}.\mathbf{r}$$

$$\mathbf{f} = \frac{-F.a^3\{a.\cos\phi.\mathbf{i} + a.\sin\phi.\mathbf{j} + (2b\phi/\pi).\mathbf{k}\}}{\{a^2 + (2b\phi/\pi)^2\}^2}$$

$$\mathbf{f}\cdot d\mathbf{r} = \frac{-Fa^3\{-(a^2/2)\sin 2\phi + (a^2/2)\sin 2\phi + (2b/\pi)^2\phi\}.d\phi}{\{a^2 + (2b\phi/\pi)^2\}^2}$$

$$= \frac{-2\pi^2F.a^3b\phi.d\phi}{\{(\pi a)^2 + (2b\phi)^2\}^2}$$

$$\int_{\mathbf{r}_1}^{\mathbf{r}_2} \mathbf{f}\cdot d\mathbf{r} = \left[\frac{-\pi^2Fa^3}{2\{(\pi a)^2 + (2b\phi)^2\}}\right]_{\phi=0}^{\phi=\pi/2}$$

$$= \frac{-F.a^3}{2(a^2 + b^2)} + \frac{Fa}{2} = \frac{-Fab^2}{2(a^2 + b^2)}.$$

Alternatively, in polar coordinates: $P = (r, \phi, z) = (a, \phi, c\phi)$, where $c = 2b/\pi$.

Therefore: $\mathbf{r} = a.\hat{\mathbf{l}}_r + c\phi.\hat{\mathbf{l}}_z$

and $|\mathbf{r}| = (a^2 + c^2\phi^2)^{\frac{1}{2}}$

and $d\mathbf{r} = (a.\,d\hat{\mathbf{l}}_r + c.\,d\phi.\hat{\mathbf{l}}_z$

that is $d\mathbf{r} = (a.\hat{\mathbf{l}}_\phi + c.\hat{\mathbf{l}}_z)d\phi$

$$\mathbf{f} = -F(a/|\mathbf{r}|)^3.\mathbf{r} = -F(a^3/|\mathbf{r}|^4).\mathbf{r}$$

$$\mathbf{f} = -Fa^3(a.\hat{\mathbf{l}}_r + c\phi.\hat{\mathbf{l}}_z)/(a^2 + c^2\phi^2)^2$$

therefore: $$\mathbf{f}\cdot d\mathbf{r} = \frac{-Fa^3c^2\phi.d\phi}{(a^2 + c^2\phi^2)^2}$$

$$\int_{\mathbf{r}_1}^{\mathbf{r}_2} \mathbf{f}\cdot d\mathbf{r} = \int_0^{\pi/2} \frac{-Fa^3c^2\phi.d\phi}{(a^2 + c^2\phi^2)^2}$$

$$= \left[\frac{Fa^3}{2(a^2 + c^2\phi^2)^2}\right]_0^{\pi/2}$$

$$= \left[\frac{Fa^3}{2\{a^2 + (2b\phi/\pi)^2\}}\right]_0^{\pi/2}$$

$$= \frac{Fa^3}{2}\left(\frac{1}{a^2 + b^2} - \frac{1}{a^2}\right) = \frac{-Fab^2}{2(a^2 + b^2)}.$$

Example 4.13

The control rod of a nuclear reactor has a mass m, and is guided by a vertical track which, in an emergency, allows the rod to enter the pile in free fall in the uniform gravitational field of strength g. After falling a vertical distance h, a

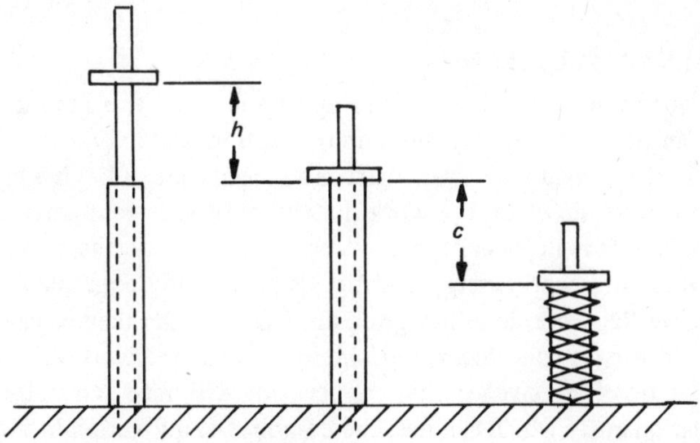

collar on the rod engages a thin tube which crumples under a constant axial force F so as to bring the rod to rest without a shock. Determine the crumpling strength required of the tube to bring the rod to rest after falling a further distance c after engagement, and the duration of the arresting action.

Solution

In free fall from rest in a uniform field, the motion is necessarily rectilinear, and it is necessary to consider only scalar motion in the vertical direction of the field. Reckoning s positive downward, the forces on the rod are:

(1) A constant gravitational force mg over the distance $h + c$,

and

(2) A constant crumpling force $-F$ over the distance c.

Over the distance $h + c$ the change of kinetic energy is zero. Therefore the net work done is zero, whence

$$mg(h + c) - F.c = 0$$

$$F = mg\left(\frac{h}{c} + 1\right) . \tag{1}$$

The duration of the arresting action is determined by equating the impulse of the force in the interval of the contact to the resulting change in the momentum of the rod.

For free fall through the distance h:

$$mg.h = (m/2)v^2_{s=h}$$

$$v^2_{s=h} = \sqrt{(2g.h)}$$

Therefore: $(mg-F)\,t = m(0-\sqrt{2gh})$

$$t = \sqrt{2gh}/\{(F/m)-g\} .$$

Substituting for F/m from (1)

$$t = c\sqrt{(2/gh)} .$$

4.6 CONSERVATIVE FIELDS

The work/energy equation is evidently appropriate to any situation in which the force can be described as a function of position, and in this case the force is said to comprise a field. However, there is a special class of fields which are said to be conservative, in which the work done in a displacement between any pair of terminal points is dependent, not on path, but only on the terminal points, and with these, the work/energy equation is particularly convenient. This is so partly because the work done in a given displacement can then be calculated for any path that proves convenient, but chiefly because the work done in moving to any point from any given datum can itself be described as a scalar field, that is, as a scalar quantity which can itself be described as a function of position.

Central Fields

In fact, we have already seen that the work done by a constant force depends only on the terminal points of the displacement, so a uniform field can be recognised as a ready example of a conservative field. However, a uniform field is only a special case of a central field in which the centre has retreated to infinity, and it can readily be shown that the whole class of central fields is, in fact, conservative. Thus, remembering that a central force is defined as one in which the force at each position is directed along the radial line to a fixed centre, and has a magnitude equal to a function of the distance from the centre, we refer to Fig. 4.15, where A and B represent a pair of arbitrary points in a field centred

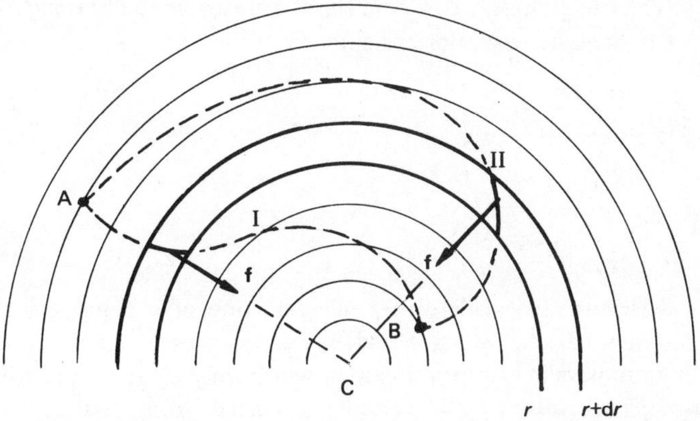

Fig. 4.15

on C, and we consider the work done by the force of the field in a displacement from A to B along each of two arbitrary paths, I and II. By their intercepts with a set of spheres centred on C, the paths are divided into infinitesimal increments in corresponding pairs, and considering one such pair, it is obvious from the definition of a central field, that since the two increments have a common distance from the centre, they would both be executed under forces of equal magnitudes. Furthermore it is obvious that the components of the displacements resolved in the radial directions of the forces would also have the same value, equal to the difference between the radii of the relevant spheres. Therefore the work done would have a common value for both increments, and, by aggregation, this implies that the work done on path I would be equal to the work done on path II. Any central field is therefore conservative, including the special case of a uniform field.

But apart from this particular, if important, case, there is a need for a general principle by which a field may be tested for conservation, and to facilitate the formulation of such a rule we shall now define three operations which may usefully be applied to fields. However, since these operations are to be applied to functions of the three coordinates of position, they themselves depend upon the principles by which the methods of calculus are extended to functions of multiple arguments, so we approach the definition of the required operations by a brief review of the basic principles of differentiation by parts.

Partial Derivatives

Let $\phi(x, y)$ represent a function of a pair of independent, scalar arguments, x and y. Since ϕ generally varies with both x and y, it is not directly amenable to differentiation with respect to either, but if y were fixed at an arbitrary parametric value $y = p =$ constant, then ϕ would vary with x alone, according to the function $\phi(x, p)$ which is determined by substituting $y = p =$ constant in $\phi(x, y)$. The function $\phi(x, p)$ would then be amenable to differentiation with respect to x in the usual way. For example

If $$\phi(x, y) = x^2y + 3xy^2 \ .$$

Then for $y = p =$ const.:

$$\phi(x, p) = px^2 + 3p^2x \ ,$$

and: $$\frac{\mathrm{d}\phi(x, p)}{\mathrm{d}x} = 2px + 3p^2 \ .$$

Now this derivative, like a derivative of a function of x alone, describes the ratio of infinitesimal changes in ϕ and x, but the present derivative is special in that the ratio is valid only for a case in which only x varies, and to mark its special significance, $\mathrm{d}\phi(x, p)/\mathrm{d}x$ is called a partial differential coefficient of ϕ with respect to x.

However, in practice both the symbolism and the procedure of the partial calculus are somewhat simplified. Thus, whereas an infinitesimal variation is usually indicated by the symbol d, in the special case in which only one of several independent variables does, in fact, vary, such a variation is represented by the symbol ∂. The partial differential coefficient of ϕ with respect to x for an arbitrary parametric value of y is then indicated by $\partial\phi/\partial x$, and the partial differential coefficient for a particular parametric value of y can be distinguished by the addition of a suitable suffix. Thus the partial differential coefficient of ϕ with respect to x, when $y =$ say, 2 may be indicated by $(\partial\phi/\partial x)_{y=2}$. The procedure of differentiation is simplified by omitting the step in which $y = p =$ const. is substituted in $\phi(x, y)$. Instead we simply differentiate $\phi(x, y)$ with respect to x while treating y as a constant, and we take it as understood that in the resulting function of x and y, the symbol y represents not the variable y itself, but any parametric value at which it may be set.

This approach can now be generalised in two ways. First, $\phi(x, y)$ could as well be differentiated partially with respect to y as to x, and the result would describe the ratio of infinitesimal changes in the ϕ and y for an arbitrary parametric value of x. Second, the method can as well be applied to functions of any number of independent arguments, provided it is understood that, taken alone, each partial coefficient is directly relevant only to an operation in which only one of the independent variables does, in fact, vary.

For example, if:

$$\phi = x^2yz^2 + xy^2z^3 \ ,$$

then $$\partial\phi/\partial x = 2xyz^2 + y^2z^3 \ ,$$

and $$\partial\phi/\partial y = x^2z^2 + 2xyz^3 \ ,$$

and $$\partial\phi/\partial z = 2x^2yz + 3xy^2z^2 \ .$$

But if a single partial coefficient is relevant only to an operation in which only one independent variable does, in fact, vary, the set of coefficients for all the independent variables, taken together, are sufficient to determine the increment in the function in the general case in which all the independent variables are given arbitrary increments simultaneously. For in saying that ϕ is a function of x and y, we mean that a unique value of ϕ is associated with each combination of values of x and y. The change in the value of ϕ therefore depends only on the net changes in the values of x and y, and not on the sequence of operations by which the changes are effected, and we are therefore free to choose a sequence in which the independent variables are varied one at a time. Thus, again considering a function $\phi(x, y)$ of x and y, let $\mathrm{d}x$ represent an arbitrary increment that is applied to x whilst holding the value of y constant. Then the ratio of the changes in ϕ and x is determined by the partial differential coefficient $\partial\phi/\partial x$, and since the actual change in x is $\mathrm{d}x$, the actual, consequential change in ϕ is given by $(\partial\phi/\partial x)\mathrm{d}x$. Now let y be varied by the arbitrary increment $\mathrm{d}y$ whilst x is held fixed. Then the ratio of the changes in ϕ and y is equal to $\partial\phi/\partial y$, and since the actual change of y is $\mathrm{d}y$, the actual change in ϕ is given by $(\partial\phi/\partial y)\mathrm{d}y$. Therefore if $\mathrm{d}\phi$ represents the actual increment of ϕ that results when x is varied by $\mathrm{d}x$, and y by $\mathrm{d}y$, we may write

$$\mathrm{d}\phi = \frac{\partial\phi}{\partial x}.\mathrm{d}x + \frac{\partial\phi}{\partial y}.\mathrm{d}y \ .$$

The extension of this rule for numbers of independent variables greater than two is obvious, and for the function $\phi(x, y, z)$ we may write

$$\mathrm{d}\phi = \frac{\partial\phi}{\partial x}.\mathrm{d}x + \frac{\partial\phi}{\partial y}.\mathrm{d}y + \frac{\partial\phi}{\partial z}.\mathrm{d}z \ .$$

The Operator ∇ (Del)

The required set of operations are defined from a vector operator ∇ (del), which, in its expanded form is presented in the appearance of a vector in which the scalar elements are replaced by the partial differential operators for each of the coordinates in turn. Thus:

$$\nabla = \frac{\partial}{\partial x}.\mathbf{i} + \frac{\partial}{\partial y}.\mathbf{j} + \frac{\partial}{\partial z}.\mathbf{k} \ . \tag{4.11}$$

The operations are then defined by treating del as the vector it resembles and applying it to the field on which it operates by one or other of the rules for the products of vectors.

The Operation Del

Given a scalar field ϕ, del is applied to ϕ by the rule for multiplying a vector by a scalar, and the result is then a vector field whose scalar elements are the derivatives of ϕ with respect to each of the coordinates in turn. This is called the gradient (Grad) of the given scalar field, and by definition:

$$\text{Grad}\ \phi = \nabla\phi = \left(\frac{\partial}{\partial x}.\mathbf{i} + \frac{\partial}{\partial y}.\mathbf{j} + \frac{\partial}{\partial z}.\mathbf{k}\right)\phi$$

that is

$$\text{Grad}\ \phi = \nabla\phi = \frac{\partial\phi}{\partial x}.\mathbf{i} + \frac{\partial\phi}{\partial y}.\mathbf{j} + \frac{\partial\phi}{\partial z}.\mathbf{k} \tag{4.12}$$

The Operation Del-dot

Since del is treated as a vector, it can be applied to a vector field either in a scalar product or in a vector product. Dealing first with the scalar product, let f represent a vector field. Then del-dot f determines a scalar field, called the divergence (Div) of the given vector field, equal to the scalar sum of the partial derivatives of the scalar elements of the given vector field, each with respect to the corresponding coordinate. Thus by definition:

$$\begin{aligned}\text{Div}\ \mathbf{f} &= \nabla\cdot\mathbf{f}\\ &= \left(\frac{\partial}{\partial x}.\mathbf{i} + \frac{\partial}{\partial y}.\mathbf{j} + \frac{\partial}{\partial z}.\mathbf{k}\right)\cdot\left(f_x.\mathbf{i} + f_y.\mathbf{j} + f_z.\mathbf{k}\right)\end{aligned}$$

$$\text{Div}\ \mathbf{f} = \nabla\cdot\mathbf{f} = \frac{\partial f_x}{\partial x} + \frac{\partial f_y}{\partial y} + \frac{\partial f_z}{\partial z} \tag{4.13}$$

The Operation Del-cross

In the operation Del-cross, the operator del is applied to a vector field according to the rule of the vector product, and the result is another vector field called the Curl of the given field. Thus, by definition:

$$\begin{aligned}\text{Curl}\ \mathbf{f} &= \nabla\times\mathbf{f}\\ &= \left(\frac{\partial}{\partial x}.\mathbf{i} + \frac{\partial}{\partial y}.\mathbf{j} + \frac{\partial}{\partial z}.\mathbf{k}\right)\times\left(f_x.\mathbf{i} + f_y.\mathbf{j} + f_z.\mathbf{k}\right)\end{aligned}$$

$$= \left(\frac{\partial f_z}{\partial y} - \frac{\partial f_y}{\partial z}\right)\mathbf{i} - \left(\frac{\partial f_z}{\partial x} - \frac{\partial f_x}{\partial z}\right)\mathbf{j} + \left(\frac{\partial f_y}{\partial x} - \frac{\partial f_x}{\partial y}\right)\mathbf{k}$$

or:

$$\text{Curl } \mathbf{f} = \nabla \times \mathbf{f} = \begin{vmatrix} \mathbf{i} & \mathbf{j} & \mathbf{k} \\ \dfrac{\partial}{\partial x} & \dfrac{\partial}{\partial y} & \dfrac{\partial}{\partial z} \\ f_x & f_y & f_z \end{vmatrix} \tag{4.14}$$

An Important Identity

It can be shown that the curl of a gradient is necessarily zero, irrespective of the scalar field to which the successive operations are applied. Thus, for any scalar field ϕ:

$$\text{Grad } \phi = \nabla\phi = \frac{\partial \phi}{\partial x}.\mathbf{i} + \frac{\partial \phi}{\partial y}.\mathbf{j} + \frac{\partial \phi}{\partial z}.\mathbf{k} \; ,$$

therefore:

$$\nabla \times (\nabla\phi) = \left(\frac{\partial}{\partial x}.\mathbf{i} + \frac{\partial}{\partial y}.\mathbf{j} + \frac{\partial}{\partial z}.\mathbf{k}\right) \times \left(\frac{\partial \phi}{\partial x}.\mathbf{i} + \frac{\partial \phi}{\partial y}.\mathbf{j} + \frac{\partial \phi}{\partial z}.\mathbf{k}\right)$$

$$= \begin{vmatrix} i & j & k \\ \dfrac{\partial}{\partial x} & \dfrac{\partial}{\partial y} & \dfrac{\partial}{\partial z} \\ \dfrac{\partial \phi}{\partial x} & \dfrac{\partial \phi}{\partial y} & \dfrac{\partial \phi}{\partial z} \end{vmatrix}$$

$$\equiv \left(\frac{\partial^2 \phi}{\partial y \partial z} - \frac{\partial^2 \phi}{\partial z \partial y}\right)\mathbf{i} - \left(\frac{\partial^2 \phi}{\partial z \partial x} - \frac{\partial^2 \phi}{\partial x \partial z}\right)\mathbf{j} + \left(\frac{\partial^2 \phi}{\partial y \partial x} - \frac{\partial^2 \phi}{\partial x \partial y}\right)\mathbf{k} \; .$$

But the sequence of successive differentiation is immaterial, so each component is equal to zero, and we conclude that the curl of the gradient of any scalar field is necessarily zero.

that is:

$$\text{Curl Grad } \phi = \nabla \times (\nabla\phi) = 0 \tag{4.15}$$

Ceneral Condition for Conservancy

A conservative force is defined as one in which the work done in a given displacement depends only on the terminal points, and this implies the existence of some scalar function ϕ of position, such that the work done by the force of the field in moving between any two positions can be equated to the difference in the values of ϕ for the two positions. This if W_{1-2} represents the work done by a conservative force $\mathbf{f}$ in a displacement from the position $\mathbf{r}_1$ to the position $\mathbf{r}_2$:

$$W_{1-2} = \int_{\mathbf{r}_1}^{\mathbf{r}_2} \mathbf{f} \cdot \mathrm{d}\mathbf{r} = \phi_2 - \phi_1$$

where ϕ is a scalar function of the coordinates which may be referred to as the work function of the field.

Equally, if $\mathrm{d}W$ represents the increment of work done in the infinitesimal increment of displacement $\mathrm{d}\mathbf{r} = \mathrm{d}x\,.\mathbf{i} + \mathrm{d}y\,.\mathbf{j} + \mathrm{d}z\,.\mathbf{k}$, then $\mathrm{d}W$ can be equated to an exact or total differential of ϕ, which can then be described in terms of the partial derivatives of ϕ as follows:

$$\mathrm{d}W = \mathbf{f} \cdot \mathrm{d}\mathbf{r} = \mathrm{d}\phi$$

or
$$\mathbf{f} \cdot \mathrm{d}\mathbf{r} = \frac{\partial\phi}{\partial x}.\mathrm{d}x + \frac{\partial\phi}{\partial y}.\mathrm{d}y + \frac{\partial\phi}{\partial z}.\mathrm{d}z\ ,$$

therefore
$$\mathbf{f} \cdot \mathrm{d}\mathbf{r} = \left(\frac{\partial\phi}{\partial x}.\mathbf{i} + \frac{\partial\phi}{\partial y}.\mathbf{j} + \frac{\partial\phi}{\partial z}.\mathbf{k}\right) \cdot (\mathrm{d}x\,.\mathbf{i} + \mathrm{d}y\,.\mathbf{j} + \mathrm{d}z\,.\mathbf{k})$$

whence
$$\mathbf{f} \cdot \mathrm{d}\mathbf{r} = \nabla\phi \cdot \mathrm{d}\mathbf{r}$$

or:
$$\mathbf{f} = \nabla\phi\ .$$

We therefore conclude that a conservative force must be the gradient of some scalar field, and since the curl of a gradient is necessarily zero, general conditions of conservation may be stated as follows:

1) If a force $\mathbf{f}$ is conservative there must exist some scalar function ϕ of position (that is, some scalar field ϕ) whose value at any position determines the work done by the force in moving to the position from a given datum, and the difference in the values of ϕ for any two positions determines the work done in a displacement from one to the other.
2) If a force $\mathbf{f}$ is conservative then its curl is necessarily zero.

Potential (Energy)

The negative of the function ϕ is called the potential (energy) of the field, and is represented symbolically by V. Thus $V = -\phi$, and we recognise that the difference of potential for any two points is the negative of the work done in moving between them. The force being conservative, the change of potential in any displacement can therefore be interpreted as the work that the force would do in the reverse displacement.

Example 4.14

The values of a scalar quantity ϕ and a vector quantity **a** vary with the rectangular coordinates of position according to:

$$\phi = x^2y + yz^2$$

and
$$\mathbf{a} = 2xz^2.\mathbf{i} + xz.\mathbf{j} + 2y^2z.\mathbf{k} .$$

Determine Grad ϕ, Div **a** and Curl **a**.

Solution

$$\text{Grad}\,\phi = \nabla\phi = \frac{\partial\phi}{\partial x}.\mathbf{i} + \frac{\partial\phi}{\partial y}.\mathbf{j} + \frac{\partial\phi}{\partial z}.\mathbf{k} .$$

The scalar elements of ϕ are therefore determined simply by differentiating ϕ partially with respect to x, y and z in turn. The partial derivative with respect to x is determined by differentiating ϕ with respect to x whilst treating y and z as constants, and so on.

Thus:
$$\text{Grad}\,\phi = 2xy.\mathbf{i} + (x^2 + z^2).\mathbf{j} + 2yz.\mathbf{k}$$

$$\text{Div}\,\mathbf{a} = \nabla \cdot \mathbf{a}$$

$$= \left(\frac{\partial}{\partial x}.\mathbf{i} + \frac{\partial}{\partial y}.\mathbf{j} + \frac{\partial}{\partial z}.\mathbf{k}\right) \cdot (a_x.\mathbf{i} + a_y.\mathbf{j} + a_z.\mathbf{k}) ,$$

that is:
$$\text{Div}\,\mathbf{a} = \frac{\partial a_x}{\partial x} + \frac{\partial a_y}{\partial y} + \frac{\partial a_z}{\partial z}$$

or:
$$\text{Div}\,\mathbf{a} = 2z^2 + 2y^2$$

$$\text{Curl}\,\mathbf{a} = \mathbf{a} = \nabla \times \mathbf{a}$$

$$\text{Curl}\,\mathbf{a} = \begin{vmatrix} \mathbf{i} & \mathbf{j} & \mathbf{k} \\ \dfrac{\partial}{\partial x} & \dfrac{\partial}{\partial y} & \dfrac{\partial}{\partial z} \\ a_x & a_y & a_z \end{vmatrix}$$

$$= \left(\frac{\partial a_z}{\partial y} - \frac{\partial a_y}{\partial z}\right)\mathbf{i} - \left(\frac{\partial a_z}{\partial x} - \frac{\partial a_x}{\partial x}\right)\mathbf{j} + \left(\frac{\partial a_y}{\partial x} - \frac{\partial a_x}{\partial y}\right)\mathbf{k}$$

$$\text{Curl}\,\mathbf{a} = (4yz - x).\mathbf{i} + 4xz.\mathbf{j} + z.\mathbf{k} .$$

Example 4.15

Repeat Example 4.12 making use of the fact that **f** is centred on the origin.

Solution

Since the force is centred on a fixed point, it is also conservative, so we can determine the work done in moving from A to B by assuming the change of position to be effected along any convenient path.

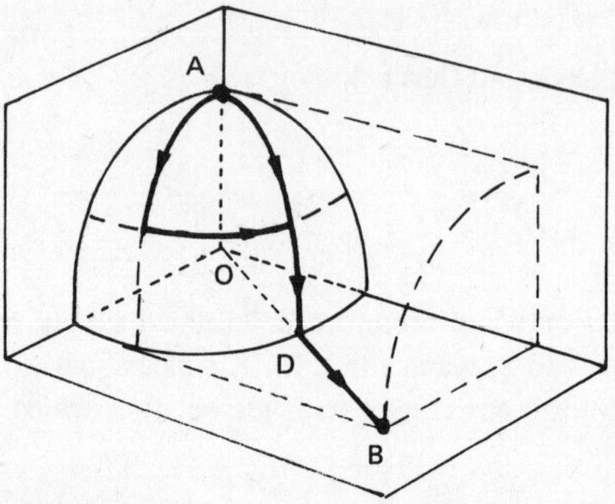

Consider a sphere of radius a centred on O, and let D be the point in which its surface cuts the line OB. We identify a path in two parts, of which AD is any path on the surface of the sphere, and DB is a straight, radial line. Since the force is centred on O, any increment of displacement on the surface of the sphere is at right-angles to the force, and it follows that the force will do work only on the path DB. At D, (that is, at distance a from the origin), $|\mathbf{f}| = F$, and its value varies inversely as s^3.

$$\text{f at } s = -\left(\frac{a}{s}\right)^3 . F$$

$$\text{Work done} = \int_D^B f \, ds$$

$$= -Fa^3 \int_a^{\sqrt{a^2+b^2}} ds/s^3 = \frac{Fa^3}{2}\left[\frac{1}{s^2}\right]_a^{\sqrt{a^2+b^2}}$$

$$= \frac{Fa^3}{2}\left(\frac{1}{a^2+b^2} - \frac{1}{a^2}\right) = -\frac{Fab^2}{2(a^2+b^2)} .$$

5

Moments of Forces and Equilibrium

5.1 MOMENTS OF FORCES

Though the lumped-parameter form of the Second Law of Motion is relevant to a distributed system, it takes no account of the locations of the various quantities which are simply treated as 'free' vectors and scalars. The equation is therefore directly applicable only to the Centre of Inertia of the System, because the motion of any other point is affected by its rotation, and this depends on both the location of the force and the distribution of the mass. But vectors being distinguished from scalars, it is found that the location of each kind of quantity is typically effective only in a specific limited way that is characteristic of the kind, and as a result, it proves convenient to define the location of a quantity by specifying, not the location itself, but some product of the quantity with an appropriate geometrical quantity which effectively determines the significant feature of its location. Such products are defined sometimes as scalars and sometimes as vectors, but in either case they are referred to as moments, and they are so defined that the rotational effects can be described by equations between moments which closely resemble the translational relationships that exist between the quantities themselves. Thus, just as the translational effect of a force is described by equating the force itself to a time-rate of change of momentum, so can its rotational effect be described by equating a moment of the force to the time-rate of change of a moment of momentum.

Here we refer specifically only to moments of forces in the context of body dynamics, but although scalars need a different treatment, the present approach is, in fact, broadly appropriate to vector quantities on a wide front.

Vector Moment About a Point

When a force acts on a 'free', rigid body, its net effect on the motion of the body can be described as the sum of two independent effects: on the translation of the Centre of Inertia of the body, and on the rotation of the body about a specific axis which passes through the Centre of Inertia. This axis is normal to the plane which contains both the line of action of the force and the Centre of

Inertia, and the turning effect is proportional to both the modulus of the force, and the perpendicular distance of its line of action from the Centre of Inertia.

Referring to Fig. 5.1, the moment of a located force **f** about an arbitrary datum point A is therefore defined as a vector quantity $\mathbf{m}_A$ whose modulus is equal to the modulus $|\mathbf{f}|$ of the force times the perpendicular distance d of its line of action from the datum, and whose direction is defined by the unit vector

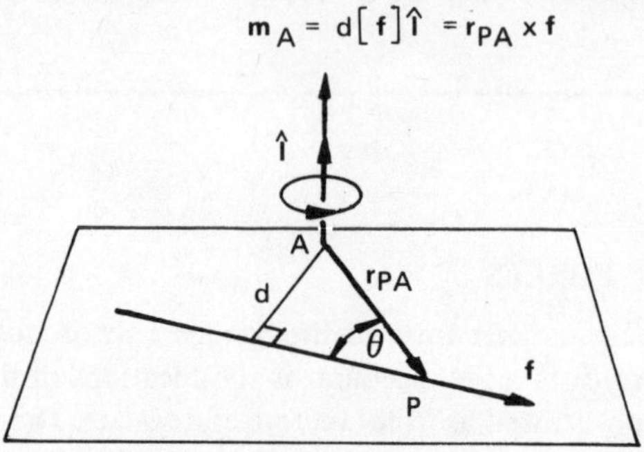

Fig. 5.1

$\hat{\mathbf{i}}$ which is normal to the plane which contains both the line of action of the force and the datum, in the sense which is arbitrarily related to the sense of the turning effect as the sense of advance is related to the sense of rotation in a right-handed screw. Thus, by definition, $\mathbf{m}_A = d\,|\mathbf{f}|\,\hat{\mathbf{i}}$.

But choosing any point P on the line of action of the force, let $\mathbf{r}_{PA}$ represent the vector which is drawn to P from the datum A, and let θ represent its inclination to the force. It is evident from Fig. 5.1 that, irrespective of the position of P on the line of action, the perpendicular distance d may be equated to $|\mathbf{r}_{PA}|\sin\theta$, and substituting accordingly gives:

$$\mathbf{m}_A = |\mathbf{r}_{PA}|\,|\mathbf{f}|\sin\theta\,\hat{\mathbf{i}}\ .$$

But the right-hand side can now be recognised as the vector product $\mathbf{r}_{PA} \times \mathbf{f}$, and we therefore specify as follows.

The moment about an arbitrary datum point A of a force **f** which is located through the point P is the vector quantity $\mathbf{m}_A$ which is alternatively, but equivalently, defined, either as the vector product $\mathbf{r}_{PA} \times \mathbf{f}$, or as the vector whose modulus is equal to the modulus of the force times the perpendicular distance of its line of action from the datum, and whose direction is that of the unit vector normal to the plane which contains both the line of action of the force and the datum, in the sense determined by the rule of the right-hand screw. Thus, by definition:

$$\mathbf{m}_A = \mathbf{r}_{PA} \times \mathbf{f} = d\,|\mathbf{f}|\,\hat{\mathbf{i}} \qquad (5.1)$$

Scalar Moment About an Axis

But commonly, a body such as the turntable in Fig. 5.2 is provided with bearings which permit its rotation only about a specific, fixed axis, and when the axis of an effect is thus determined, it remains to specify only its modulus and its sense.

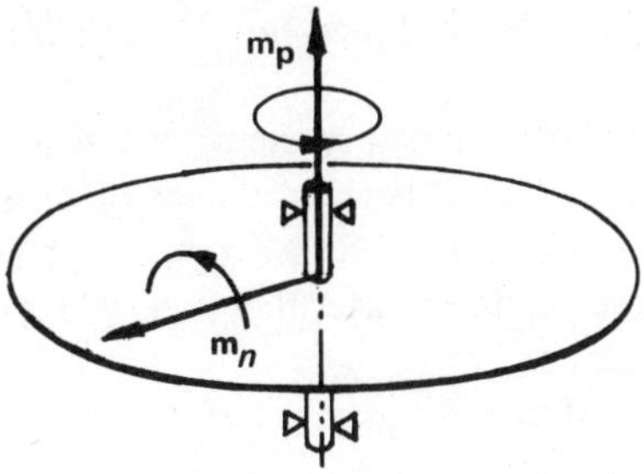

Fig. 5.2

So, as well as a vector moment about a point, we also define a scalar moment about an axis.

Evidently a body which is so mounted is free to respond to any moment like $\mathbf{m}_p$ whose axis coincides with the permitted axis. But any moment which, like $\mathbf{m}_n$, has an axis which is normal to the permitted axis, calls-up at the bearings the forces of reaction that are required to hold the axis in its fixed position, and the moment generated by these forces entirely frustrates the effect of the applied moment. So, supposing a turntable to be subjected to a given force **f** applied at a given point P, and choosing a datum point A at any position on the permitted axis, consider the vector moment of the force about the datum. By definition: $\mathbf{m}_A = \mathbf{r}_{PA} \times \mathbf{f}$. But, referring to Fig. 5.3, let $\mathbf{m}_A$, $\mathbf{r}_{PA}$ and **f** be described as sums of pairs of rectangular components, of which $\mathbf{m}_p$, $\mathbf{r}_p$ and $\mathbf{f}_p$

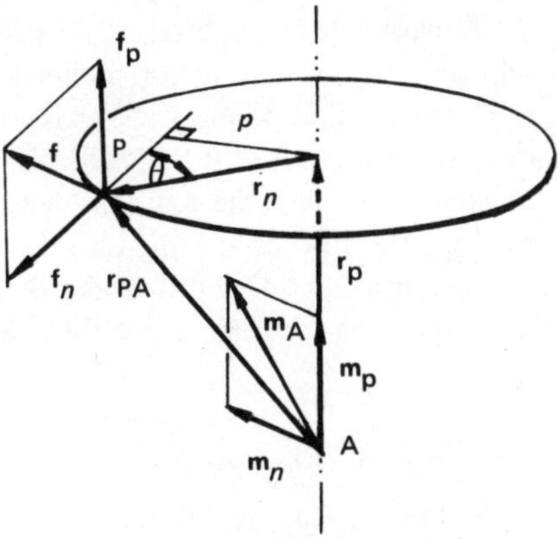

Fig. 5.3

represent components parallel to the permitted axis, when the remaining components $\mathbf{m}_n$, $\mathbf{r}_n$ and $\mathbf{f}_n$ will generally lie in different directions normal to the axis. Substituting accordingly:

$$(\mathbf{m}_p + \mathbf{m}_n) = (\mathbf{r}_p + \mathbf{r}_n) \times (\mathbf{f}_p + \mathbf{f}_n)$$

or

$$(\mathbf{m}_p + \mathbf{m}_n) = (\mathbf{r}_p \times \mathbf{f}_p) + (\mathbf{r}_p \times \mathbf{f}_n) + (\mathbf{r}_n \times \mathbf{f}_p) + (\mathbf{r}_n \times \mathbf{f}_n) \ .$$

But since $\mathbf{r}_p$ and $\mathbf{f}_p$ are both parallel to the permitted axis, they are also parallel to one another, and $\mathbf{r}_p \times \mathbf{f}_p$ is therefore equal to zero. Furthermore, by the definition of a vector product, $\mathbf{r}_p \times \mathbf{f}_n$ is normal to $\mathbf{r}_p$, and $\mathbf{r}_n \times \mathbf{f}_p$ is normal to $\mathbf{f}_p$. Therefore both these products have directions which are normal to the permitted axis, and it follows that $\mathbf{r}_n \times \mathbf{f}_n$ alone has a direction which coincides with that of the permitted axis. We may therefore write:

$$\mathbf{m}_p = (\mathbf{r}_n \times \mathbf{f}_n)$$

and

$$\mathbf{m}_n = (\mathbf{r}_p \times \mathbf{f}_n) + (\mathbf{r}_n \times \mathbf{f}_p)$$

We note that $\mathbf{m}_p$ is independent of $\mathbf{r}_p$, and since $\mathbf{r}_n$ is evidently independent of the position of the datum A on the permitted axis it follows that $\mathbf{m}_p$ also is independent of the position of A along the axis.

We also note that $\mathbf{m}_p$ is independent of $\mathbf{f}_p$ also, and depends only on $\mathbf{f}_n$, the component of the force projected on a plane normal to the permitted axis. Furthermore, if θ represents the inclination of $\mathbf{f}_n$ to $\mathbf{r}_n$, as indicated in Fig. 5.3, the modulus of $\mathbf{m}_p$ may be equated to $|\mathbf{r}_n|\,|\mathbf{f}_n| \sin\theta$. But $|\mathbf{f}_n| \sin\theta$ can evidently be equated to the perpendicular distance p between the line of action of $\mathbf{f}_n$ and the permitted axis, so the modulus of $\mathbf{m}_p$ can alternatively be equated to $p\,|\mathbf{f}_n|$, and gathering these observations together we specify as follows.

The scalar moment about an axis $\overrightarrow{AB}$ which passes through the points A and B, of a force **f** which is located through the point P is indicated symbolically by m_{AB} and it is alternatively, but equivalently defined, either as the component resolved along the axis, of the vector moment about any point which lies on the axis, or as the product of the scalar component of the force projected on a plane normal to the axis with the perpendicular distance between the line of action of this component and the axis, the sense of the moment being arbitrarily determined by the rule of the right-hand screw. Thus, if P is any point on the line of action of a force **f**, and if A and B are any two points on the axis $\overrightarrow{AB}$, the scalar moment of the force about the axis is, by definition,, given by

$$m_{AB} = \hat{AB} \cdot \mathbf{m}_A = \hat{AB} \cdot (\mathbf{r}_{PA} \times \mathbf{f})$$

$$\text{or } m_{AB} = \hat{AB} \cdot \mathbf{m}_B = \hat{AB} \cdot (\mathbf{r}_{PB} \times \mathbf{f}) \qquad (5.2)$$

$$\text{or } |m_{AB}| = p\,|\mathbf{f}_n|$$

We note, firstly, that since the scalar elements of the vector moment about any point are the coefficients of its components in the directions of the co-ordinate axes, they can be recognised as the scalar moments about axes which pass through the datum point parallel to the x, y and z axes in turn.

Secondly, we note that a force exerts no moment about any axis which is either parallel to the force, or is intersected by its line of action.

Resultant Moment

Whether for vector moments or for scalar, the resultant moment of a set of distributed forces is defined as the sum of the moments of the several forces, and provided it is recognised that a summation of vectors is always to be performed in the vectorial manner, such a summation may be indicated by the symbol Σ (sigma). For example, the resultant moment of a set of discrete forces about a given datum point A is indicated by $\Sigma \mathbf{m}_A$, and if P represents any point on the line of action of a typical member $\mathbf{f}$ of the set, the resultant moment is given by:

$$\Sigma \mathbf{m}_A = \Sigma(\mathbf{r}_{PA} \times \mathbf{f}) \tag{5.3}$$

But consider a force which is distributed over a continuum, as the gravitational force is distributed over a body. The strength of the 'field' must then be described by specifying the intensity of the force per unit volume, and, first describing the elemental moment of the force which acts on a typical element of volume, the resultant moment of the force on the whole continuum is determined by integrating over its volume. Thus if ρ represents the force per unit volume (ρ is commonly a constant but may generally be a function of position), the force on an element of volume $\mathrm{d}V$ at P may be equated to $\rho\, \mathrm{d}V$; the elemental moment of this force about a given datum point A is given by

$$\mathrm{d}\mathbf{m}_A = \mathbf{r}_{PA} \times \mathrm{d}\mathbf{f} = (\mathbf{r}_{PA} \times \rho)\, \mathrm{d}V,$$

and integrating over the volume V of the continuum gives the resultant moment of the forces on the whole continuum:

$$\mathbf{m}_A = \int_0^V \mathrm{d}\mathbf{m}_A = \int_0^V (\mathbf{r}_{PA} \times \rho)\, \mathrm{d}V \tag{5.3a}$$

The Resultant of a Couple

A pair of equal but opposite forces having different lines of action are said to comprise a couple. This is a special case having distinctive characteristics, in that, not only is the resultant of the forces themselves evidently zero, but also it can be shown that their resultant moment has a unique value which is common to all datum points.

Thus in Fig. 5.4, let $\mathbf{f}$ and $-\mathbf{f}$ represent a pair of equal to opposite forces,

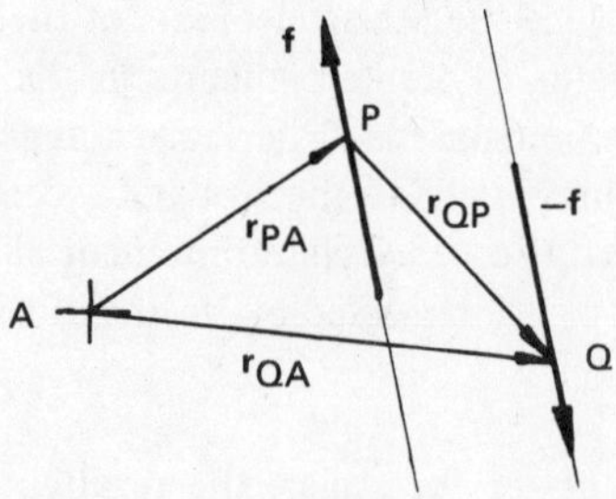

Fig. 5.4

let P and Q represent any points which lie on their lines of action, and let A represent an arbitrary datum point. Then calculating the resultant moment of the couple as the sum of the moments of the several forces we have:

$$\Sigma \mathbf{m}_A = (\mathbf{r}_{PA} \times \mathbf{f}) + (\mathbf{r}_{QA} \times -\mathbf{f})$$
$$= (\mathbf{r}_{PA} - \mathbf{r}_{QA}) \times \mathbf{f} = \mathbf{r}_{PQ} \times \mathbf{f}\ .$$

Evidently this product is independent of the datum A. Furthermore, by definition, its axis, being normal to both $\mathbf{r}_{PQ}$ and $\mathbf{f}$, is, in fact, normal to the plane which contains the pair of forces, whilst in its modulus $|\mathbf{r}_{PQ}|\ |\mathbf{f}|\ \sin\theta$, the product $\mathbf{r}_{PQ}\ \sin\theta$ may be equated to the perpendicular distance p between the lines of action of the forces. We may therefore conclude as follows.

A couple (which comprises a pair of equal but opposite forces for which the resultant of the forces themselves is evidently zero) has a unique resultant moment for all datum points; its axis is normal to the plane which contains the couple, whilst its modulus is equal to the modulus of one of the forces times the perpendicular distance between the two.

Shift of Datum

Given the resultant moment of a set of forces about one datum point, together with the resultant of the forces themselves, the resultant moment about an alternative datum can readily be determined without recalculating the moments of the several forces about the new datum.

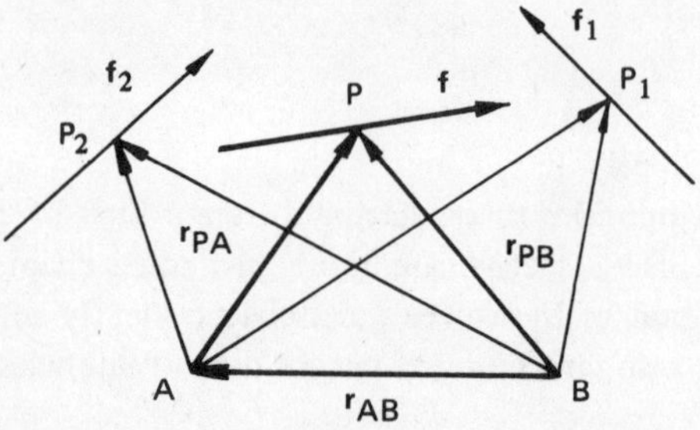

Fig. 5.5

To derive the necessary relationship, let f represent a typical member of the set of forces, let P represent any point on its line of action, and let A and B represent an arbitrary pair of alternative datum points.

By definition:

$$\Sigma \mathbf{m}_B = \Sigma(\mathbf{r}_{PB} \times \mathbf{f}) \ .$$

But it is obvious from Fig. 5.5 that $\mathbf{r}_{PB} = \mathbf{r}_{PA} + \mathbf{r}_{AB}$, and substituting accordingly gives:

$$\Sigma \mathbf{m}_B = \Sigma\{(\mathbf{r}_{PA} + \mathbf{r}_{AB}) \times \mathbf{f}\}$$

or

$$\Sigma \mathbf{m}_B = \Sigma(\mathbf{r}_{PA} \times \mathbf{f}) + \Sigma(\mathbf{r}_{AB} \times \mathbf{f}) \ .$$

But the first of the sums on the right can be recognised as the resultant moment about A, whilst in the second, $\mathbf{r}_{AB}$ is a common factor which may therefore be extracted from under the summation. We may therefore write

$$\Sigma \mathbf{m}_B = \Sigma \mathbf{m}_A + (\mathbf{r}_{AB} \times \Sigma \mathbf{f}) \qquad (5.4)$$

Thus, given the resultant moment of a set of forces about one datum A, the moment about a new datum B can be determined simply by adding to the given moment, the moment about the new datum of the resultant of the set of forces located through the old datum.

The Centroidal Axis

In Fig. 5.6, $\mathbf{f}_1$, $\mathbf{f}_2$ and $\mathbf{f}_3$ represent a general set of forces which are located through P_1, P_2 and P_3, whilst $\Sigma\mathbf{f}$ represents a single force equal to their resultant, acting on an arbitrary line of action which passes through the point C.

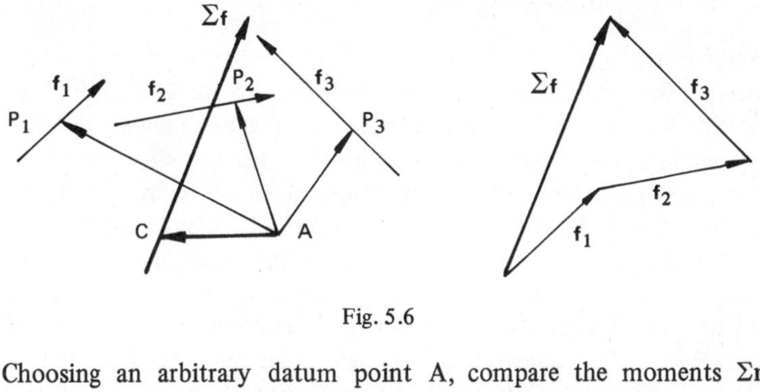

Fig. 5.6

Choosing an arbitrary datum point A, compare the moments $\Sigma \mathbf{m}_A = \Sigma(\mathbf{r}_{PA} \times \mathbf{f})$, and $(\mathbf{r}_{CA} \times \Sigma\mathbf{f})$, that is, compare the resultant of the moments of the forces, with the moment of the resultant of the forces, both about the same arbitrary datum. Irrespective of the positions of A and C, the moment of the resultant of the forces will, by definition, be at right-angles to the resultant force itself. But the resultant moment will similarly be at right-angles to the resultant force only in the special cases in which the forces of the set are either concurrent, parallel, coplanar, or parallel and coplanar, and it follows that, although it is not generally possible to find a line of action on which the resultant force gives a moment equal to the resultant moment, for any of the special cases it is possible.

Furthermore, we note, firstly that the moment of the resultant force about any point on its own line of action is necessarily zero, and secondly, that a scalar moment about an axis may be equated to a resolved component of the vector moment about any point which lies on the axis, and gathering these observations together we conclude as follows.

For any set of forces which are concurrent, parallel, coplanar, or parallel and coplanar, there exists a unique line of action on which the resultant of the forces gives a moment about any datum point equal to the resultant of the moments of the several forces about the same datum, and this line is called the Centroidal Axis of the set of forces. As a corollary, the resultant moment of a set of forces about any datum point which lies on their centroidal axis, or about any axis which intersects the centroidal axis, is necessarily zero, and we may therefore summarise as follows.

$$\Sigma \mathbf{m}_A = \Sigma(\mathbf{r}_{PA} \times \mathbf{f}) = \mathbf{r}_{CA} \times \Sigma \mathbf{f}$$

$$\Sigma \mathbf{m}_C = \Sigma m_{Cl} = 0$$

where $\mathbf{f}$ represents a typical member of a set of forces which are either concurrent, parallel, coplanar or parallel and coplanar, and where A is any arbitrary datum point, P is any point on the line of action of $\mathbf{f}$, and C is any point on the centroidal axis of the set of forces.

Centre of Gravity

Consider a given distribution of mass which may rotate in a uniform gravitational field. Irrespective of the orientation of the assembly in the field, the gravitational forces on its parts would all lie in the uniform direction of the field. Therefore, for each orientation, there would be in the body a corresponding centroidal axis, and the point in which the centroidal axes for different orientations intersect is called the Centre of Gravity of the assembly. Thus the Centre of Gravity of a given distribution of mass is the point about which the resultant moment of a uniform gravitational field is zero, irrespective of the orientation of the assembly, and it follows that, for any other datum point, the resultant moment of the distributed forces may be equated to the moment of the resultant of the forces, acting through the Centre.

Example 5.1

The force $\mathbf{f} = (10\mathbf{i} + 20\mathbf{j} + 30\mathbf{k})$ N acts through the point P = (3, 1, 2) m, whilst a point A has the coordinates (1, −2, 2) m. Determine the moments of $\mathbf{f}$

about: (a) the origin O of coordinates, and (b) the coordinate axis Ox, and (c) the axis $\vec{OA}$.

Solution

(a) By definition, the moment of **f** about the point O is the vector quantity given by:

$$\mathbf{m}_O = \mathbf{r}_{PO} \times \mathbf{f} = (3\mathbf{i} + \mathbf{j} + 2\mathbf{k}) \times (10\mathbf{i} + 20\mathbf{j} + 30\mathbf{k})\ \text{N m}\,,$$

$$\mathbf{m}_O = (-10\mathbf{i} - 70\mathbf{j} + 50\mathbf{k})\ \text{N m}\ .$$

(b) The moment of **f** about the axis Ox can be recognised as the **i** element of the vector moment about the point O.

Thus $$m_{Ox} = -10\ \text{N m}\ .$$

Alternatively, we note that **f** is described as a sum of components, each of which is either or normal or parallel to Ox, and that the perpendicular distance between their lines of action and the axis is obvious by inspection. Thus, the **i**

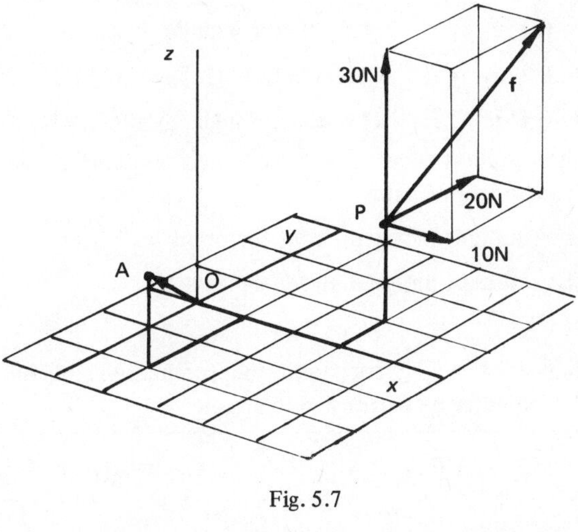

Fig. 5.7

component of **f** is parallel to Ox, so its contribution to the required moment is zero. However, the **j** and **k** components of **f** are both normal to Ox, and the perpendicular distances between their lines of action and Ox are 2 m and 1 m respectively. Furthermore, by the rule of the right-handed screw, the moment generated by the **j** component is negative whereas the moment generated by the **k** component is positive,

whence: $$m_{Ox} = -(2 \times 20) + (1 \times 30)\ \text{N m}$$

that is: $$m_{Ox} = -10\ \text{N m}\ .$$

(c) It is relatively simple to determine the component of **f** projected on a plane normal to OA, simply by subtracting the component resolved in the direction of the axis. However, it is relatively difficult to determine both the perpendicular distance between the line of action of the component and the axis, and the sense of the moment. It is therefore simplest to determine m_{OA} as the resolved component of the vector moment about either O or A. Thus, using the value of $\mathbf{m}_O$ determined in (a),

$$m_{OA} = \hat{OA} \cdot \mathbf{m}_O = |\mathbf{OA}|^{-1} \mathbf{OA} \cdot \mathbf{m}_O$$

Thus $$m_{OA} = \frac{1}{3}(\mathbf{i} - 2\mathbf{j} + 2\mathbf{k}) \cdot (-10\mathbf{i} - 70\mathbf{j} + 50\mathbf{k})\ \mathrm{N\ m}\ ,$$

whence: $$m_{OA} = 76\tfrac{2}{3}\ \mathrm{N\ m}\ .$$

Example 5.2

The members f of a set of discrete forces have lines of action which pass through points P as listed, and two points A and B have the coordinates: A = (−2, 3, −1) m, B = (−1, 5, 1) m.

$$\mathbf{f}_1 = (\ \mathbf{i} + 2\mathbf{j} + 3\mathbf{k})\ \mathrm{N} \text{ acts through } P_1 = (1, -2, \ 4)\ \mathrm{m}$$
$$\mathbf{f}_2 = (2\mathbf{i} + 3\mathbf{j} \qquad\quad)\ \mathrm{N} \text{ acts through } P_2 = (2, \ 0, \ 2)\ \mathrm{m}$$
$$\mathbf{f}_3 = (\ \mathbf{i} + 3\mathbf{j} - \ \mathbf{k})\ \mathrm{N} \text{ acts through } P_3 = (3, \ 1, -2)\ \mathrm{m}$$
$$\mathbf{f}_4 = (-\mathbf{i} + \ \mathbf{j} + 2\mathbf{k})\ \mathrm{N} \text{ acts through } P_4 = (1, -1, \ 0)\ \mathrm{m}$$

Determine: a) The resultant of the forces and their resultant moment about the origin of coordinates.

and b) The resultant moment about the point A.

and c) The resultant moment about the axis $\vec{AB}$.

Solution

a) The calculation of the moments, and the summations of the forces and moments, can conveniently be performed in a table.

FORCE	$\mathbf{r}_{PO}$			$\mathbf{f}$			$\mathbf{m}_O = \mathbf{r}_{PO} \times \mathbf{f}$					
	i	j	k	i	j	k	i		j		k	
	a	b	c	d	e	f	bf	−ce	cd	−af	ae	−bd
$\mathbf{f}_1$	1	−2	4	1	2	3	−6	−8	4	−3	2	2
$\mathbf{f}_2$	2	0	2	2	3	0	0	−6	4	0	6	0
$\mathbf{f}_3$	3	1	−2	1	3	−1	−1	6	−2	3	9	−1
$\mathbf{f}_4$	1	−1	0	−1	1	2	−2	0	0	−2	1	−1
Σ				3	9	4	−17		4		18	

$$\Sigma \mathbf{f} = (3\mathbf{i} + 9\mathbf{j} + 4\mathbf{k})\ \mathrm{N}$$
$$\Sigma \mathbf{m}_O = (-17\mathbf{i} + 4\mathbf{j} + 18\mathbf{k})\ \mathrm{N\ m}\ .$$

b) Given the resultant forces and their resultant moment about O, the resultant moment about A can be determined from the rule for a shift of datum:

$$\Sigma \mathbf{m}_A = \Sigma \mathbf{m}_O + \mathbf{r}_{OA} \times \Sigma \mathbf{f}\ ,$$

that is: $\Sigma \mathbf{m}_A = -17\mathbf{i} + 4\mathbf{j} + 18\mathbf{k} + (2\mathbf{i} - 3\mathbf{j} + \mathbf{k}) \times (3\mathbf{i} + 9\mathbf{j} + 4\mathbf{k})$

whence $\Sigma \mathbf{m}_A = -38\mathbf{j} - \mathbf{j} + 45\mathbf{k}\ \mathrm{N\ m}\ .$

c) The scalar moment about $\overrightarrow{AB}$ can be determined as the component resolved in the direction AB, of the vector moment about either A or B.

Thus: $m_{AB} = \hat{\mathbf{AB}} \cdot \mathbf{m}_A = |\mathbf{AB}|^{-1}\mathbf{AB} \cdot \mathbf{m}_A\ ,$

that is: $m_{AB} = \frac{1}{3}(\mathbf{i} + 2\mathbf{j} + 2\mathbf{k}) \cdot (-38\mathbf{i} - \mathbf{j} + 45\mathbf{k})\ \mathrm{N\ m}$

whence $m_{AB} = 16\frac{2}{3}\ \mathrm{N\ m}\ .$

Example 5.3

Fig. 5.8 shows a plane triangular sheet of uniform thickness t, and uniform density ρ which lies in the xy plane, in a uniform gravitational field of strength·

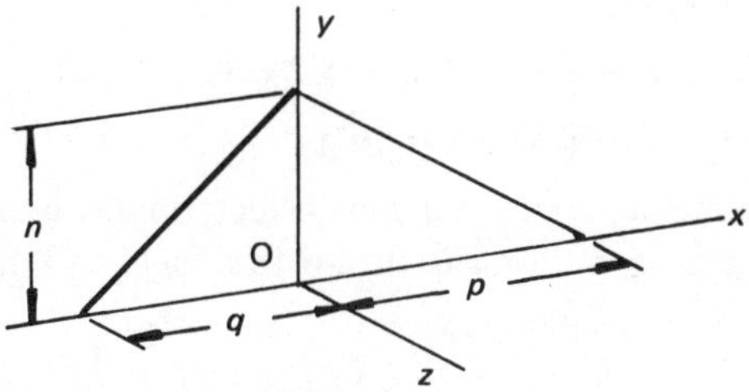

Fig. 5.8

$-g\,\mathbf{k}$ per unit mass. Determine: (a) the resultant force on the sheet; and (b) the resultant moment about the origin O of coordinates, and (c) the coordinates of the Centre of Gravity, and (d) the resultant moment about A = (1, 2, 3) m.

Solution

(a) The elemental force on the element $dx \times dy \times t$ at $P = (x, y, 0)$ is given by:

$$d\mathbf{f} = -\rho\, g\, t\, dx\, dy\, \mathbf{k}\ ,$$

and the resultant force on the sheet can be determined by integrating $d\mathbf{f}$, first with respect to x from A to B, and then with respect to y from the x axis to D.

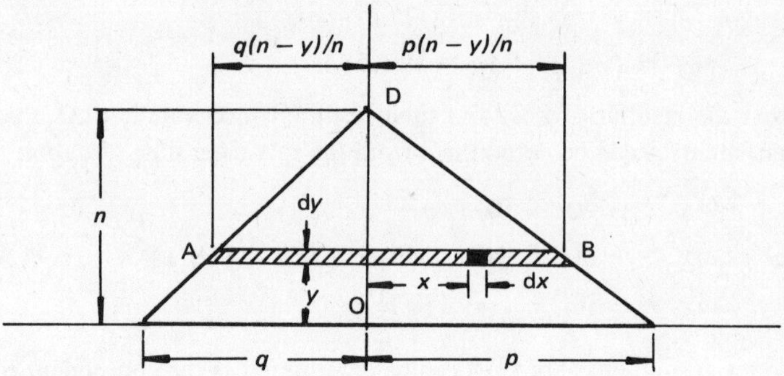

Thus: $$\mathbf{f} = \int_0^{y_D}\int_{x_A}^{x_B} \mathrm{d}\mathbf{f}\,\mathrm{d}x\,\mathrm{d}y = -\rho g t \int_0^n \int_{-q(n-y)/n}^{p(n-y)/n} \mathbf{k}\,\mathrm{d}x\,\mathrm{d}y$$

$$= \frac{-\rho g t(p+q)}{n}\int_0^n (n-y)\,\mathbf{k}\,\mathrm{d}y\ ,$$

$$\mathbf{f} = \frac{-\rho g t\, n(p+q)}{2}\mathbf{k}\ .$$

(b) By definition:

$$\mathrm{d}\mathbf{m}_O = \mathbf{r}_{PO} \times \mathrm{d}\mathbf{f}\ ,$$

that is: $$\mathrm{d}\mathbf{m}_O = (x\mathbf{i} + y\mathbf{j}) \times -\rho g t \mathbf{k}\,\mathrm{d}x\,\mathrm{d}y$$

$$= -\rho g t\,(y\mathbf{i} - x\mathbf{j})\,\mathrm{d}x\,\mathrm{d}y\ .$$

The resultant moment about O can now be determined by again integrating, first with respect to x from A to B, and then with respect to Y from the x axis to D.

Thus: $$\mathbf{m}_O = -\rho g t \int_0^n \int_{-q(n-y)/n}^{p(n-y)/n} (y\,.\,\mathbf{i} - x\,.\,\mathbf{j})\,.\,\mathrm{d}x\,.\,\mathrm{d}y$$

$$= -\rho g t \int_0^n \left[xy\,.\,\mathbf{i} - \frac{x^2}{2}.\mathbf{j}\right]_{-q(n-y)/n}^{p(n-y)/n}.\,\mathrm{d}y$$

$$= \frac{-\rho g t(p+q)}{2n}\int_0^n \left\{2(n-y).\mathbf{i} - \frac{(p-q)}{n}(n-y)^2.\mathbf{j}\right\}.\,\mathrm{d}y$$

$$= \frac{-\rho g t(p+q)}{2n}\left[\left(ny^2 - \frac{2y^3}{3}\right).\mathbf{i} - \right.$$

$$\left. - \frac{(p-q)}{n}\left(n^2 y - ny^2 + \frac{y^3}{3}\right).\mathbf{j}\right]_0^n$$

whence: $$\mathbf{m}_O = \frac{-\rho g t n(p+q)}{6}\left\{n.\mathbf{i} - (p-q).\mathbf{j}\right\} .$$

(c) Let $c = (x_c, y_c)$ be the Centre of Gravity. Then equating the resultant moment about O to the moment of the resultant force acting through the Centre of Gravity gives:

$$\Sigma\mathbf{m}_O = \mathbf{r}_{co} \times \Sigma\mathbf{f}$$

that is: $$\tfrac{1}{3}\{(p-q)\,\mathbf{i} - n\mathbf{j}\} = (x_c\mathbf{i} - y_c\mathbf{j}) \times \mathbf{k} .$$

Thus: $$\tfrac{1}{3}\{(p-q)\,\mathbf{i} - n\mathbf{j}\} = y_c\mathbf{i} - x_c\mathbf{j}$$

whence, by equating components,

$$(x_c, y_c) = \tfrac{1}{3}\{n, (p-q)\} .$$

(d) The resultant moment of the distributed forces about any point can be equated to the moment of the resultant force, acting through the Centre of Gravity.

$$\mathbf{m}_A = \mathbf{r}_{CA} \times \mathbf{f} = \left\{\left(1 - \frac{n}{3}\right)\mathbf{i} + \left(2 - \frac{p-q}{3}\right)\mathbf{j}\right\} \times \frac{\rho g t}{2} n(p+q)\,\mathbf{k} ,$$

that is: $$\mathbf{m}_A = \frac{-\rho g t n(p+q)}{6}\{(6-p+q)\,\mathbf{i} - (3-n)\,\mathbf{j}\} .$$

Example 5.4

A set of forces **f** which lie in the xy plane act through points P as listed. Determine the value of x in which the centroidal axis of the set intersects the x axis.

$$\mathbf{f}_1 = (2\mathbf{i} - \mathbf{j})\text{ N acts through } P_1 = (4, \ 2, 0)\text{ m}$$
$$\mathbf{f}_2 = (\mathbf{i} + \mathbf{j})\text{ N acts through } P_2 = (1, -2, 0)\text{ m}$$
$$\mathbf{f}_3 = (2\mathbf{j})\text{ N acts through } P_3 = (1, \ 1, 1)\text{ m} .$$

Solution

By the method as before:

$$\Sigma\mathbf{f} = (3\mathbf{i} + 2\mathbf{j})\text{ N; and } \Sigma\mathbf{m}_O = -2\mathbf{k}\text{ N m} .$$

Letting $c = (x_c, 0, 0)$ be the point in which the centroidal axis cuts the x axis, and equating the moment of the resultant force to the resultant moment of the forces:

$$x_c\mathbf{i} \times (3\mathbf{i} + 2\mathbf{j}) = -2\mathbf{k}\text{ N m: whence } x_c = -1 .$$

Example 5.5

A set of parallel forces **f** act through points P as listed. Determine the coordinates of the point in which the centroidal axis of the set of forces intersects the xy plane.

$$\mathbf{f}_1 = 2(2\mathbf{i} - \mathbf{j} + \mathbf{k})\text{ N acts through } P_1 = (1, 1, \ \ 1)\text{ m}$$

$$\mathbf{f}_2 = -3(2\mathbf{i} - \mathbf{j} + \mathbf{k})\text{ N acts through } P_2 = (2, 0, -1)\text{ m}$$

$$\mathbf{f}_3 = 4(2\mathbf{i} - \mathbf{j} + \mathbf{k})\text{ N acts through } P_3 = (0, 0, \ \ 2)\text{ m}$$

Solution

$$\Sigma\mathbf{f} = 3(2\mathbf{i} - \mathbf{j} + \mathbf{k})\text{ N}$$

$$\Sigma\mathbf{m}_O = \{2(\mathbf{i},\mathbf{j},\mathbf{k}) - 3(2\mathbf{i} - \mathbf{k}) + 4.2\,\mathbf{k}\} \times (2\mathbf{i} - \mathbf{j} + \mathbf{k})\text{ N m}$$

$$= (-4, 2, 13) \times (2, -1, 1) = (15\mathbf{i} + 30\mathbf{j})\text{ N m} .$$

Let $\mathbf{c} = (c_x, c_y, O)$ be the point in which the centroidal axis intersects the xy plane. Then equating the moment of the resultant force to the resultant moment of the forces:

$$(x_c\mathbf{i} + y_c\mathbf{j}) \times 3(2\mathbf{i} - \mathbf{j} + \mathbf{k}) = 15\mathbf{i} + 30\mathbf{j}$$

whence, by completing the product and equating components:

$$x_c = 10\text{ m}; \text{ and } y_c = 5\text{ m} .$$

5.2 EQUILIBRIUM OF FORCES

Let $\Sigma\mathbf{f}$ represent the resultant of the (unlocated) vectors of the forces that act on a body, and let $\Sigma\mathbf{m}_c$ represent the resultant moment of the forces about its Centre of Inertia. Then the moments of forces have been so defined that in much the same way as $\Sigma\mathbf{f}$ determines the translation of the Centre of Inertia, so does $\Sigma\mathbf{m}_c$ determine the rotation of the body about the Centre, and it follows, that in the context of rigid body mechanics, a set of forces are generally equivalent, not to the set of force vectors alone, but to the set of their moments about the Centre of Inertia also.

In particular, a body will remain at rest, only if both the resultant of the force vectors and their resultant moment about the Centre of Inertia are zero. However, by the rule for a shift of datum from A to B:

$$\Sigma\mathbf{m}_B = \Sigma\mathbf{m}_A + \mathbf{r}_{AB} \times \Sigma\mathbf{f} .$$

This implies that if both the resultant of the forces and their resultant moment about one point A are zero, then their moment about any other point B is also zero, and we conclude that for the equilibrium of a set of located forces, the resultants of both the force vectors themselves, and their moments about an arbitrary datum point must be zero.

Vector Equations of Equilibrium

In general, the conditions for the equilibrium of a set of located forces can therefore be expressed in a pair of vector equations, as follows:

$$\text{For equilibrium:} \qquad \Sigma\mathbf{f} = 0$$

$$\text{And:} \qquad \Sigma\mathbf{m}_{\mathrm{A}} = 0 \tag{5.6}$$

Where A is any arbitrary datum point.

But again referring to the rule for a shift of datum, we note that if the moments about two arbitrary datum points A and B are zero, then provided only that $\mathbf{r}_{\mathrm{AB}}$ is not parallel to $\Sigma\mathbf{f}$, we can infer that $\Sigma\mathbf{f}$ also is zero. The conditions for equilibrium can therefore be expressed in an alternative pair of vector equations in which the force equation is replaced by a second moment equation for a second datum B, as follows:

$$\text{For equilibrium} \qquad \Sigma\mathbf{m}_{\mathrm{A}} = 0$$

$$\text{and} \qquad \Sigma\mathbf{m}_{\mathrm{B}} = 0 \tag{5.6a}$$

where A and B are any pair of datum points not on a line parallel to $\Sigma\mathbf{f}$.

We note, however, that there is no relationship that allows the moment equation of equilibrium to be replaced by a second vector force equation, because only the moments contain the necessary information about the locations of the forces as well as their vector properties.

Scalar Equations of Equilibrium

Since vectors are equal only when all three of their scalar elements are equal, a single vector equation is equivalent to a set of three scalar equations. The conditions of equilibrium can therefore be expressed in a set of six scalar equations, and a suitable set is commonly prepared simply by describing the vector quantities in terms of their components in given rectangular coordinates. Thus, we may write:

$$\Sigma\mathbf{f} = (\Sigma f_x)\,\mathbf{i} + (\Sigma f_y)\,\mathbf{j} + (\Sigma f_z)\,\mathbf{k}$$

and

$$\Sigma\mathbf{m}_{\mathrm{A}} = (\Sigma m_{\mathrm{A}x})\,\mathbf{i} + (\Sigma m_{\mathrm{A}y})\,\mathbf{j} + (\Sigma m_{\mathrm{A}z})\,\mathbf{k}$$

where Σf_x, Σf_y and Σf_z represent the sums of the scalar components of the forces in the x, y and z directions in turn, whilst $\Sigma m_{\mathrm{A}x}$, $\Sigma m_{\mathrm{A}y}$ and $\Sigma m_{\mathrm{A}z}$ represent the sums of the scalar moments of the forces about axes through A parallel to x, y and z in turn, and since a vector is zero only when each of its components is zero, the conditions of equilibrium can be expressed in a set of six scalar equations as follows.

For equilibrium: $\Sigma f_x = \Sigma f_y = \Sigma f_z = 0$

and: $\Sigma m_{Ax} = \Sigma m_{Ay} = \Sigma m_{Az} = 0$ (5.6b)

where A is any arbitrary datum point.

Alternatively, the set of force equations can be replaced by the equations for the scalar moments about a set of axes through a second datum point B, but in fact, the two vector equations need not be resolved in the same set of directions, and provided they are not coplanar, neither set of directions need be rectangular. Indeed, since equilibrium implies zero moment about every datum point, it is not even necessary to choose the moment axes in sets which intersect in a common point. In preparing scalar equations of equilibrium there is therefore a great deal of freedom in the choice of axes of which it is commonly convenient to take advantage. However, it is then necessary to ensure the independence of the set, and in this connection it will be helpful to note the particular characteristics of a number of special cases.

These arise when the forces of the set are concurrent (a set of forces are said to be concurrent when they have a common intersection), or coplanar, or parallel, or parallel and coplanar. In a sense, these are simpler than general sets of forces because the special conditions imply that some of the conditions of equilibrium are satisfied automatically. Some of the equations of equilibrium then reduce to zero equations, and so drop out of consideration. However, with the reduction in the number of independent equations there is also a characteristic reduction of the number of force equations that may be included, and we now consider these effects in the various special cases, remembering that a set of forces is equivalent to a set of force vectors together with a set of moment vectors, and noting, that since the forces of a general set of vectors may differ in all three elements, such a set has three degrees of freedom, whereas a coplanar set has only two, and a parallel set only one.

Concurrent Forces

Since the moment of a force about any point on its line of action is zero, it is obvious that the moment of a concurrent set of forces about their common intersection is zero, irrespective of the values of the forces. If therefore the resultant of the vectors of the concurrent forces themselves is zero, their moment about any point will automatically be zero, and it follows that for a set of concurrent forces we can write only three independent scalar equations of equilibrium, and all three may be force equations.

Coplanar Forces

Fig. 5.9 illustrates a coplanar set of forces $\mathbf{f}_1$, $\mathbf{f}_2$ etc., and since the vector moment of a force about a point is normal to the plane which contains both the force and the datum, it is obvious that the moments $\mathbf{m}_1$, $\mathbf{m}_2$ etc. of the set of forces

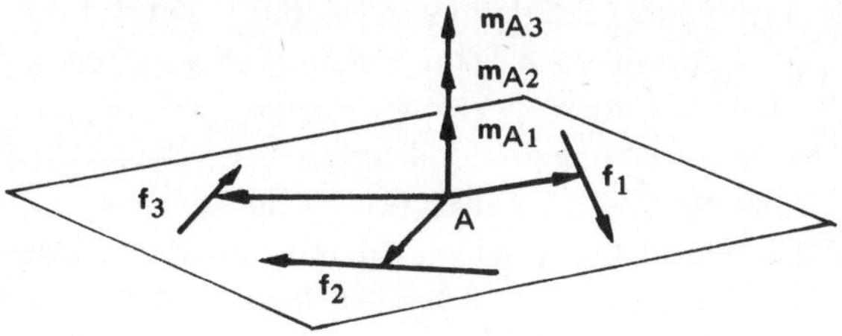

Fig. 5.9

which itself has two degrees of freedom can be described by a parallel set of vectors which has only one. Therefore, for a parallel set of forces we can again write only three independent scalar equations of equilibrium, but in this case only two may be force equations.

Parallel Forces

Fig. 5.10 indicates that, irrespective of the datum, the moments of a parallel set of forces are coplanar. Thus, the set of forces which itself has one degree of

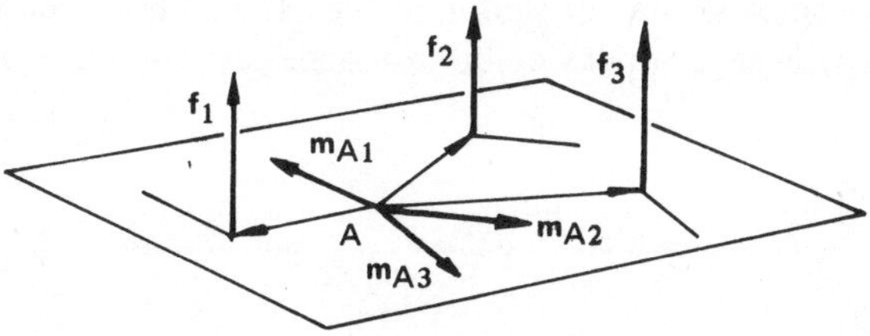

Fig. 5.10

freedom has a set of vector moments which has two, so for a parallel set of forces we can yet again write only three independent scalar equations of equilibrium of which only one may be a force equation.

Parallel and Coplanar Forces

Finally, the moments of a set of forces which is coplanar as well as parallel can be defined by a set of vectors which also are parallel, and it follows that for such a set of forces we can write only two independent scalar equations of equilibrium, of which only one may be a force equation.

5.3 STATICALLY DETERMINATE STRUCTURAL SUPPORT REACTIONS

The equations of equilibrium are directly applicable in the analysis of static structures where they are used in the determination of the 'internal' forces and moments that the externally-reacted forces induce in the various parts of the structure. However, the specification of a structural requirement does not usually provide a complete description of all the external forces. Rather it specifies that a structure of given form is required to support a given set of loads when it is supported in an appropriate manner, and we first use the equations of equilibrium to complete the description of the external forces by determining the reactions that the loads induce at the supports.

For the equilibrium of the structure, the resultant of the support reactions must be equal but opposite to the resultant of the known loads, but the values of the individual reactions may vary substantially according to the nature of the restraints which the various supports impose, and it is therefore necessary that each support should be so designed that it reacts to the loads in some clearly determined manner. Now, in principle, a support may have either or both of two effects. Thus it may oppose a force with a force so as to inhibit a displacement, or it may oppose a moment with a couple so as to inhibit a rotation, and supposing the characteristics of each support to be specified in these terms we now postulate a situation as follows.

Given that a set of known loads are applied to known points in a structure, when other known points are provided with supports which are capable of reacting the loads with forces along known lines and couples about known axes, we require to complete the description of the external forces on the structure by determining the magnitudes of the reactions that the loads induce at the supports.

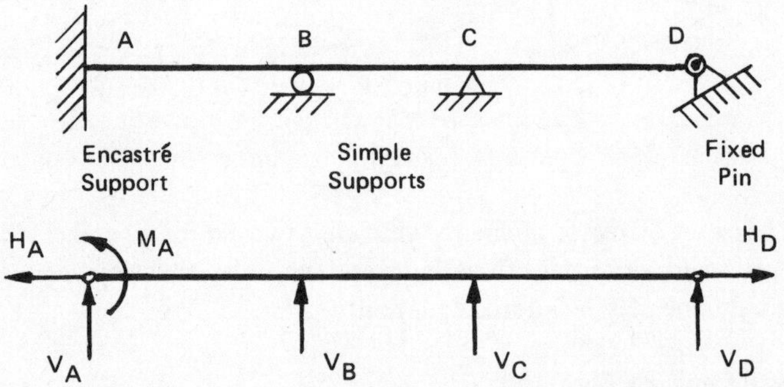

Fig. 5.11

For the present we are concerned only with the three types of support indicated in Fig. 5.11, and their characteristics are as follows.

Simple Support

In general, a simple support is characterised by a rigid ball which is interposed between the point of support and a rigid foundation, though for a plane system it may be characterised by a cylindrical roller whose axis is normal to the plane of the case. It is designed completely to prevent displacement of the point of support in the direction normal to the mutual surface of contact, but to offer no restraint either to displacements in the plane of the surface of contact or to its rotation. The reaction at a simple support is therefore represented as a single component of force normal to the surface of the structure at the point of support.

Fixed Pin Support

Fixed pin supports are relevant only to two-dimensional cases, where they are characterised by a frictionless pin which fixes the position of the point of support in the plane of the problem, but offers no restraint to in-plane rotation. The reaction at a fixed pin is therefore represented as a pair of in-plane components of force.

Encastré Supports

An encastré support is represented as one that is rigidly built-into a rigid foundation, and it is designed completely to prevent, not only the displacement of the structure in the immediate vicinity of the support, but also its rotation. In general, the reactions of an *encastré* support must therefore be represented as a combination of three components of force with three of moment. However, we are usually concerned only with two-dimensional cases, when only the two in-plane components of force, and the one couple about the axis normal to the plane are relevant.

Static Determinacy

By preparing a set of scalar equations of equilibrium for forces in different directions, and moments about different axes, we can formulate a set of equations which relate the unknown magnitudes of the support reactions to the known loads, but whereas the number of independent equations is restricted to the number of degrees of freedom of the system, the number of possible support reactions is unlimited. We can therefore distinguish three cases, according as the number of unknown support reactions is less than, equal to, or greater than the number of degrees of freedom in the set of external forces that the structure sustains.

When the number of support reactions is less than the number of degrees of freedom, the supports are not sufficient completely to determine the positions of the forces. The problem is then one, not of statics, but of dynamics, and as such it falls outside the range of present interest.

When the number of support reactions is greater than the number of degrees of freedom, they are more than sufficient to determine the location of the system, but the number of independent equations of equilibrium is then less than the number of unknown support reactions which they contain. It is therefore not possible completely to determine the support reactions, simply by the solution of the equations of equilibrium, and such a case is said to be statically indeterminate. In fact, the superfluous or redundant support reactions then impose additional constraints on the deformation of the structure, and provided these are known, the reactions of a redundant system can be determined by considering the deformation of the structures simultaneously with its equilibrium, but the treatment of problems in this category will necessarily be deferred until we have considered the question of deformation.

Between the dynamical case on the one hand, and the statically indeterminate case on the other, there lies the case in which the number of support reactions is equal to the number of degrees of freedom of the external forces. The support reactions are then just sufficient to determine the location of the system, and the number of independent equations of equilibrium is equal to the number of unknown support reactions they contain. The support reactions can therefore be determined simply by the simultaneous solution of an appropriate set of equations of equilibrium, and since the support reactions can thus be determined from considerations of equilibrium alone they are said to be statically determinate. Here we consider only such cases as fall within this category.

Choice of Equilibrium Equations

Commonly the calculation of the support reactions is only the prelude to the larger task of determining the forces and moments that the loads and support reactions induce in the individual members of the structure, and since the support reactions are typically large, it is evidently important that their values are checked before passing on to the next stage. This can conveniently be done by substituting the calculated values in additional equations of equilibrium, which should then reduce to an obvious identity. In practice, we therefore prepare more equations than are required for an independent set, and in selecting which equations to use for the different purposes, it is helpful to note that whereas the equations of a simultaneous set are most easily solved when each equation contains as few unknowns as possible, an equation that is to be used for checking the results is most useful when it serves to verify the values of as many of the results as possible. Thus for the independent set we aim to select equations which contain few unknowns, whilst for checking we aim to select equations that contain as many of the unknowns as possible, and in this connection it is helpful to remember that no force will appear in the moment equation for any axis which either intersects the line of action of the force, or is parallel to it.

Example 5.6

The diagram represents a straight, horizontal beam which is simply supported at A and C, and which carries a load of 20 kN at B, and a co-planar couple of 40 kNm at D. Determine the support reactions.

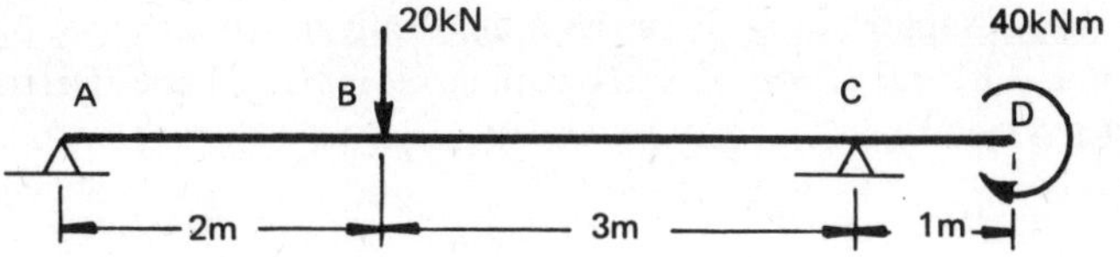

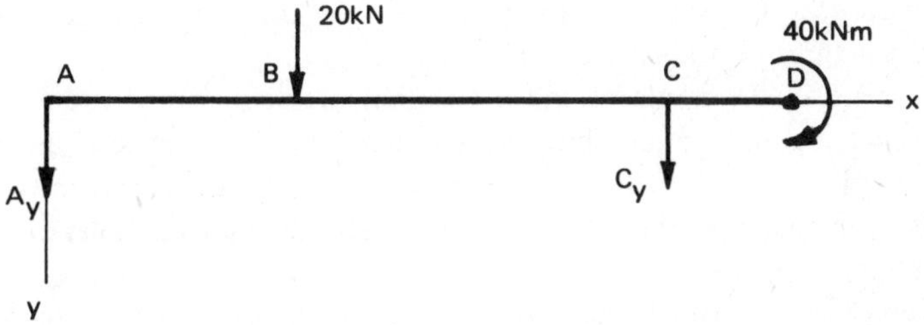

Solution

Identifying a set of rectangular directions by the axes x, y, z indicated, we note that the supports at A and C react with vertical forces which we represent by A_y and C_y. The plane parallel system of forces has two degrees of freedom. Consider the equilibrium of forces in the direction y, and of moments about axes parallel to z through A and C.

In the direction y: $$20 + A_y + C_y = 0 \qquad \text{(i)}$$

For moments about A: $$(20.2) + 40 + 5.C_y = 0 \qquad \text{(ii)}$$

For moments about C: $$40 - 5.A_y - (20.3) = 0 \qquad \text{(iii)}$$

From (ii): $C_y = -16$ kN; From (iii): $A_y = -4$ kN.

When these values are substituted in (i) it reduces to the obvious identity $0 = 0$, and this checks the calculations.

Example 5.7

The diagram represents a straight, horizontal beam having a fixed pin support at

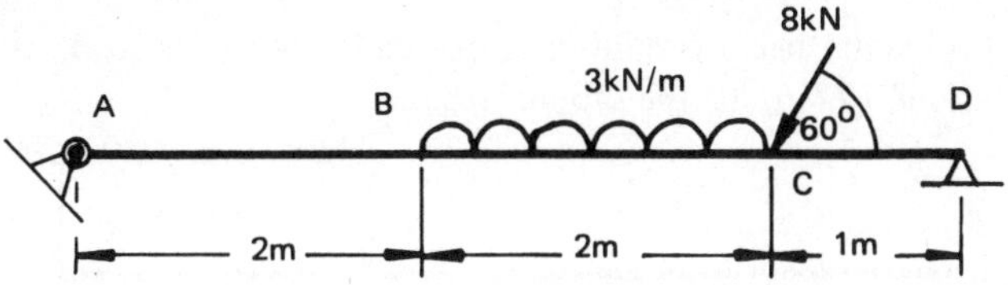

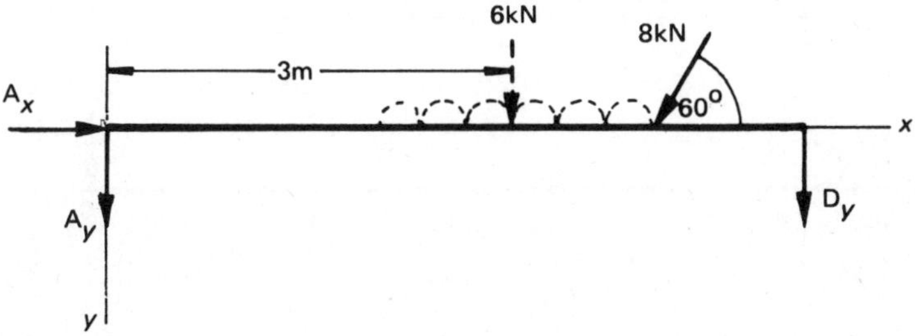

A, and a simple support at D. It carries a uniformly distributed load of 3 kN/m between B and C, and a load of 8 kN, inclined as indicated at C. Determine the support reactions.

Solution

The fixed pin at A reacts with two components of force, and the simple support at D with one. The forces on the beam are then coplanar but not parallel, and the set has three degrees of freedom.

In general, the resultant and a resultant moment of a distributed load is determined by integration, but the resultant moment can be equated to the moment of the resultant force located through its centroid, and in the present case it is obvious that the resultant of the distributed load is equal to 6 kN, and that the centroid is 3 m from A.

The equations for the equilibrium of forces in the x and y directions, and for moments about axes normal to the plane of the system, through A and D are:

For x direction: $$A_x - 8\cos 60^\circ = 0 \quad \text{(i)}$$

For y direction $$6 + 8\sin 60^\circ + A_y + D_y = 0 \quad \text{(ii)}$$

About A: $$(3.6) + (4.8\sin 60^\circ) + 5D_y = 0 \quad \text{(iii)}$$

About D: $$-(1.8\sin 60^\circ) - (2.6) - 5A_y = 0 \quad \text{(iv)}$$

From (i): $A_x = 4$ kN

From (ii): $A_y = -(2{\cdot}4 + 0{\cdot}8\sqrt{3})$ kN

From (iii): $D_y = -(3{\cdot}6 + 3{\cdot}2\sqrt{3})$ kN .

When the values of A_y and D_y are substituted in (ii), that equation reduces to an obvious identity. This still leaves the value of A_x unchecked, but since this is equal to the resolved component of a single load it can as well be checked by verifying the original calculation.

Example 5.8

A straight, horizontal beam is built-in at one end, and carries loads of 4 kN and 5 kN as indicated. Determine the support reactions.

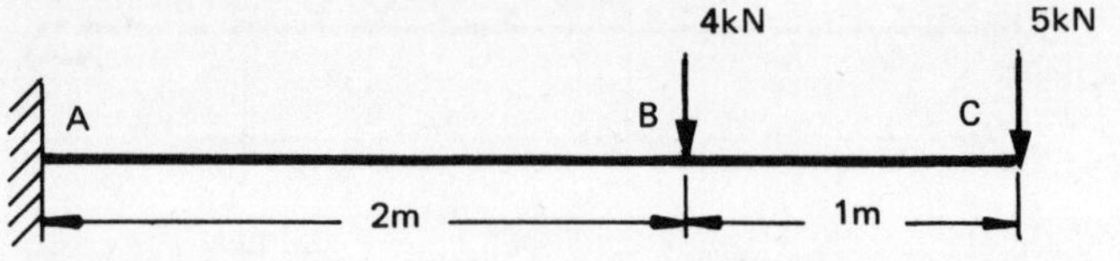

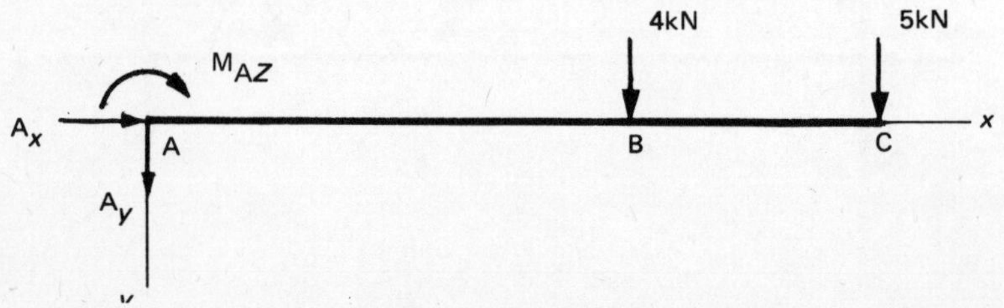

Solution

The plane, parallel loads are reacted by the coplanar components of reaction A_x, A_y and m_{Az}, and, considering the equilibrium of forces in the x and y directions, and of moments about axes parallel to z through A and C, we have

For x direction: $A_x = 0$ (i)

For y direction: $4 + 5 + A_y = 0$ (ii)

About A: $(2.4) + (3.5) + m_{Az} = 0$ (iii)

About C: $-3.A_y) - (1.4) + m_{Az} = 0$ (iv)

From (ii): $A_y = -9$ kN; From (iii): $m_{Az} = -23$ kN m.

When these results are substituted in (iv) they lead to an obvious identity, and this checks the calculation.

Example 5.9

A rod is bent to the form of a U, with the distance of $2R$ between the parallel legs. One leg is inserted into a frictionless, horizontal sleeve in a vertical wall.

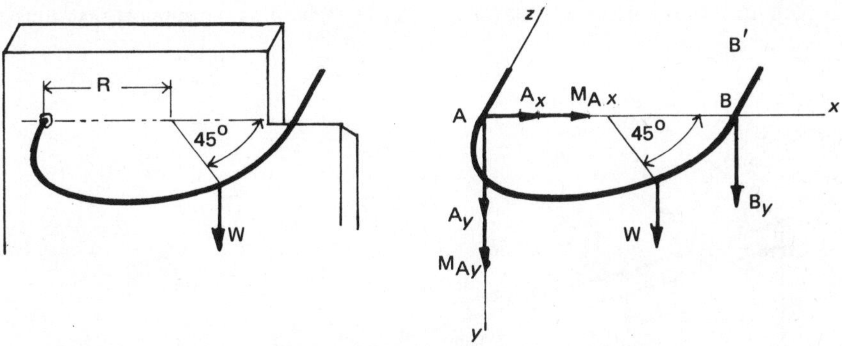

The free end rests on a simple support at the same level as the sleeve, so that the U lies in a horizontal plane with only the semicircle projecting from the wall. A weight W is suspended at a position 45° from the simple support. Determine the support reactions.

Solution

The sleeve resists displacement in any direction in the plane of the wall, and rotation about any axis in the plane. It is therefore capable of two components of force reaction A_x and A_y, and two components of moment reaction m_{Ax} and m_{Ay}. The simple support is capable of one component of force reaction B_y.

Considering the equilibrium of forces in the x and y directions, and of moments about the x, y and z axes, and an axis parallel to z through B:

For x direction: $A_x = 0$ (i)

For y direction: $W + A_y + B_y = 0$ (ii)

About x axis: $m_{Ax} + W.R/\sqrt{2} = 0$ (iii)

About y axis: $m_{Ay} = 0$ (iv)

About z axis: $R(1 + \sqrt{2})W + 2R.B_y = 0$ (v)

About BB′: $-R\sqrt{2}W - 2R.A_y = 0$ (vi)

From (iii): $m_{Ax} = -WR/\sqrt{2}$

From (v): $B_y = W(1 + \sqrt{2})/2$

From (vi): $A_y = -W/\sqrt{2}$.

Substituting the calculated values of B_y and A_y, checks these values by reducing to an obvious identity.

Example 5.10

A rod ABCDE lies along four edges of a rectangular prism as indicated. It is simply supported at A, F and G, and carries loads of 5 N at E, 10 N at C, and 4 N/m uniformly distributed between F and G. Determine the support reactions.

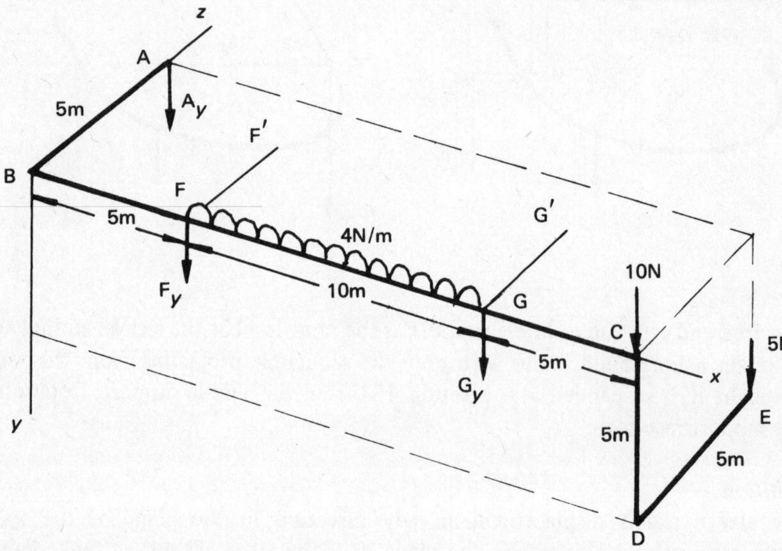

Solution

The forces are parallel, but not coplanar. They therefore have three degrees of freedom, and we consider the equilibrium of forces in the y direction, and of moments about BC′, and axes through F and G in the z direction.

Vertically	$A_y + F_y + G_y + 40 + 10 + 5$	$= 0$	(i)
About BC′	$5A_y + 5.5$	$= 0$	(ii)
About FF′	$-5A_y + 10G_y + 5.40 + 15.15$	$= 0$	(iii)
About GG′	$-15A_y - 10F_y - 5.40 + 5.15$	$= 0$	(iv)
From (ii):	$A_y = -5$ N		(v)
(v) in (iii):	$G_y = -45$ N		
(v) in (iv):	$F_y = -5$ N .		

These values may be checked by substituting in (i) which then reduces to an obvious identity.

Example 5.11

The illustration represents a structure in which the members DA, DB and BC are each 5 m long. The structure is supported at the joints D, A and B in the *xy* plane, and the joint C is vertically above B. Support at D is by a fixed, frictionless ball joint. The support at B allows free movement in the *y* direction only. The support of A allows free movement in the *xy* plane, but not normal to it. Determine the support reactions when the joint C supports a force, $\mathbf{W}_c = 8j - 10k$ N, as shown.

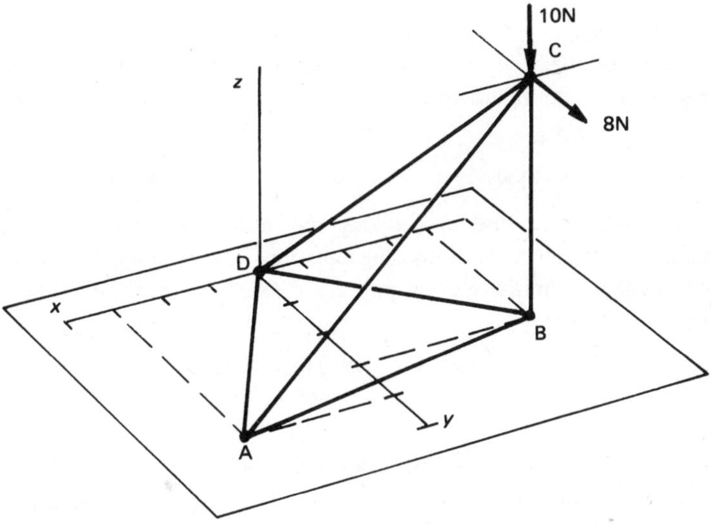

Solution

The support at D reacts with three components of force: D_x, D_y and D_z; that at B with two components: B_x and B_z, and that at A with one: A_z. The forces of the system are neither parallel nor coplanar, and the system therefore has six degrees of freedom.

It is still simple to write down the scalar equations of force equilibrium. Thus:

For x direction: $$B_x + D_x = 0 \tag{1a}$$

For y direction: $$D_y + 8 = 0 \tag{1b}$$

For z direction: $$A_z + B_z + D_z + 10 = 0 \tag{1c}$$

However, it becomes progressively more difficult to write down the scalar equations of moment equilibrium without error, and in this example it is probably more reliable to determine vector moments about a point. Thus, for moments about D:

$$(\mathbf{r}_{AD} \times \mathbf{f}_A) + (\mathbf{r}_{BD} \times \mathbf{f}_B) + (\mathbf{r}_{CD} \times \mathbf{f}_C) = 0$$

$$\left.\begin{matrix}(3,4,0\ )\\ \times(0,0,A_z)\end{matrix}\right\} + \left.\begin{matrix}(-4,3,0\ )\\ \times(B_x,0,B_z)\end{matrix}\right\} + \left.\begin{matrix}(-4,3,5\ )\\ \times(0,8,-10)\end{matrix}\right\} = \mathbf{0}$$

$$(4A_z, -3A_z, 0) + (3B_z, 4B_z, -3B_z) + (-70, -40, -32) = 0$$

But a vector is zero only when each of its components is zero,

Therefore: $$4A_z + 3B_z - 10 = 0 \tag{2a}$$

$$-3A_z + 4B_z - 40 = 0 \tag{2b}$$

$$-3B_x - 32 = 0 \tag{2c}$$

You are left to satisfy yourself that these are the scalar equations of moment equilibrium about the x, y and z axes in turn, and their simultaneous solution with the set of force equations gives:

$$A_z = 6{\cdot}4\text{ N} \quad B_x = -(32/3)\text{ N}; \quad B_z = 14{\cdot}8\text{ N}$$

$$D_x = (32/3)\text{ N}; \quad D_y = -8\text{ N}; \quad D_z = -11{\cdot}2\text{ N}.$$

Additional equations for use in checking the results, or for inclusion in the independent set instead of the foregoing equations, can be similarly determined by taking moments about any other point. For example, for the equilibrium of moments about B:

$$(\mathbf{r}_{AB} \times \mathbf{f}_A) + (\mathbf{r}_{DB} \times \mathbf{f}_D) + (\mathbf{r}_{CB} \times \mathbf{f}_C) = 0$$

and so on.

5.4 PRIMARY FORCES IN THE MEMBERS OF TRUSSES

Structures are load-bearing assemblies of sensibly fixed geometries which are commonly formed by the interconnection of a number of long, straight members, and supposing the loads and support reactions on such a structure to be determined, we now consider the determination of the 'internal' interactions that the 'external' forces induce between its parts. However, we can now deal only with cases which are statically determinate, so we first distinguish between 'frames' and 'trusses', for it is only the latter that are commonly determinate.

A structural frame, as typified by the simple portal indicated in Fig. 5.12, is an assembly which depends essentially for its stiffness on the stiffness in bending, partly of the members themselves, but particularly of the joints by which they are connected, and if these allowed free relative rotation between the members

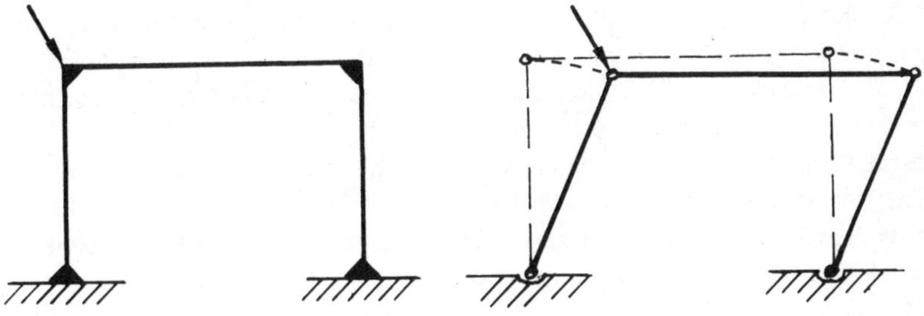

Fig. 5.12

the assembly would degenerate to a mechanism of variable geometry. A structural frame can therefore be regarded as an assembly of beams which are effectively connected by rigid joints into a single continuous whole, but with only one or two exceptions, the internal forces and moments in such an assembly are statically indeterminate. This is therefore a question which, for the present, we necessarily defer.

On the other hand, a truss is a structure whose stiffness depends essentially on its triangulated form. Consider, for example, the three-dimensional space truss, and the two-dimensional plane truss indicated in Fig. 5.13. These are evidently of fixed geometry, irrespective of the stiffness of the joints, so suppose

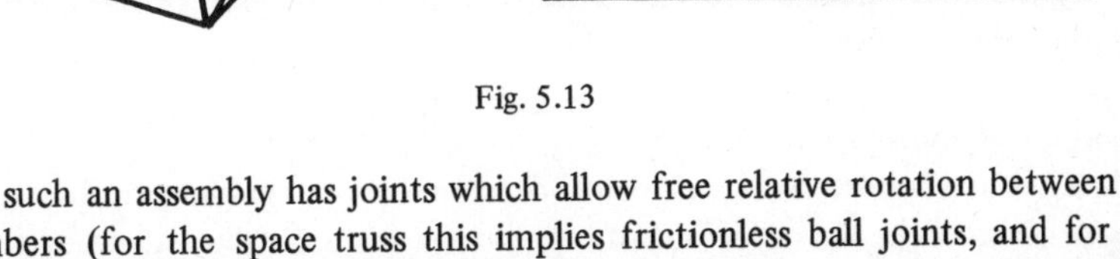

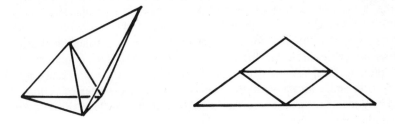

Fig. 5.13

that such an assembly has joints which allow free relative rotation between the members (for the space truss this implies frictionless ball joints, and for the plane truss frictionless pins), and suppose also that external forces are applied to the structure only at its joints. Then whether they arise from the external forces or from its interactions with other members, forces are applied to a member only through the joints at its ends, and since the frictionless joints are incapable of transmitting a couple, it follows from the equilibrium of the member that the forces applied to its ends must be equal, opposite, and colinear with the member itself. A truss is therefore conceived as a structure which depends essentially for its stiffness on its triangulated form; its members are conceived, not as beams in bending, but as 'ties' in simple tension, and as 'struts' in simple compression, and it is understood that external forces are to be applied to the truss only at its joints.

Member forces which are calculated on the assumption that the joints are frictionless are called the primary forces in the members. It is true that the assumption of frictionless joints is commonly contrary to the fact, and as a result secondary stresses that result from the stiffness of the joints are added to those that result from the primary forces. However, the secondary stresses are so called, not merely on account of their magnitude (in fact they may well be larger than the primary stresses), but rather because they have only a secondary effect on the ultimate load-carrying capacity of the structure. To justify this assertion in detail, it would be necessary to consider the deformation of the structure, but briefly, we note that whereas the primary forces are essential to the equilibrium of the loads, the secondary stresses arise only from a lack of compatibility in the deformations that the primary forces impose, and from this it falls out that the ultimate load-carrying capacity of a truss largely depends on the primary forces alone. As to the stiffness of the truss, this can only be enhanced by stiffness in the joints, and thus it is that the primary forces in a truss are indeed of primary importance, even though they are based on an assumption which might appear to be an unreasonable idealisation of the fact.

So, in summary, we now postulate a truss of known form which is maintained in equilibrium, as a single whole, by a set of externally reacted forces which are applied to the structure only at its joints. These external forces comprise the applied loads, together with the reactions they induce at the supports, and supposing them to be completely determined, we now consider their relationship to the 'internal' forces which they induce in the individual members. However, the joints being frictionless, it follows from the equilibrium of each of the individual members, that the forces applied to its ends must be equal, opposite, and colinear with the member itself, so the lines of action of the primary member forces are in fact, already known, and it remains to determine only their magnitudes and senses.

Equations of Joint Equilibrium

In brief, the support reactions being determined by applying to the set of external forces, the six equations of equilibrium for a distributed set of forces, Fig. 5.14a), the external forces are related to the internal by applying to the mixed set of forces that act on the ball of each joint, the three equations of equilibrium for a concurrent set of forces. Thus, each joint being identified by a letter, and each member by the pair of letters associated with the two joints to which it

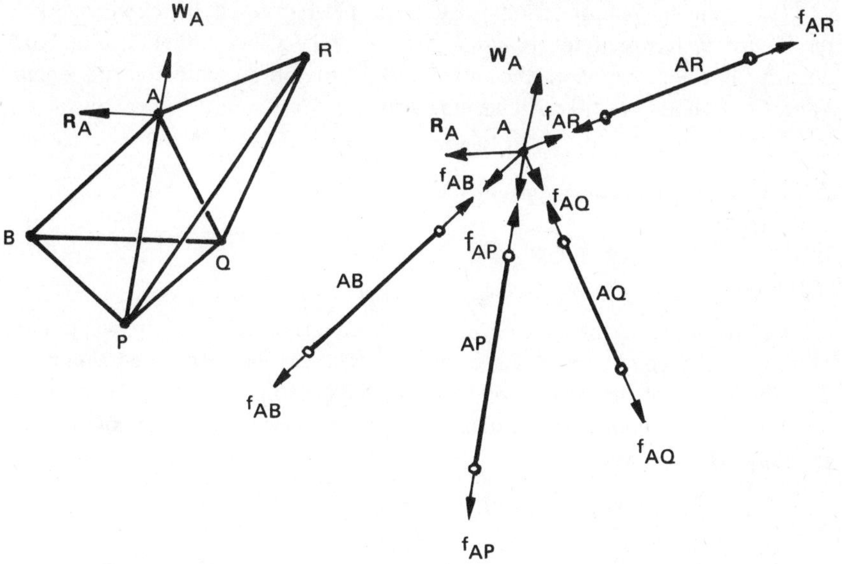

Fig. 5.14

connects, consider the forces on the ball of a joint A which is connected to the joints B, P, Q etc. by the members AB, AP, AQ etc. (Fig. 5.14b). As well as any external forces, these include the reactions of the 'internal' forces which the joint imposes on the ends of all the members it connects, and the pair of equal but opposite forces on the ends of each member being reckoned positive when they place the member in tension, it follows that the reactions on each joint must be reckoned positive in the sense which is directed along the members, outward from the joint. So let $\hat{AB}$, $\hat{AP}$, $\hat{AQ}$ etc. represent the unit vectors which radiate from the joint A along the members AB, AP AQ etc., and let f_{AB}, f_{AP}, f_{AQ} etc. represent the magnitudes of the forces in the members. Then the internal forces on the joint may be described as vectors by $f_{AB}\hat{AB}$, $f_{AP}\hat{AP}$, $f_{AQ}\hat{AQ}$, etc., and if ΣW_A and $\Sigma \mathbf{R}_A$ respectively represent the resultants of any external loads and support reactions on the joint, the condition for the equilibrium of the joint may be expressed in a single vector equation which equates the resultant force on the joint to zero:

$$f_{AB}\hat{AB} + f_{AP}\hat{AP} + f_{AQ}\hat{AQ} + \ldots . + \Sigma \mathbf{R}_A + \Sigma \mathbf{W}_A = 0 \ ,$$

Alternatively, if AB is regarded as a typical member of the set which radiates from the joint A, and if $\sum_A$ is taken to indicate a summation over all the members of the set, the equation for the equilibrium of the joint A may be represented as:

$$\sum_A (f_{AB}\hat{AB}) + \Sigma \mathbf{R}_A = -\Sigma \mathbf{W}_A \tag{5.7}$$

However, the computations may be somewhat eased by expressing this result, not in terms of the forces f in the members, but rather in terms of a quantity ϕ which expresses the ratio of the force in a member to its length. Thus, for each member like AB we may write:

$$f_{\mathrm{AB}}\hat{\mathbf{AB}} = f_{\mathrm{AB}}|\mathbf{AB}|^{-1}\mathbf{AB} = \phi_{\mathrm{AB}}\mathbf{AB} \ ,$$

and substituting accordingly gives:

$$\sum_{\mathrm{A}}\phi_{\mathrm{AB}}\mathbf{AB} + \Sigma\mathbf{R}_{\mathrm{A}} = -\Sigma\mathbf{W}_{\mathrm{A}} \tag{5.7a}$$

Of course, either of the foregoing vector equations can be expressed in a set of three scalar equations, and a convenient set may be formulated simply by describing each of the vectors **AB**, $\mathbf{R}_{\mathrm{A}}$ and $\mathbf{W}_{\mathrm{A}}$ in terms of its components in an arbitrary set of rectangular coordinates, and then equating the three components severally. Thus we write:

$$\begin{aligned} \mathbf{AB} &= x_{\mathrm{AB}}\mathbf{i} + y_{\mathrm{AB}}\mathbf{j} + z_{\mathrm{AB}}\mathbf{k} \\ \mathbf{R}_{\mathrm{A}} &= R_{\mathrm{A}x}\mathbf{i} + R_{\mathrm{A}y}\mathbf{j} + R_{\mathrm{A}z}\mathbf{k} \\ \mathbf{W}_{\mathrm{A}} &= W_{\mathrm{A}x}\mathbf{i} + W_{\mathrm{A}y}\mathbf{j} + W_{\mathrm{A}z}\mathbf{k} \ , \end{aligned}$$

and substituting accordingly, and equating the x, y and z components in turn, we obtain the result

$$\left.\begin{aligned} \sum_{\mathrm{A}}(\phi_{\mathrm{AB}}x_{\mathrm{AB}}) + \Sigma R_{\mathrm{A}x} &= -\Sigma W_{\mathrm{A}x} \\ \sum_{\mathrm{A}}(\phi_{\mathrm{AB}}y_{\mathrm{AB}}) + \Sigma R_{\mathrm{A}y} &= -\Sigma W_{\mathrm{A}y} \\ \sum_{\mathrm{A}}(\phi_{\mathrm{AB}}z_{\mathrm{AB}}) + \Sigma R_{\mathrm{A}z} &= -\Sigma W_{\mathrm{A}z} \end{aligned}\right\} \tag{5.7b}$$

Evidently a similar set of equations can be prepared for each joint, and where the joint connects only three members, the three equations will contain only three unknown values of ϕ, and the member forces on the individual joint will be statically determinate. But even though some of the joints connect more than three members, the truss as a whole is not necessarily redundant. This is so because the unknown value of ϕ for each member will appear in the equations of equilibrium for both of the two joints to which the member connects. For example, if a member AB connects to joint A of three members with a joint B of four members, the unknown value of ϕ_{AB} will appear in the sets of equations for both of the joints A and B, and its value being determined from the one, the result may be substituted in the other so as to reduce the number of unknowns remaining to three. It therefore remains to determine the condition for statical determinacy, and in this it is necessary to distinguish between ball-jointed space trusses on the one hand, and pin-jointed plane trusses on the other.

Statical Determinacy of a Ball-Jointed Space Truss

In the basic space truss indicated in Fig. 5.15a, six members are interconnected by four joints, and since each joint connects three members, the sets of equations

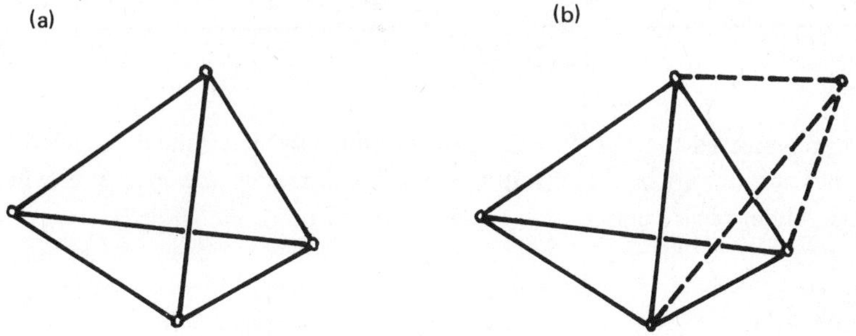

Fig. 5.15

of joint equilibrium are all statically determinate severally. But suppose that the truss is extended (Fig. 5.15b). To fix the position of each additional joint in its three degrees of freedom it is necessary to provide three additional (non-coplanar) members, and if the truss is extended in this way, it is obvious that the sets of equations of equilibrium for the joints of the truss as a whole will remain statically determinate, even though some of the joints would then connect more than three members. So taking it as understood that no joint may connect less than two members, and recognising that the four joints of the basic truss are connected by six members, it follows that the form of a space truss having B ball joints will be completely determined provided that the number of members M is not less than $6 + 3(B - 4) = 3(B - 2)$, but that the primary forces in the members will be statically determinate only if the number of members M is equal to $3(B - 2)$.

Static Determinacy of a Pin-Jointed Plane Truss

A plane truss is one in which both the truss itself, and the external forces to which it is subjected are contained in a plane, and the analysis of such an assembly is only the two-dimensional counterpart of the three-dimensional analysis of a space truss. However, to maintain the out-of-plane stability it is necessary to replace the ball-joints of a space truss with frictionless pins which allow relative rotation between the members only in the plane of the case, and as a result, different conditions of determinacy apply.

In brief, the basic plane truss is a triangle of three members, connected through three joints, and it may be extended by adding two additional members for each additional joint. It follows that, it being understood that no joint may connect less than two members, the form of a plane truss in which M members

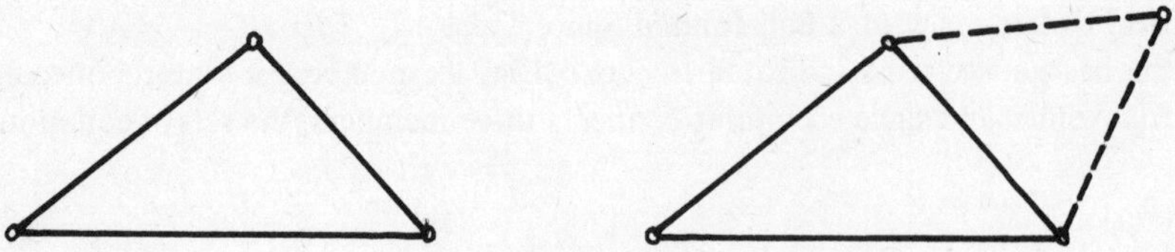

Fig. 5.16

are interconnected through P pins will be completely determined provided that M is not less than $(2P - 3)$, but the primary forces in the members will be statically determinate only if M is equal to $(2P - 3)$.

Example 5.12

The diagram illustrates a symmetrical space truss whose joints have the coordinates:

$$\begin{aligned} O &= (0,\ \ 0, 0)\ \text{m} \\ A &= (4,\ \ 3, 0)\ \text{m} \\ B &= (4, -3, 0)\ \text{m} \\ C &= (3,\ \ 0, 4)\ \text{m} \\ D &= (6,\ \ 0, 7)\ \text{m}\ . \end{aligned}$$

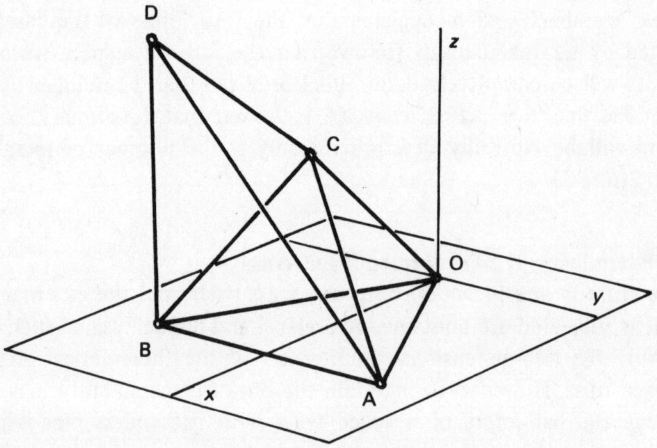

The joint O is fixed in position; the joint B is free to move only along a line parallel to the y axis; the joint C is free to move in any direction in the xy plane.

Determine the magnitudes and senses of the primary forces induced in the members by a load $\mathbf{W}_D = 2\mathbf{i} + \mathbf{j} - 10\mathbf{k}$ kN applied to the joint D.

Solution

By the statement of the problem, the joint O is constrained in all three coordinate directions, whilst B is constrained in the x and z directions, and C in the z direction only. Therefore, as well as the single known load W_D, the external forces comprise the six unknown components of support reaction R_{Ox}, R_{Oy}, R_{Oz}, R_{Bx}, R_{Bz}, and R_{Cz}, and since the general set of forces has six degrees of freedom, it follows that the support reactions are statically determinate.

Furthermore, since $M = 9$ members are interconnected through $B = 5$ ball joints, $M = 3(B - 2)$, and it follows that the primary member forces are statically determinate.

Considering the equilibrium of the distributed set of external forces, the radius vectors to the various joints, the vectors of external forces that act upon them, and the moments of these forces about the origin, are listed in the following table.

JOINT	r			F			$\mathbf{m}_O = \mathbf{r} \times \mathbf{F}$		
	i	j	k	i	j	k	i	j	k
O	0	0	0	R_{Ox}	R_{Oy}	R_{Oz}	0	0	0
A	4	3	0	R_{Ax}	0	R_{Az}	$3R_{Az}$	$-4R_{Az}$	$-3R_{Ax}$
B	4	−3	0	0	0	R_{Bz}	$-3R_{Bz}$	$-4R_{Bz}$	0
C	3	0	4	0	0	0	0	0	0
D	7	0	8	2	1	−10	−8	86	7

For the equilibrium of the distributed set of external forces, the resultant of the forces, and their resultant moment about an arbitrary datum (say the origin O), must be zero, and considering the **i**, **j** and **k** components in turn we have:

$$
\left.
\begin{aligned}
\Sigma F_x &= R_{Ox} + R_{Ax} + 2 = 0 \\
\Sigma F_y &= R_{Oy} + 1 = 0 \\
\Sigma F_z &= R_{Oz} + R_{Az} + R_{Bz} - 10 = 0 \\
\Sigma M_{Ox} &= 3R_{Az} - 3R_{Bz} - 8 = 0 \\
\Sigma M_{Oy} &= -4R_{Az} - 4R_{Bz} + 86 = 0 \\
\Sigma M_{Oz} &= -3R_{Ax} + 7 = 0
\end{aligned}
\right\}
$$

The simultaneous solution of these six equations gives the six components of support reaction as follows.

$$R_{Ox} = -4\tfrac{1}{3}\text{kN};\quad R_{Oy} = -1\text{ kN};\quad R_{Oz} = -11\tfrac{1}{2}\text{kN}$$
$$R_{Ax} = 2\tfrac{1}{3}\text{kN};\quad R_{Az} = 12\tfrac{1}{12}\text{kN};\quad R_{Bz} = 9\tfrac{5}{12}\text{ kN}.$$

The external forces now being entirely known, they may be related to the member forces by considering the equilibrium of forces on each joint in turn, starting with the joints which connect only three members. For example, for the joint O which connects the members OA, OB and OC we have:

$$\phi_{OA}x_{OA} + \phi_{OB}x_{OB} + \phi_{OC}x_{OC} = -W_{Ox}$$
$$\phi_{OA}y_{OA} + \phi_{OB}y_{OB} + \phi_{OC}y_{OC} = -W_{Oy}$$
$$\phi_{OA}z_{OA} + \phi_{OB}z_{OB} + \phi_{OC}z_{OC} = -W_{Oz}.$$

where (x_{OA}, y_{OA}, z_{OA}), (x_{OB}, y_{OB}, z_{OB}) and (x_{OC}, y_{OC}, z_{OC}) are the scalar elements of the vectors **OA**, **OB** and **OC**, and may be determined by subtracting the coordinates of the initial point O from those of the final points A, B and C. The values of these components for the present truss are as listed in the following table.

	OA −AO	OB −BO	OC −CO	AB −BA	AC −CA	BC −CB	AD −DA	BD −DB	CD −DC
x	4	4	3	0	−1	−1	2	2	3
y	3	−3	0	−6	−3	3	−3	3	0
z	0	0	4	0	4	4	7	7	3

Substituting accordingly for **OA, OB** and **OC** in the foregoing equations for the joint O then gives the set of three equations whose simultaneous solution determines the values of ϕ_{OA}, ϕ_{OB} and ϕ_{OC} as follows:

$$\left.\begin{aligned} 4\phi_{OA} + 4\phi_{OB} + 3\phi_{OC} &= 13/3 \\ 3\phi_{OA} - 3\phi_{OB} &= 1 \\ 4\phi_{OC} &= 23/2 \end{aligned}\right\}$$

whence: $\phi_{OA} = -\dfrac{103}{24}$ kN/m; $\phi_{OB} = -\dfrac{111}{24}$ kN/m, $\phi_{OC} = \dfrac{23}{8}$ kN/m.

Similarly for the joint D which connects the three members DA, DB and DC:

$$\left.\begin{aligned} -2\phi_{DA} - 2\phi_{DB} - 3\phi_{DC} &= -2 \\ 3\phi_{DA} - 3\phi_{DB} \qquad\quad &= -1 \\ -7\phi_{DA} - 7\phi_{DB} - 3\phi_{DC} &= 10 \end{aligned}\right\}$$

whence: $\phi_{DA} = -\dfrac{41}{30}$ kN/m; $\phi_{DB} = -\dfrac{31}{30}$ kN/m; $\phi_{DC} = 6{\cdot}8$ kN/m .

Since each of the remaining joints A, B and C connects four members, the three equations of equilibrium for the joint will contain four values of ϕ. However, in each case, these will include at least one of the values already determined, and the known values being substituted, the number of unknowns remaining will be at most three. Thus, substituting the values of $\phi_{CO} = -\phi_{OC}$ and of $\phi_{CD} = -\phi_{DC}$ already determined in the equations of equilibrium for the joint C gives:

$$\begin{aligned} -3\,.\frac{23}{8} + \phi_{CA} + \phi_{CB} + 3.6.8 &= 0 \\ + 3\phi_{CA} - 3\phi_{CB} \qquad\quad &= 0 \\ -4\,.\frac{23}{8} - 4\phi_{CA} - 4\phi_{CB} + 3.6.8 &= 0\ . \end{aligned}$$

The solution of any two then gives the values of ϕ_{CA} and ϕ_{CB}:

$$\phi_{CA} = \phi_{CB} = 471/80 \text{ kN/m}\ .$$

It now remains to determine the values of ϕ for the member AB only, and for this we require only one of the equations of equilibrium for either of the joints A and B. Thus, taking the y equation for the joint B, with the values of ϕ_{AO}, ϕ_{AC} and ϕ_{AD} substituted from the previous results gives:

$$3.\frac{103}{24} - 6\phi_{AB} - 3.\frac{471}{80} + 3.\frac{41}{30} = 0$$

whence $\phi_{AB} = -\dfrac{11}{96}$ kN/m .

The forces in the members may now be determined by multiplying the values of ϕ by the lengths of the members, the lengths being determined from the tabulated components of the vectors of the members, as follows.

	OA	OB	OC	AB	AC	BC	AD	BD	CD
ϕ, kN/m	$-\frac{103}{24}$	$-\frac{111}{22}$	$\frac{23}{8}$	$-\frac{11}{96}$	$\frac{471}{80}$	$\frac{471}{80}$	$-\frac{41}{30}$	$-\frac{31}{30}$	6·8
l, m	5	5	5	6	$\sqrt{26}$	$\sqrt{26}$	$\sqrt{62}$	$\sqrt{62}$	$3\sqrt{2}$
f, kN	−21·5	−23·1	14·4	−0.69	30·0	30·0	−10·8	−8·1	28.9

The positive results indicate ties in tension, whilst the negative results indicate struts in compression.

A Matrix Presentation

In the foregoing analysis of a statically determinate truss, the determination of the member forces from the equations of joint equilibrium was preceded by a separate calculation in which the support reactions were determined by applying the laws of equilibrium for a distributed set of forces to the external forces that act on the truss as a single whole. But in a statically determinate truss, B ball joints are interconnected by $(3B - 6)$ members, and when this number of unknown member forces is added to the six unknown components of support reaction, the total of $3B$ unknowns is, in fact, only just equal to the number of relationships that are available in the equations of joint equilibrium alone. The separate reference to the equilibrium of the external forces is therefore not strictly essential, but it has the effect of partitioning the equations into a number of small sub-sets which, by a simple choice of sequence, may be treated as though they were independent; and where the equations have to be solved by hand, this greatly eases the task of computation. But where an automatic computer is available, this consideration is only of minor significance, so we now consider an alternative presentation of the equations of joint equilibrium which treats both the support reactions and the member forces in a single calculation, and has a form which is suitable for direct entry to the machine. It is based on a few of the elementary principles of matrix algebra.

A matrix is a set of quantities which are logically arranged in a rectangular array of rows and columns, such that the rows distinguish the quantities in sub-sets according to one criterion, and the columns according to another. There is no limitation on either the number of rows or the number of columns, but where the quantities are distinguished in only one respect, they are necessarily presented either in a single row or in a single column, and such a matrix is referred to either as a row vector or as a column vector, as appropriate.

$$[\mathbf{a}] = \begin{bmatrix} a_{11} & a_{12} & a_{13} \\ a_{21} & a_{22} & a_{23} \\ a_{31} & a_{32} & a_{33} \end{bmatrix}$$

In general, a matrix as a single whole may be indicated algebraically by a bold symbol enclosed within square brackets, when its elements may be indicated by the same symbol, qualified by a pair of numerical suffices which identify, first the row in which the element appears, and then the column. Thus a_{23} indicates the term which appears in the second row and the third column of a matrix $[\mathbf{a}]$. However, the scalar elements of a geometrical vector may evidently be presented either as a row vector or a column vector, and these we will continue to distinguish as in vector algebra: by bold type for a general vector, or by the addition of a circumflex for a unit vector.

A product of matrices [a] . [b] is defined only for a pair in which the number of terms in each row of the pre-factor [a] is equal to the number of terms in each column of the post-factor [b], and it may be regarded as a generalisation of the dot product of vector algebra which is adapted to matrices by regarding each row of the pre-factor as a row vector, and each column of the post-factor as a column vector. Then, the dot product of two vectors being determined as the sum of the products of the corresponding pairs of their elements, the product of matrices [a] · [b] is defined as another matrix, each of these rows is generated from the corresponding row of the pre-factor [a] by taking its vectorial dot product with each column of the post-factor [b] in turn.

In particular, a product in which a matrix is post-multiplied by a column vector is another column vector each of whose elements is determined as the vectorial dot product of the corresponding row of the pre-factor with the post-factor. Thus, algebraically:

$$\begin{bmatrix} a_{11} & a_{12} \\ a_{21} & a_{22} \\ a_{31} & a_{32} \end{bmatrix} \cdot \begin{bmatrix} b_{11} \\ b_{21} \end{bmatrix} = \begin{bmatrix} (a_{11}b_{11} + a_{12}b_{21}) \\ (a_{21}b_{11} + a_{22}b_{21}) \\ (a_{31}b_{11} + a_{32}b_{21}) \end{bmatrix},$$

or numerically:

$$\begin{bmatrix} 1 & 2 \\ 3 & 2 \\ 5 & 6 \end{bmatrix} \cdot \begin{bmatrix} 7 \\ 8 \end{bmatrix} = \begin{bmatrix} (1.7 + 2.8) \\ (3.7 + 4.8) \\ (5.7 + 6.8) \end{bmatrix} = \begin{bmatrix} 23 \\ 53 \\ 83 \end{bmatrix}$$

You are left to satisfy yourself, that although a vectorial dot product is commutative, a dot product of matrices is not.

Given this definition, consider a set of simultaneous, linear, algebraic equations, such as:

$$\left.\begin{aligned} a_{11}x_1 + a_{12}x_2 &= b_1 \\ a_{21}x_1 + a_{22}x_2 &= b_2 \end{aligned}\right\},$$

where the a's and b's represent known constants. By adopting the matrix conventions, the equations may alternatively be presented as follows.

$$\begin{bmatrix} a_{11} & a_{12} \\ a_{21} & a_{22} \end{bmatrix} \cdot \begin{bmatrix} x_1 \\ x_2 \end{bmatrix} = \begin{bmatrix} b_1 \\ b_2 \end{bmatrix},$$

or $$[\mathbf{a}] \cdot [\mathbf{x}] = [\mathbf{b}],$$

where [a] represents the known matrix of the constant coefficients of the x's, and where [x] and [b] represent the column vectors of the unknown x's and the known b's.

Solving the equations simultaneously by first eliminating x_2 and then x_1, determines the values of the unknown x's in terms of the known a's and b's as follows:

$$\left.\begin{aligned} x_1 &= (a_{22}/|[\mathbf{a}]|)b_1 - (a_{12}/|[\mathbf{a}]|)b_2 \\ \text{and} \quad x_2 &= -(a_{22}/|[\mathbf{a}]|)b_1 + (a_{11}/|[\mathbf{a}]|)b_2 \end{aligned}\right\}$$

where $|[\mathbf{a}]| = (a_{11}a_{22} - a_{12}a_{21})$

or

$$\begin{bmatrix} x_1 \\ x_2 \end{bmatrix} = \frac{1}{|[\mathbf{a}]|}\begin{bmatrix} a_{22} & -a_{12} \\ -a_{21} & a_{11} \end{bmatrix} \cdot \begin{bmatrix} b_1 \\ b_2 \end{bmatrix}$$

Evidently then, if the b's can be described as linear functions of the x's, so can the x's be described as linear functions of the b's. In other words, if the known vector [**b**] can be equated to the product of the known matrix [**a**] times the vector of x's, then so can the vector of the unknown x's be equated to the product of another matrix with the vector of the b's, and this matrix, which derives from [**a**] by an operation known as inversion, is indicated symbolically by $[\mathbf{a}]^{-1}$. By this convention, we may therefore write:

If: $[\mathbf{a}] \cdot [\mathbf{x}] = [\mathbf{b}]$,

then: $[\mathbf{x}] = [\mathbf{a}]^{-1} \cdot [\mathbf{b}]$.

We note that the symbolism is so designed that it appears as though the matrix [**a**] can be divided through the matrix equation in much the same way as a scalar can be divided through an ordinary algebraic equation, but whereas a^{-1} only indicates the operation of inverting a scalar by simple division, $[\mathbf{a}]^{-1}$ indicates the operation of inverting a matrix, and this is effectively that of solving a set of linear simultaneous equations. Nevertheless, in an automatic computer, the matrix operations of inversion and multiplication are usually both available directly on demand, so a simultaneous set of linear algebraic equations being expressed in matrix convention as follows:

$$[\mathbf{a}] \cdot [\mathbf{x}] = [\mathbf{b}] \ ,$$

the solution of the equations is adequately indicated by:

$$[\mathbf{x}] = [\mathbf{a}]^{-1} \cdot [\mathbf{b}] \ ,$$

and a and b being entered as known data, the program for the solution of the equations will usually need only to call directly for the inversion of [**a**], followed by the multiplication of the result by [**b**].

Consider, for example, the application of the equations of joint equilibrium (5.7a) to a joint A which is connected to joints B, C, D and E, and is liable to two components of support reaction R_{Ax} and R_{Az}, and two components of load W_{Ay} and W_{Az}. Writing the equations for the x, y and z directions in turn, we have:

$$\begin{aligned}
\phi_{AB}x_{AB} + \phi_{AC}x_{AC} + \phi_{AD}x_{AD} + \phi_{AR}x_{AR} + R_{Ax} &= 0 \\
\phi_{AB}y_{AB} + \phi_{AC}y_{AC} + \phi_{AD}y_{AD} + \phi_{AR}y_{AR} &= -W_{Ay} \;, \\
\phi_{AB}z_{AB} + \phi_{AC}z_{AC} + \phi_{AD}z_{AD} + \phi_{AR}z_{AR} + R_{Az} &= -W_{Az}
\end{aligned}$$

or, in matrix form:

$$\begin{bmatrix} x_{AB} & x_{AC} & x_{AD} & x_{AR} & 1 & 0 \\ y_{AB} & y_{AC} & y_{AD} & y_{AR} & 0 & 0 \\ z_{AB} & z_{AC} & z_{AD} & z_{AR} & 0 & 1 \end{bmatrix} \cdot \begin{bmatrix} \phi_{AB} \\ \phi_{AC} \\ \phi_{AD} \\ \phi_{AR} \\ R_{Ax} \\ R_{Az} \end{bmatrix} = - \begin{bmatrix} 0 \\ W_{Ay} \\ W_{Az} \end{bmatrix}$$

This equation may be formed, by rote, as follows. The unknown values of ϕ for the members which radiate from the joint, and the unknown values of any component of support reaction being entered in a column vector in arbitrary order, the vector is applied as a post-factor to a three-row matrix whose columns are allocated to the members and the support reactions in the same order. The rows of the matrix are allocated to x, y and z components in turn, and the matrix is formed by entering in the column allocated to a member the three components of its vector, whilst in each column allocated to a component of support reaction, unity is added in the appropriate row.

The equation is completed by equating the product to the negative of the column vector of the components of the load applied to the joint.

Such a set of equations for a single joint is not generally determinate, but only for those joints which connect only three members and which are not liable to support reactions. However, if the equations for all B joints of a statically determinate truss are taken as a single whole, there are $3B$ equations containing $3B$ unknowns, and the solution for the truss can be written in the inverse form in which the column vector of the unknowns is equated to the product of the inverse of the matrix with the column vector of the loads, and the following illustrates the matrix formulation for the truss of Example 5.12.

$$
\begin{bmatrix} \phi_{OA} \\ \phi_{OB} \\ \phi_{OC} \\ \phi_{AB} \\ \phi_{AC} \\ \phi_{BC} \\ \phi_{AD} \\ \phi_{BD} \\ \phi_{CD} \\ R_{Ox} \\ R_{Oy} \\ R_{Oz} \\ R_{Bx} \\ R_{Bz} \\ R_{Cz} \end{bmatrix}
= -
\begin{array}{c}
\begin{array}{ccccccccccccccc} \text{OA} & \text{OB} & \text{OC} & \text{AB} & \text{AC} & \text{BC} & \text{AD} & \text{BD} & \text{CD} & R_{Ox} & R_{Oy} & R_{Oz} & R_{Bx} & R_{Bz} & R_{Cz} \end{array} \\
\begin{bmatrix}
4 & 4 & 3 & . & . & . & . & . & . & 1 & . & . & . & . & . \\
3 & -3 & 0 & . & . & . & . & . & . & . & 1 & . & . & . & . \\
0 & 0 & 4 & . & . & . & . & . & . & . & . & 1 & . & . & . \\
. & . & . & . & . & . & -2 & -2 & -3 & . & . & . & . & . & . \\
. & . & . & . & . & . & 3 & -3 & 0 & . & . & . & . & . & . \\
. & . & . & . & . & . & -7 & -7 & -3 & . & . & . & . & . & . \\
. & . & -3 & . & 1 & 1 & . & . & 3 & . & . & . & . & . & . \\
. & . & 0 & . & 3 & -3 & . & . & 0 & . & . & . & . & . & . \\
. & . & -4 & . & -4 & -4 & . & . & 3 & . & . & . & . & . & 1 \\
-4 & . & . & 0 & -1 & . & 2 & . & . & . & . & . & . & . & . \\
-3 & . & . & -6 & -3 & . & -3 & . & . & . & . & . & . & . & . \\
0 & . & . & 0 & 4 & . & 7 & . & . & . & . & . & . & . & . \\
. & -4 & . & 0 & . & -1 & . & 2 & . & . & . & . & 1 & . & . \\
. & 3 & . & 6 & . & 3 & . & 3 & . & . & . & . & . & . & . \\
. & 0 & . & 0 & . & 4 & . & 7 & . & . & . & . & . & 1 & .
\end{bmatrix}^{-1}
\end{array}
\cdot
\begin{bmatrix} . \\ . \\ . \\ 2 \\ 1 \\ -10 \\ . \\ . \\ . \\ . \\ . \\ . \\ . \\ . \\ . \end{bmatrix}
\begin{matrix} x \\ y \\ z \\ x \\ y \\ z \\ x \\ y \\ z \\ x \\ y \\ z \\ x \\ y \\ z \end{matrix}
\begin{matrix} \text{JOINT O} \\ \\ \\ \text{JOINT D} \\ \\ \\ \text{JOINT C} \\ \\ \\ \text{JOINT A} \\ \\ \\ \text{JOINT B} \\ \\ \\ \end{matrix}
$$

In this, the extension of the vector of unknowns and of the columns of the matrix so as to provide for all $(3B - 6)$ members and the six components of support reaction is obvious. As to the rows of both the matrix and the vector of applied loads, these are allocated in sets of three to each of the B joints in any arbitrary order, and we note, that since each member connects to two joints there are two sets of entries in the column allocated to each member. Thus just as **OA** is the vector of the member OA which radiates from the joint O, so is **AO** the vector of the member AO which radiates from the joint A, and we note that the column allocated to this member contains the component of **OA** in the rows allocated to the joint O, and their negatives in the rows allocated to the joint A.

6

Continuum Mechanics I – Stress

6.1 INTRODUCTION

In all branches of mechanics we are particularly concerned with the effects of forces on the motions of the bodies on which they act, and among these effects we invariably find relative motions between the parts of the body that result from its deformation. It is true that the effects of the deformation are often small compared with the motions as a whole, so that where we are concerned only with the overall motion of the body, we are often justified in ignoring the effects of the deformation by treating the body as rigid. Yet in other situations, and in other respects, the effects of the deformation may be crucial, and it is situations of this kind that fall within the special province of Continuum Mechanics. In other words, Continnum Mechanics is the branch of Mechanics that is particularly concerned with the consequences of deformation, and it is developed by focusing on the manner in which forces that are applied to a body are transmitted and diffused through it as stresses.

It does not adopt the particulate model of matter. Rather, as the title implies, it treats a body as a continuum of infinitesimal elements, in which stresses and strains are smoothly distributed in a manner which justifies the use of the calculus.

6.2 STRESS ON A PLANE SECTION

Let a body which supports an equilibrium set of externally reacted forces be conceived to comprise the two domains that lie on either side of any plane section that completely divides it.

Although the external forces as a whole are in equilibrium, in general those that are applied to any one domain are not. Yet the parts must be in equilibrium just like the whole, and by considering the equilibrium, first of the whole (Fig. 6.1a), and then of each part in turn (Fig. 6.1b), we can conclude, first that the resultants of the external forces on each of the two domains must be equal but opposite, and second, that the resultant of the external forces on each

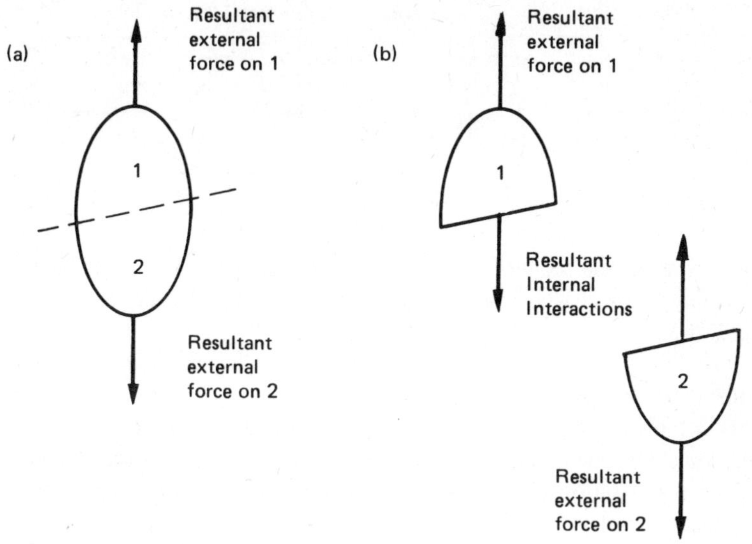

Fig. 6.1

domain must be offset by internal forces which each domain exerts upon the other, across the section that divides them.

In principle, these internal interactions should be considered to derive from the action of each element of each domain on every element of the other, but in practice, it is found that, even in the aggregate, the effects are significant only between immediate neighbours. It is therefore realistic to consider the interactions across any section to occur only between the elements that lie immediately on either side of the section. If, therefore, we imagine the body to be parted on the section, the internal interactions across it can reasonably be regarded as contact forces that are exchanged between the faces of the two parts so as to maintain the equilibrium of each, and it is this exchange across a section of a pair of equal but opposite internal interactions that is said to place the section under stress.

Average Stress on a Plane Section

It is a matter of common experience that larger bodies can usually support larger forces, or, more specifically, that in situations that are geometrically similar, bodies of different sizes can usually support forces that are broadly proportional to their sectional areas. This suggests that a stress should be gauged, not by the net force of interaction itself, but by its average intensity per unit area of the section on which it acts. So let $\mathbf{f}$ represent the resultant internal interaction that acts on either of the faces exposed by a section whose superficial area is represented by A. Then it is the ratios $\mathbf{f}/A$ that is defined as the

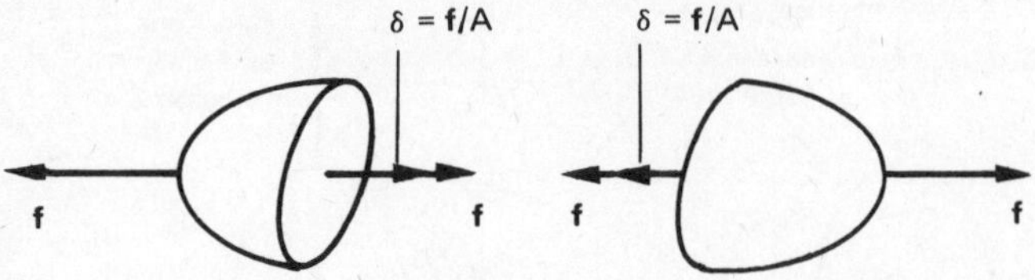

Fig. 6.2

average stress on the section, and, A being a scalar, the ratio may be recognised as a vector quantity which shares the same direction as the net force of interaction f, itself.

Dilation v. Distortion

By this definition, stress is a vector quantity which is much like any other, but it requires a treatment different from that usually applied to vectors, because experiment and rational argument both suggest that the directions of stresses are significant, not in relation to some common reference, but in each relation to the particular section on which it acts. In fact, the arguments which lead to this conclusion vary from one kind of material to another, but taking a single example, we now outline the argument for the important class of materials known as crystalline solids, for which purpose, we temporarily revert to the atomic model.

Atomic theory conceives a crystalline solid as an assembly of atoms or ions which are tightly bound in a highly regular geometrical lattice by the forces of interaction that exist between their charges. These forces are distinguished in two kinds, for some are attractive whilst others are repulsive, and it is found, that although the net attractive and repulsive forces on each atom both increase substantially as the distance between the atoms reduces, they do so according to different laws. These are such, that whereas the attractive forces are greater than the repulsive at larger spacings, as the spacing reduces toward the value at which the electron shells of the atoms overlap, there is such a sharp change in the rate of increase of the repulsive force, that thereafter, this rapidly becomes the greater. Thus when account is taken of both effects together, the variation with the lattice spacing of the net force on the atoms has the general form indicated in Fig. 6.3. This shows that when the spacing is very small, the net force on each atom is repulsive, but that as the spacing increases, the net force first rises rapidly, through zero, up to a maximum attractive value, and then slowly falls away again. Of course, in the absence of external forces, the normal equilibrium spacing is that at which the internal interactions balance, so that the net internal force on each atom is zero.

The application of external forces disturbs this balance of internal interactions and obliges the atoms to seek new equilibrium positions at which the net internal force on each balances the net external force, and in general this may

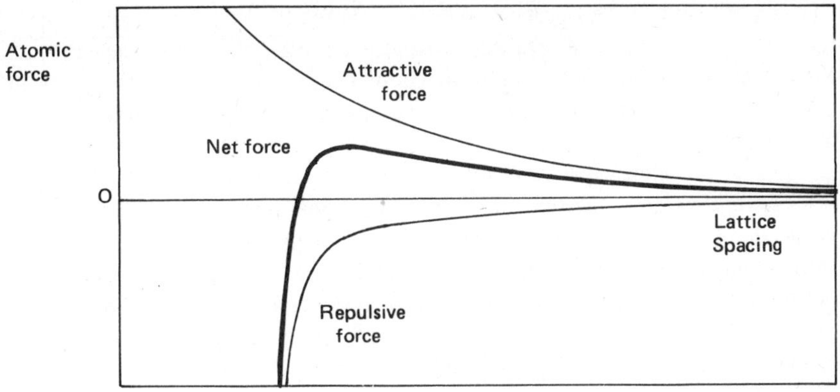

Fig. 6.3

require a pattern of point displacements which is very complex. But no matter how complex it may be, the resulting pattern of deformation at any position can always be described as a sum of deformations in only two basic modes (Fig. 6.4). One is called the dilational mode, and in this there is a simple extension or contraction of the lattice without any change in the form of the assembly. The other is called the shear mode, and in this, sheets of atoms slide past one another so as to change the form of the assembly without any change in the lattice spacing. Clearly then, dilation results in a change of volume without change of shape, whilst shear results in its converse – a change of shape without change of volume, and it follows that any general pattern of deformation can be described as a combination of the two.

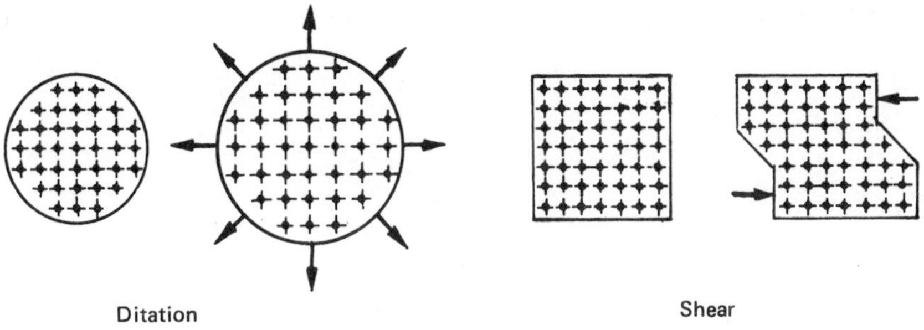

Fig. 6.4

In comparing the characteristics of these two modes, it is crucial that the spacing at which the binding forces typically have their maximum value, differs from the normal equilibrium spacing by only a fractional amount, say, about one fifth. This implies that dilational forces large enough to cause proportional changes of length of this order would be sufficient to extend the lattice indefinitely. On the other hand, if a dilational effect smaller than this were removed, the spacing would revert to its normal equilibrium value, and it follows that, in crystalline materials, dilational deformations are generally both small and impermanent.

But in the relative slide of the shear mode, one neighbour approaches as another recedes, so each atom takes up a new bond as it surrenders an old, and by accumulation from layer to layer, shear can result in large deformations without impairing the integrity of the material. In addition we note that a new equilibrium configuration is achieved at regular intervals of displacement, so that whereas dilational deformations (in solids and liquids) are generally both small and impermanent, in shear they may be both large and permanent.

Direct Stress and Shear

Arguments like these suggest that there is a crucial difference between one pair of forces which, acting as right-angles to a section, tend to pull the two domains apart, and another which, acting parallel to the section, tend to promote the relative slide of shear. Therefore we elect to describe the stresses on different sections, not in terms of their components in the directions of a fixed set of axes, but each in terms of its components normal and parallel to the particular section on which it acts. Thus for each section we choose a set of rectangular axes of which one is normal to the section. Then the others are necessarily parallel to the section, and resolving the stress on the section in the directions of these axes, we describe the stress in terms of its normal or 'direct' component, and two shear components in an arbitrary pair of rectangular directions parallel to the section (Fig. 6.5).

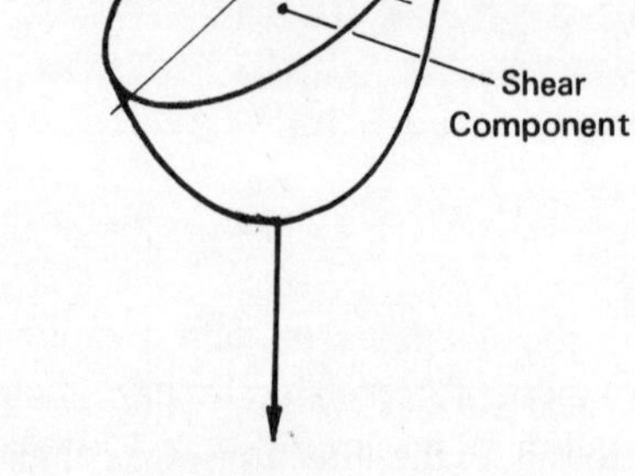

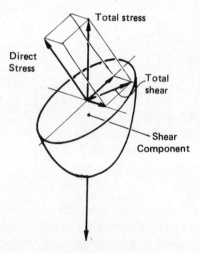

Fig. 6.5

Of course, the total shear stress on the section is the vector sum of the two shear components. Furthermore, the components being rectangular, the square of the total shear stress can be equated to the sum of the squares of the two shear components, whilst the square of the total stress can be equated, either to the sum of the squares of the direct and total shear stresses, or to the sum of the squares of all three of the original components.

Example 6.1

A long cylindrical rod of radius a is subjected to equal but opposite axial tensile forces of modulus F applied to its ends, and an imaginary section MM is identified by the angle θ which its normal makes with the axis of the rod. Determine the magnitudes of the direct and total shear stresses on the section as functions of θ.

Solution

The net internal interaction on MM can be equated to the resultant of the external forces applied to either side of the section, and in the present case,

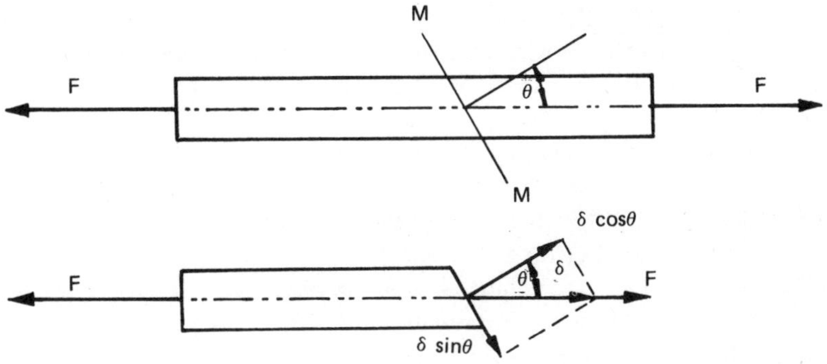

this is a constant. However, the average stress on the section varies with θ because the area of the section varies, whilst the direct and shear components are further affected by the variation of their directions of resolution.

The stress on the section, like the applied forces, lies in the axial direction, and since the section is an ellipse whose semi-axes are equal to a and $a/\cos\theta$, its magnitude is given by:

$$|\sigma| = \frac{|\mathbf{f}|}{A} = \frac{F}{\pi a^2/\cos\theta} = \frac{F.\cos\theta}{\pi a^2} .$$

The direct and total shear stresses on the section can now be determined as the components of σ resolved in the directions normal and parallel to the section, that is:

$$\text{Direct stress} = |\sigma| \,.\, \cos\theta = (F/\pi a^2) \,.\, \cos^2\theta$$
$$= (F/2\pi a^2)\,(1 + \cos 2\theta)$$
$$\text{Total shear} = |\sigma| \,.\, \sin\theta = (F/\pi a^2) \,.\, \sin\theta \,.\, \cos\theta$$
$$= (F/2\pi a^2) \,.\, \sin 2\theta \;.$$

Stress at a Point in a Plane

By its definition, an average stress takes account only of the resultant of the internal interactions that are transmitted across a section, and not of their distribution across the section. Yet experiment confirms what intuition suggests – that the distribution may, in fact, be crucial, so that whereas the concept of average stress may be useful for comparing situations that are geometrically similar, more generally it is only of limited value. We therefore require to refine the concept in a manner which distinguishes the stresses at different positions in any given section, and this we do by invoking the continuum assumptions.

In brief, these assume that a body can be treated as a continuum of infinitesimal elements in which stresses and strains are smoothly distributed in a manner which justifies their analysis by infinitesimal calculus. This may appear to be strongly in conflict with the alternative, atomic theory, but we are now concerned not with the effects on individual particles, but only with their measurable, macroscopic aggregates and averages, and sensibly, these may well be treated as smoothly distributed even though the discrete effects may not. But the continuum assumptions require more, namely, that the aggregates and averages should still retain their significance even when they are taken over an infinitesimal domain. Yet even this does not necessarily imply any significant conflict between the two points of view, because there is such a wide gulf between the scales of atomic events on the one hand, and of macroscopic measurement on the other, that a domain that is regarded as large on the atomic scale may yet be regarded as indefinitely small on the macroscopic scale. Thus it is that two approaches which stem from assumptions as different as those of atomic and continuum theories can both lead to reasonable interpretations of real effects in real materials.

In the present context, the continuum assumptions may be stated more specifically as follows:

1) Considering any point on a plane section drawn through a body, let δA represent the area of an arbitrary part of the section that contains the given point, and let $\delta \mathbf{f}$ represent the resultant of those of the internal interactions that are transmitted across that part of the section. Then the stress on the zone δA of the section may be taken as the ratio $\delta \mathbf{f}/\delta A$, and the continuum assumptions postulate that as δA tends to zero, the value of this ratio tends to a specific limiting value for each point in each section. This we identidy as the stress on the particular section at the particular point, so for materials which satisfy the

continuum assumptions the stress at a given point on a given section may be defined as:

$$\text{Stress} = \underset{\delta A \to 0}{\text{Limit}} \frac{\delta f}{\delta A} = \frac{df}{dA},$$

2) It is also assumed that the stresses at different points on a given section vary smoothly with the positions of the points, and that the stresses on different sections at a given point vary smoothly with the orientations of the sections.

6.3 THE STATE OF STRESS AT A POINT

The concept of stress being now defined in a manner which distinguishes the stresses at each point in each section, we now consider the totality of the stresses at any one point. Thus, considering a particular point in a particular body which is subjected to a particular set of forces, we recognise that, in general, a different value of the stress vector would be associated with each section through the point. Each point is therefore associated with an indefinite number of different stress vectors. But although their number is unlimited, the set of stresses associated with the particular point would all fall within some specific range, and they are conceived to comprise the state of stress at the point. The state of stress at a point is thus conceived to comprise the set of stress vectors for all sections through the point, and this can be recognised as a set of limited range, but unlimited number.

But only a few of the vectors of any one state can be included in an independent set. To show that this is so, we note that no three-dimensional figure has less than four plane faces. On the other hand, given three planes through a point, we can always complete a three-dimensional figure having its fourth face at any arbitrary orientation, provided only that no three of the faces intersect in parallel lines (Fig. 6.6).

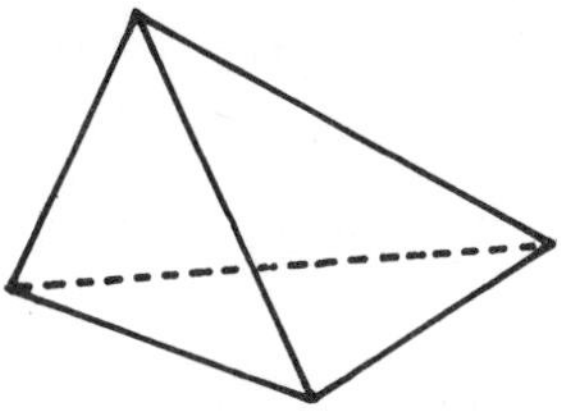

Fig. 6.6

So excepting the special case (it is relatively trivial) suppose that the stress vectors for any three sections at a point are known, and consider an element which is isolated from the body by parting on the three sections on which the stresses are known, and a fourth section of arbitrary orientation. The force on each face can be equated to the stress times the area, and by applying the six equations of equilibrium to the element, we could establish between the twelve components of stress that act on the four faces, six fixed relationships that are universally imposed by the operation of the natural laws. Of these, three would suffice to determine the unknown components of stress on the arbitrary face in terms of the given components, and later we shall prepare these equations in order to reveal the essential features of a state of stress. However, for the present it is sufficient merely to note that the relationships exist, for this implies that the stress vectors for any three sections at a point are sufficient to determine the stress on any other, and it follows that not more than three of the stress vectors at any point can be independent. In fact the existence of the three further fixed relationships implies that the nine components of the three stress vectors will be thrice redundant, so that only six of the components of any three stress vectors may be recognised as independent; but it can be shown that the set must include components from at least three stress vectors, because not more than five of the components of any two vectors can be independent.

The Tensors of a State of Stress

Thus we recognise that a state of stress can be determined completely by the stress vectors for only three sections at the point (in fact their nine scalar components are thrice redundant), and in principle, each of the vectors could be described by its components in a set of directions which, like the orientations of the sections themselves, are arbitrary. But it would obviously be convenient if all these directions were defined by a single set of axes, so in practice, a standard form of description is prepared by first identifying a convenient set of rectangular stress axes (x_1, x_2, x_3) at the point, and by then describing the stress vectors for the three coordinate planes, each in terms of its components in the three coordinate directions.

In a diagram, this set of components can be represented as the stresses on the faces of an infinitesimal, rectangular element whose edges are parallel to the stress axes, and such a diagram is shown in Fig. 6.7. However, we shall be able to describe the state algebraically only when we have resolved the question of a sign convention.

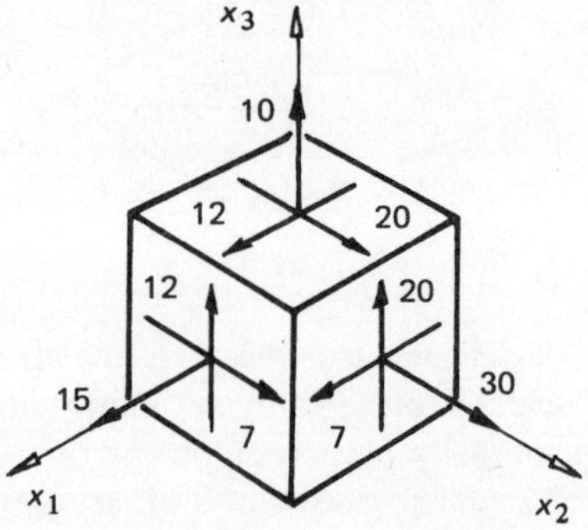

Fig. 6.7

Sign Convention

The sense of a force is determined by comparing its direction with a reference direction – usually that of a coordinate. However, since a stress implies a pair of interactions whose senses, by the force convention, are opposed, it follows that stresses require a convention which is different from that applied to forces.

To illustrate the point by a simple example, consider the case of a straight, slender rod which supports a pair of equal but opposite axial forces, applied to its ends (Fig. 6.8). Clearly, the pair of opposed forces can be so arranged as to place the rod either in tension, or, merely by interchanging the forces end to end, in compression, and confining attention to transverse sections, we require to distinguish the tensile case as positive, and the compressive case as negative. Yet in either case, the pair of faces exposed by a transverse section are subjected to the same pair of interactions which are simply switched from fact to face when the external forces are switched from end to end, so it is clear that we shall be able to make the required distinction only if the directions of the interactions are referred, not to some fixed frame of reference, but each to the particular face on which it acts.

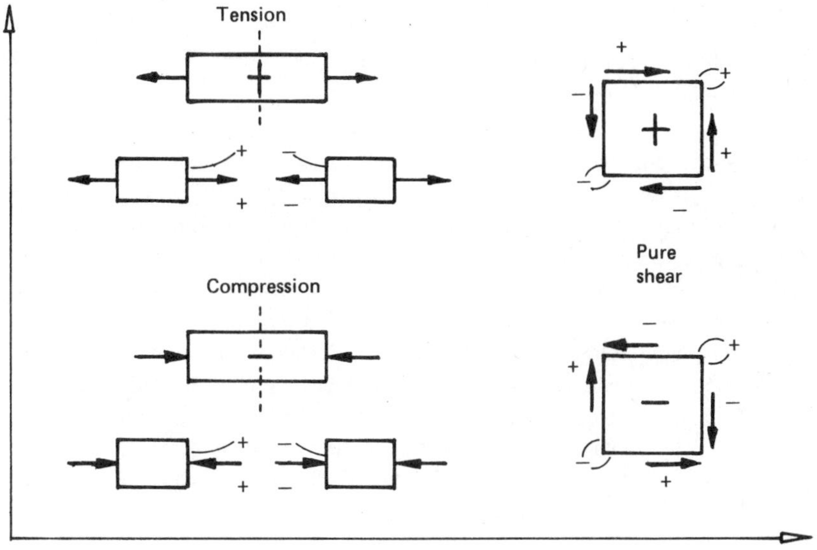

Fig. 6.8

This, unfortunately, is a requirement that cannot be satisfied by a single convention that is convenient in all cases; but in a standard description of a state of stress, we are concerned only with components that act in coordinate directions on coordinate sections, and for these we can conveniently determine the sense of a stress by separately distinguishing the senses, first of the pair of faces exposed by the section, and then of the associated pair of interactions.

To distinguish the senses of the pair of faces exposed by a coordinate section, we erect the outward-facing normal to each. These are necessarily opposed, with one facing in the positive sense of coordinates, and the other in the negative, and the face which takes the positive normal we reckon to be positive, and the other negative (Fig. 6.9).

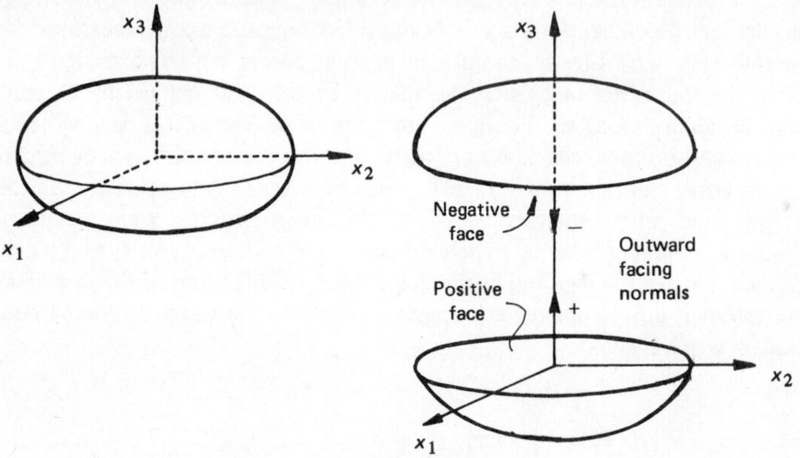

Fig. 6.9

As to the forces of interaction, we note that we are concerned only with components parallel to the coordinate axes, so for these we can retain the usual convention that reckons a force positive when it acts in the positive sense of coordinates, and applying these conventions to the situation shown in Fig. 6.8, we note that the tensile case is one in which the positive face takes the positive interaction and the negative face the negative interaction. Conversely, in the compressive case, it is the negative face which takes the positive interaction, and the positive face the negative interaction, and generalising this observation in a manner that allows it to be applied to any component of a standard description, we reckon a positive component as one in which the sense of each of the forces of interaction agrees with the sense of the face on which it acts, whilst a negative component is one in which the sense of each of the forces of interaction disagrees with the sense of the face on which it acts.

With this convention, a standard description of a given state of stress in a given set of stress axes can be specified purely numerically, without recourse to a diagram, simply by specifying the magnitudes and senses of the nine components in a square array or matrix in which each component is allocated its own position. Thus the vectors of the stresses on the sectional normal to x_1, x_2, and x_3 are entered as the first, second, and third rows of the array in turn, and in each case, the components which act in the directions of x_1, x_2, and x_3 are entered in the first, second, and third positions in the row respectively. Note that each row then contains the three components that act on the same coordinate section, whilst each column contains the three components that act in the same coordinate direction on different sections.

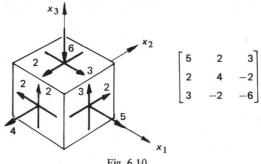

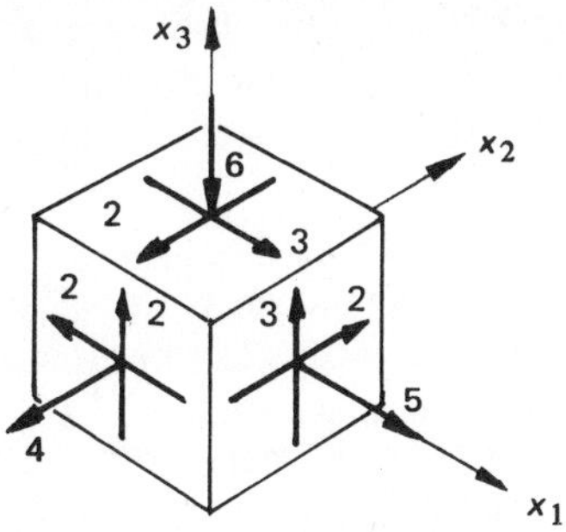

$$\begin{bmatrix} 5 & 2 & 3 \\ 2 & 4 & -2 \\ 3 & -2 & -6 \end{bmatrix}$$

Fig. 6.10

As an example, consider the state of stress defined in Fig. 6.10. The element has three pairs of parallel faces, each of which includes one positive face and one negative, and depending on the point of view, an illustration of such an element might reveal any mixture of positive and negative faces. In fact, of the pair of faces parallel to x_1, Fig. 6.10 reveals the one whose outward-facing normal is directed in the positive sense of x_1. This is therefore a positive face, and since each of the components which act upon it is also directed in the positive sense of coordinates, they are all reckoned positive, and, taking the components parallel to x_1, x_2 and x_3 in turn, we enter as the first row of the array, the set of scalar components: (5, 2, 3). But of the two faces normal to x_2, Fig. 6.10 reveals the one whose outward-facing normal is directed in the negative sense of x_2. This is therefore a negative face on which a stress is reckoned positive when it acts in the negative sense of coordinates, so again taking the components parallel to x_1, x_2 and x_3 in turn, we enter as the second row of the array: (2, 4, −2). Finally, of the faces normal to x_3, the figure reveals the positive, and as the third row of the array, we enter (3, −2, −6).

It may be observed that a particular state of stress, like a particular vector, can be described by its components in any set of axes. Thus a given state is not uniquely associated with a particular set of components, but has a different set of components in each set of axes. Yet all the different descriptions of any one state must, of course, be equivalent, for, as we have argued, each of the components of one set must be related to the components of the same state in any other set of axes, by a set of relationships that are universally imposed by the operation of the natural laws.

In the event, we shall find that these relationships can be expressed as a particularly convenient rule of transformation by which the complete set of components for one set of axes can be resolved into the corresponding set of components for a different set of axes, but more than this, we shall find that this rule of transformation is characteristic not of states of stress alone, but of a whole class of similar quantities. In other words, just as we found that there are some quantities which resolve according to the rule of vector transformation, so shall we find that a state of stress resolves according to a similar but different rule of transformation that characterises another class of quantities. These are known as tensors, and in future, we shall anticipate this outcome by referring to the standard description of a given state of stress as a stress tensor.

Symbolism

Algebraically, tensors, vectors and scalar components of stress are all represented by suitable qualifications of the symbol σ (sigma). For a tensor the symbol is printed in bold, and is enclosed in square brackets, thus: $[\boldsymbol{\sigma}]$. For a vector the symbol is printed in bold and is qualified by a single suffix which nominates the normal to the section on which the stress acts. For a scalar component the symbol is printed in italic and is qualified by a pair of suffices, of which the first nominates the normal to the section on which the stress acts, whilst the second nominates the direction in which the scalar component is resolved. Thus σ_{ll}, represents a scalar component of the stress vector $\boldsymbol{\sigma}_l$ which can itself be regarded as a vector component of the stress tensor $[\boldsymbol{\sigma}]$. However, the stress axes can be distinguished by their numerical suffices alone, so numerical suffices are reserved for this purpose, and the stress vector for the section normal to x_1 is indicated by $\boldsymbol{\sigma}_1$, the coefficients of its components in the directions of x_1, x_2 and x_3 are indicated by σ_{11}, σ_{12} and σ_{13}, and these are the terms that appear in the first row of the tensor of the state, in the given axes.

Since the direct stress is one that acts in the direction of the normal to the section, it always takes a repeated suffix, and it follows that the direct components of a tensor are those that fall on the leading diagonal. This is the diagonal which runs from the top, left-hand corner of the array, to the bottom right. But a shear stress, being parallel to the section, always acts in a direction different from that of the normal, so a shear stress always takes a pair of suffices which differ, and the shear components of a tensor are those that are entered at positions away from the leading diagonal.

$$[\boldsymbol{\sigma}] = \begin{bmatrix} \sigma_{11} & \sigma_{12} & \sigma_{13} \\ \sigma_{21} & \sigma_{22} & \sigma_{23} \\ \sigma_{31} & \sigma_{32} & \sigma_{33} \end{bmatrix}$$

6.4 ELEMENTARY MATRIX ALGEBRA

We have argued that the three vectors of a stress tensor are more than sufficient completely to determine the state of stress at any point, but although a stress tensor thus implies all the features of a given state, they do not directly reveal them. We therefore require to establish the relationships that exist between the stresses of any state, first, so as to reveal those of its features that best characterise its physical significance, and second, so as to determine how the relevant parameters may be determined from any given description of the state. However, this process of stress resolution, which is unreasonably tedious in scalar analysis, can be vastly simplified by only a few of the simpler concepts of matrix algebra, and these can themselves be presented as simple generalisations of operations with which we are already familiar in vector algebra. For although a tensor is evidently more extensive and more complex than a vector, the two kinds of quantity still have much in common. Indeed, if a tensor is seen as the set of vectors that are associated with a given set of rectangular axes, whilst a vector is seen as the set of scalars that are associated with a given set of rectangular axes, then tensors, vectors and scalars could well be regarded as progressively simpler orders of a single class of quantities, and a vector being regarded as a simple form of tensor, we are led to question whether the rules of vector algebra cannot similarly be regarded as special cases of a set of rules which are applicable to the whole class. The rules of vector algebra have therefore been generalised in a manner which allows them to be applied to matrices, whilst still retaining the application to geometrical vectors as a special case, and we find that these generalised rules can, in fact, be usefully applied, not only to the special class of matrices known as tensors, but commonly to matrices in general. Such is the genesis of matrix algebra, and in what follows we present, as arbitrary definitions, those few of its simpler rules that are relevant to stress resolution.

Transposition

We commonly need to rewrite a matrix so as to present its rows and columns and its columns as rows. This operation is known as transposition, and it is indicated by the application to a matrix of the index T.

Thus:
$$\begin{bmatrix} 1 & 2 & 3 \\ 10 & 20 & 30 \end{bmatrix}^T = \begin{bmatrix} 1 & 10 \\ 2 & 20 \\ 3 & 30 \end{bmatrix}$$

Addition and Subtraction

In the sum (or difference) of two geometrical vectors, each scalar element of the sum (or difference) is equal to the sum (or difference) of the corresponding elements of the given vectors.

A sum (or difference) of matrices is defined only for matrices which have equal numbers of rows and equal numbers of columns, when each element of the sum or difference is again determined as the sum (or difference) of the corresponding elements of the given matrices. Thus:

$$\begin{bmatrix} 0 & 10 \\ 1 & 20 \\ 2 & 30 \end{bmatrix} + \begin{bmatrix} 3 & 40 \\ 4 & 50 \\ 5 & 60 \end{bmatrix} = \begin{bmatrix} (0+3) & (10+40) \\ (1+4) & (20+50) \\ (2+5) & (30+60) \end{bmatrix} = \begin{bmatrix} 3 & 50 \\ 5 & 70 \\ 7 & 90 \end{bmatrix}$$

Evidently addition is again commutative, whereas subtraction is not.

Thus: $[\mathbf{a}] + [\mathbf{b}] = [\mathbf{b}] + [\mathbf{a}]$

But: $[\mathbf{a}] - [\mathbf{b}] = -([\mathbf{b}] - [\mathbf{a}])$.

The Dot Product

The product of matrices **[a]** · **[b]** is defined only for the case when the number of elements in each row of the pre-factor **[a]** is equal to the number of elements in each column of the post-factor **[b]** when the product is defined as the matrix, each of whose rows is determined by taking the (vectorial) dot product of the corresponding row of the pre-factor **[a]** with each column of the post-factor **[b]** in turn. Thus:

$$\begin{bmatrix} 1 & 0 & 1 & 0 \\ 2 & 0 & 2 & 0 \end{bmatrix} \cdot \begin{bmatrix} 3 & 4 \\ 3 & 4 \\ 0 & 0 \\ 3 & 4 \end{bmatrix} = \begin{bmatrix} [1\ 0\ 1\ 0] \cdot \begin{bmatrix} 3 \\ 3 \\ 0 \\ 3 \end{bmatrix} & [1\ 0\ 1\ 0] \cdot \begin{bmatrix} 4 \\ 4 \\ 0 \\ 4 \end{bmatrix} \\ [2\ 0\ 2\ 0] \cdot \begin{bmatrix} 3 \\ 3 \\ 0 \\ 3 \end{bmatrix} & [2\ 0\ 2\ 0] \cdot \begin{bmatrix} 4 \\ 4 \\ 0 \\ 4 \end{bmatrix} \end{bmatrix}$$

$$= \begin{bmatrix} 3 & 4 \\ 6 & 8 \end{bmatrix}$$

Alternatively, the product **[a]** · **[b]** can be defined as a matrix having as many rows as the pre-factor **[a]**, and as many columns as the post-factor **[b]**, in which the term in the ith row and the jth column is equal to the dot product of the ith row of the pre-factor with the jth column of the post-factor.

Since a product is valid only when the number of columns in the pre-factor is equal to the number of rows in the post-factor, both **[a]** · **[b]** and **[b]** · **[a** can be valid only when **[a]** and **[b]** are square matrices of the same order.

However, even then, **[a]** · **[b]** is not generally equal to **[b]** · **[a]**. We therefore note, that even though the dot product is commutative when applied to a pair of vectors, in application to matrices in general, it is not commutative. That is:

$$[\mathbf{b}] \cdot [\mathbf{a}] \neq [\mathbf{a}] \cdot [\mathbf{b}] \ .$$

On the other hand a dot product of matrices is associative. That is:

$$[\mathbf{a}] \cdot [\mathbf{b}] \cdot [\mathbf{c}] = [\mathbf{a}] \cdot ([\mathbf{b}] \cdot [\mathbf{c}]) = ([\mathbf{a}] \cdot [\mathbf{b}]) \cdot [\mathbf{c}]$$

It will be helpful to note the effect of applying the product rule to particular combinations of matrices as follows:

When a matrix is pre-multiplied by a row vector the result is a row vector whose elements may be determined by multiplying each element of the vector into the corresponding row of the matrix, and adding the products in each column. For example:

$$[1\ \ 2\ \ 3] \cdot \begin{bmatrix} 11 & 12 & 13 \\ 21 & 22 & 23 \\ 31 & 32 & 33 \end{bmatrix} = \left[\begin{array}{c|c|c} 1\text{X}11 & 1\text{X}12 & 1\text{X}13 \\ + & + & + \\ 2\text{X}21 & 2\text{X}22 & 2\text{X}23 \\ + & + & + \\ 3\text{X}31 & 3\text{X}32 & 3\text{X}33 \end{array}\right]$$

$$= [146 \qquad 152 \qquad 158]$$

When a matrix is pre-multiplied by a row vector and post-multiplied by a column vector the result is a scalar which can be determined by multiplying each element of the pre-factor into the corresponding row of the matrix, and each element of the post-factor into the corresponding column, and adding all the triple products. For example:

$$[1\ \ 2\ \ 3] \cdot \begin{bmatrix} 11 & 12 \\ 21 & 22 \\ 31 & 32 \end{bmatrix} \cdot \begin{bmatrix} 4 \\ 5 \end{bmatrix} = \begin{array}{l} \ 1\text{X}4\text{X}11 + 1\text{X}5\text{X}12 \\ + 2\text{X}4\text{X}21 + 2\text{X}5\text{X}22 \\ + 3\text{X}4\text{X}31 + 3\text{X}5\text{X}32 \end{array}$$

$$= 1344$$

When a Matrix is pre-multiplied by one matrix and post-multiplied by another, the result is a matrix having as many rows as the pre-factor and as many columns as the post-factor, and the term in the *i*th row and the *j*th column is equal to the given matrix pre-multiplied by the *i*th row of the pre-factor and post-multiplied by the *j*th column of the post-factor. For example:

$$\begin{bmatrix} 1 & 1 & 0 \\ 2 & 0 & 2 \end{bmatrix} \cdot \begin{bmatrix} 5 & 0 \\ 0 & 7 \\ 6 & 8 \end{bmatrix} \cdot \begin{bmatrix} 3 & 4 \\ 3 & 0 \end{bmatrix} =$$

$$= \begin{bmatrix} [1\ 1\ 0]\cdot\begin{bmatrix}5&0\\0&7\\6&8\end{bmatrix}\cdot\begin{bmatrix}3\\3\end{bmatrix} & [1\ 1\ 0]\cdot\begin{bmatrix}5&0\\0&7\\6&8\end{bmatrix}\cdot\begin{bmatrix}4\\0\end{bmatrix} \\ [2\ 0\ 2]\cdot\begin{bmatrix}5&0\\0&7\\6&8\end{bmatrix}\cdot\begin{bmatrix}3\\3\end{bmatrix} & [2\ 0\ 2]\cdot\begin{bmatrix}5&0\\0&7\\6&8\end{bmatrix}\cdot\begin{bmatrix}4\\0\end{bmatrix} \end{bmatrix}$$

$$= \begin{bmatrix}36 & 20\\ 114 & 88\end{bmatrix}$$

Determinants

A matrix in which the number of columns is equal to the number of rows is said to be square, and each such matrix is associated with a quantity called its determinant, which, like a modulus, is indicated by enclosing the array within vertical strokes.

For a 2 X 2 matrix, the determinant is defined, (as in the example), as the difference between the products of the pairs of elements on the leading and trailing diagonals, the leading diagonal being the one which runs from the top left-hand corner to the bottom right.

$$\begin{vmatrix}1 & 2\\ 3 & 4\end{vmatrix} = 1 \times 4 - 2 \times 3 = 4 - 6 = -2$$

To evaluate the determinant of a 3 X 3 matrix we first take the product of each element of the first row with the determinant of the 2 X 2 array that remains when we ignore both the row and the column in which the element resides. Then the required determinant is defined as the first product, minus the second, plus the third. For example:

$$\begin{vmatrix}1 & 2 & 3\\ 8 & 7 & 6\\ 0 & 4 & 5\end{vmatrix} = 1\begin{vmatrix}7 & 6\\ 4 & 5\end{vmatrix} - 2\begin{vmatrix}8 & 6\\ 0 & 5\end{vmatrix} + 3\begin{vmatrix}8 & 7\\ 0 & 4\end{vmatrix}$$

$$= 1(35 - 24) - 2(40 - 0) + 3(32 - 0)$$

$$= 11 - 80 + 96 = 27\ .$$

Alternative Rectangular Coordinates

A ready application of the foregoing principles is found in the equation for the transformation of a vector for a rotation of coordinates. In this, we first use a matrix to describe the inclination of one set of rectangular coordinates to another, when the matrix may be used in a simple equation which relates the scalar elements of any given vector in the alternative coordinates.

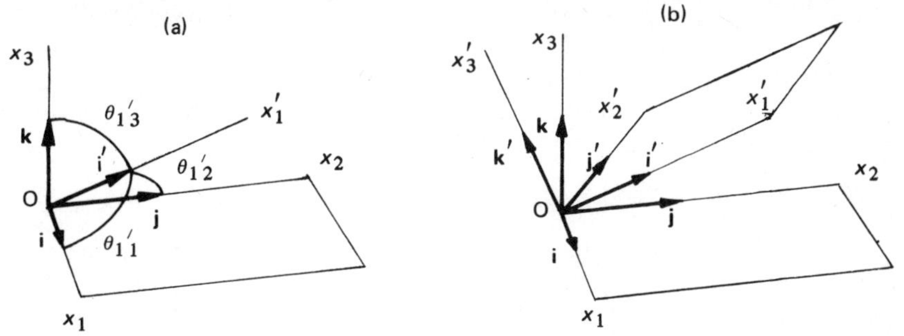

Fig. 6.11

Referring to Fig. 6.11a, let $Ox_1x_2x_3$ represent a set of rectangular coordinates, let $\mathbf{i}'$ represent the unit vector which coincides with an arbitrary axis x_1', and let $(l_{1'1}, l_{1'2}, l_{1'3})$ represent the cosines of the angles $\theta_{1'1}$, $\theta_{1'2}$ and $\theta_{1'3}$ which x_1' makes with x_1, x_2 and x_3 in turn. Then the set of cosines will serve to define the direction of x_1' in $Ox_1x_2x_3$, and we recognise that they may be considered to comprise either the direction cosines of x_1', or the scalar elements of the coincident unit vector $\mathbf{i}'$, each in the coordinates $Ox_1x_2x_3$.

Referring to Fig. 6.11b, now let x_1' be one of an alternative set of rectangular coordinates $Ox_1'x_2'x_3'$. Evidently, the directions of x_2' and x_3' in $Ox_1x_2x_3$ may similarly be defined by sets of direction cosines $(l_{2'1}, l_{2'2}, l_{2'3})$ and $(l_{3'1}, l_{3'2}, l_{3'3})$ respectively, and we recognise that these may alternatively be considered to comprise the scalar elements of the coincident unit vectors $\mathbf{j}'$ and $\mathbf{k}'$, again in terms of their components in $Ox_1x_2x_3$.

Taken together, the three sets of direct cosines will serve to define the directions of the alternative coordinates $Ox_1'x_2'x_3'$ in $Ox_1x_2x_3$, and, being presented in a matrix $[\mathbf{l}_{x'x}]$ in which the direction cosines of x_1', x_2' and x_3' are entered in the first, second and third rows respectively, we recognise that the rows may alternatively be regarded as the unit vectors $\mathbf{i}'$, $\mathbf{j}'$, and $\mathbf{k}'$, described in terms of their components in $Ox_1x_2x_3$.

$$[\mathbf{l}_{x'x}] = \begin{bmatrix} l_{1'1} & l_{1'2} & l_{1'3} \\ l_{2'1} & l_{2'2} & l_{2'3} \\ l_{3'1} & l_{3'2} & l_{3'3} \end{bmatrix} = \begin{bmatrix} \longleftarrow \mathbf{i}' \longrightarrow \\ \longleftarrow \mathbf{j}' \longrightarrow \\ \longleftarrow \mathbf{k}' \longrightarrow \end{bmatrix} = \begin{bmatrix} \uparrow & \uparrow & \uparrow \\ \mathbf{i} & \mathbf{j} & \mathbf{k} \\ \downarrow & \downarrow & \downarrow \end{bmatrix}$$

However, reflection will show that the first column of $[\mathbf{l}_{x'x}]$ then contains the cosines of the angles which x_1', x_2' and x_3' make with x_1. It may therefore be considered to comprise, either the direction cosines of x_1 in $Ox'_1x'_2x'_3$, or the unit vector **i** in the same, and since parallel observations can be made of the second and third columns, we can conlude as follows:

a) In the matrix $[\mathbf{l}_{x'x}]$ whose rows comprise the direction cosines of x'_1, x'_2 and x'_3 in $Ox_1x_2x_3$, the columns comprise the direction cosines of x_1, x_2 and x_3 in $Ox'_1x'_2x'_3$, and it follows that the transpose of $[\mathbf{l}_{x'x}]$, (the matrix of direction cosines of $Ox'_1x'_2x'_3$ in $Ox_1x_2x_3$) may be equated to $[\mathbf{l}_{xx'}]$ (the matrix of direction cosines of $Ox_1x_2x_3$ in $Ox'_1x'_2x'_3$).

Thus: $$[\mathbf{l}_{xx'}] = [\mathbf{l}_{x'x}]^{\mathrm{T}} \,.$$

b) In the matrix of direction cosines of $Ox'_1x'_2x'_3$ in $Ox_1x_2x_3$, the rows may be considered to describe the unit vectors $\mathbf{i}', \mathbf{j}'$ and $\mathbf{k}'$ in $Ox_1x_2x_3$, whilst the columns may be considered to comprise the unit vectors **i**, **j** and **k** in terms of their components in $Ox'_1x'_2x'_3$, and since **i**, **j**, **k** and $\mathbf{i}', \mathbf{j}', \mathbf{k}'$ are both right-handed sets of rectangular unit vectors, we may conclude as follows.

In the matrix of direction cosines which describes the inclination of one set of rectangular coordinates to another:

a) The sum of the squares of the elements in each row and in each column is equal to unity.

b) The dot product of any two rows or of any two columns is equal to zero.

c) The rows or columns being taken in cyclical order 1, 2, 3, 1, 2, 3, then for any consecutive trio, the vector product of the first with the second will be equal to the third.

The matrix is therefore greatly redundant, so that given three elements of any two rows or columns, together with the senses only of two other elements, the remaining elements may be determined from the foregoing relationships.

The Equation of Vector Transformation

A given vector is not uniquely associated with a particular set of scalar elements. Rather it may be described in terms of its components in any set of rectangular coordinates that proves convenient, and Fig. 6.12 shows a vector which is described in terms of its components in both $Ox_1x_2x_3$ and $Ox'_1x'_2x'_3$. Let $\mathbf{r} = (r_1\,r_2\,r_3)$ represent the description of the vector in terms of its scalar elements in $Ox_1x_2x_3$, let $\mathbf{r}' = (r'_1\,r'_2\,r'_3)$ represent the description in $Ox'_1x'_2x'_3$, and let $[\mathbf{l}_{x'x}]$ represent the matrix of direction cosines which defines the inclination of $Ox'_1x'_2x'_3$ to $Ox_1x_2x_3$. Since r'_1, r'_2 and r'_3 are the components of the given vector resolved in the directions of x'_1, x'_2 and x'_3 in turn, they can be equated to the dot products of the vector with the unit vectors $\mathbf{i}', \mathbf{j}'$ and $\mathbf{k}'$, and since these are given in the

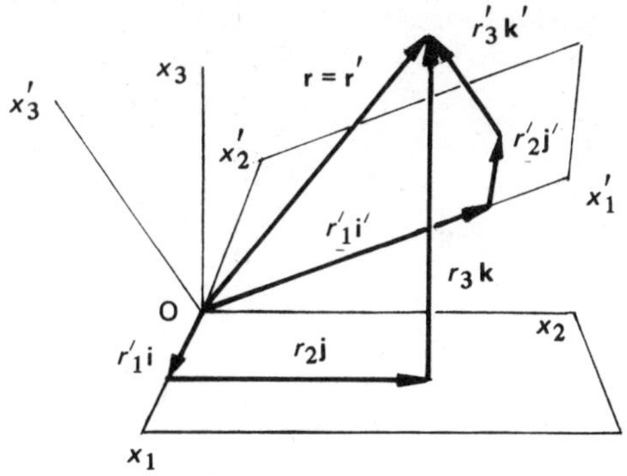

Fig. 6.12

rows of $[\mathbf{l}_{x'x}]$, we conclude that the column vector of $\mathbf{r}'$ can be equated to the matrix $[\mathbf{l}_{x'x}]$ times the column vector of $\mathbf{r}$. Thus

$$\begin{bmatrix} r'_1 \\ r'_2 \\ r'_3 \end{bmatrix} = \begin{bmatrix} \mathbf{i}' \cdot \mathbf{r} \\ \mathbf{j}' \cdot \mathbf{r} \\ \mathbf{k}' \cdot \mathbf{r} \end{bmatrix} = \begin{bmatrix} \longleftarrow \mathbf{i}' \longrightarrow \\ \longleftarrow \mathbf{j}' \longrightarrow \\ \longleftarrow \mathbf{k}' \longrightarrow \end{bmatrix} \cdot \begin{bmatrix} \uparrow \\ \mathbf{r} \\ \downarrow \end{bmatrix}$$

or

$$\mathbf{r}' = [\mathbf{l}_{x'x}] \cdot \mathbf{r} \qquad (6.1)$$

Example 6.2

$Ox_1x_2x_3$ and $Ox'_1x'_2x'_3$ are two sets of rectangular coordinates in which x'_1 passes throught the point P whose coordinates in $Oxyz$ are (2, 2, 1), whilst x'_2 lies in the x_1x_2 plane in the sense of x_2 increasing. The scalar elements of a vector $\mathbf{r}$ in $Ox_1x_2x_3$ are (2 3 4). Determine

a) The matrix of direction cosines of $Ox'_1x'_2x'_3$ in $Ox_1x_2x_3$.

and b) The scalar elements of $\mathbf{r}$ in $Ox'_1x'_2x'_3$.

Solution

a) The rows of the required matrix are the scalar elements of $\mathbf{i}', \mathbf{j}'$ and $\mathbf{k}'$ in $Ox_1x_2x_3$, and since $\mathbf{i}'$ coincides with OP.

$$\mathbf{i}' = \hat{\mathbf{OP}} = |\mathbf{OP}|^{-1}\mathbf{OP} = (1/3)\,(2 \quad 2 \quad 1)$$

Since $\mathbf{j}'$ lies in the x_1x_2 plane its $\mathbf{k}$ component is zero, and it can therefore be represented by $(l_{2'1}\; l_{2'2}\; 0)$. Furthermore, it is at right-angles to $\mathbf{i}' = (1/3)(2\,2\,1)$, and its modulus is unity, whence:

$$2l_{2'1} + 2l_{2'2} = 0; \quad \text{and:} \quad l^2_{2'1} + l^2_{2'2} = 1\ .$$

Solving simultaneously gives:

$$l_{2'1} = \pm \frac{1}{\sqrt{2}}; \quad l_{2'2} = \mp \frac{1}{\sqrt{2}} .$$

But we require $\mathbf{j}'$ to be directed in the sense of x_2 increasing. Its $\mathbf{j}$ component must therefore be positive, whence

$$\mathbf{j}' = \frac{1}{\sqrt{2}}(-\mathbf{i} + \mathbf{j}) .$$

Finally: $$\mathbf{k}' = \mathbf{i}' \times \mathbf{j}' = \frac{1}{3\sqrt{2}}(2 \quad 2 \quad 1) \times (-1 \quad 1 \quad 0)$$

$$\mathbf{k}' = \frac{1}{3\sqrt{2}}(-1 \quad -1 \quad 4) .$$

Thus: $$[\mathbf{l}_{x'x}] = \begin{bmatrix} \longleftarrow \mathbf{i}' \longrightarrow \\ \longleftarrow \mathbf{j}' \longrightarrow \\ \longleftarrow \mathbf{k}' \longrightarrow \end{bmatrix} = \frac{1}{3\sqrt{2}} \begin{bmatrix} 2\sqrt{2} & 2\sqrt{2} & \sqrt{2} \\ -3 & 3 & 0 \\ -1 & -1 & 4 \end{bmatrix}$$

b) $$\mathbf{r}' = [\mathbf{l}_{x'x}] \cdot \mathbf{r}$$

$$\mathbf{r}' = \frac{1}{3\sqrt{2}} \begin{bmatrix} 2\sqrt{2} & 2\sqrt{2} & \sqrt{2} \\ -3 & 3 & 0 \\ -1 & -1 & 4 \end{bmatrix} \cdot \begin{bmatrix} 2 \\ 3 \\ 4 \end{bmatrix}$$

$$\mathbf{r}' = \frac{1}{3\sqrt{2}} \begin{bmatrix} 14\sqrt{2} \\ 3 \\ 11 \end{bmatrix} .$$

6.5 STRESS RESOLUTION

Stress resolution is concerned only with relationships between the stresses of a given state, so it deals only with the variation of stress from section to section at a given point, and not with the variation in the state of stress from point to point. In establishing the rule of resolution it is therefore sufficient to assume the specially simple, 'homogeneous' case in which there is no variation in the state of stress from point to point, and the stress at any point on any section varies only with the orientation of the section, and not with the position of the point.

We also note that the forces generated by the stresses which act on the faces of an infinitesimal element may be treated as an equilibrium system, irrespective of any body forces or of any acceleration of the element. For whereas the forces generated by the stresses vary with the areas of the faces on which they act, any body forces which arise either from some field effect or from the acceleration of the element would be proportional to its volume. Thus the forces that arise from the stresses are proportional to the square of the linear dimensions of the element, whereas any body forces would be proportional to their cube, and in the limit, as the linear dimensions tend to zero, the body forces, including any force associated with the acceleration of the element, would be negligible compared with those that arise from the stresses which act on the faces of the element.

Given that the state of stress in a homogeneous case is defined by its tensor $[\sigma]$ in a given set of stress axes (Fig 16.3a), and that $\hat{1}$ is an arbitrary unit vector which also is defined in terms of its components (l_1, l_2, l_3) in the same coordinates (Fig. 6.13b), consider the equilibrium of the forces generated by the

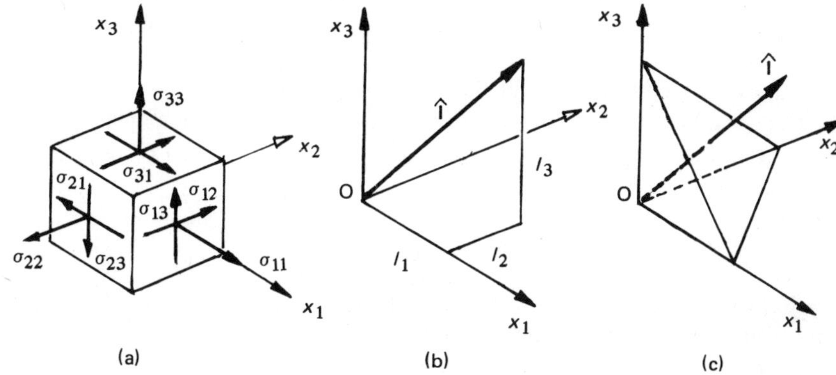

Fig. 6.13

stresses which act on the faces of an infinitesimal element isolated between the three coordinate planes through one point, and a fourth, general section normal to the arbitrary unit vector $\hat{1}$, through an adjacent point (Fig. 6.13c).

Since the force generated by a stress is determined by multiplying the stress by the area of the face on which it acts, we first require to determine the proportional areas of the faces of the element; so let the superficial area of the general face be taken as the unit of area, when its vector area is its unit normal vector $\hat{1}$. Then we note that the faces which lie in the coordinate planes are the projections of the general face whose areas are equal to the scalar elements of its vector area, and it follows, that the area of the general face being taken as unity, the areas of the faces normal to x_1, x_2 and x_3 may be equated to l_1, l_2 and l_3 in turn.

Consider the forces that act on the three faces that lie in the coordinate planes, noting that they are all negative faces on which positive stresses therefore give negative forces. The stress vectors for the faces normal to x_1, x_2 and x_3 are given in the first, second and third rows of $[\sigma]$ respectively, and the areas of the faces on which these stresses act are equal to l_1, l_2 and l_3 in turn. The negatives of the forces can therefore be determined by multiplying the terms in the first, second and third rows of $[\sigma]$ by l_1, l_2 and l_3 respectively, and since the negative of their resultant can then be determined by adding the products in each column, we recognised that it can be determined by the product $\hat{\mathbf{l}} \cdot [\sigma]$.

For the equilibrium of the element, this resultant of the negative forces on the coordinate faces can now be equated to the positive of the force generated by the unknown stress σ_l acting on the unit area of the general face, whence:

$$\sigma_l = \hat{\mathbf{l}} \cdot [\sigma] \ .$$

We note that since $\hat{\mathbf{l}}$ and $[\sigma]$ are both described in the given stress axes, their product determines σ_l in terms of its components in the same directions, and not in terms of the usual set of direct and shear components normal and parallel to the section. However once σ_l is determined in any set of coordinates, it may readily be resolved in any required direction, simply by taking its dot product with the unit vector that lies in that direction. Thus if $\hat{\mathbf{l}}'$ represents a second arbitrary unit vector which is again described in terms of its components in the given stress axes, the scalar component of σ_l resolved in the direction of $\hat{\mathbf{l}}'$ is indicated by $\sigma_{ll'}$, and is given by

$$\sigma_{ll'} = \sigma_l \cdot \hat{\mathbf{l}}' = \hat{\mathbf{l}} \cdot [\sigma] \cdot \hat{\mathbf{l}}' \ .$$

Concerning the sequence of terms in the products $\hat{\mathbf{l}} \cdot [\sigma]$ and $\hat{\mathbf{l}} \cdot [\sigma] \cdot \hat{\mathbf{l}}'$, we remember that a matrix product is not generally commutative. However, we shall show that a stress tensor is necessarily symmetrical about its leading diagonal, and reflection will show that $\hat{\mathbf{l}} \cdot [\sigma]$ is then equal to $([\sigma] \cdot \hat{\mathbf{l}})^{\mathrm{T}}$ (i.e., that $[\sigma] \cdot \hat{\mathbf{l}}$ determines the row vector $\hat{\mathbf{l}} \cdot [\sigma]$ as a column vector), and that $\hat{\mathbf{l}} \cdot [\sigma] \cdot \hat{\mathbf{l}}'$ is equal to $\hat{\mathbf{l}}' \cdot [\sigma] \cdot \hat{\mathbf{l}}$. On the other hand, it is obvious that neither $\hat{\mathbf{l}} \cdot \hat{\mathbf{l}}' \cdot [\sigma]$ nor $[\sigma] \cdot \hat{\mathbf{l}} \cdot \hat{\mathbf{l}}'$ is equal to $\hat{\mathbf{l}} \cdot [\sigma] \cdot \hat{\mathbf{l}}'$ because whereas the latter determines a scalar, either of the former determines a matrix. In stress resolution it is therefore immaterial which of the unit vectors is applied to the tensor as a pre-factor, and which as a postfactor, provided only that no two unit vectors are multiplied directly.

Summarising the foregoing conclusions for the resolution of a stress tensor in comparison with the resolution of a vector, we remark as follows.

A vector a is defined by specifying its scalar components in a given set of rectangular directions, and it is resolved into its scalar component in the direction of an arbitrary unit vector $\hat{\mathbf{l}}$ by the product $\hat{\mathbf{l}} \cdot \mathbf{a}$.

A tensor $[\sigma]$ of the state of stress at any point is defined by specifying the vectors of the stresses which act on the sections normal to a given set of rectangular directions, and the stress vector for the general section normal to an arbitrary unit vector $\hat{l}$ is given by the product $\hat{l} \cdot [\sigma]$.

The scalar component of the stress on the section normal to $\hat{l}$, resolved in the direction $\hat{l}$,, is indicated by $\sigma_{ll'}$, and is therefore given by $\hat{l} \cdot [\sigma] \cdot \hat{l}'$.

$$\sigma_l = [\sigma_{l1}\ \sigma_{l2}\ \sigma_{l3}] = \hat{l} \cdot [\sigma] = ([\sigma] \cdot \hat{l})^T \tag{6.2}$$

$$\sigma_{ll'} = \hat{l} \cdot [\sigma] \cdot \hat{l}' = \hat{l}' \cdot [\sigma] \cdot \hat{l} \tag{6.3}$$

Direct and Total Shear Stresses on a General Section

In particular, the direct stress on the section normal to $\hat{l}$ is the scalar component resolved along the normal $\hat{l}$, and is therefore given by:

$$\sigma_{ll} = \sigma_l \cdot \hat{l} = \hat{l} \cdot [\sigma] \cdot \hat{l} \,.$$

Furthermore, since the direct stress σ_{ll} and the total shear stress τ comprise a pair of rectangular components of the total stress vector σ_l, the magnitude (but not the direction) of the total shear stress on the section normal to **l** may be determined by equating its square to the difference between the squares of the total stress σ_l, and the direct stress σ_{ll}.

$$\tau^2 = |\sigma_l|^2 - \sigma_{ll}^2 \tag{6.4}$$

Example 6.3

A tensor $[\sigma]$ of the state of stress at a point has the value indicated. For the section normal to OP, where O is the origin of the tensor axes and P has co-ordinates (1, 2, 2) determine

a) The modulus of the total stress
b) The direct stress
c) The modulus of the total shear
d) The scalar component of shear parallel to the x_1x_2 plane of the stress axes.

$$[\sigma] = \begin{bmatrix} 4 & 6 & 0 \\ 6 & 9 & 0 \\ 0 & 0 & 2 \end{bmatrix}$$

Solution

The unit vector $\hat{l}$ normal to the specified section is given by:

$$\hat{l} = |\mathbf{OP}|^{-1}\mathbf{OP} = (1/3)\,(1 \quad 2 \quad 2) \,.$$

a) The components of σ_l resolved in the directions of the stress axes are given by $\hat{l} \cdot [\sigma]$, and the modulus of σ_l is equal to the square root of the sum of their squares.

$$[\sigma_{l1} \quad \sigma_{l2} \quad \sigma_{l3}] = \hat{\mathbf{l}} \cdot \underline{\sigma} = \frac{1}{3}[1 \quad 2 \quad 2] \cdot \begin{bmatrix} 4 & 6 & 0 \\ 6 & 9 & 0 \\ 0 & 0 & 2 \end{bmatrix}$$

$$= (4/3)\,[4 \quad 6 \quad 1]$$

$$|\tilde{\sigma}_l| = (4/3)\sqrt{(16 + 36 + 1)}$$

$$|\tilde{\sigma_l}| = 4\sqrt{53}/3\ .$$

b) The direct stress is the component of σ_l resolved in the direction l.

$$\sigma_{ll} = \sigma_l \cdot \hat{\mathbf{l}} = (4/9)\,[4 \quad 6 \quad 1] \cdot \begin{bmatrix} 1 \\ 2 \\ 2 \end{bmatrix}$$

$$\sigma_{ll} = 8\ .$$

c) The square of the total stress is equal to the sum of the squares of the direct and total shear components. Thus, the modulus of the total shear being indicated by τ:

$$\tau^2 = |\sigma_l|^2 - \sigma_{ll}^2 = (4\sqrt{53}/3)^2 - 8^2$$

$$\tau = 4\sqrt{17}/3\ .$$

d) Let $\hat{\mathbf{l}}' = (l_1\, l_2\, l_3)$ represent the unit vector in the required direction of resolution.

Since $\hat{\mathbf{l}}'$ is parallel to x_1x_2:

$$\hat{\mathbf{l}}'_3 = 0 \tag{i}$$

Since $\hat{\mathbf{l}}'$ is normal to $\hat{\mathbf{l}}$, their dot product is zero, that is:

$$\frac{1}{3}(l'_1 + 2l'_2 + 2l'_3) = 0\ . \tag{ii}$$

Since the modulus of $\hat{\mathbf{l}}'$ is unity:

$$(l'_1)^2 + (l'_2)^2 + (l'_3)^2 = 1\ . \tag{iii}$$

Between (i), (ii) and (iii):

$$l'_1 = \pm\frac{2}{\sqrt{5}};\quad l'_2 = \mp\frac{1}{\sqrt{5}};\quad l'_3 = 0\ .$$

The alternative solutions arise because we have not defined the positive sense along the direction of resolution. However, we require to determine only the modulus of the component, so either of the two combinations of signs will suffice.

Take: $\hat{\mathbf{l}}' = \dfrac{1}{\sqrt{5}}(2, \ -1, \ 0)$

Then $\sigma_{ll'} = \sigma_l \cdot \hat{\mathbf{l}} = \dfrac{4}{3\sqrt{5}}\begin{bmatrix} 4 & 6 & 1 \end{bmatrix} \cdot \begin{bmatrix} 2 \\ -1 \\ 0 \end{bmatrix},$

that is: $\sigma_{ll'} = \dfrac{8}{3\sqrt{5}}$.

6.6 PROPERTIES OF A STRESS TENSOR

Provided that a state of stress is defined by its tensor in a given set of axes, we are now able to resolve the tensor so as to determine either the vector of the stress on any arbitrary section at the point, or the scalar component of this vector in any arbitrary direction, and now we use this facility to investigate the chief properties of a given state.

Tensor Symmetry and Complementary Shears

Earlier we saw that the nine components of a stress tensor are thrice redundant because they are liable to three fixed relationships that are universally imposed by the laws of motion, and we now show that these relationships imply that any stress tensor must be symmetrical about its leading diagonal. To this end, consider the equilibrium of the forces generated by the stresses on the faces of an infinitesimal, cubical element of edge length a, isolated from a body in a state of homogeneous stress.

In Fig. 6.14 the several components of a stress tensor are shown in separate sets, though all the components do, of course, act together. Specifically Fig. 6.14a shows the direct stresses that act on all six faces, whilst each of the Figs. (b), (c) and (d) shows the 'belt' of shear stresses parallel to one of the coordinate planes.

Evidently the direct stresses which act on each pair of parallel faces generate a pair of equal but opposite forces which share the same line of action, and these satisfy the requirements of equilibrium in every respect.

Now consider the two pairs of shear components σ_{12} and σ_{21} indicated in Fig. 6.14b. Again the shears on each pair of parallel faces generate a pair of forces which are equal in magnitude, and opposite in direction, so again they automatically satisfy the requirements of force equilibrium. However, the forces of each pair do not share a common line of action, but are separated by the distance a, so each of the two pairs of forces generates a couple about an axis parallel to x_3. Similarly, the shear stresses σ_{13} and σ_{31} shown in Fig. 6.14c, generate a pair of couples about an axis parallel to x_2, whilst the shear stresses σ_{23} and σ_{32} shown in Fig. 6.14d generate a pair of couples about an axis parallel

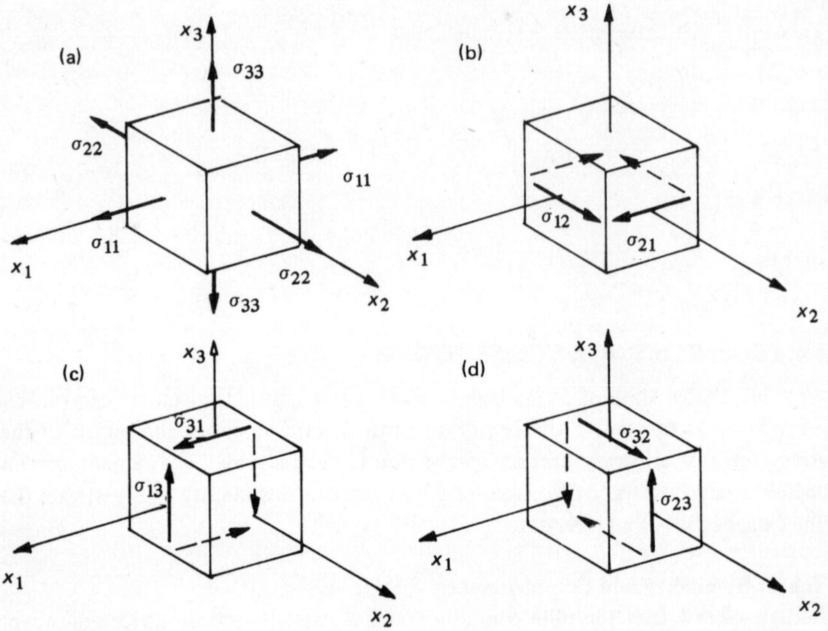

Fig. 6.14

to x_1, and equilibrium requires that the resultant of each pair of couples must be equal to zero. Thus, remembering that the force generated by a stress is equal to the stress times the area of the face on which it acts, and that the couple generated by a pair of equal but opposite forces is equal to one of the forces times the perpendicular distances between, we note that, with positive stresses, the couples generated by the two pairs of forces are opposed, and for moment equilibrium about x_1, we have:

$$a(\sigma_{12}\,a^2) = a(\sigma_{21}\,a^2)$$

or: $$\sigma_{12} = \sigma_{21}\,.$$

Similarly: $$\sigma_{13} = \sigma_{31}$$

and $$\sigma_{23} = \sigma_{32}\,.$$

We therefore conclude that any stress tensor is necessarily symmetrical about its leading diagonal, but more than this, reflection will show that $\hat{1}\cdot[\sigma]\cdot\hat{1}'$ must then be equal to $\hat{1}'\cdot[\sigma]\cdot\hat{1}$, that is, that the stress on the section normal to $\hat{1}$ resolved in the direction $\hat{1}'$ must be equal to the stress on the section normal to $\hat{1}'$ resolved in the direction $\hat{1}$, no matter what directions are defined by $\hat{1}$ and $\hat{1}'$.

The equal shears of any rectangular pair, such as σ_{12} and σ_{21}, or σ_{23} and x_{32}, are known as complementary shears, and we note that by the stress convention, the shears of a complementary pair are equal in sense as well as in magnitude, though by the moment convention they generate couples which are opposed.

Transformation of Tensor

A tensor, like a vector, can be described in terms of its components in any set of coordinates that proves convenient, so let $[\sigma]$ represent the tensor of a given state of stress in one set of axes $(x_1\,x_2\,x_3)$, and let $[\sigma']$ represent the tensor of the same state in terms of its components in an alternative set of axes $(x'_1\,x'_2\,x'_3)$, when the inclination of the alternative axes to the original is defined by the usual matrix of direction cosines by $[\mathbf{l}_{x'x}]$. The unit vectors $\mathbf{i}', \mathbf{j}'$ and $\mathbf{k}'$ parallel to x'_1, x'_2 and x'_3 are given in the first, second and third rows of $[\mathbf{l}_{x'x}]$, and we could evidently determine the components of $[\sigma']$ in terms of the components of $[\sigma]$ and $[\mathbf{l}_{x'x}]$ by the repeated application of the rule of resolution. For example, σ'_{12} is the component of stress on the section normal to x'_1, resolved in the direction of x'_2, and since the unit vectors in the directions of x'_1 and x'_2 are $\mathbf{i}'$ and $\mathbf{j}'$, it follows that σ'_{12} could be determined by pre-multiplying $[\sigma]$ by the first row of $[\mathbf{l}_{x'x}]$, and post-multiplying by a column vector equal to its second row. More generally, therefore, the term in the pth row and the $q$the column of $[\sigma']$ may be determined by pre-multiplying $[\sigma]$ by the pth row of $[\mathbf{l}_{x'x}]$, and post-multiplying by the transpose of its qth row, and it follows that the alternative tensor $[\sigma']$ as a whole can be determined by pre-multiplying $[\sigma]$ by the matrix $[\mathbf{l}_{x'x}]$, and post-multiplying by its transpose. In brief, the rule for the transformation of a tensor for a rotation of axes is:

$$[\sigma'] = [\mathbf{l}_{x'x}] \cdot [\sigma] \cdot [\mathbf{l}_{x'x}]^{\mathrm{T}} \tag{6.5}$$

where $[\mathbf{l}_{x'x}]$ represents the matrix of direction cosines which describes the orientation of the axes of $[\sigma']$ to the axes of $[\sigma]$.

Example 6.4

A tensor of a state of stress has the value $[\sigma]$ indicated below. Determine the tensor of the state in an alternative set of stress axes whose inclination to the axes of $[\sigma]$ is defined in the given matrix $[\mathbf{l}_{x'x}]$.

$$[\sigma] = \begin{bmatrix} 2 & 3 & 0 \\ 3 & -4 & 0 \\ 0 & 0 & 5 \end{bmatrix} ; \qquad [\mathbf{l}_{x'x}] = \frac{1}{\sqrt{6}} \begin{bmatrix} \sqrt{2} & \sqrt{2} & \sqrt{2} \\ -2 & 1 & 1 \\ 0 & -\sqrt{3} & \sqrt{3} \end{bmatrix} .$$

Solution

By the rule for the transformation of a tensor for a rotation of axes:

$$[\sigma'] = [\mathbf{l}_{x'x}]\,[\sigma]\,[\mathbf{l}_{x'x}]^{\mathrm{T}} = \frac{1}{6} \begin{bmatrix} \sqrt{2} & \sqrt{2} & \sqrt{2} \\ -2 & 1 & 1 \\ 0 & -\sqrt{3} & \sqrt{3} \end{bmatrix} \cdot \begin{bmatrix} 2 & 3 & 0 \\ 3 & -4 & 0 \\ 0 & 0 & 5 \end{bmatrix} \cdot \begin{bmatrix} \sqrt{2} & -2 & 0 \\ \sqrt{2} & 1 & -\sqrt{3} \\ \sqrt{2} & 1 & \sqrt{3} \end{bmatrix}$$

$$= \frac{1}{6}\begin{bmatrix} \sqrt{2} & \sqrt{2} & \sqrt{2} \\ -2 & 1 & 1 \\ 0 & -\sqrt{3} & \sqrt{3} \end{bmatrix} \cdot \begin{bmatrix} 5\sqrt{2} & -1 & -3\sqrt{3} \\ -\sqrt{2} & -10 & 4\sqrt{3} \\ 5\sqrt{2} & 5 & 5\sqrt{3} \end{bmatrix}$$

$$= \frac{1}{2}\begin{bmatrix} 6 & -2\sqrt{2} & 2\sqrt{6} \\ -2\sqrt{2} & -1 & 5\sqrt{3} \\ 2\sqrt{6} & 5\sqrt{3} & 1 \end{bmatrix}.$$

Principal Axes and Principal Stresses

In any state of stress, each axis is associated with a vector (the vector of the stress on the section normal to the axis), but the vector does not generally coincide with the axis itself. In general, therefore, when a state of stress is defined by specifying the vectors associated with an arbitrary set of rectangular axes, the vectors are inclined to the axes, and so have shear components, as well as normal components (Fig. 6.15a). But, if the stress axes are rotated, the inclinations of the stress vectors to the axes vary, and it can be shown that for every

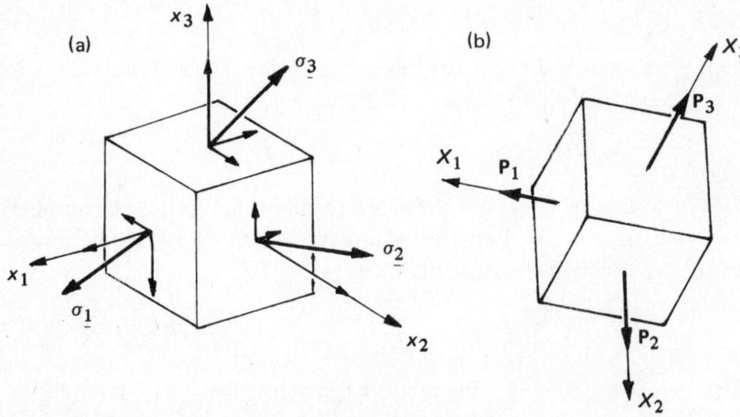

Fig. 6.15

state of stress there is one set of 'principal' axes for which each of the three stress vectors coincides with the axis normal to the section on which it acts (Fig. 6.15b), and the wholly normal stresses which act on the principal planes are called the principal stresses of the state. Symbolically, the principal axes are indicated by letters in capital case, i.e. by $(X_1\, X_2\, X_3)$, and to mark their special character, the magnitudes of the principal stresses which act in these directions are indicated, not by σ_{11}, σ_{22} and σ_{33}, but by p_1, p_2 and p_3 and since the principal tensor is the one for which all the shear components are zero simultaneously, it follows that such a tensor takes the form **[p]** indicated.

$$[\mathbf{p}] = \begin{bmatrix} p_1 & 0 & 0 \\ 0 & p_2 & 0 \\ 0 & 0 & p_3 \end{bmatrix}.$$

To prove the invariable existence of principal axes, we formulate the equation that would relate the value p of any principal stress to the components of a general tensor of the state, when the existence of real roots of p would imply

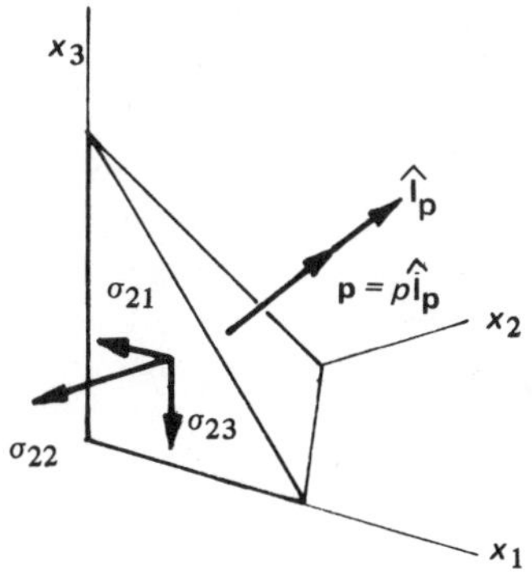

Fig. 6.16

the existence of real principal stresses. So supposing that a state of stress is defined by its tensor $[\sigma]$ in a general set of stress axes $(x_1 x_2 x_3)$, we assume the existence of a principal stress p which acts along the principal axes X whose inclination to the given stress axes is defined by the unit vector $\hat{\mathbf{l}}_p$. Then the total stress on the principal plane has the modulus p in the direction $\hat{\mathbf{l}}_p$ and it is therefore described vectorially by $p\,\hat{\mathbf{l}}_p$. But by resolving the given tensor $[\sigma]$ in the direction $\hat{\mathbf{l}}_p$, the value of this vector can alternatively by determined by the product $\hat{\mathbf{l}}_p \cdot [\sigma]$, and equating the difference of the alternative descriptions to zero, we have:

$$[\sigma] \cdot \hat{\mathbf{l}}_p - p\,\hat{\mathbf{l}}_p = 0 .$$

The second term of this equation can now be rewritten by making use of a special matrix $[\mathbf{I}]$ in which each term on the leading diagonal is unity whilst each off-diagonal term is zero. This is called a unit or identity matrix because its product with any (valid) matrix leaves the value of that matrix unaltered. For example, taking the product of $[\mathbf{I}]$ with the vectorial matrix $\hat{\mathbf{l}}_p$, we have:

$$[\mathbf{I}] = \begin{bmatrix} 1 & 0 & 0 \\ 0 & 1 & 0 \\ 0 & 0 & 1 \end{bmatrix} .$$

$$[\mathbf{I}] \cdot \hat{\mathbf{l}}_p = \begin{bmatrix} 1 & 0 & 0 \\ 0 & 1 & 0 \\ 0 & 0 & 1 \end{bmatrix} \cdot \begin{bmatrix} l_{p1} \\ l_{p2} \\ l_{p3} \end{bmatrix} = \begin{bmatrix} l_{p1} \\ l_{p2} \\ l_{p3} \end{bmatrix} = \hat{\mathbf{l}}_p .$$

Substituting accordingly gives:

$$[\sigma] \cdot \hat{\mathbf{l}}_p - p\,[\mathbf{I}] \cdot \hat{\mathbf{l}}_p = 0$$

or:

$$[[\sigma] - p\,[\mathbf{I}]] \cdot \hat{\mathbf{l}}_p = 0 .$$

Now in matrix algebra it is shown that this equation can be satisfied by non-trivial values of $\hat{\mathbf{l}}_p$ only if the determinant of $[[\sigma] - p\,[\mathbf{I}]]$ is zero, and it follows that the value of any principal stress p must be related to any general tensor of the state according to

$$|\,[\sigma] - p\,[\mathbf{I}]\,| = 0 .$$

This result can be expressed as a relationship between p and the components of $[\sigma]$ by expressing $[\sigma]$ and $[\mathbf{I}]$ in terms of their components, completing the subtraction and expanding the determinant of the resulting matrix. Thus:

$$[[\sigma] - p\,[\mathbf{I}]] = \begin{bmatrix} \sigma_{11} & \sigma_{12} & \sigma_{13} \\ \sigma_{21} & \sigma_{22} & \sigma_{23} \\ \sigma_{31} & \sigma_{32} & \sigma_{33} \end{bmatrix} - p \begin{bmatrix} 1 & 0 & 0 \\ 0 & 1 & 0 \\ 0 & 0 & 1 \end{bmatrix}$$

or:

$$[[\sigma] - p\,[\mathbf{I}]] = \begin{bmatrix} (\sigma_{11} - p) & \sigma_{12} & \sigma_{13} \\ \sigma_{21} & (\sigma_{22} - p) & \sigma_{23} \\ \sigma_{31} & \sigma_{32} & (\sigma_{33} - p) \end{bmatrix} .$$

Expanding the determinant of $[[\sigma] - p\,[\mathbf{I}]]$ and equating to zero then results in a cubic equation of the form:

$$p^3 - a_1 p^2 + a_2 p - a_3 = 0 \tag{6.6}$$

where:

$$a_1 = \sigma_{11} + \sigma_{22} + \sigma_{33}$$

$$a_2 = \sigma_{11}\sigma_{22} + \sigma_{11}\sigma_{33} + \sigma_{22}\sigma_{33} - \sigma_{12}^2 - \sigma_{13}^2 - \sigma_{23}^2$$

$$a_3 = \sigma_{11}\sigma_{22}\sigma_{33} + 2\,\sigma_{12}\sigma_{13}\sigma_{23} - \sigma_{11}\sigma_{23}^2 - \sigma_{22}\sigma_{13}^2 - \sigma_{33}\sigma_{12}^2 .$$

Alternatively, this relationship can be determined directly from eqn. (a), by again substituting for $[\sigma]$, $[\mathbf{I}]$ and $\hat{\mathbf{l}}_p$ in terms of their components, and by then eliminating l_{p1}, l_{p2} and l_{p3} from the resulting set of simultaneous equations.

Thus: $$[[\sigma] - p\,[\mathbf{I}]] \cdot \hat{\mathbf{l}} = 0 \,,$$

that is:
$$\begin{bmatrix} (\sigma_{11}-p) & \sigma_{12} & \sigma_{13} \\ \sigma_{22} & (\sigma_{22}-p) & \sigma_{23} \\ \sigma_{23} & \sigma_{32} & (\sigma_{33}-p) \end{bmatrix} \cdot \begin{bmatrix} l_{p1} \\ l_{p2} \\ l_{p3} \end{bmatrix} = 0$$

or:
$$\begin{aligned} (\sigma_{11}-p)l_{p1} + \sigma_{12}l_{p2} + \sigma_{13}l_{p3} &= 0 \\ \sigma_{21}l_{p1} + (\sigma_{22}-p)l_{p2} + \sigma_{23}l_{p3} &= 0 \\ \sigma_{31}l_{p1} + \sigma_{32}l_{p2} + (\sigma_{33}-p)l_{p3} &= 0 \;. \end{aligned}$$

Eliminating l_{p3} between the first and second equations gives:

$$\{\sigma_{23}(\sigma_{11}-p) - \sigma_{13}\sigma_{21}\}l_{p1} + \{\sigma_{23}\sigma_{12} - \sigma_{13}(\sigma_{22}-p)\}l_{p2} = 0$$

whence: $$\frac{l_{p1}}{l_{p2}} = \frac{\sigma_{12}\sigma_{23} - \sigma_{13}(\sigma_{22}-p)}{\sigma_{13}\sigma_{21} - \sigma_{23}(\sigma_{11}-p)} \;.$$

Similarly, the elimination of l_{p3} between the second and third equations gives:

$$\frac{l_{p1}}{l_{p2}} = \frac{(\sigma_{33}-p)(\sigma_{22}-p) - \sigma_{23}\,\sigma_{32}}{\sigma_{23}\sigma_{31} - \sigma_{12}(\sigma_{33}-p)} \,,$$

and by equating the two expressions for l_{p1}/l_{p2}, cross-multiplying, and collecting terms, we again arrive at eqn. (6.6).

The relationship between the value p of a possible principal stress and the components of any general tensor of the state is therefore seen to be described by a cubic equation in p, and we know that at least one of the three roots of any cubic equation must be real. We can therefore infer that in every state of stress there is at least one principal stress, though it can in fact be shown that for real states of stress all three of the roots of the cubic equation are real, and the directions of the three principal axes are rectangular. We may therefore summarise as follows.

In every state of stress there is a 'principal' set of rectangular stress axes for which each of the stress vectors coincides with the axis itself, so that all the shear terms of the principal tensor are zero simultaneously. The wholly normal stresses that act on the principal planes are called the principal stresses of the state.

Given a general tensor $[\sigma]$ of a state of stress, the values p of the principal stresses of the state are given by the roots of the cubic equation $|[\sigma] - p\,[\mathbf{I}]| = 0$, and for each principal stress, the unit vector $\hat{\mathbf{l}}_p$ that determines the direction of the corresponding principal axis is given by $[[\sigma] - [\mathbf{I}]] \cdot \hat{\mathbf{l}}_p = 0$. That is:

$$|[\sigma] - p\,[\mathbf{I}]| = 0 \tag{6.7}$$

and $$[[\sigma] - p\,[\mathbf{I}]] \cdot \hat{\mathbf{l}}_p = 0$$

Example 6.5

The tensor of a given state of stress in a given set of stress axes has the value $[\sigma]$ indicated.

$$[\sigma] = \begin{bmatrix} 6 & 2 & 2 \\ 2 & 5 & 0 \\ 2 & 0 & 7 \end{bmatrix}$$

Determine the principal stresses of the state, and the set of direction cosines which define the direction, relative to the axes of the given tensor, of the smallest.

Solution

The values of the principal stresses p, are the roots of the cubic equation:

$$\begin{vmatrix} (6-p) & 0 & 2 \\ 2 & (5-p) & 0 \\ 2 & 2 & (7-p) \end{vmatrix} = 0\ ,$$

that is: $(6-p)\,\{(5-p)\,(7-p)-0\} - 2\{2(7-p)-0\} + 2\{0-2(5-p)\} = 0$

or $(6-p)\,(5-p)\,(7-p) - 4(7-p) - 4(5-p) = 0$

that is: $(6-p\,(5-p)\,(7-p) - 8(6-p) = 0$

that is: $(6-p)\,\{p^2-12p+27\} = 0$

that is: $(6-p)\,(p-9)\,(p-3) = 0$

therfore $p = 3, 6$ and 9.

For each value of p:

$$\begin{bmatrix} (6-p) & 2 & 2 \\ 2 & (5-p) & 0 \\ 2 & 0 & (7-p) \end{bmatrix} \begin{bmatrix} l_{11} \\ l_{12} \\ l_{13} \end{bmatrix} = \begin{bmatrix} 0 \\ 0 \\ 0 \end{bmatrix}$$

When $p = 3$:

$$\begin{bmatrix} 3 & 2 & 2 \\ 2 & 2 & 0 \\ 2 & 0 & 4 \end{bmatrix} \begin{bmatrix} l_{11} \\ l_{12} \\ l_{13} \end{bmatrix} = \begin{bmatrix} 0 \\ 0 \\ 0 \end{bmatrix} \quad \begin{matrix} \text{(i)} \\ \text{(ii)} \\ \text{(iii)} \end{matrix}$$

From (ii) and (iii):

$$l_{12} = -l_{11} \quad \text{and} \quad l_{13} = -l_{11}/2 \ .$$

But $$l_{11}^2 + l_{12}^2 + l_{13}^2 = 1, \quad \text{that is: } l_{11}^2 + l_{11}^2 + \frac{1}{4} l_{11}^2 = 1 \ ,$$

therefore: $$l_{11} = \pm \frac{2}{3}; \quad l_{12} = \mp \frac{2}{3}; \quad l_{13} = \mp \frac{1}{3} \ .$$

Invariants

For purposes of analysis, the tensorial manner of describing a state of stress is evidently very convenient, yet in other respects it is awkward, chiefly because it does not readily reveal those parameters of the state that are particularly significant in determining its physical significance. To overcome this deficiency, we will show that the tensor of any state can be related to an alternative description which, though it is analytically awkward, is more revealing. But in the meantime we note that the difficulty with a tensor is aggravated by the fact that the tensor of any state varies with the orientation of the axes in which it is determined, so there is an obvious interest in invariant properties which are independent of the axes.

In this connection consider eqn. (6.6). This is a cubic in the principal stresses p of any state, whose coefficients are functions of the components of any general tensor of the state, so that if values of the coefficients are determined from any such tensor, then the roots of the equation will identify the values of the principal stresses of the state. Now for any given state these must comprise a unique set of values, yet the roots of the cubic can be unique only if its coefficients are unique, and it follows that, for any given state of stress, the functions a_1, a_2 and a_3 must have the same set of values, no matter from which tensor of the state they are determined. They are called the first, second and third variants of the state, and:

$$a_1 = \sigma_{11} + \sigma_{22} + \sigma_{33}$$

$$a_2 = \sigma_{11}\sigma_{22} + \sigma_{11}\sigma_{33} + \sigma_{22}\sigma_{33} - \sigma_{12}^2 - \sigma_{13}^2 - \sigma_{23}^2$$

$$a_3 = \sigma_{11}\sigma_{22}\sigma_{33} + 2\sigma_{12}\sigma_{13}\sigma_{23} - \sigma_{11}\sigma_{23}^2 - \sigma_{22}\sigma_{13}^2 - \sigma_{33}\sigma_{12}^2 \ .$$

We are immediately concerned with a_1, and with an additional quantity a_2', which is so defined as a function of a_1 and a_2, that it, also, is obviously an invariant.

The invariance of a_1 implies that the sum (or, of course, the average) of the three direct stresses that act on all sets of rectangular sections at any point have a common value. In other words, in all the tensors of any one state, the sum (or average) of the three direct stresses that lie on the leading diagonal have the same value.

The additional invariant a_2' is defined as follows:

$$a_2' = a_1^2 - 3a_2 \; .$$

It can be expressed in terms of the components of any general tensor of the state, simply by substituting appropriately for a_1 and a_2, when it is found that:

$$a_2' = (\sigma_{11}-\sigma_{22})^2 + (\sigma_{11}-\sigma_{33})^2 + (\sigma_{22}-\sigma_{33})^2 + 6(\sigma_{12}^2+\sigma_{13}^2+\sigma_{23}^2) \; .$$

Alternatively, a_2' can be described in terms of the principal stresses of the state by substituting p_1, p_2 and p_3 for σ_{11}, σ_{22} and σ_{33}, and by setting all the shear components to zero.

Thus: $$a_2' = (p_1-p_2)^2 + (p_1-p_3)^2 + (p_2-p_3)^2 \; .$$

6.7 OCTAHEDRAL STRESSES

In seeking an alternative to the tensorial description that will better reveal the physical significance of a state of stress, we are drawn by considerations of symmetry to the four 'octahedral' axes each of which is equally inclined to the three principal axes of the state. Thus Fig. 6.17 indicates a cubical element whose faces lie in the principal planes, at right-angles to the principal axes. The octahedral axes of the state then coincide with the body diagonals of the cube, and within the cube, Fig. 6.17 identifies the octahedron having a pair of faces normal to each of the four octahedral axes of the state.

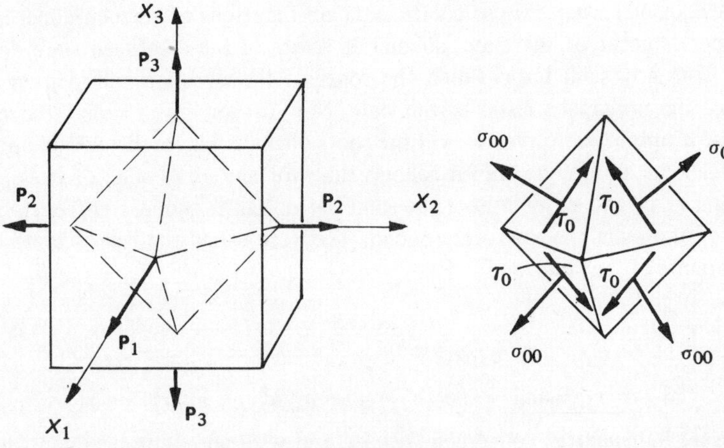

Fig. 6.17

Imagine the octahedron to be parted on each of the principal planes. This would isolate eight elements which, apart from their orientation in space, would be identical in every respect. It follows that the eight octahedral faces must all be subjected to the same value of direct stress, σ_{00}, and the same value of total shear stress τ_0, so that whereas we require six parameters to describe the nine components of stress that act on the faces of a rectangular element, to describe the direct stress and the magnitude (but not the direction) of the total shear stresses on the faces of the octahedral element we require only two. Moreover, the uniform direct stress and the uniform shear on the faces of the octahedron can be seen to correspond to what we recognised as the two basic modes of uniform dilation and uniform shear, and we can reasonably anticipate that the two parameters of the octahedral description may well prove as physically significant as they are evidently convenient.

To determine σ_{00} and τ_0 in terms of the components of a standard tensor, let $\hat{\mathbf{l}} = (l_1\ l_2\ l_3)$ represent the unit vector which lies along one of the octahedral axes in terms of its components in the principal axes. Though the senses of the direction cosines l_1, l_2 and l_3 vary from axis to axis, their magnitudes must be equal because, by definition, each octahedral axis is equally inclined to the principal axes. The squares of the three direction cosines must therefore, in each case, be equal, and since the sum of the squares is unity, it follows that for each axis we may write:

$$l_1^2 = l_2^2 = l_3^2 = 1/3 \ .$$

Resolving the principal tensor to determine the total stress σ_0 and the direct stress σ_{00} on the octahedral sections, we have

$$\sigma_0 = \sigma_l = \hat{\mathbf{l}} \cdot [\mathrm{p}] \ ,$$

that is: $$\sigma_0 = [l_1 \quad l_2 \quad l_3] \cdot \begin{bmatrix} p_1 & 0 & 0 \\ 0 & p_2 & 0 \\ 0 & 0 & p_3 \end{bmatrix}$$

or: $$\sigma_0 = l_1p_1 + l_2p_2 + l_3p_3 \ .$$

Then: $$\sigma_{00} = \sigma_0 \cdot \hat{\mathbf{l}}$$

that is: $$\sigma_{00} = [l_1p_1 \quad l_2p_2 \quad l_3p_3] \cdot \begin{bmatrix} l_1 \\ l_2 \\ l_3 \end{bmatrix} ,$$

or: $$\sigma_{00} = l_1^2p_1 \quad l_2^2p_2 \quad l_3^2p_3 \ .$$

But $l_1^2 = l_2^2 = l_3^2 = 1/3$, and substituting accordingly:

$$\sigma_{00} = (p_1 + p_2 + p_3)/3 \ .$$

But $(p_1 + p_2 + p_3)$ is the invariant sum of the direct stresses of any tensor of the state, and we may therefore write:

$$\sigma_{00} = (\sigma_{11} + \sigma_{22} + \sigma_{33})/3 \ .$$

To determine the total octahedral shear stress τ_0, we equate its square to the difference in the squares of the total stress σ_0, and the direct stress σ_{00}.

Thus: $\tau_0^2 = |\sigma_0|^2 - \sigma_{00}^2$.

But: $|\sigma_0|^2 = (p_1 l_1)^2 + (p_2 l_2)^2 + (p_3 l_3)^2$,

and again substituting $l_1^2 = l_2^2 = l_3^2 = 1/3$, we have

$$|\sigma_0|^2 = (p_1^2 + p_2^2 + p_3^2)/3 \ .$$

Therefore: $\tau_0^2 = (p_1^2 + p_2^2 + p_3^2)/3 - (p_1 + p_2 + p_3)^2/9$

whence: $\tau_0 = \dfrac{1}{3}\sqrt{((p_1 - p_2)^2 + (p_1 - p_3)^2 + (p_2 - p_3)^2)}$.

But the quantity under the square root can be recognised as the invariant a_2', and expressing this in terms of the components of a general tensor, we have:

$$\tau_0 = \frac{1}{3}\sqrt{((\sigma_{11} - \sigma_{22})^2 + (\sigma_{11} - \sigma_{33})^2 + (\sigma_{22} - \sigma_{33})^2 + 6(\sigma_{12}^2 + \sigma_{13}^2 + \sigma_{23}^2))} \ .$$

To summarise, in any state of stress, the four octahedral sections which are equally inclined to the three principal axes are subjected to the same value of direct stress σ_{00}, and the same value of total shear stress τ_0, and these can be described in terms of the components of any general tensor of the state as follows:

$$\sigma_{00} = \frac{1}{3}(\sigma_{11} + \sigma_{22} + \sigma_{33})$$

$$\tau_0 = \frac{1}{3}\sqrt{((\sigma_{11} - \sigma_{22})^2 + (\sigma_{11} - \sigma_{33})^2 + (\sigma_{22} - \sigma_{33})^2 + 6(\sigma_{12}^2 + \sigma_{13}^2 + \sigma_{33}^2))} \ . \qquad (6.8)$$

6.8 THE HYDROSTATIC AND DEVIATORIC TENSORS

Since common materials can usually be treated as isotropic, there are many situations in which we are concerned only with the magnitudes of the stresses of a state, and in this case the foregoing relationships allow a tensorial description of the state to be assessed in the two parameters of the octahedral mode. However, in other situations it is necessary to take account of the directions of the octahedral shears; and to avoid analysis in the non-independent octahedral axes, we now show that any total tensor of a given state of stress can readily be parted into two, such that one tensor describes the effect of the uniform direct stresses on the octahedral planes, whilst the other describes the effect of the uniform octahedral shears, and this allows the octahedral interpretation to be analysed in the convenient tensorial manner.

Hydrostatic Stress

The ease with which the total tensor can be suitably partitioned depends upon the spherical symmetry of a particular state of stress which is known as a hydrostatic state, for this is a state whose tensor is not only of particularly simple form, but is also invariant to the orientation of the stress axes. The concept stems from fluid mechanics where a fluid is defined as a substance in which any level of shear causes progressive deformation which continues to develop for so long as the stress is applied. It follows that a fluid will remain at rest only when shears are entirely absent, and by analogy, any such state is said to be hydrostatic, irrespective of the nature of the material.

To investigate the case, consider a state in which the three principal stresses have a common value p, so that its principal tensor can conveniently be described in terms of the unit matrix [I] as follows:

$$[\mathrm{p}] = \begin{bmatrix} p & 0 & 0 \\ 0 & p & 0 \\ 0 & 0 & p \end{bmatrix} = p\,[\mathrm{I}] \ .$$

Then if $\hat{\imath}$ and $\hat{\imath}'$ are arbitrary unit vectors, the component of stress on the arbitrary section normal to $\hat{\imath}$, resolved in the arbitrary direction of $\hat{\imath}'$, is given by:

$$\sigma_{\hat{\imath}\hat{\imath}'} = \hat{\imath} \cdot \underline{\mathrm{p}} \cdot \hat{\imath}' = \hat{\imath} \cdot p\,[\mathrm{I}] \cdot \hat{\imath}' \ ,$$

and since p is a scalar, and $[\underline{\mathrm{I}}] \cdot \hat{\imath}' = \hat{\imath}'$, we may write

$$\sigma_{\hat{\imath}\hat{\imath}'} = p\,\hat{\imath} \cdot \hat{\imath}' \ .$$

Now irrespective of its direction in the section, the direction of resolution of a shear stress is always at right-angles to the normal to the section, and when $\hat{\imath}$ and $\hat{\imath}'$ are rectangular, their scalar product is zero. On the other hand, for the direct stress the direction of resolution coincides with the normal to the section, and when $\hat{\imath}$ and $\hat{\imath}'$ coincide, the scalar product is equal to unity. A state of stress in which the three principal stresses are equal is therefore hydrostatic in that every section at the point is entirely free of shear, and the case is spherically symmetrical in that every section is a principal section, and the principal stress has a common value on them all.

The Hydrostatic and Deviatoric Tensors

It follows from the spherical symmetry of the state that the faces of any infinitesimal element at a position of hydrostatic stress would be uniformly free of shear, and subjected throughout to a uniform value of direct stress, irrespective of either the shape of the element or its orientation. But for any state of stress, the faces of the octahedral element are subjected to a uniform direct stress σ_{00} with a uniform shear τ_0. Any state of stress may therefore be considered to comprise two essentially different effects. One is a spherically symmetrical hydrostatic stress of magnitude σ_{00}; the other is represented by a uniform shear of magnitude τ_0 on the four octahedral sections.

Finally, the value of σ_{00} for any state of stress can be equated to the invariant average of the direct stresses of any tensor of the state, and a hydrostatic state of stress of magnitude σ_{00} can be described by the invariant tensor σ_{00} [I] irrespective of the orientation of the axes. So suppose that a general state of stress is defined by its tensor $[\sigma]$ for an arbitrary set of axes. Then the tensor of the hydrostatic stress of the state in the same axes is described by σ_{00} [I], where σ_{00} is equal to the average of the direct stresses of the given tensor $[\sigma]$, and it follows that the difference $[\sigma] - \sigma_{00}$ [I] must be the description, in tensorial form, of the effect which appears in the octahedral mode as the uniform shears on the octahedral sections. This is called a deviatoric tensor of the state, and we may therefore summarise as follows.

i) In each state of stress there is a principal set of rectangular stress axes for which the vector of the total stress on each of the coordinate planes coincides with the normal to the plane. The principal tensor of a state is therefore the tensor for those axes for which all the shear components are zero simultaneously.

ii) The four axes which are equally inclined to the three principal axes are called the octahedral axes of the state. It follows from symmetry that the four octahedral sections normal to the octahedral axes are subjected to a common value of direct stress represented by σ_{00}, and to a common value of total shear stress represented by τ_0.

iii) The values of σ_{00} and τ_0 for any state of stress are invariants of the tensors of the state. This means that each can be described as a specific function of the components of any tensor of the state, irrespective of the orientation of the axes. In particular, the value of σ_{00} can be equated to the invariant average of the three direct stresses of any tensor of the state.

iv) A hydrostatic state of stress is a spherically symmetrical special case in which every section is a principal section, and in which all the sections are subjected to the same value of direct, principal stress.

v) The octahedral mode reveals any state of stress as a combination of two effects of essentially different kinds. One is the spherically symmetrical, hydrostatic stress of magnitude σ_{00}, whilst the other is described by uniform shears of magnitude τ_0 on the octahedral sections.

vi) Irrespective of the orientation of the axes, the hydrostatic stress of any state is described as a tensor by the invariant σ_{00} [I].

vii) If $[\sigma]$ represents any general tensor of a given state, the difference $[\sigma] - \sigma_{00}$ [I] is called a deviatoric tensor of the state, and this is a description, in tensorial form, of the effect which appears as the uniform shears τ_0 on the octahedral sections.

6.9 MOHR'S CIRCULAR DIAGRAM OF STRESS

Provided that the description of a state of stress is given in the standard form of a tensor, we are now able to determine for the point, both the vector of the stress on any arbitrary section, and its scalar component in any arbitrary direction, and we have used this facility both to determine the chief characteristics of the tensor, and to reveal its features in either the principal or the octahedral mode. However, we are not yet able to deal with descriptions which are presented in non-standard forms, and it is chiefly for this purpose that we now introduce a graphical presentation of a state of stress. In this, the total shear stress τ on each section at the point is plotted against the direct stress σ, when all the points define a domain whose form, size and location give an immediate indication of the general characteristics of the state of stress at the point (Fig. 6.18). But more, it can be shown that if the sections are taken in appropriate sets, the points for each set take the particularly convenient form of a circular arc, and thus presented, the graph, which is known as Mohr's Circular Diagram of Stress, is of considerable analytical utility.

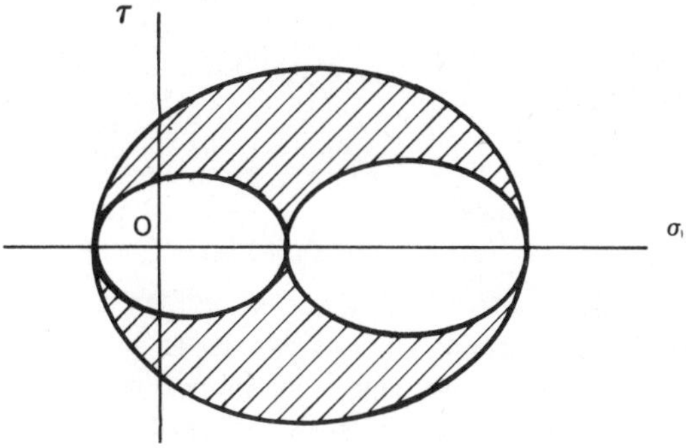

Fig. 6.18

In the sets of interest, each section bears a common inclination to one of the principal axes (X_1, X_2, X_3) of the state. Such a set may be visualised by imagining a cone whose axis coincides with one of the principal axes, say X_3, as indicated in Fig. 6.19. Then it is obvious, first, that a plane which rolls around the surface of the cone defines a set of sections which bear a common inclination to X_3, and second, that if the directions of the sections are defined by describing their unit normal vectors $\hat{l}$ in terms of their components $(l_1\ l_2\ l_3)$ in

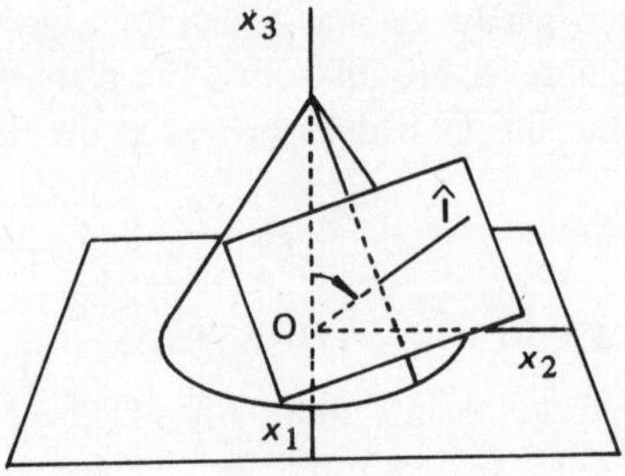

Fig. 6.19

the principal axes, then l_3 is equal to the cosine of the angles which the normals make with X_3, and for the set of sections which bear a common inclination to X_3, this has a common value. Similarly, for sets of sections which bear a common inclination to X_2 or X_1, it is l_2 or l_1 respectively which has a common value.

A limiting case of particular interest is that in which the ratio of the height of the cone to its base tends to infinity, when the cone becomes a cylinder (Fig. 6.20). The plane which rolls around the cylinder then identifies a set of

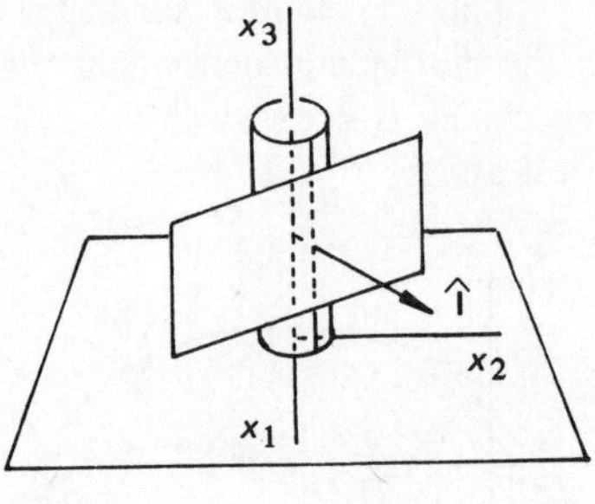

Fig. 6.20

planes which are parallel to the axis, and since their normals are at 90° to the axis, it follows that the sets of sections parallel to X_3, or X_2 or X_1, are the sets for which l_3 or l_2 or l_1 respectively has a common value of zero.

So supposing that the state of stress at a point is defined by its principal tensor [p] in its principal axes (X_1, X_2, X_3), let ($l_1\, l_2\, l_3$) represent the scalar components (in the same axes) of the unit vector $\hat{1}$ normal to an arbitrary section. The vector of the stress on the section is given by

$$\sigma_l = \hat{1} \cdot \underline{p} = [l_1 \quad l_2 \quad l_3] \cdot \begin{bmatrix} p_1 & 0 & 0 \\ 0 & p_2 & 0 \\ 0 & 0 & p_3 \end{bmatrix} ,$$

whence $\quad \sigma_l = [l_1 p_1 \quad l_2 p_2 \quad l_3 p_3] ,$

and equating the square of the vector to the sum of the squares of its scalar components gives

$$|\vec{\sigma_l}|^2 = l_1^2 p_1^2 + l_2^2 p_2^2 + l_3^2 p_3^2 \ .$$

The direct stress σ on the section is then given by

$$\sigma = \sigma_{ll} = \mathbf{1} \cdot \sigma_l = l_1^2 p_1 + l_2^2 p_2 + l_3^2 p_3 \ ,$$

and although its direction remains unknown, the modulus of the total shear stress τ may be determined by equating the square of the stress vector to the sum of the squares of its direct and total shear components, as follows:

$$\sigma^2 + |\tau|^2 = \sigma_l^2 \ .$$

This relationship between σ and $|\tau|$ can be modified by adding equally to both sides as follows

$$\left(\sigma - \frac{p_1 + p_2}{2}\right)^2 + |\tau|^2 = \sigma_l^2 - \sigma(p_1 + p_2) + \left(\frac{p_1 + p_2}{2}\right)^2 ,$$

and substituting for σ_l and σ, and rearranging, gives

$$\left(\sigma - \frac{p_1 + p_2}{2}\right)^2 + |\tau|^2 = l_3^2 \left\{p_3^2 + p_1 p_2 - p_3(p_1 + p_2)\right\} + \left(\frac{p_1 + p_2}{2}\right)^2 .$$

Now for a given state of stress the principal stresses p_1, p_2 and p_3 have a specific set of values, and for any set of sections for which l_3 has a common value, the relationship between σ and $|\tau|$ therefore has the form:

$$(\sigma - a)^2 + |\tau|^2 = b^2$$

where $a = (p_1 + p_2)/2 = \text{const.},$

and $b = \sqrt{\left(l_3^2 \left\{p_3^2 + p_1 p_2 - p_3(p_1 + p_2)\right\} + \left(\frac{p_1 + p_2}{2}\right)^2\right)} = \text{Const.}$

But this can be recognised as the equation of a circle whose radius b varies with the value of l_3 for the set, but whose centre falls on the axis of σ at the position $\sigma = (p_1 + p_2)/2$ which is independent of the value of l_3, and since parallel conclusions hold for sets of sections which have a common value of either l_2 or l_1, we may conclude as follows.

In a diagram of a state of stress in which the total shear stress τ is plotted against the direct stress σ, the graph for each set of sections which bear a common inclination to any one of the three principal axes is a circular arc whose centre falls at a specific point on the axis of σ, but whose radius varies with the inclination. For the sets of sections which bear a common inclination to X_3, and which therefore have a common value of l_3, the common centre falls at $\sigma = (p_1 + p_2)/2$, whilst for the sets which have a common value of l_2 and l_1, the centres fall at $\sigma = (p_1 + p_3)/2$, and at $\sigma = (p_2 + p_3)/2$ respectively, as shown in Fig. 6.21.

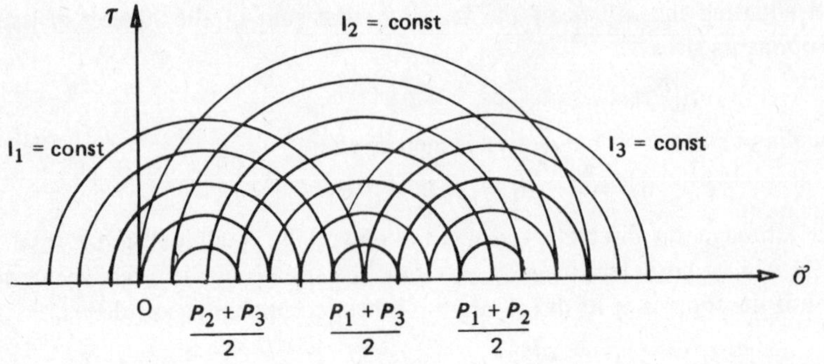

Fig. 6.21

However, each of the graphs is valid only over the arc that falls within the domain of the particular state of stress, and to determine the bounds of the domain, we first consider the graphs for the sets of sections that are parallel to one of the principal axes. Thus, the set of sections for which l_3 has a common value of zero is the set parallel to X_3, and these include the two principal sections on which the direct stress has the values p_1 and p_2, whilst the shear stress τ is zero. The circle for the set of sections for which l_3 is zero will therefore cut the axis of σ at $\sigma = p_1$ and $\sigma = p_2$, and by parallel arguments, the circle for $l_2 = 0$ will cut the axis at $\sigma = p_1$ and $\sigma = p_3$, whilst the circle for $l_1 = 0$ will cut at $\sigma = p_2$ and $\sigma = p_3$, as shown in Fig. 6.22.

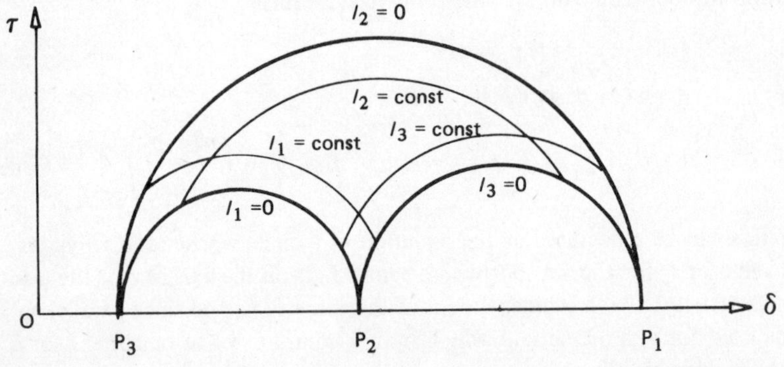

Fig. 6.22

Now consider the graph for a set of sections for which, say, l_3 has an arbitrary common value. Each point on the graph indicates the stresses on a section for which the values of l_1 and l_2 vary from point to point, but the variation must be such that the sum of the squares of l_1, l_2 and l_3 retains a constant unit value, when it follows that l_1 will have its largest value when l_2 is zero, whilst l_2 will have its greatest value when l_1 is zero. Furthermore, parallel conclusions will hold for the sets of sections of which l_2 and l_1 have a common value, and we conclude that the circular graphs are valid only over the arcs which fall within the domain bounded by the three basic circles for $l_3 = 0, l_2 = 0$ and $l_1 = 0$ (Fig. 6.22).

By thus defining the bounds of the (σ, τ) domain of a given state of stress, the three basic circles are sufficient of themselves to reveal a number of significant features of the state. For example, it is obvious by inspection: that of all the direct stresses of any state, one of the principal stresses is the algebraic greatest whilst another is the algebraic least; that the greatest shear stress of the state is determined by the radius of the largest of the three circles, and that this is equal to half the greatest difference between any two of the principal stresses; and (since a variation in the hydrostatic stress affects all the principal stresses equally) that the position of the circles along the σ axis indicates the hydrostatic stress of the state. However, to determine the basic circles from a non-standard description of the state, it is necessary to associate each point in the diagram with the section to which it applies, and in this, as so far presented, the diagram is somewhat deficient. In essence, the deficiency stems from a lack of compatibility with the tensorial conventions, for it follows from the symmetry of a stress tensor that, special cases apart, the stresses on one section are repeated on another, and as a result, each point on the diagram corresponds, not to a single section, but to a pair which are symmetrical about the principal sections. However, this deficiency can be overcome by adopting a sign convention which distinguishes between the senses of the shear stresses which act on each such pair, so in using Mohr's circle we adopt a special sign convention which is reserved solely for this purpose, and which differs essentially from the tensorial convention only by distinguishing between the senses of a pair of complementary shears.

Referring to Fig. 6.23, consider, for example, the set of sections parallel to X_3, for which l_3 has a common value of zero. These include the principal sections on which p_1 and p_2 act, and on which the shear stress is zero in every direction.

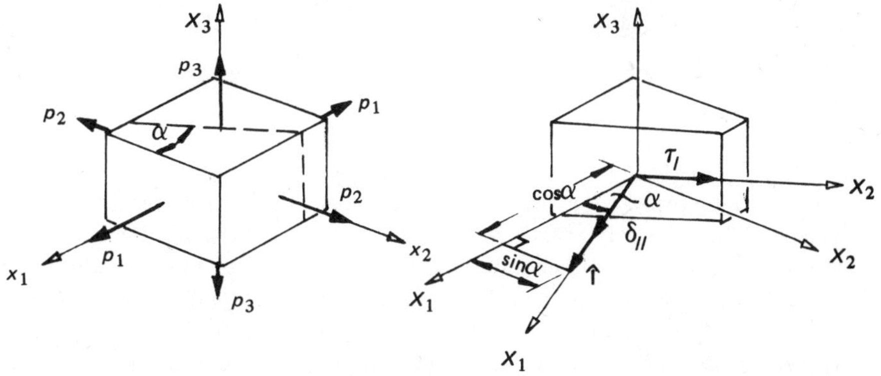

Fig. 6.23

But any of the set is normal to the principal section on which p_3 acts, and it follows from the equality of complementary shears, that the component of shear in the X_3 direction is zero on all the sections. Thus, for the set of sections parallel to X_3, both the direct stress σ and the total shear stress τ are uniformly parallel to the X_1X_2 plane, and it follows, that even though every state of stress is necessarily three-dimensional, the analysis of the stresses on the set of sections parallel to X_3 can be treated as a problem in only the two dimensions of the X_1X_2 plane.

To define a sign convention for the application of Mohr's circle to such a set, the circular diagram is associated with a physical diagram which views the set of sections edge on, by looking, in either sense, directly along the principal axis to which the sections are parallel (Fig. 6.24). Then each section such as AA or BB exposes a pair of faces, and, as in the tensor convention, the direct stress on either is reckoned positive in the sense of the outward-facing normal. A shear stress is reckoned positive in the direction for which the sense of rotation from $+ \sigma$ to $+ \tau$ in the physical diagram is opposed to that in the circular diagram. Here, in the circular diagram, $+ \tau$ will be chosen $\pi/2$ counterclockwise (CCW) from $+ \sigma$, so in the physical diagram, τ will be reckoned positive in the direction $\pi/2$ clockwise (CW) from the outward facing normal.

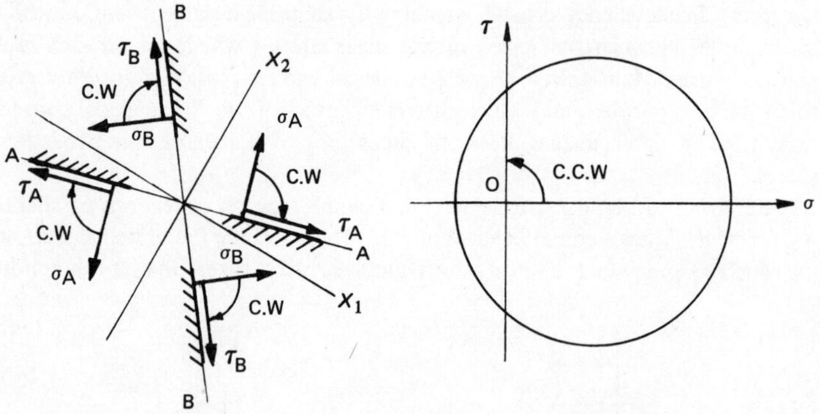

Fig. 6.24

Let X_3 represent the principal axis to which the sections of the set are parallel. Then the set includes the principal sections on which the principal stresses p_1 and p_2 act, and choosing the section on which p_1 acts as an arbitrary datum, let α represent the magnitude and sense of the angle from the datum to an arbitrary member of the set. Such a section is shown in Fig. 6.25, where $\hat{\mathbf{l}}$ represents one of the two unit vectors normal to the section, whilst $\hat{\mathbf{l}}'$ is the unit vector $\pi/2$ clockwise from $\hat{\mathbf{l}}$.

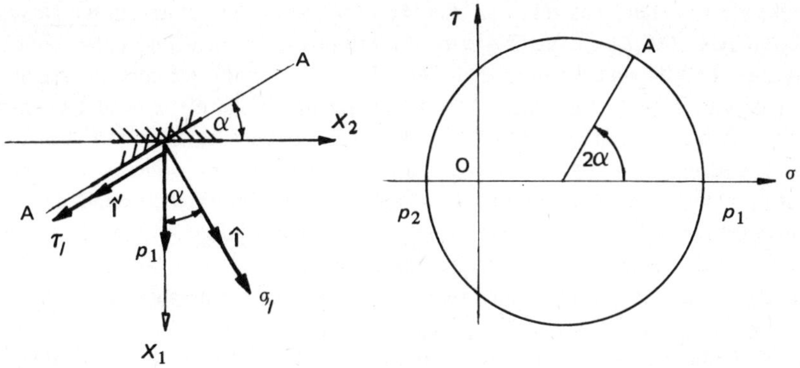

Fig. 6.25

Evidently, the components of $\hat{\imath}$ and $\hat{\imath}'$ in the principal axis can be described in terms of α as follows:

$$\hat{\imath} = [\cos\alpha \quad \sin\alpha \quad 0] \ ,$$

and $$\hat{\imath}' = [\sin\alpha \ -\cos\alpha \quad 0] \ ,$$

and $\hat{\imath}$ being treated as the outward-facing normal to one of the faces exposed by the arbitrary section, the direct and total shear stresses on the face can be determined from the principal tensor $[\mathbf{p}]$ of the state as follows:

$$\sigma_l = \hat{\imath} \cdot [\mathbf{p}] = [\cos\alpha \quad \sin\alpha \quad 0] \begin{bmatrix} p_1 & 0 & 0 \\ 0 & p_2 & 0 \\ 0 & 0 & p_3 \end{bmatrix} ,$$

$$= p_1 \cos\alpha + p_2 \sin\alpha \ .$$

Therefore $\sigma = \sigma_{ll} = \sigma_l \cdot \hat{\imath} = p_1\cos^2\alpha + p_2\sin^2\alpha \ ,$

and $\tau = \sigma_{ll'} = \sigma_l \cdot \hat{\imath}' = (p_1 - p_2)\cos\alpha\sin\alpha \ .$

But $2\cos^2\alpha = 1 + \cos 2\alpha \ ,$

and $2\sin^2\alpha = 1 - \cos 2\alpha \ ,$

and $2\cos\alpha\sin\alpha = \sin 2\alpha \ ,$

and substituting accordingly gives

$$\left.\begin{aligned} \sigma &= \frac{p_1 + p_2}{2} + \frac{p_1 - p_2}{2}\cos 2\alpha \\ \text{and} \qquad \tau &= \frac{p_1 - p_2}{2}\sin 2\alpha \end{aligned}\right\} .$$

These are the equations that describe the variation of the stresses on the general section AA with its inclination α to the principal section whose (σ, τ) coordinates are $(p_1, 0)$, and we note that since the circle for the set cuts the σ axis at $\sigma = p_1$ and $\sigma = p_2$, the term $(p_1 + p_2)/2$ determines the distance of its centre along the axis, whilst $(p_1 - p_2)/2$ determines its radius. Reference to Fig. 6.25 will therefore show that the equations determine that the point on the circle that corresponds to the section AA falls at A, where the angle subtended at the centre by the arc drawn from the point $(p_1, 0)$ is equal to 2α, and reflection will show that this observation can be generalised as follows.

If AA and BB identify any two sections of the set parallel to one of the principal axes, and if A and B represent the corresponding points on the Mohr's circle for the set, while C represents the centre of the circle, the magnitude and sense of the angle from CA to CB is twice the angle from section AA to section BB (Fig. 6.26).

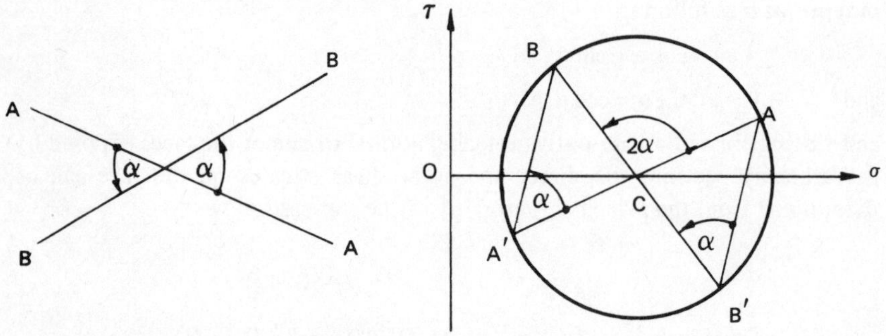

Fig. 6.26

Plane Stress

A special case of particular interest is one which is idealised as a plane lamina of uniform thickness in which there is no effect which tends to distort the lamina out of its plane. This requires that the faces of the lamina be free of shear so that the plane of the lamina is a principal plane, that the direct stresses on the faces be uniform on each and equal on the two, and that all the 'external' forces and stresses act in the plane of the lamina. This case is convenient because it can be analysed as a problem in only two dimensions, but it is important to remember that every state of stress is necessarily three-dimensional, and in assessing such a state it is important to take account of the third principal stress in the direction at right-angles to the plane of the analysis, even though its value is zero. The case is important because it is a reasonable idealisation of many practical problems involving flat plates or, indeed, any 'free', unloaded surface which is sensibly plane.

Example 6.6

The diagram indicates the stresses at a point O on a pair of sections normal to a 'free', unloaded surface. For the point, determine the principal stresses, the

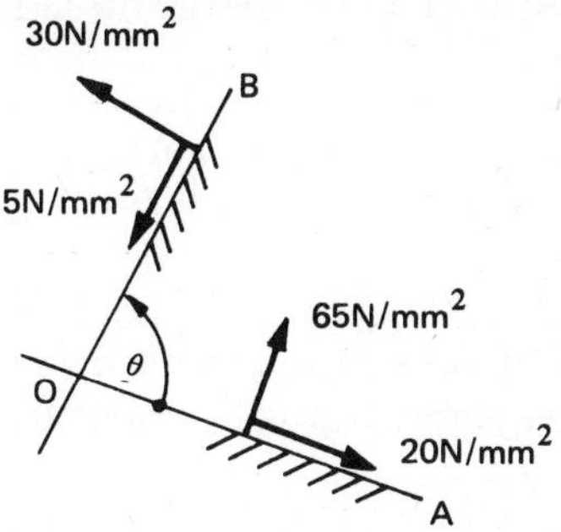

greatest shear stress, and the angle θ, and indicate the locations of the principal sections and the stresses which act on them in a reproduction of the given diagram.

Solution

By the convention for Mohr's Circle, the (σ, τ) coordinates for the sections A and B are:

$$A = (65, 20); \quad B = (30, -15) .$$

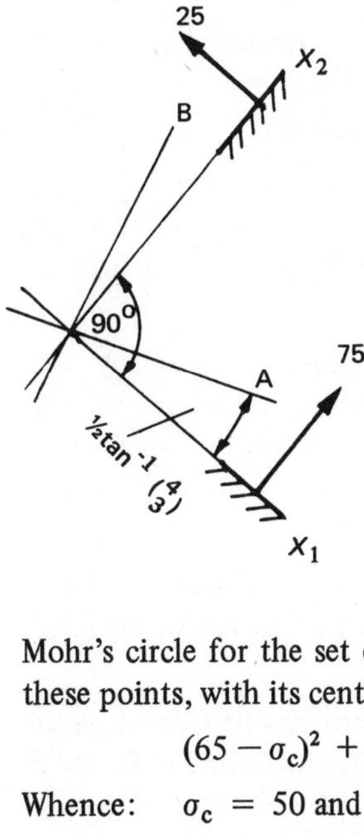

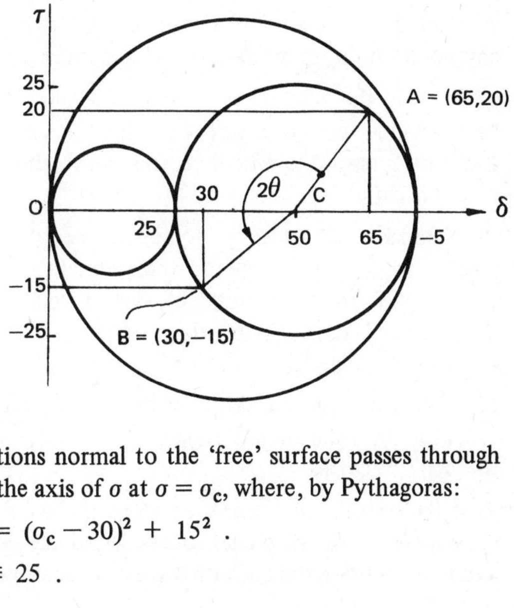

Mohr's circle for the set of sections normal to the 'free' surface passes through these points, with its centre on the axis of σ at $\sigma = \sigma_c$, where, by Pythagoras:

$$(65 - \sigma_c)^2 + 20^2 = (\sigma_c - 30)^2 + 15^2 .$$

Whence: $\sigma_c = 50$ and $AC \equiv 25$.

The sections normal to the free surface therefore include the principal sections on which the principal stresses have the values $p = 50 \pm 25$, whence $p_1 = 75$ and $p_2 = 25$ N/mm^2. The third principal section is the free surface on which $p_3 = 0$.

Since the inclination of CB to CA is twice the inclination of B to the section A,

$$\theta = \left(\pi + \tan^{-1}\frac{3}{4} - \tan^{-1}\frac{4}{3}\right)/2 \ .$$

Example 6.7

The figure indicates the direct stresses only for a set of three sections normal to a 'free' surface. The shear stresses are unknown. Determine the principal stresses of the state.

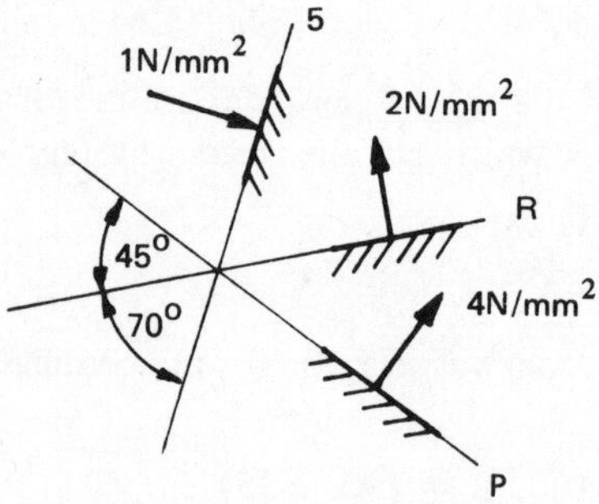

Solution

Mohr's Circle can be drawn from the given data by remembering that the angle subtended by a chord of a circle at the centre is twice the angle subtended at any point on the circumference. The method is then as follows:

a) Treating the zero of τ as unknown, choose a zero of σ, and draw the ordinates through the given values of direct stress. (We know that the points in the graph of τ against σ which correspond to the sections P, Q and R must lie on these ordinates).

b) Choosing an arbitrary point on one of the ordinates (it is usually convenient to choose a point on the ordinate of intermediate value, in this case the point A$'$ on the R ordinate), draw through it the lines A$'$P$'$ and A$'$Q$'$ which indicate the relative inclinations of the given sections, but with A$'$R$'$ aligned along the R ordinate.

c) Extend A$'$P$'$ to cut the P ordinate in P; extend A$'$Q$'$ to cut the Q ordinate in Q; and draw the circle through A$'$, P and Q, and draw the abscissa of the graph through its centre.

d) If the point A$'$ is transferred along the ordinate to the opposite point on the circumference R, the angles subtended at the centre, by the points P, Q and R will then be twice the angles between the sections P, Q and R.

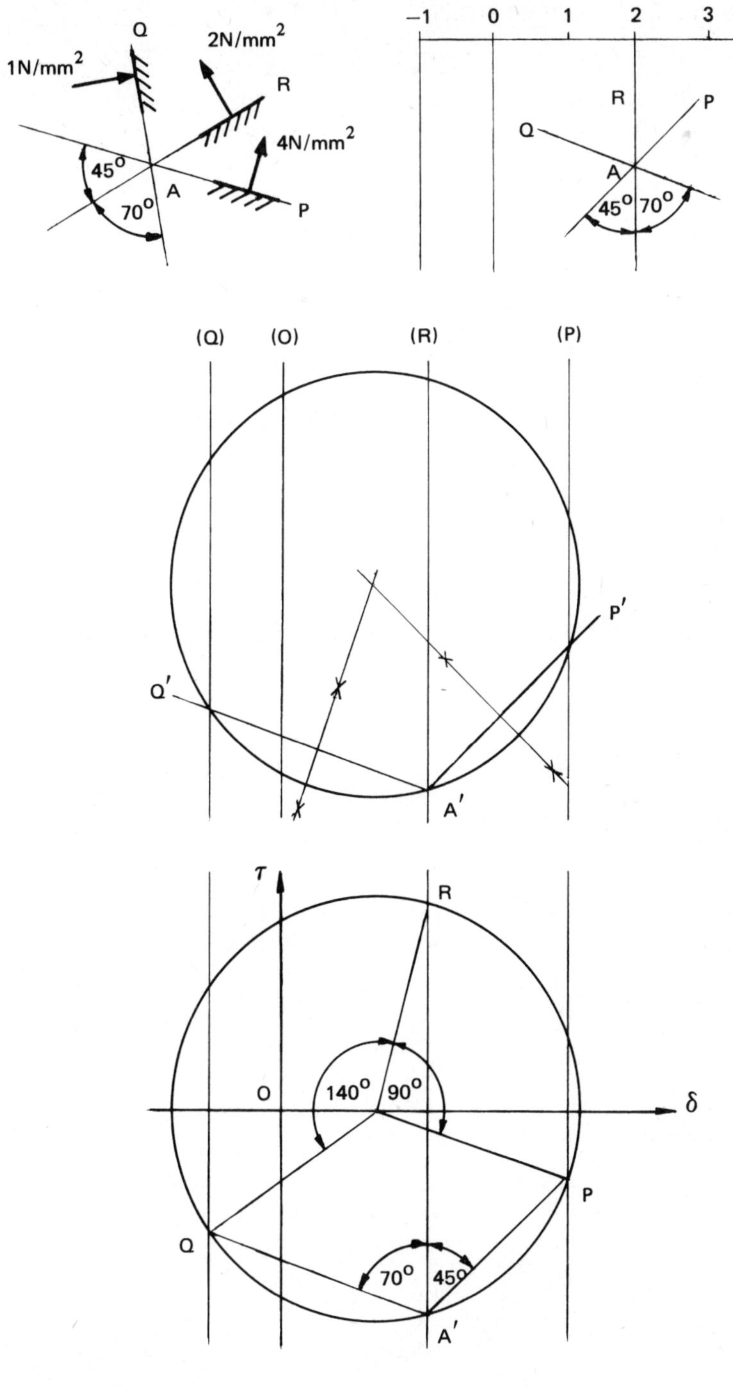
Q
2N/mm²
1N/mm²
R
4N/mm²
45°
A
70°
P
−1
0
1
2
3
4
R
P
Q
A
45°
70°
(Q)
(O)
(R)
(P)
P′
Q′
A′
τ
R
O
140°
90°
δ
P
Q
70°
45°
A′

Thus, by drawing to scale, we obtain:

$$p_1 = 4{\cdot}10$$

$$p_2 = -1{\cdot}6$$

$$p_3 = \quad 0 \text{ (given)}.$$

7

Continuum Mechanics II – Strain

7.1 INTRODUCTION

In Stress Resolution we were essentially concerned only with the different stresses of a single state, and we have yet to consider the methods of Stress Analysis and Fluid Mechanics by which we determine how the state of stress varies from point to point. But it is a matter of common experience, that such problems are not generally statically determinate. Rather, their solution must ensure, not only that the stresses on each element are consistent with the laws of motion, but also that the associated deformations of the elements are consistent with the physical continuity of the body as a whole, and since the effect of a given state of stress may vary from material to material, it is necessary to characterise the properties of the various materials in 'constitutive equations' which relate the states of stress to the patterns of deformation to which they give rise. Moreover, in practice, our direct concern is as much with questions of deformation and flexibility as of strength. On both these counts it is therefore necessary now to consider the description of the pattern of deformation at a point, in much the same way as we have already considered the description of the state of stress, and we approach this question by first considering the relative displacements between the points of the body.

7.2 POINT DISPLACEMENTS IN A CONTINUUM

In so doing, we shall assume throughout, that the body of interest can be treated as a continuum of infinitesimal 'point' elements, whose essential continuity is not impaired by its deformation.

The Continuum Assumption

Thus let P represent a particular arbitrary point in a body, let Q represent an adjacent point at a variable position adjacent to P, and supposing the body to be deformed, consider the variation in the displacement of Q relative to P as the initial position of Q tends to that of P. If for any point P the relative displacement did not tend to zero as the difference between the initial positions tended to zero, this would imply the opening of a gap between elements which were initially contiguous. It follows that the deformation of the body will leave its continuity unimpaired only if the relative displacements tend to zero as the initial relative positions tend to zero, and this implies that, in a frame of reference which is independent of the body, the displacements of the points of the body must be a smoothly continuous function of their initial positions.

The Matrix $[\partial \mathbf{u}]$

So assuming a set of rectangular coordinates which are independent of the body, let **u** represent the vector function of the coordinates which describes the displacements of the points of the body as a function of their initial positions, or, in brief, let **u** represent the function of the displacement field. Then each of its scalar components $[u_1\ u_2\ u_3]$ is a scalar function of the coordinates, so each may be differentiated partially with respect to each of the three coordinates in turn, and each such coefficient may conveniently be represented by the symbol ∂u, qualified by a pair of suffices which identify, first the component of **u** and then the coordinate. Thus, ∂u_{11} represents $\partial u_1/\partial x_1$; ∂u_{12} represents $\partial u_1/\partial x_2$, and so on, and the nine partial derivatives may then be conveniently presented in a matrix

$$[\partial \mathbf{u}] = \begin{bmatrix} \partial u_{11} & \partial u_{12} & \partial u_{13} \\ \partial u_{21} & \partial u_{22} & \partial u_{23} \\ \partial u_{31} & \partial u_{32} & \partial u_{33} \end{bmatrix}$$

wherein the derivatives of u_1, u_2 and u_3 appear in the first, second and third rows respectively, with the derivatives with respect to x_1, x_2 and x_3 in the first, second and third positions in each case.

Infinitesimal Relative Displacements

Noting that the displacement of one point relative to another is determined by the difference in their individual displacements, again let $\mathbf{u} = [u_1\, u_2 u_3]$ represent the vector function which describes the displacement of a general point P in terms of its initial coordinates (x_1, x_2, x_3), and let $\mathbf{u} + \mathbf{du}$ represent the displacement of an adjacent point Q whose initial coordinates $(x_1 + \mathrm{d}x_1,\ x_2 + \mathrm{d}x_2,\ x_3 + \mathrm{d}x_3)$ differ from those of P by the arbitrary increments of $\mathrm{d}x_1$ in x_1, of $\mathrm{d}x_2$ in x_2, and of $\mathrm{d}x_3$ in x_3. Then the initial position of Q relative to P is defined by the infinitesimal vector $\mathrm{dr} = [\mathrm{d}x_1\ \mathrm{d}x_2\ \mathrm{d}x_3]$, whilst the displacement of Q relative to P is determined by the difference $\mathrm{du} = [\mathrm{d}u_1\ \mathrm{d}u_2\ \mathrm{d}u_3]$, and the rule of partial differentiation being applied to each component of **u** in turn, we may write:

$$\left.\begin{aligned} du_1 &= \partial u_{11}dx_1 + \partial u_{12}dx_2 + \partial u_{13}dx_3 \\ du_2 &= \partial u_{21}dx_1 + \partial u_{22}dx_2 + \partial u_{23}dx_3 \\ du_3 &= \partial u_{31}dx_1 + \partial u_{32}dx_2 + \partial u_{33}dx_3 \end{aligned}\right\}.$$

In matrix form, this set of equations appears as follows:

$$\begin{bmatrix} du_1 \\ du_2 \\ du_3 \end{bmatrix} = \begin{bmatrix} \partial u_{11} & \partial u_{12} & \partial u_{13} \\ \partial u_{21} & \partial u_{22} & \partial u_{23} \\ \partial u_{31} & \partial u_{32} & \partial u_{33} \end{bmatrix} \cdot \begin{bmatrix} dx_1 \\ dx_2 \\ dx_3 \end{bmatrix}$$

or

$$\mathbf{du} = [\partial \mathbf{u}] \cdot \mathbf{dr} \tag{7.1}$$

Thus, the relative displacement between two adjacent points whose initial relative position is defined by the infinitesimal vector **dr** is given by the product of the matrix $[\partial \mathbf{u}]$ with the column vector of **dr**.

Transformation of the Matrix $[\partial \mathbf{u}]$

However, the orientation of the coordinates is arbitrary, so the matrix of partial derivatives of a given displacement field is not unique, but varies with the orientation of the axes. Yet each of the matrices is sufficient completely to determine the pattern of relative displacements, so all the different matrices must be related in some characteristic manner, and it can be shown that the different matrices for a given displacement field are, in fact, liable to the same rule of transformation for a rotation of axes as the different tensors of a given state of stress.

Thus, let $Ox_1'x_2'x_3'$ be an alternative set of coordinates whose orientation to $Ox_1x_2x_3$ is defined by the usual matrix of direction cosines [l] and let quantities which are referred to the alternative axes $Ox_1'x_2'x_3'$ be distinguished by a

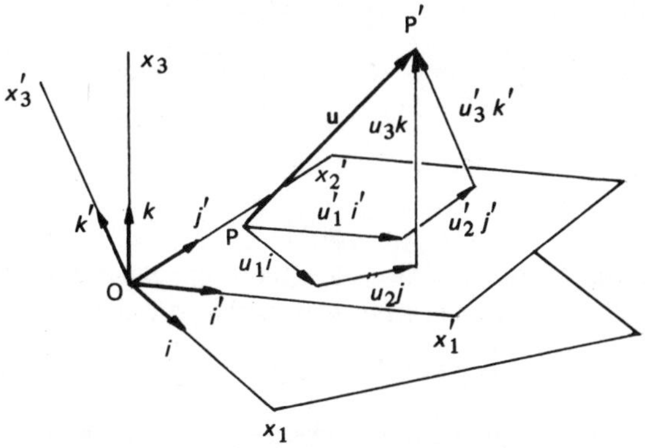

Fig. 7.1

dash. Then the position of the general point P may alternatively be described either by its coordinates (x_1, x_2, x_3) in $Ox_1x_2x_3$ or by its coordinates (x'_1, x'_2, x'_3) in $Ox'_1x'_2x'_3$. Equally, any vector function may alternatively be defined either by describing its components parallel to the x_1, x_2 and x_3 axes as functions of the coordinates (x_1, x_2, x_3), or by describing its components parallel to the x'_1, x'_2 and x'_3 axes as functions of the coordinates (x'_1, x'_2, x'_3). In particular, the point displacements may be defined either by

$$\mathbf{u} = u_1\mathbf{i} + u_2\mathbf{j} + u_3\mathbf{k} \ ,$$

or by

$$\mathbf{u}' = u'_1\mathbf{i}' + u'_2\mathbf{j}' + u'_3\mathbf{k}' \ ,$$

and we note, that just as the components of $\mathbf{u}$ may be differentiated partially with respect to their arguments x_1, x_2 and x_3 to determine the matrix $[\partial\mathbf{u}]$, so may the components of $\mathbf{u}'$ be differentiated with respect to x'_1, x'_2 and x'_3 to determine the alternative matrix $[\partial'\mathbf{u}']$. Furthermore, just as we have seen that the relative displacement can be determined in terms of its components in $Ox_1x_2x_3$ from the equation

$$\mathrm{d}\mathbf{u} = [\partial\mathbf{u}] \cdot \mathrm{d}\mathbf{r} \ ,$$

so it may be determined in terms of its components in $Ox'_1x'_2x'_3$ from

$$\mathrm{d}\mathbf{u}' = [\partial'\mathbf{u}'] \cdot \mathrm{d}\mathbf{r}' \ .$$

But these alternative descriptions of the relative displacement are related by the rule for the transformation of a vector:

$$\mathrm{d}\mathbf{u}' = [\mathrm{l}] \cdot \mathrm{d}\mathbf{u} \ .$$

Therefore

$$[\partial'\mathbf{u}'] \cdot \mathrm{d}\mathbf{r}' = [\mathrm{l}] \cdot [\partial\mathbf{u}] \cdot \mathrm{d}\mathbf{r} \ ,$$

and by first again making use of the rule of vector transformation to substitute $\mathrm{d}\mathbf{r} = [\mathrm{l}]^{\mathrm{T}} \cdot \mathrm{d}\mathbf{r}'$, and then cancelling the dot product with $\mathrm{d}\mathbf{r}'$ on both sides of the equation, we arrive at a rule of transformation which is identical with that for a stress tensor, namely:

$$[\partial'\mathbf{u}'] = [\mathrm{l}] \cdot [\partial\mathbf{u}] \cdot [\mathrm{l}]^{\mathrm{T}} \qquad (7.2)$$

The matrix of partial derivatives of a given displacement field is therefore another tensor of the same kind as a stress tensor, except that it is not generally symmetrical.

Example 7.1

In given rectangular coordinates $Ox_1x_2x_3$, the displacements $\mathbf{u}$ of points in a continuum vary with the coordinates of their initial positions according to:

$$\mathbf{u} \times 10^4 = (x_1x_2 + x_3^2)\,\mathbf{i} + (2x_1 + x_3^3)\,\mathbf{j} + x_1x_2x_3\,\mathbf{k} \ .$$

a) Determine the tensor of partial derivatives of relative point displacements in the given axes

$$[\mathbf{l}] = \frac{1}{3}\begin{bmatrix} 2 & 2 & 1 \\ 1 & -2 & 2 \\ 2 & -1 & -2 \end{bmatrix}.$$

b) Determine the tensor of partial differential coefficients of the relative point displacements for the particular point whose initial coordinates are (1, 2, −1), both in the original axes, and in an alternative set of axes $Ox'_1x'_2x'_3$ whose orientations in the original axes are defined by the matrix of direction cosines [l] indicated.

Solution

a) Each row of the tensor [∂u] is formed by entering the partial derivatives of the corresponding component of **u** with respect to x_1, x_2 and x_3 in turn. Thus, the first row of [∂u] contains the partial derivatives of $u_1 = (x_1x_2 + x_3^2)$, with respect to x_1, x_2 and x_3 in order, and so on.

Thus:

$$[\partial \mathbf{u}] = 10^{-4}\begin{bmatrix} x_2 & x_1 & 2x_3 \\ 2 & 0 & 3x_3^2 \\ x_2x_3 & x_1x_3 & x_1x_2 \end{bmatrix},$$

b) The tensor in the given axes of the partial differential coefficients for the point whose initial coördinates are (1, 2, −1) is determined simply by substituting accordingly in the tensor of partial derivatives above.

Thus:

$$[\partial \mathbf{u}] = 10^{-4}\begin{bmatrix} 2 & 1 & -2 \\ 2 & 0 & 3 \\ -2 & -1 & 2 \end{bmatrix}.$$

The corresponding tensor in the alternative axes is then determined by applying the rule of tensor transformation for a rotation of axes. That is:

$$[\partial'\mathbf{u}'] = [\mathbf{l}] \cdot [\partial\mathbf{u}] \cdot [\mathbf{l}]^{\mathrm{T}}.$$

Thus:

$$[\partial'\mathbf{u}'] = \frac{10^{-4}}{9}\begin{bmatrix} 2 & 2 & 1 \\ 1 & -2 & 2 \\ 2 & -1 & -2 \end{bmatrix} \cdot \begin{bmatrix} 1 & 1 & -2 \\ 2 & 0 & 3 \\ -2 & -1 & 2 \end{bmatrix} \cdot \begin{bmatrix} 2 & 1 & 2 \\ 2 & -2 & -1 \\ 1 & 2 & -2 \end{bmatrix}$$

$$[\partial'\mathbf{u}'] = \frac{10^{-4}}{9}\begin{bmatrix} 2 & 2 & 1 \\ 1 & -2 & 2 \\ 2 & -1 & -2 \end{bmatrix} \cdot \begin{bmatrix} 4 & -4 & 7 \\ 7 & 8 & -2 \\ -4 & 4 & -7 \end{bmatrix}$$

$$[\partial' u'] = \frac{10^{-4}}{3}\begin{bmatrix} 6 & 4 & 1 \\ -6 & -4 & -1 \\ 3 & -8 & 10 \end{bmatrix} .$$

7.3 STRAIN

Though the relative displacements between the points of a body determine its deformation, they also reflect its rotation, and to focus attention on the deformation alone, we now consider the effects of the relative displacements on both the lengths of the straight lines drawn between arbitrary points, and on the angles between them. There is an obvious special case which is spherically symmetrical in that every line suffers the same proportional change of length. In this case there is no change in any of the angles between the lines, and the body suffers a simple dilation without any change in its geometrical form. However, in general the extensions of the lines may be disproportionate, in which case there will also be changes in the angles between them. So recognising a general deformation as one in which a change of shape is superimposed on a simple change of size, we define a shear strain to quantify the one, and a linear strain the other.

Linear Strain

When the distance between two points varies, the straight line which joins them is said to suffer a linear strain, and, in principle, this is quantified as a non-dimensional ratio by describing the extension as a proportion of the length. However, when the change of length is significant in comparison with the length itself, the total extension must be regarded as a succession of infinitesimal increments. In each of these the length may be treated as a constant, and the total strain is determined as the sum of the increments of strain that occur in all the increments of extension that make up the total extension. Thus, the initial length of the line (Fig. 7.2) being represented by l_1, and its final length by l_2, let $\mathrm{d}l$ represent an infinitesimal increment of the extension which occurs when the instantaneous length of the line is l. Then, by definition, the increment of strain which occurs in the increment of extension $\mathrm{d}l$ is equal to $\mathrm{d}l/l$,

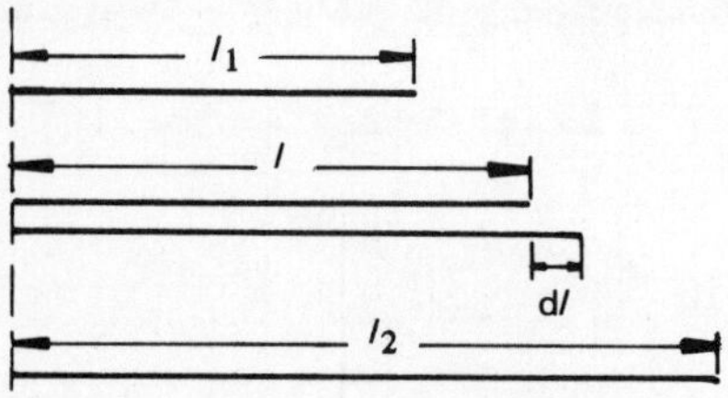

Fig. 7.2

and the total strain in the total extension is determined by integrating the increments of strain from the initial length l_1 to the final length l_2. Thus, by definition:

$$\text{Linear strain} = \int_{l_1}^{l_2} \frac{dl}{l} = \log_e \frac{l_2}{l_1} .$$

Alternatively, this result may be expressed in terms of the ratio ϵ of the total extension to the original length, for it is obvious that:

$$\log_e \frac{l_2}{l_1} = \log_e \left(1 + \frac{l_2 - l_1}{l_1}\right) = \log_e(1 + \epsilon) .$$

Thus: $$\text{Linear strain} = \log_e \frac{l_2}{l_1} = \log_e(1 + \epsilon)$$

where $\epsilon = (l_2 - l_1)/l_1$ is the ratio of the change of length to the original length. (7.3)

Shear Strain

In principle, any pair of straight lines which suffer a relative rotation may be said to suffer shear, but in practice the concept can usefully be developed only for pairs of lines which are initially rectangular. Furthermore, to assign a sense to the effect it is necessary first to assign a sense to each of the lines, so in

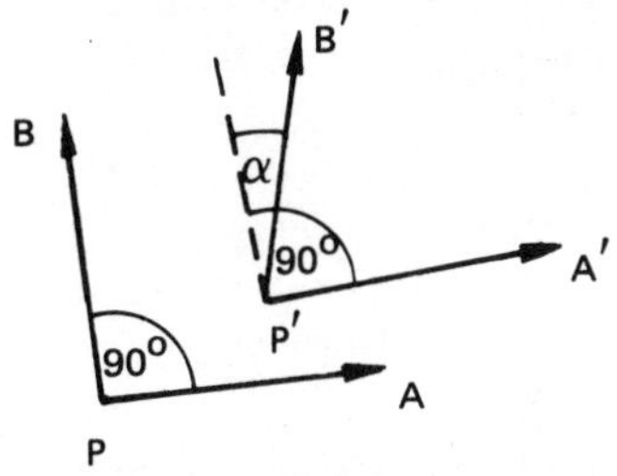

Fig. 7.3

practice a shear strain is defined only for pairs of directed lines which are initially rectangular, and for these it is quantified by the tangent of the angle by which the right-angle in the positive quadrant is reduced (Fig. 7.3). Thus, by definition:

$$\text{Shear strain} = \tan\gamma$$

where γ is the angle by which the right-angle in the positive quadrant between a rectangular pair of directed lines is reduced. (7.4)

7.3 SMALL STRAINS

The analysis of strains in general on the basis of the foregoing definitions is undoubtedly complex, but where the extensions of the lines are negligible compared with their lengths, both the definitions and the subsequent analysis may be greatly simplified. It is therefore significant that, with one notable exception, it is generally necessary, by design, to limit the strains to which engineering components are subjected in normal operation to very small values. Thus, in manufacture the specific objective is commonly that of imposing a significant change of shape, and for this it is commonly necessary to impose large strains; but for normal operation, considerations of both safety and deformation invariably require that the strains be limited to values which are generally less than 1×10^{-3}. So, recognising that the approach will commonly not be applicable to the study of manufacturing processes, we now restrict consideration to small strains of the order appropriate to normal operation only, and these may alternatively, but equivalently, be regarded, either as strains which are negligible compared with unity, or as strains whose squares are negligible compared with the strains themselves.

Small Linear Strains

When the extension of the line is negligible compared with its length, it is immaterial whether the strain is defined as the ratio of the extension to the original length or to the final length. Equally, we may argue mathematically, that as ϵ tends to zero, $\log_e(1 + \epsilon)$ tends to equality with ϵ itself, so for small strains in which ϵ is negligible compared with unity, the linear strain of a line is redefined as the ratio of the extension to the original length.

$$\text{Small linear strain} = \epsilon = \frac{\text{Extension}}{\text{Original length}} \tag{7.5}$$

Volumetric Strain

However, in the analysis of small strains it is sometimes convenient to describe the change of size, not by specifying the proportional changes of length, but rather by directly specifying the proportional change of volume. This is called the volumetric strain, and it is indicated symbolically by Δ. Thus, by definition:

$$\text{Volumetric strain} = \Delta = \frac{\text{Change of volume}}{\text{Original volume}} \tag{7.6}$$

Small Shear Strains

As the angle of shear γ tends to zero, its tangent tends to equality with the angle itself, so for small strains, the shear strain is redefined as follows.

Small shear strain $= \gamma$

where γ is the (small) angle by which the right-angle in the positive quadrant between a directed pair of rectangular lines is reduced. (7.7)

7.5 SETS OF SMALL STRAINS AT A POINT

Deformation is a problem in the three-dimensional geometry of relative point displacements which, in a continuum, are a smoothly continuous function of the initial positions. The pattern of deformation in the vicinity of any point is therefore determined completely by the displacements relative to the point of any set of three adjacent points which are not coplanar with it. However, it is obviously convenient if the three adjacent points are so chosen that the straight lines which join them to the point of interest form a rectangular triad, when the effect of the rotation of the body can be eliminated by specifying, not the relative displacements of the points, but the linear strains of the three lines, together with the shear strains of the three right-angles which lie between them.

So choosing an arbitrary set of rectangular coordinates which are independent of the body, let (x_1, x_2, x_3) represent the initial coordinates of a general point P in the body, and let A, B and C be a set of adjacent points whose initial coordinates are:

$$A = \{(x_1 + dx_1),\ x_2,\ x_3\}$$
$$B = \{x_1,\ (x_2 + dx_2),\ x_3\}$$
$$C = \{x_1,\ x_2,\ (x_3 + dx_3)\}\ .$$

Then PA, PB and PC (Fig. 7.4) form a rectangular triad of infinitesimal lines through P, of which PA is of length dx_1 parallel to x_1, PB is of length dx_2 parallel to x_2, and PC is of length dx_3 parallel to x_3, and we define the pattern of deformation at P in the given coordinates by specifying the linear strains of the lines, together with the shear strains of the right-angles which lie between them.

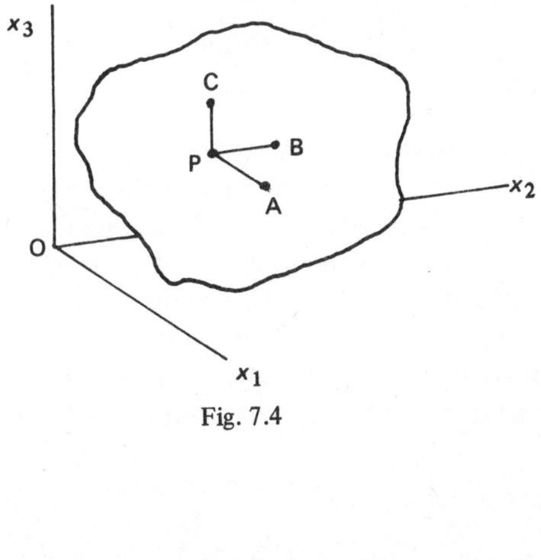

Fig. 7.4

Each of the three linear strains is indicated by the symbol ϵ (epsilon), qualified by a repeated suffix which identifies the initial direction of the line. Thus the linear strains of the lines PA, PB and PC which were initially parallel to x_1, x_2 and x_3 are indicated symbolically by ϵ_{11}, ϵ_{22} and ϵ_{33}, respectively.

Each of the three shear strains is indicated by the symbol γ, qualified by a pair of suffices which identify the initial directions of the pair of lines which form the relevant right-angle. Thus for the right-angles between the pairs of lines initially parallel to x_1 and x_2, to x_1 and x_3, and to x_2 and x_3, the shear strains are indicated symbolically by γ_{12}, γ_{13} and γ_{23} respectively.

The Set of Small Strains v the Partial Derivatives of the Components of the Point Displacements

To establish the relationships of the set of small strains to the point displacements, we consider the effects on each of the lines PA, PB and PC of the relative displacement between its ends. The latter may be described in terms of the partial derivatives of the point displacements, either by applying eqn. (7.1) to PA, PB and PC in turn, or, more directly, by noting that for points which lie on any one of these lines, only one of the three coordinates does, in fact, vary. It follows, that for points on PA the actual rate of change of the displacement **u** with respect to x_1 may be equated to the partial derivative $\partial\mathbf{u}/\partial x_1$, that for points on PB the actual rate of change of **u** with respect to x_2 may be equated to $\partial\mathbf{u}/\partial x_2$, and that for points along PC, the actual rate of change of **u** with respect to x_3 may be equated to $\partial\mathbf{u}/\partial x_3$.

For example, let the displacement of A relative to P be represented by $\mathrm{d}\mathbf{u}_\mathrm{A}$, and let the scalar components of $\mathrm{d}\mathbf{u}_\mathrm{A}$ in the directions of x_1, x_2 and x_3 be represented by $[\mathrm{d}u_{\mathrm{A}1}, \mathrm{d}u_{\mathrm{A}2}, \mathrm{d}u_{\mathrm{A}3}]$. Then between P and A, the actual rate of change of **u** with respect to x_1 may be equated to the partial derivative of the field function **u** with respect to x_1, and the operation being distributed to the components of **u**, it follows that:

$$\frac{\mathrm{d}u_{\mathrm{A}1}}{\mathrm{d}x_1} = \frac{\partial u_1}{\partial x_1} = \partial u_{11}; \qquad \frac{\mathrm{d}u_{\mathrm{A}2}}{\mathrm{d}x_1} = \frac{\partial u_2}{\partial x_1} = \partial u_{21};$$

$$\frac{\mathrm{d}u_{\mathrm{A}3}}{\mathrm{d}x_1} = \frac{\partial u_3}{\partial x_1} = \partial u_{31}.$$

Of these three derivatives, consider the two in which the suffices differ, that is, ∂u_{21} and ∂u_{31}. Since $\mathrm{d}u_{\mathrm{A}2}$ and $\mathrm{d}u_{\mathrm{A}3}$ are components of displacement which are both normal to the line PA itself, each implies a rotation of the line, and it is obvious from Fig. 7.5, that whereas ∂u_{21} determines the angle of rotation of PA (the line parallel to x_1) toward PB (the line parallel to x_2), ∂u_{31} determines the tangent of the angle by which PA turns toward PC (the line parallel to x_3). In addition, $\mathrm{d}u_{\mathrm{A}2}$ and $\mathrm{d}u_{\mathrm{A}3}$ imply a lengthening of the line PA. However, it follows from Pythagoras' theorem that the extension is only of the

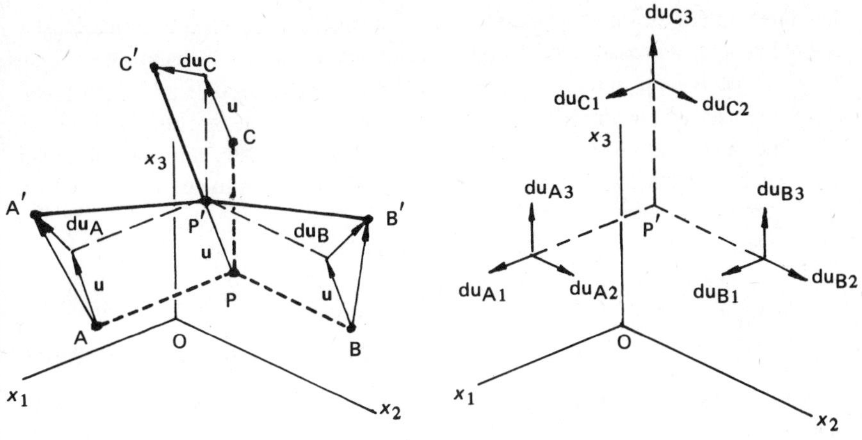

Fig. 7.5

order of the square of displacement, and for small strains, the effects of the squares are assumed to be negligible compared with the effects of the displacements themselves. Furthermore, it then follows that the tangents of the angles of rotation may be equated to the angles themselves, and we conclude that the partial derivatives ∂u_{21} and ∂u_{31} determine the angles of rotation of the line parallel to x_1 towards the lines parallel to x_2 and x_3, respectively, but that the implied lengthening of the line is negligible.

Effectively, therefore, the extension of the line PB derives solely from the component of relative displacement du_{A1} which is colinear with PA itself, and since $du_{A1}/dx_1 = \partial u_{11}$ can then be recognised as the proportional extension of an infinitesimal line parallel to x_1, it follows that the partial derivative ∂u_{11} can be equated to the linear strain ϵ_{11}.

Evidently, parallel results will hold for the variation of **u** with respect to x_2 along PB, and for the variation of **u** with respect to x_3 along PC. Furthermore, we note that the shear strain of the right-angle between any pair of the lines PA, PB and PC may be equated to the sum of the angles through which each of the lines rotates toward the other, and we conclude that, both being analysed in the same coordinates, the small strains can be described in terms of the partial derivatives of the components of the displacement field as follows:

$$\epsilon_{11} = \partial u_{11}; \quad \epsilon_{22} = \partial u_{22}; \quad \epsilon_{33} = \partial u_{33}$$
$$\gamma_{12} = \partial u_{12} + \partial u_{21}; \quad \gamma_{13} = \partial u_{13} + \partial u_{31}; \quad \gamma_{23} = \partial u_{23} + \partial u_{32} \tag{7.8}$$

7.6 TENSORS OF SMALL STRAINS AT A POINT

If each of the shear strains γ_{12}, γ_{13} and γ_{23} is treated as the sum of two equal shear elements, the three pairs of equal shear elements may be combined with the three linear strains ϵ_{11}, ϵ_{22} and ϵ_{33} in a symmetrical matrix which closely resembles a stress tensor, and is, in fact, another tensor of the same kind. Thus let $\epsilon_{12} = \epsilon_{21} = \frac{1}{2}\gamma_{12}$, let $\epsilon_{13} = \epsilon_{31} = \frac{1}{2}\gamma_{13}$, let $\epsilon_{23} = \epsilon_{32} = \frac{1}{2}\gamma_{23}$, and let $[\epsilon]$ represent the matrix in which the linear strains are entered on the leading diagonal, with the three pairs of equal shear elements symmetrically disposed on either side. Then substituting for the elements of $[\epsilon]$ in terms of the partial derivatives of the point displacements, we have:

$$[\epsilon] = \begin{bmatrix} \epsilon_{11} & \epsilon_{12} & \epsilon_{13} \\ \epsilon_{21} & \epsilon_{22} & \epsilon_{23} \\ \epsilon_{31} & \epsilon_{32} & \epsilon_{33} \end{bmatrix} = \frac{1}{2} \begin{bmatrix} 2\partial u_{11} & (\partial u_{12} + \partial u_{21}) & (\partial u_{13} + \partial u_{31}) \\ (\partial u_{12} + \partial u_{21}) & 2\partial u_{22} & (\partial u_{23} + \partial u_{32}) \\ (\partial u_{13} + \partial u_{31}) & (\partial u_{23} + \partial u_{32}) & 2\partial u_{33} \end{bmatrix}$$

But the matrix on the right may be partitioned to write:

$$\begin{bmatrix} \epsilon_{11} & \epsilon_{12} & \epsilon_{13} \\ \epsilon_{21} & \epsilon_{22} & \epsilon_{23} \\ \epsilon_{31} & \epsilon_{32} & \epsilon_{33} \end{bmatrix} = \frac{1}{2} \left(\begin{bmatrix} \partial u_{11} & \partial u_{12} & \partial u_{13} \\ \partial u_{21} & \partial u_{22} & \partial u_{23} \\ \partial u_{31} & \partial u_{32} & \partial u_{33} \end{bmatrix} + \begin{bmatrix} \partial u_{11} & \partial u_{21} & \partial u_{31} \\ \partial u_{12} & \partial u_{22} & \partial u_{32} \\ \partial u_{13} & \partial u_{23} & \partial u_{33} \end{bmatrix} \right)$$

whence: $$[\epsilon] = \tfrac{1}{2}([\partial \mathbf{u}] + [\partial \mathbf{u}]^{\mathrm{T}}) \qquad (7.9)$$

Thus the relationships between a set of small strains at a point, and the partial derivatives of the point displacements in the same axes can be described by a simple relationship in which the matrix of small strains is equated to half the sum of the matrix of partial derivatives $[\partial \mathbf{u}]$ and its transpose.

Moreover, we have already seen that $[\partial \mathbf{u}]$ transforms for a rotation of axes in the same manner as a stress tensor, and reflection will show that $[\partial \mathbf{u}] + [\partial \mathbf{u}]^{\mathrm{T}}$ must then be a symmetrical matrix which obeys the same rule. The matrix of small strains $[\epsilon]$ can therefore be recognised as another tensor of precisely the same kind as a stress tensor, and so shares all its properties of resolution, as outlined in the following.

Strain Resolution

Given a tensor $[\epsilon]$ of small strains for any point, together with a rectangular pair of unit vectors $\hat{\mathbf{l}}$ and $\hat{\mathbf{l}}'$ which are described in terms of their scalar elements (or direction cosines) in the axes of the tensor, the linear strain in the direction $\hat{\mathbf{l}}$, and the shear element for the pair of rectangular directions $\hat{\mathbf{l}}$ and $\hat{\mathbf{l}}'$ are given by:

$$\epsilon_{ll} = \hat{l} \cdot [\epsilon] \cdot \hat{l}$$
$$\epsilon_{ll'} = \frac{1}{2}\gamma_{ll'} = \hat{l} \cdot [\epsilon] \cdot \hat{l}' \tag{7.10}$$

Tensor Transformation

Given the tensor $[\epsilon]$ for a given point in one set of axes, together with the matrix of direction cosines [l] which describes the orientation of an alternative set of axes relative to the axes of $[\epsilon]$, the matrix $[\epsilon']$ of small strains at the point in the alternative axes is given by the characteristic rule of transformation:

$$[\epsilon'] = [l] \cdot [\epsilon] \cdot [l]^T \tag{7.11}$$

Principal Axes of Strain and Principal Strains

At every point in the continuum there exists a set of rectangular axes which suffer no shear strains. These are called the principal axes of strain at the point, and the values of the principal strains ϵ_p at the point (i.e. the linear strains of the principal axes) are related to the elements of any general tensor $[\epsilon]$ for the point by the cubic equation:

$$|[\epsilon] - \epsilon_p [I]| = 0 \tag{7.12}$$

Furthermore, for each of the three values of ϵ_p, the unit vector $\hat{l}$ which defines the direction of the corresponding principal axis in the axes of the general tensor $[\epsilon]$ is determined by substituting the value of ϵ_p in:

$$[[\epsilon] - \epsilon_p [I]] \cdot l = 0 \tag{7.13}$$

In an isotropic material, the principal axes of strain coincide with the principal axes of stress.

Octahedral and Volumetric Strains

In any state of strain, the linear strain along each of the four octahedral axes which are equally inclined to the three principal axes has a value equal to the invariant average of the three linear strains of any tensor for the point.

$$\epsilon_{00} = \frac{1}{3}(\epsilon_{11} + \epsilon_{22} + \epsilon_{33}) .$$

However, it can be shown that, for small strains, the octahedral strain is one third of the volumetric strain, and results are usually described in terms of the latter rather than the former. To demonstrate the relationship, consider a rectangular element having edges which are parallel to the principal axes X_1,

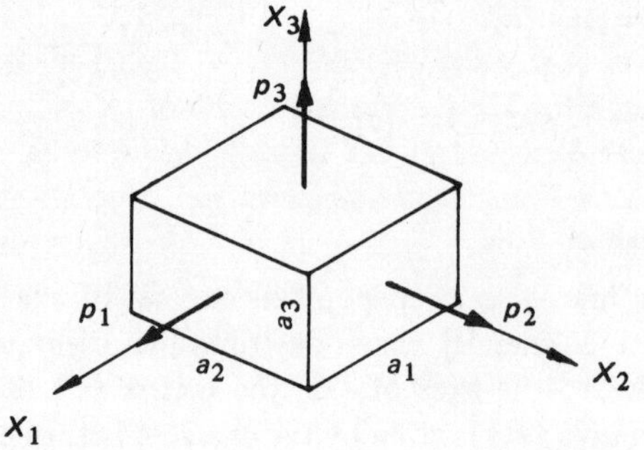

Fig. 7.6

X_2 and X_3 at the point, and whose initial lengths are indicated by a_1, a_2 and a_3. Then if ϵ_{p1}, ϵ_{p2} and ϵ_{p3} represent the principal strains at the point, the final lengths of the edges are $a_1(1+\epsilon_{p1}), a_2(1+\epsilon_{p2})$ and $a_3(1+\epsilon_{p3})$, and by definition:

$$\text{Volumetric strain} = \Delta = \frac{\text{Change of volume}}{\text{Original volume}}$$

that is:
$$\Delta = \frac{a_1a_2a_3(1+\epsilon_{p1})\,(1+\epsilon_{p2})\,(1+\epsilon_{p3}) - a_1a_2a_3}{a_1a_2a_3} \,,$$

and ignoring terms of second order, this gives the result

$$\Delta = \epsilon_{p1} + \epsilon_{p2} + \epsilon_{p3} \,.$$

But the sum of the direct strains is an invariant of the tensors of any given state, whence:

$$\Delta = 3\epsilon_{00} = \epsilon_{11} + \epsilon_{22} + \epsilon_{33} \tag{7.14}$$

The Volumetric and Deviatoric Strain Tensors

In parallel with the hydrostatic state of stress, it can be shown that in pure dilation when there are no changes in the angle between any pair of lines in a body, all lines in the body suffer the same linear strain. A strain tensor, like a stress tensor, can therefore readily be split into its dilational and deviatoric parts, as follows:

$$\begin{aligned} [\epsilon] &= \epsilon_{00}\,[\mathrm{I}] + ([\epsilon] - \epsilon_{00}\,[\mathrm{I}]) \\ \text{or } [\epsilon] &= \frac{\Delta}{3}[\mathrm{I}] + [\epsilon] - \frac{\Delta}{3}[\mathrm{I}]) \end{aligned} \tag{7.15}$$

Mohr's Circular Diagram of Strain

The tensor of small strains for any given point can be resolved by Mohr's Circular Diagram in much the same way as the state of stress, but it is important to note that the shear elements of the strain tensor are only one half the corresponding shear strains. Thus, each line through a given point is associated with a linear strain represented by ϵ, and a shear strain represented by γ, and if the value of $\gamma/2$ for each line is plotted against its linear strain ϵ, the plot for all lines through any point will occupy a specific domain which is bounded by three circles whose centres lie on the axis of linear strain ϵ (Fig. 7.7). Each of these boundary circles is the graph of $\gamma/2$ *vs* ϵ for the set of lines which lie in one of the principal planes at the point, and it cuts the axis of ϵ in two of the principal strains.

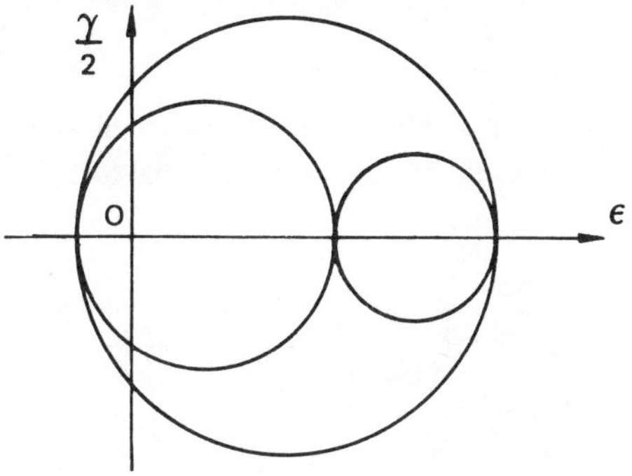

Fig. 7.7

For each of the boundary circles, a suitable sign convention can be formulated by interpreting the circular diagram in an associated physical diagram which views the principal plane in which the set of lines lie, face-on, from either direction. Then retaining the convention whereby the shear strain between a pair of rectangular directions is reckoned positive when the right-angle is reduced, the shear strain that is associated with each line in the physical diagram is taken as that of the right-angle which lies to that side of the line whose sense is opposed to the sense of rotation (through $\pi/2$) from the axis $+\epsilon$ to the axis $+\gamma/2$ in the circular diagram.

Here we shall adopt the convention of choosing the axis of $+\gamma/2$ at the angle $\pi/2$ counterclockwise from the axis $+\epsilon_1$, so the shear that is associated with any line will be interpreted in the physical diagram as that of the quadrant that lies to the clockwise side of the line. For example, Fig. 7.8 shows four lines through P, of which PQ$'$ and PR$'$ are $\pi/2$ clockwise from PQ and PR respectively, and the shear strain associated with PQ is taken as that of the right-angle $\angle$QPQ$'$, whilst the shear strain associated with PR is taken as that of the right-angle $\angle$RPR$'$. The points Q and R on the circle which correspond to the lines PQ and PR in the physical diagram, subtend at the centre of the circle an angle which, with due regard to sense, is twice the angle between the lines in the physical diagram.

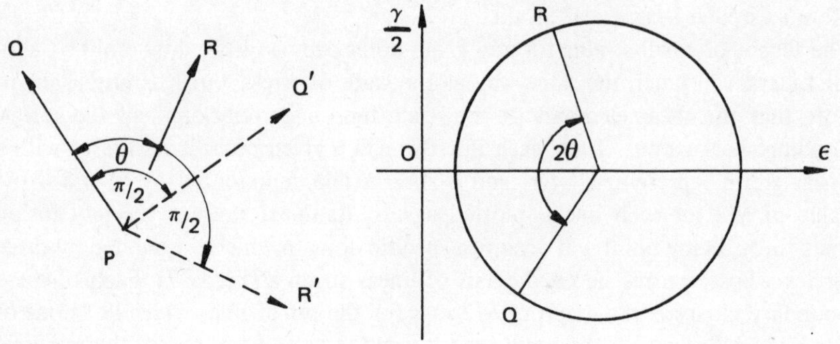

Fig. 7.8

Example 7.2

The diagram shows an elemental cube having an edge length of 2 units which suffers a plane deformation in planes parallel to one pair of faces, so that its

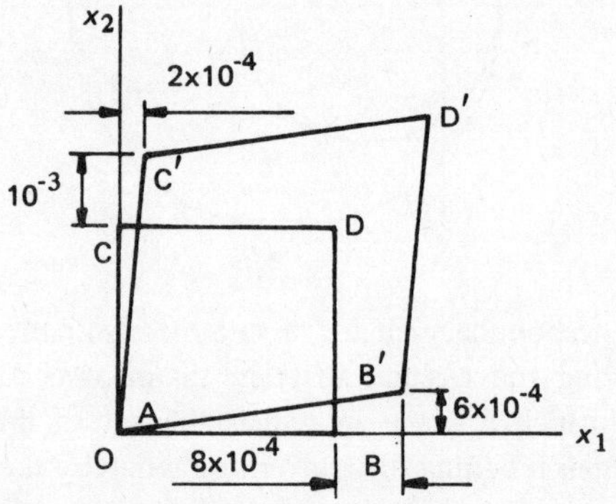

elevation changes from ABCD to A′B′C′D′. In the direction normal to the plane of the figure the edges suffer a uniform reduction of length of 2×10^{-4} units. Treating the strains as small, determine:

a) The strain tensor for the element in the axes (x_1, x_2, x_3) which coincide with its edges.

b) The strain tensor for the alternative axes (x_1', x_2', x_3') whose direction cosines relative to (x_1, x_2, x_3) are given by the matrix [l].

$$[l] = \frac{1}{3}\begin{bmatrix} 1 & 2 & -2 \\ 2 & 1 & 2 \\ 2 & -2 & 2 \end{bmatrix} .$$

c) The change in the right-angle between x_1' and x_2'.

Solution

a) Since the strains are small, the linear strain ϵ_{11} is given by the ratio to the initial length of AB, of the component in the direction of x_1 of the displacement of B relative to A:

$$\epsilon_{11} = 8/2 \times 10^{-4} = 4 \times 10^{-4} .$$

Similarly: $\epsilon_{22} = 10/2 \times 10^{-4} = 5 \times 10^{-4}$

and $\epsilon_{33} = -2/2 \times 10^{-4} = -1 \times 10^{-4}$.

The shear strain γ_{12} is the reduction in the right-angle $\angle APB$.

Thus: $\epsilon_{12} = \epsilon_{21} = \frac{1}{2}\gamma_{12} = \frac{1}{2}(6+2)/2 \times 10^{-4} = 2 \times 10^{-4}$.

Since the edges parallel to x_3 remain at right-angles to the x_1x_2 plane:

$$\epsilon_{13} = \epsilon_{31} = \epsilon_{23} = \epsilon_{32} = 0 .$$

Thus:

$$[\epsilon] = \begin{bmatrix} 4 & 2 & 0 \\ 2 & 5 & 0 \\ 0 & 0 & -1 \end{bmatrix} \times 10^{-4}$$

b) By the rule of tensor transformation:

$$[\epsilon'] = [l] \cdot [\epsilon] \cdot [l]^T$$

that is:

$$[\epsilon'] = 10^{-4}/9 \begin{bmatrix} 1 & 2 & -2 \\ 2 & 1 & 2 \\ 2 & -2 & -1 \end{bmatrix} \cdot \begin{bmatrix} 4 & 2 & 0 \\ 2 & 5 & 0 \\ 0 & 0 & -1 \end{bmatrix} \cdot \begin{bmatrix} 1 & 2 & 2 \\ 2 & 1 & -2 \\ -2 & 2 & 1 \end{bmatrix}$$

whence:

$$[\epsilon'] = \frac{10^{-4}}{9} \begin{bmatrix} 28 & 32 & -10 \\ 32 & 25 & 4 \\ -10 & 4 & 19 \end{bmatrix}$$

c) The shear element $\epsilon'_{12} = \epsilon'_{21}$ is equal to half the reduction of the right-angle between the axes x'_1 and x'_2. The right-angle is therefore reduced by $(64/9) \times 10^{-4}$ rad.

Example 7.3

The displacement $\mathbf{u}$ of the points of a continuum vary with the initial coordinates (x_1, x_2, x_3) according to:

$$u \times 10^4 = 0{\cdot}5x_1^2x_2^2\,\mathbf{i} - x_1^3x_2\,\mathbf{j} + 2x_1x_3\,\mathbf{k} .$$

a) Determine the strain tensor for any point (in the given axes) as a function of its initial coordinates.

b) For the point whose initial coordinates are (1, 2, 0) determine:

i) The strain tensor in the given axes.

ii) The proportional change of density.

iii) The linear strain along the line whose direction cosines in the given axes are $\mathbf{l} = \frac{1}{3}(2\ 1\ 2)$.

iv) The principal strains.

Solution

a) The matrix $[\partial \mathbf{u}]$ is formed by entering the partial derivatives of $u_1 = 0{\cdot}5x_1^2x_2^2$ in the first row, of $u_2 = -x_1^3x_2$ in the second row, and of $u_3 = 2x_1x_3$ in the third row. Thus:

$$[\partial \mathbf{u}] = \begin{bmatrix} x_1x_2^2 & x_1^2x_2 & 0 \\ -3x_1^2x_2 & -x_1^3 & 0 \\ 2x_3 & 0 & 2x_1 \end{bmatrix} \times 10^{-4}$$

Then:
$$[\epsilon] = \tfrac{1}{2}([\partial \mathbf{u}] + [\partial \mathbf{u}]^{\mathrm{T}}) = \begin{bmatrix} x_1x_2^2 & -x_1^2x_2 & x_3 \\ -x_1^2x_2 & -x_1^3 & 0 \\ x_3 & 0 & 2x_1 \end{bmatrix} \times 10^{-4}$$

b(i)) To determine the tensor for the point whose initial coordinates are (1, 2, 0) we substitute accordingly in $[\epsilon]$ above.

$$[\epsilon] = \begin{bmatrix} 4 & -2 & 0 \\ -2 & -1 & 0 \\ 0 & 0 & 2 \end{bmatrix} \times 10^{-4}$$

b(ii)) Let m represent the mass of an element whose volumes before and after deformation are represented by V and $V + \mathrm{d}V$.

$$\begin{matrix}\text{Proportional change} \\ \text{of density}\end{matrix} = \frac{m/(V + \mathrm{d}V) - m/V}{m/V} = \frac{-\mathrm{d}V}{V + \mathrm{d}V} .$$

But when $\mathrm{d}V$ is negligible compared with V, this may be equated to $-\mathrm{d}V/V$, and since this is the negative of the volumetric strains, it can be equated to the negative of the invariant sum of the linear strains of any tensor for the point. Thus:

$$\begin{matrix}\text{Proportional change} \\ \text{of density}\end{matrix} = -(4\ -1\ +2) \times 10^{-4} = -5 \times 10^{-4} .$$

b(iii)) By the rule of tensor resolution:

$$\epsilon_{ll} = \hat{\mathbf{l}} \cdot [\epsilon] \cdot \hat{\mathbf{l}}$$

that is:

$$\epsilon_{ll} = \frac{10^{-4}}{9} [2 \;\; 1 \;\; 2] \cdot \begin{bmatrix} 4 & -2 & 0 \\ -2 & -1 & 0 \\ 0 & 0 & 2 \end{bmatrix} \cdot \begin{bmatrix} 2 \\ 1 \\ 2 \end{bmatrix}$$

$$\epsilon_{ll} = (7/9) \times 10^{-4} \, .$$

b(iv)) By inspection of the matrix $[\epsilon]$ in (b(i)), one of the principal strains is 2×10^{-4}, and the remaining pair may be determined as the roots of the quadratic equation

$$\left| \begin{bmatrix} 4 & -2 \\ -2 & -1 \end{bmatrix} - p \, 10^4 \, [\mathrm{I}] \right| = 0 \; ,$$

whence: $p_1 = 1{\cdot}70 \times 10^{-4}; \;\; p_2 = -4{\cdot}70 \times 10^{-4}; \;\; p_3 = 2 \times 10^{-4}$.

Example 7.4

A strain gauge rosette is used to measure the linear strains along three lines q, r and s through a point P on a plane unloaded surface, with the results indicated. Determine the two principal strains in the plane of the surface at the

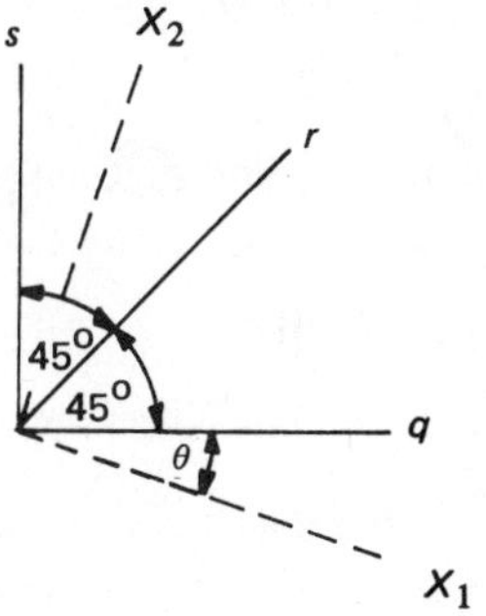

point, and the angle θ which defines the direction of the set of axes relative to the principal axes X_1 and X_2.

Solution

Let S represent half the sum of the principal strains ϵ_{p1} and ϵ_{p2}, and let D represent half their difference. Then S determines the position of the centre of Mohr's circle, whilst D determines its radius.

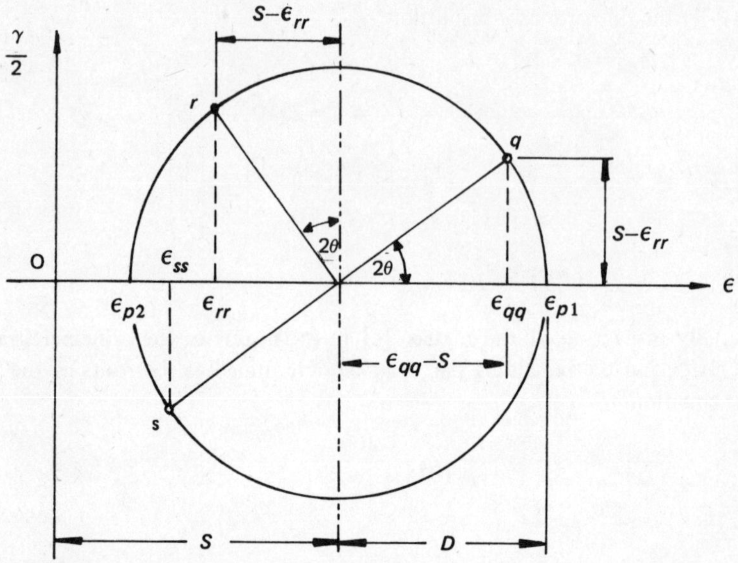

Since the sum of the linear strains in any pair of rectangular directions in the principal plane of the surface is invariant:

$$S = (\epsilon_{qq} + \epsilon_{ss})/2 \ . \tag{a}$$

Furthermore, by Pythagoras' Theorem:

$$D^2 = (\epsilon_{qq} - S)^2 + (S - \epsilon_{rr})^2 \ . \tag{b}$$

Also: $$\tan\theta = (S - \epsilon_{rr})/(\epsilon_{qq} - S) \ . \tag{c}$$

Substituting the given values of ϵ_{qq}, ϵ_{rr} and ϵ_{ss} then gives

$$S = 1{\cdot}95 + 10^{-4}; \quad D = 1{\cdot}08 \times 10^{-4}; \quad \theta = 13{\cdot}4^\circ \ .$$

Then: $\epsilon_{p1} = S + D = 3{\cdot}03 \times 10^{-4}$

and $\epsilon_{p2} = S - D = 0{\cdot}87 \times 10^{-4}$.

Alternatively, the linear strain at the angle α to X_1 is given by $\epsilon = S + D\cos 2\alpha$, and applying this to the lines p, q and r in turn, we have:

$$\epsilon_{qq} = \qquad\qquad = S + D\sin 2\theta \tag{d}$$

$$\epsilon_{rr} = S + D\cos(2\theta + \frac{\pi}{2}) = S - D\sin 2\theta \tag{e}$$

$$\epsilon_{ss} = S + D\cos(2\theta + \pi) = S - D\cos 2\theta \ . \tag{f}$$

(d) + (f) gives:

$$S = (\epsilon_{qq} + \epsilon_{rr}) \tag{a'}$$

whilst from (d) and (e) we obtain:

$$D^2 = (\epsilon_{qq} - S)^2 + (S - \epsilon_{rr})^2 \qquad (b')$$

and $$\tan\theta = (S - \epsilon_{rr})/(\epsilon_{qq} - S) \ . \qquad (c')$$

The equations (a′), (b′) and (c′) are then identical with (a), (b) and (c).

8

Continuum Mechanics III – Hookean Solids and Newtonian Fluids

8.1 INTRODUCTION

When real materials are stressed they invariably deform, but the nature of the deformation that results from a given state of stress varies from one kind of material to another, and this affords one way of distinguishing between them. Thus one material may be classified as a solid, and another as a liquid, or one as ductile, and another as brittle, but this does not imply that a particular material can be assigned, once for all, to a particular category. On the contrary, the nature of a given material invariably depends on a variety of circumstances. For example, a change of temperature may change a solid to a liquid; rocks would generally be classified as brittle, yet folds in rocks that are now exposed reveal that they have exhibited considerable ductility under the extreme hydrostatic pressures of their geological histories; and mild steel would usually be classified as ductile, yet there are, in fact, several circumstances that can rob the material of its ductility. It is therefore clear that any statement which purports to characterise a material must assume some particular conditions, and here we shall always assume conditions which are typical of those in which the material is commonly used. Thus the picture that we now present is only a simple, first view of a situation whose details are really somewhat complex, so that observations that are now made as categoric statements may well require a measure of amplification, and even of modification, as the study of the subject develops.

The materials considered fall into two classes which are distinguished as Hookean solids and Newtonian fluids. The Hookean properties are defined as a generalised idealisation of the properties in simple tension of the great majority of structural solids, whilst the Newtonian properties reflect those of a range of common liquids and gases.

8.2 UNIFORM SIMPLE TENSION

We have argued that, in ductile solids, a state of stress can best be regarded as a combination of two effects of essentially different kinds. One is the dilational or hydrostatic effect, and since this is spherically symmetrical it can be treated as a scalar, whereas the other is the deforming or deviatoric effect and this has all the directional properties of a general tensor. But the common structural solids, in bulk, are sensibly isotropic. This implies that the effect of a given stress is independent of the direction in which it is applied, and when the direction of the deviatoric effect is irrelevant, its magnitude can reasonably be characterised, as a scalar, by the octahedral shear stress. This suggests, first, that we should seek to describe the stress/strain properties of the common structural solids in a pair of relationships of which one relates the dilation to the hydrostatic stress whilst the other relates the distortion to the deviatoric effect, and second, that we should investigate these relationships in a series of tests in which the hydrostatic stress and the octahedral shear stress are varied systematically. But such a test is not easy to devise. Furthermore, it is difficult to develop ideas in terms of a quantity like the octahedral shear stress which is not readily to be recognised in everyday experience. So instead we adopt a type of test which is limited in that it does not permit the independent variation of the significant variables, but which does afford the alternative virtues of conceptual and experimental simplicity, and this is the test in uniform, simple tension.

A stress field is said to be uniform when each point suffers the same state of stress, and in a state of simple tension (or of simple compression) two of the principal stresses are, by definition, zero. For example, Fig. 8.1 illustrates an

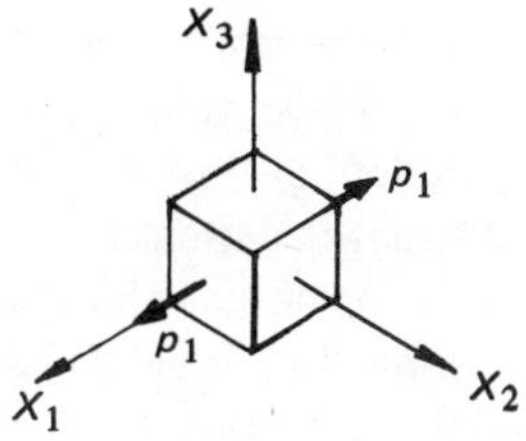

$$[\sigma] = \begin{bmatrix} p_1 & 0 & 0 \\ 0 & 0 & 0 \\ 0 & 0 & 0 \end{bmatrix}$$

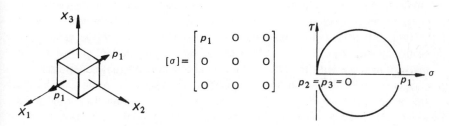

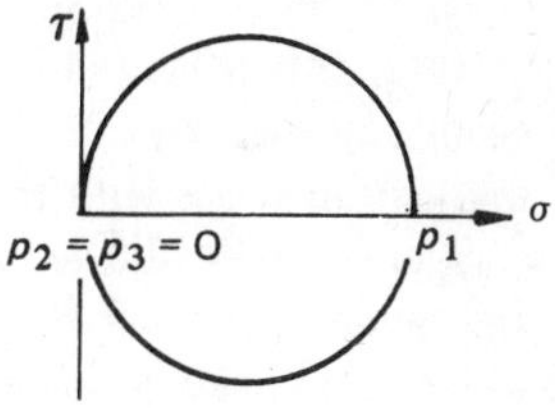

Fig. 8.1

element which is placed in simple tension of magnitude p_1 on the axis X_1, when the principal tensor, and Mohr's Circular Diagram of the state, take the particularly simple forms indicated.

Evidently, the hydrostatic stress of the state is equal to $p_1/3$, whilst the octahedral shear stress is equal to $\sqrt{2}(p_1/3)$, so the state of simple tension is, in fact, a complex state in that it incorporates both a hydrostatic effect and a deviatoric effect, but it is a special case in that the ratio of the octahedral shear stress to the hydrostatic stress retains a fixed value of $\sqrt{2}$.

As to the circular diagram, since p_2 and p_3 fall into coincidence at the origin, the basic circle for the set of sections parallel to X_1 shrinks to a point, and it follows that every axis normal to X_1 is a principal axis on which the principal stress is zero, whilst every section parallel to X_1 is entirely free of stress. Furthermore, the circles for the sets of sections parallel to X_2 and X_3 then coalesce, and it follows that all the combinations of direct stress and shear that arise in simple tension are indicated by the coordinates of the points that lie on the single remaining circle.

The Tensile Test

A tensile testing machine is designed to apply to the ends of a long slender member of uniform section, a pair of equal but opposite forces whose common line of action is accurately aligned to the straight line which passes through the centres of area of the transverse sections, and this gives rise to states of stress and strain which, up to a certain level, are sensibly uniform.

Gauge Length

Though the strains at a point are defined in terms of proportional changes in infinitesimal lines, in a uniform strain field the strains along any straight line have a value at each point equal to that of the line as a whole. In a uniform field the strains may therefore be measured on lines of finite length, and in the tensile test they are usually measured on gauge lengths of several centimetres.

'True' and Nominal Stresses

Similarly, when the stresses are uniform, the stress at each point on a section is equal to the average stress on the section as a whole, and in the test in simple tension, the tensile stress on transverse sections can therefore be determined simply by dividing one of the forces by the area of the section. But the area of the section varies with the stress. It is true, that within the range of stresses that can generally be allowed in normal operation, the effect on the stress is hardly significant, but otherwise it may be considerable, so we distinguish a 'true' stress which does take account of the variation in the area of the section from a nominal stress which does not, and these are formally defined as follows:

$$\begin{aligned} \text{True stress} &= \frac{\text{Force}}{\text{Current sectional area}} \\ \text{Nominal stress} &= \frac{\text{Force}}{\text{Original section area}} \end{aligned} \tag{8.1}$$

8.3 MILD STEEL IN SIMPLE TENSION

Prominent among the materials of engineering are the structural metals, and among these mild steel is preeminent. It is therefore reasonable that, in turning to a consideration of the properties of particular materials, we first consider mild steel. However, that should not be taken to imply that this material is typical of the others. Rather, the properties of mild steel are in some ways unique, but they are distinctive in ways which are of particular practical convenience, so we start with mild steel partly because of its prevalence, but even more because its distinctive features suggest how we can conveniently interpret the less distinctive features of other materials.

The Stress/Strain Graph

If a virgin specimen of mild steel is steadily pulled in simple tension, progressively and without reversal, up to fracture, a graph in which the nominal tensile stress is plotted against the linear 'axial' strain in the direction of the stress itself takes the general form indicated in Fig. 8.2. However, to show the essential features

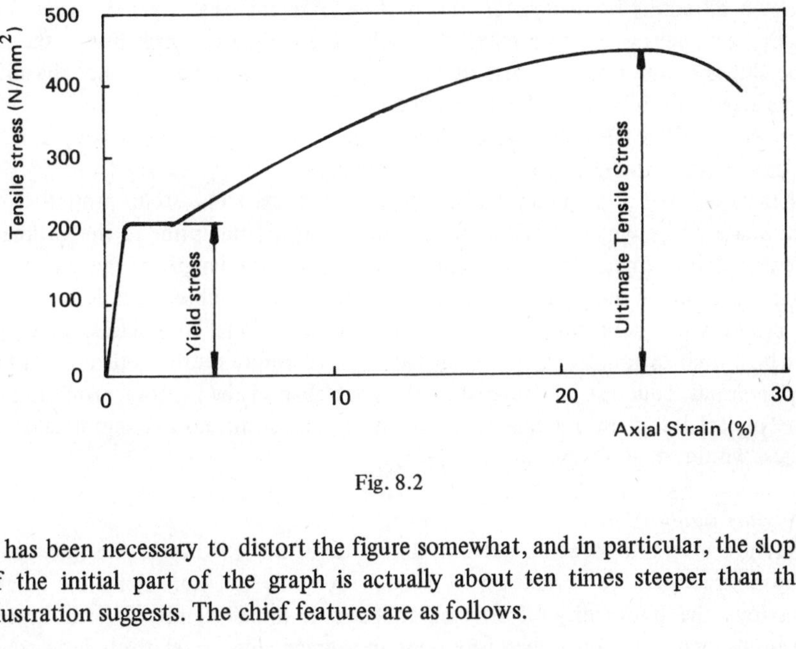

Fig. 8.2

it has been necessary to distort the figure somewhat, and in particular, the slope of the initial part of the graph is actually about ten times steeper than the illustration suggests. The chief features are as follows.

The Proportional Range and the Limit of Proportionality

Initially, as the stress increases from zero, the axial strain increases in proportion to the stress. The point in which the proportional range ends is called the Limit of Proportionality, and this typically occurs at a stress of about 200 N/mm^2, and a strain of about 0·001, or 0·1%. However, in special grades of mild steel the values may be somewhat greater.

Yield Stress

Shortly thereafter there occurs a phenomenon known as yield. Here a sudden increment of strain of about 2% occurs, and the value of tensile stress at which it occurs is known as the Yield Stress σ_y of the material. Since the other principal stresses are zero, the octahedral shear stress at yield is then equal to $\sigma_y\sqrt{2}/3$, whilst the greatest value of shear stress at yield is equal to $\sigma_y/2$.

With appropriate equipment it can be shown that the variation of stress with strain during yield is really more complex than the diagram suggests, but the details are seldom of practical significance, and for the present, yield can be treated simply as a phenomenon in which the incremental stiffness of the material falls to zero over an increment of strain of about 2%.

Strain Hardening and the Ultimate Tensile Stress

The strains at yield are limited because they arise from a shearing of the crystals, and this restores a measure of stiffness to the material through a phenomenon known as strain hardening. In the post-yield region the stress can therefore again be increased, up to a maximum called the Ultimate Tensile Stress, though the slopes of the relevant parts of the stress/strain graph indicate that the stiffness after yield is very much less than before.

Beyond the Ultimate Tensile Stress, the slope of the stress/strain graph is, of course, negative, but whilst this truly indicates the force/extension properties of the specimen as a whole, it is misleading as to the stress/strain properties of the material of which it is composed. In fact, at a point prior to the Ultimate Tensile Stress a neck forms at some position along the length of the specimen, and as a result of the progressive local reduction of area, subsequent deformation is progressively concentrated in the weakened neck. In its later stages, the shape of the graph of nominal stress against strain is therefore more a reflection of the preferential reduction of the area of the neck than of the intrinsic properties of the material, and in a graph of true stress against strain no maximum value of stress would generally appear.

Fracture ductility

Fracture is indicated at an axial strain of about 30%. However, we note that, before fracture, the formation of a neck at some position along the length destroys the uniformity of stresses and strains in its vicinity. The indicated fracture strain of 30% is therefore only an average along a relatively long gauge length, of a strain which varies from point to point, and from observations on the area of the fracture surface we can infer that in the immediate vicinity of the fracture, the longitudinal strain at fracture is about twice the value indicated on the long gauge length.

Elasticity and the Elastic Limit

Strains which can be recovered by the removal of the stress are said to be elastic, and it is commonly found that the material has a characteristic range of stress within which the associated strains are wholly elastic. The value of simple tensile stress at the upper end of the elastic range is called the elastic limit, and in mild steel this falls in the small interval between the Proportional Limit and the Yield Stress.

Hooke's Law and Young's Modulus

It follows that for stresses which are maintained within the elastic range, a particular value of axial strain is associated with each value of the simple tensile stress. Within the elastic range the one is therefore a function of the other, and within the proportional range, the function takes the simple proportional form known as Hooke's Law. The constant of proportionality is known as Young's Modulus, E, and for mild steel in simple tension within the proportional range we can therefore write

$$\begin{aligned} &\text{Tensile stress} = E.\ \text{Axial strain} \\ \text{or:}\quad &\text{Axial strain} = (1/E).\ \text{Tensile stress} \end{aligned} \qquad (8.2)$$

Although the value of the Yield Stress may vary from one grade of mild steel to another, there is little variation in their values of Young's Modulus, and in all grades and conditions of mild steel its value falls close to $210 \times 10^3\ \text{N/mm}^2$, that is, to $2{\cdot}1 \times 10^6$ bar.

Lateral Strains and Poisson's Ratio

In the great majority of the common structural solids, the linear strain along lines which are normal to the axis of simple tension has a value which is sensibly uniform in all directions, and the ratio of the lateral linear strains to the axial is known as Poisson's Ratio, ν. However, the sense of the lateral strain is invariably opposed to that of the axial. In other words, an extension along the axis is invariably associated with a contraction in directions at right-angles to the axis (or vice versa), and since Poisson's Ratio is invariably specified as a positive fraction, it should strictly be defined as the negative of the ratio of the lateral strain to the axial. Thus, by definition:

$$\text{Poisson's ratio} = \nu = \frac{-\text{Lateral strain}}{\text{Axial strain}} \qquad (8.3)$$

In mild steel in simple tension within the proportional range, the lateral strain, like the longitudinal, is proportional to the tensile stress, and Poisson's Ratio is then a constant whose value falls close to $\frac{1}{3}$. In other words, the magnitude of the lateral strain is about one third that of the axial, and it is taken as understood that its sense is always opposed to that of the axial strain.

Evidently, within the proportional range, where the axial strain can be equated to $1/E$ times the tensile stress, the lateral strain can be related to the tensile stress as follows:

$$\text{Lateral strain} = -\frac{\nu}{E} \cdot \text{Tensile stress} \tag{8.4}$$

Plastic Strains

Strains which are not recovered when the stress is removed, and which therefore result in a permanent deformation, are said to be plastic.

If a test in simple tension is interrupted at any point beyond the elastic limit, and the load is reduced, the graph of tensile stress against axial strain for the unloading of the specimen no longer retraces back along the graph for the loading, and there is a residual strain which remains when all the stress has been removed. It follows that beyond yield the strain can no longer be regarded as a function of the stress, and can only be determined from the complete stress/strain history of the material. Furthermore, the total strain will then include both an elastic part which is recoverable, and a plastic part which implies a permanent deformation.

More specifically, the stress/strain graph for the unloading of the specimen is a straight line parallel to the initial, straight part of the loading graph, as indicated in Fig. 8.3. At any tensile stress σ, the amount of the recoverable elastic strain is therefore equal to σ/E, and the remainder is the plastic strain.

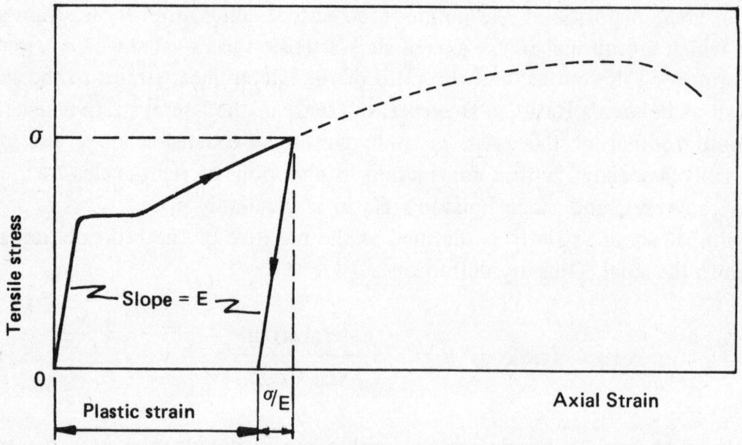

Fig. 8.3

Evidently, in a body which is loaded cyclically, plastic strains could accumulate from cycle to cycle in an incremental type of failure, and this is one of the reasons why working stresses must usually be confined to the elastic range.

The Elimination of Yield and Enhancement of the Elastic Limit

If, after unloading from a point beyond yield, the specimen is reloaded, the stress/strain graph for the reloading retraces back along the unloading graph, until it rejoins the graph of virgin loading. This implies that once the specimen has been loaded beyond yield, the phenomenon is absent from any subsequent loading. Furthermore, the Elastic Limit of the material is raised, virtually to a value of stress equal to that from which it was unloaded, and cold working in this way is sometimes used as a means of enhancing the Elastic Range of the material, and so of extending its working range. However, cold working modifies the atomic structure of the material in a way which both impairs its ductility, and renders it liable to a catastrophic type of failure known as brittle fracture, and more commonly the effects of any incidental cold working of a material are removed by a suitable heat treatment. This is known as annealing, and it effectively restores the material to its virgin state.

8.4 OTHER MATERIALS IN SIMPLE TENSION

The crucial properties of mild steel in simple tension are the simple law of stress/strain proportionality which holds within the normal working range, and the yield, which conveniently sets an upper bound to the range. However, in other materials (Fig. 8.4), the proportional range is less clear cut, whilst the yield

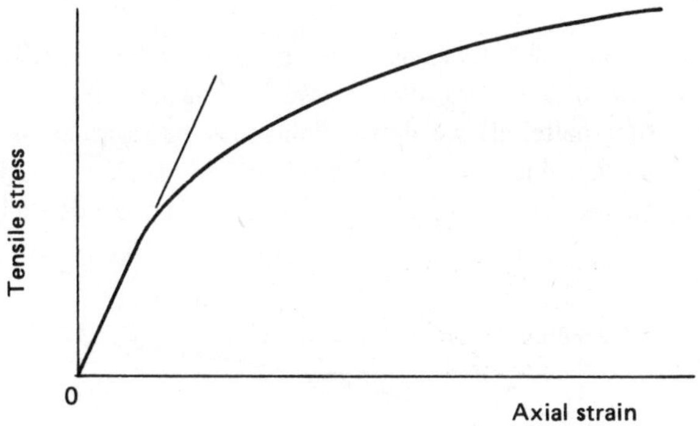

Fig. 8.4

is entirely absent. Nevertheless, for the great majority of materials within their normal working ranges, the departures from proportionality can usually be ignored, and in stress analysis the common materials are usually assumed to obey Hooke's Law.

Typical, approximate values of Young's Modulus for various materials in various units are listed in Table 8.1.

Table 8.1

	Young's Modulus		
	GN/m^2	bar	N/mm^2
Tungsten	350	$3{\cdot}5 \times 10^6$	350×10^3
Mild steel	210	$2{\cdot}1 \times 10^6$	210×10^3
Copper Cast iron	110	$1{\cdot}1 \times 10^6$	110×10^3
Brass, bronze	100	$1{\cdot}0 \times 10^6$	100×10^3
Aluminium	70	$0{\cdot}7 \times 10^6$	70×10^3
Concrete (compr$^{\underline{n}}$.)	20	$0{\cdot}2 \times 10^6$	20×10^3
Timber	10	$0{\cdot}1 \times 10^6$	10×10^3

Proof Stress

As to the absence of yield, this phenomenon is significant chiefly because it marks the point at which a significant level of plastic strain of about 0·2% occurs, so for other materials we define what may be regarded as an artificial yield stress, as one at which a specified level (usually 0·1% or 0·2%) of plastic strain accumulates, and such is called a Proof Stress, or an Offset Yield Stress.

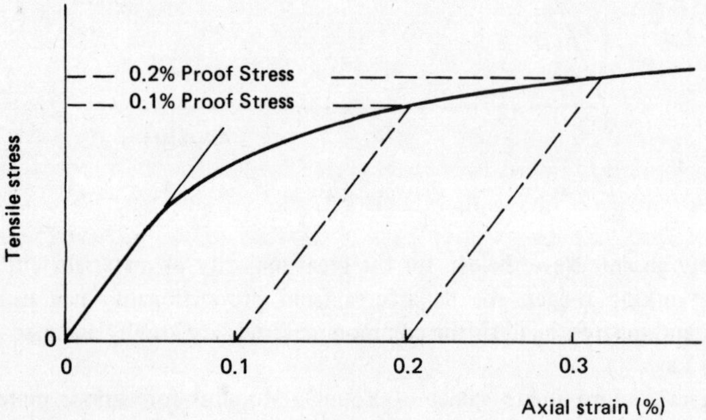

Fig. 8.5

The Proof Stress of a given material at a given value of proof strain is determined by first determining the tensile stress v axial strain graph for the material in simple tension (Fig. 8.5). Then the specified value of proof strain being offset along the strain axis, a straight line is drawn through the point, parallel to the initial portion of the graph, when the value of stress in which the line cuts the graph is the required Proof Stress at the specified proof strain.

8.5 THERMAL STRAINS

The strains in a body depend, not only on the stresses in the body, but on its temperature also. However, for small strains, the two effects are sensibly independent, and we now consider the deformation at a point due to a change of temperature only, the state of stress at the point being assumed not to vary.

Reckoning the deformation from an arbitrary datum temperature, and assuming a set of rectangular axes which are independent of the body, let $[\epsilon_T]$ represent the tensor in the given axes of the thermal strains at a point in the body at a general temperature T. In general, the strains will vary with the temperature, so the components of $[\epsilon_T]$ will be functions of temperature, and reflection will show that if $[\alpha] = \mathrm{d}[\epsilon_T]/\mathrm{d}T$ represents the matrix whose components are the derivatives of the components of $[\epsilon_T]$, then $[\alpha]$ will be another symmetrical tensor of the same kind as $[\epsilon_T]$, and it will describe the deformation at the point per unit change of temperature. The deformation due to a change of temperature from T to $T + \delta T$ will therefore be given by $[\alpha] \, . \, \delta T$, and it follows that to describe the properties of a material in thermal expansion, it will generally be necessary to determine the six functions which describe the variation with temperature of the six independent parameters of $[\alpha]$.

However, the common structural solids, in bulk, are often sensibly isotropic. In the present context, this implies linear strains which have a common value in all directions, with shears which are zero throughout, and it follows that the thermal properties of an isotropic material are determined by the single scalar function α, called the Coefficient of Thermal Expansion of the material, which describes the variation with temperature of the common value of linear strain per unit change of temperature. So where a single Coefficient of Thermal Expansion is quoted, it is to be supposed that the material is thermally isotropic, and reflection will show that this implies a pure dilation, without distortion, in a pattern of deformation which precisely parallels a hydrostatic state of stress. The deformation due to a change of temperature from T to $T + \delta T$ can therefore be described completely by the invariant tensor $\alpha \, . \, \delta T \, . \, [\mathbf{I}]$, where $[\mathbf{I}]$ is the unit matrix.

But although α is strictly to be regarded as a function of temperature, the effect of the variation is often small enough to be ignored, and the Coefficients of Thermal Expansion of the common solids are therefore commonly quoted as constants. In that event, the Coefficient of Thermal Expansion is a constant of proportionality which relates the linear thermal strains to the change of temperature, in much the same way as Young's Modulus relates the tensile stress to the axial strain in simple tension. However, whereas the patterns of deformation produced by stresses are complex, the pattern of thermal strains in an isotropic material is particularly simple. Thus the tensor of strains produced by a state of stress will generally contain both linear strains and shears, and the tensor will vary with the orientation of the axes. But in any case, and irrespective of the orientation of the axes, the effect of a change of temperature δT can be accounted for, simply by adding $\alpha \,.\, \delta T$ to the linear terms of the tensor of strains that arise from the stresses, and this can be expressed in tensor form, simply by the addition of the invariant tensor $\alpha \,.\, \delta T \,.\, [\mathbf{I}]$.

8.6 EXAMPLES IN SIMPLE TENSION

In the following examples it is to be assumed that the materials are isotropic, and that each transverse section is in uniform simple tension or compression.

Example 8.1

A cylindrical specimen of mild steel having a diameter of 2·5 cm is tested in simple tension, the extension of the specimen being measured on a gauge length of 8 cm. Within the proportional range an increase of 20 kN in the load causes an increase in the gauge length of 16×10^{-3} mm and a reduction in the diameter of the specimen of $1{\cdot}7 \times 10^{-3}$ mm. The load at yield is 108 kN. Determine: Young's Modulus and Poisson's Ratio for the material, and the values at yield, of the tensile stress, the greatest shear stress, and the octahedral shear stress.

Solution

The variation in the tensile stress is determined by dividing the variation in the load by the area of the section, and the consequential linear strains in the axial and lateral directions are determined as the ratios of the changes, (i.e., *increases*) in length to the original lengths. Thus, for the increase of load of 20 kN.

$$\text{Variation of stress} = \frac{20 \times 10^3}{\pi \times 12{\cdot}5^2} = 40{\cdot}74\ \text{N/mm}^2$$

$$\text{Resulting axial strain} = \frac{16 \times 10^{-3}}{80} = 2 \times 10^{-4}$$

$$\text{Resulting lateral strain} = \frac{-1{\cdot}7 \times 10^{-3}}{25} = -0{\cdot}68 \times 10^{-4}$$

Then within the proportional range, Young's Modulus may be determined as the ratio of the change of stress to the resulting axial strain, whilst Poisson's Ratio may be determined as the negative of the ratio of the lateral strain to the longitudinal. Thus:

$$\text{Young's Modulus} \quad = E = \frac{40{\cdot}74}{2 \times 10^{-4}} = 203{\cdot}7 \times 10^3\ \text{N/mm}^2$$

$$\text{Poisson's Ratio} \quad = \nu = -\frac{-0{\cdot}68 \times 10^{-4}}{2 \times 10^{-4}} = 0{\cdot}34\ .$$

The value of simple tensile stress at yield σ_y is determined by dividing the load at yield by the area of the section, and since the remaining principal stresses are both zero, it follows that the greates shear stress at yield is equal to $\sigma_y/2$, whilst the octahedral shear stress at yield is equal to $\sqrt{2}\sigma_y/3$

$$\text{Tensile stress at yield} \quad = \frac{108 \times 10^3}{\pi \times 12{\cdot}5^2} = 220\ \text{N/mm}^2$$

$$\text{Greatest shear stress at yield} \quad = \frac{220}{2} = 110\ \text{N/mm}^2$$

$$\text{Octahedral shear stress at yield} \quad = \frac{220\sqrt{2}}{3} = 104\ \text{N/mm}^2$$

Example 8.2

The following concurrent values of tensile force F and extension Δ are taken from a test in simple tension over a gauge length of 25 mm on a light alloy specimen having a sectional area of 20 mm². Plot the load/extension graph for the test and thus determine the values of Young's Modulus, and the 0·1% and 0·2% Proof Stresses for the material.

F (kN)	0·2	0·4	0·6	0·8	1·0	1·2	1·4	1·6	1·8
Δ (mm × 10^{-3})	3	6	10	12	17	22	28	41	90

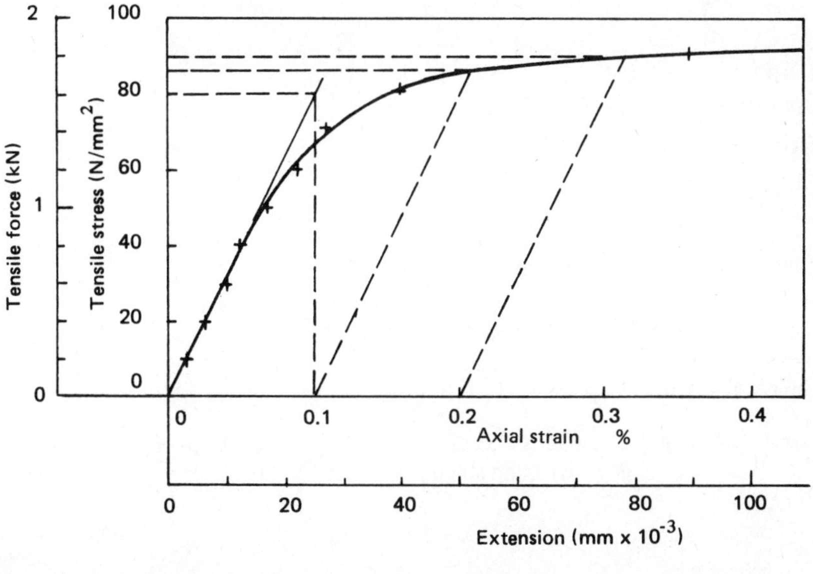

Solution

From the slope of the initial portion of the graph:

$$\text{Young's Modulus} = E = \frac{80\ \text{N/mm}^2}{0{\cdot}1\%} = 80 \times 10^3\ \text{N/mm}^2\ .$$

From the intercepts with the graph of the parallel lines through the 0·1% and 0·2% strain offsets:

$$0{\cdot}1\%\ \text{Proof Stress} = 82\ \text{N/mm}^2$$

$$0{\cdot}2\%\ \text{Proof Stress} = 84\ \text{N/mm}^2\ .$$

Example 8.3

A baulk of timber BC is to be formed into a hoist for a load of 12 kN by the addition of a horizontal steel tie, AB, 2 m long. The ties available have diameters in multiples of 5 mm, and the allowable tensile working stress, and Young's Modulus for the material have values of 50 N/mm^2 and 210 $\times$ 10^3 N/mm^2 respectively. Select a suitable diameter of tie, and calculate its extension under the load.

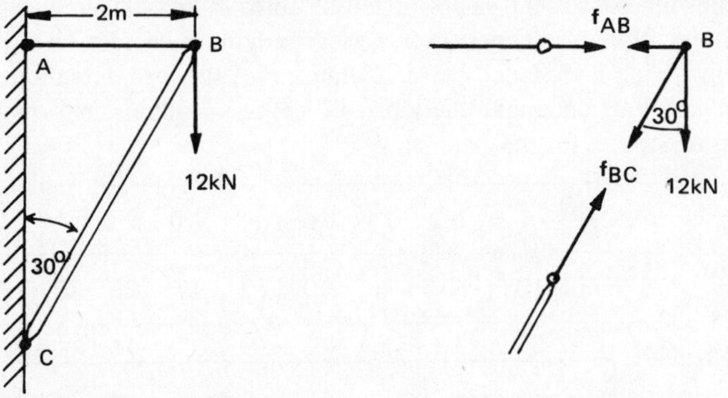

Solution

From the equilibrium of joint B:

$$\left.\begin{aligned} &\text{Vertically: } f_{BC}\cos 30° + 12\,000 = 0 \\ &\text{Horizontally:} \\ &\qquad f_{BC}\sin 30° + f_{AB} = 0 \end{aligned}\right\}$$

Whence: $f_{AB} = 12\,000 \tan 30° = 6929\ \text{N}\ .$

$$\text{Allowable working stress} = \frac{\text{Force}}{\pi d^2/4}\ .$$

The necessary diameter is therefore given by

$$d = \left(\frac{4 \times \text{force}}{\pi \times \text{working stress}}\right)^{\frac{1}{2}} = \left(\frac{4 \times 6929}{\pi \times 50}\right)^{\frac{1}{2}} \simeq 13{\cdot}3\ \text{mm}\ .$$

Of the available diameters the next largest is of 15 mm.

$$\text{Extension} = \text{axial strain} \times \text{initial length}$$

$$= \frac{\text{tensile stress}}{\text{Young's Modulus}} \times \text{initial length}$$

$$= \frac{\text{tensile force}}{\text{section area}} \times \frac{\text{initial length}}{\text{Young's Modulus}}$$

$$= \frac{6929}{\pi.15^2/4} \times \frac{2 \times 10^3}{210 \times 10^3}\,\text{mm} = 0{\cdot}373\ \text{mm}\ .$$

Example 8.4

A solid right cone of axial length l is made of a material having specific weight w and Young's Modulus E. Show that when the cone is suspended from its circular base, its elongation δ under its own weight is given by $\delta = wl^2/6E$.

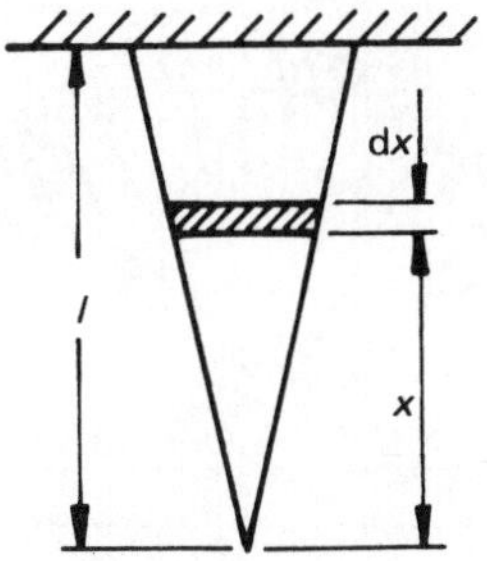

Solution

Let A represent the sectional area of an elemental slice of thickness dx at a distance x from the apex. The tensile force on the section at x is the weight of a cone of base area A and height $x = w\,x\,A/3$, and the tensile stress on the element is therefore equal to $w\,x/3$. The axial strain is therefore equal to $w\,x/3E$, and the axial extension is equal to $w\,x\,\mathrm{d}x/3E$. The axial extension of the whole cone can then be determined by integrating the extension of the elements from $x = 0$ to $x = l$:

$$\delta = (w/3E)\int_0^l x\,\mathrm{d}x = wl^2/6E\ .$$

Example 8.5

A uniform bar of material has a coefficient of thermal expansion of $1 \times 10^{-5}/°\mathrm{C}$ and Young's Modulus of $200 \times 10^3\ \mathrm{N/mm^2}$. Determine the stress in the bar when its ends are rigidly clamped and its temperatuee is then raised by 50°C.

Solution
Let σ be the tensile stress induced by the change of temperature. Then the total axial strain is the sum of the strains due to σ and σT, and since, by the statement of the problem, both the elongation and the total strain are zero:

$$\text{Total axial strain} = \sigma/E + \alpha.\delta T = 0 \; ,$$

that is: $\sigma = -E\,\alpha\,\delta T$

or $\sigma = -200 \times 10^3.10^{-5}.50 = -100\ \text{N/mm}^2$

Example 8.6
A length of a linear spring is stretched around the circumference of a disc of 50 cm diameter so as to give a radial force per unit length around the periphery of 4 N per centimetre. a) Determine the necessary tension in the spring. b) Given that a 10 cm length of the spring extends by 0·5 cm under a load of 36 N, determine the free length of spring required.

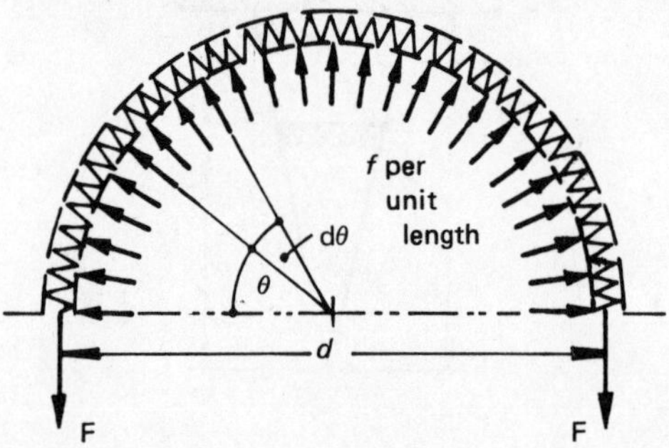

Solution
a) Let F be the tensile force in the spring for a radial force f per unit length of circumference, on the diameter d. For the equilibrium of the spring around one half of the circumference:

$$2F = (fd/2)\int_0^{\pi} \sin\theta\, d\theta = fd$$

but $f = 400$ N/m

and: $d = 0{\cdot}5$ m .

Therefore: $F = 100$ N .

b) Stiffness of spring 0·1 m long = 36/0·005 = 7200 N/m. But the extensions under a given load of different lengths of a given spring would be proportional to their lengths. The stiffnesses of the different lengths would therefore be in the inverse proportion of the lengths, and the stiffness K of a spring l metres long is therefore given by:

$$K = 7{\cdot}2 \times 10^3 \,.\, (0{\cdot}1/l) = 720/l \text{ N/m} \,.$$

$$\text{Extension of stretched spring} = \pi d - l = (0{\cdot}5\pi - l) \text{ m} \,.$$

$$\text{Tensile force in spring} = (0{\cdot}5\pi - l)720/l = 100 \text{ N} \,.$$

Thus: $0{\cdot}5\pi \times 7{\cdot}2 = 8{\cdot}2\,l$

whence: $l = 1{\cdot}379$ m .

Example 8.7

A load W is supported by two equal ties PQ and PR, of length l, sectional area a, and Young's Modulus E, suspended from a rigid horizontal as indicated.

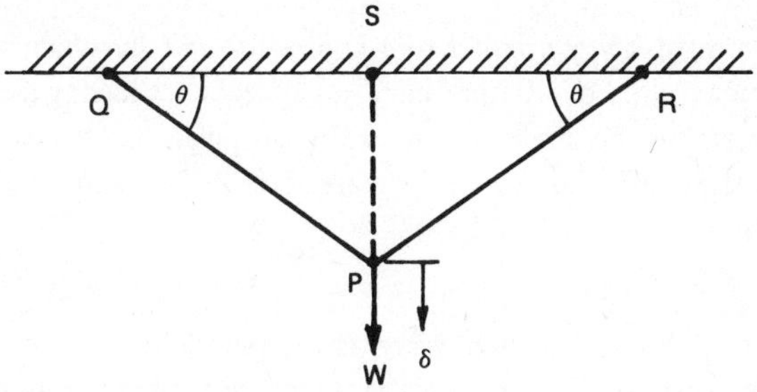

a) show that the stiffness of the joint P, (i.e., the force W per unit displacement δ) is given by:

$$W/\delta = 2Ea\sin^2\theta/l \,.$$

b) An additional tie of the same material but of area a' is to be added between P and S so as to double the stiffness of the joint P. Show that the required ratio of a' to a is:

$$a'/a = 2\sin^3\theta \,.$$

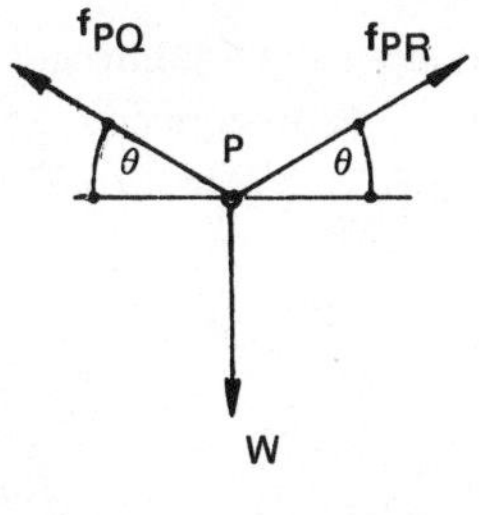

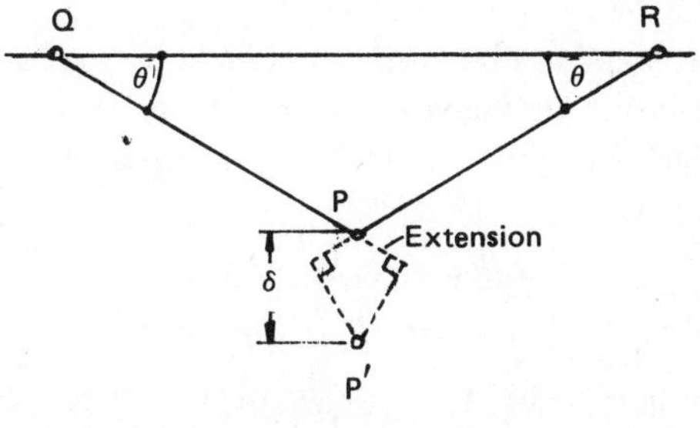

Solution

a) From the equilibrium of the joint P it is obvious that the force f in each of PQ and PR is equal to $W/2\sin\theta$. The tensile stress is therefore $W/2a\sin\theta$, the axial strain is $W/2Ea\sin\theta$, and the extension is $Wl/2Ea\sin\theta$. The displacement δ is therefore given by:

$$\delta = \text{Ext}^{\text{n}}/\sin\theta = Wl/2Ea\sin^2\theta$$

whence $$W/\delta = 2Ea\sin^2\theta/l\ .$$

b) The extension of PS alone due to a load W is given by $\delta = W\,l\sin\theta/Ea'$, and the stiffness of PS alone is therefore given by $W/\delta = Ea'/l\sin\theta$. But if the addition of PS doubles the stiffness of PQ and PR, then the stiffness of PS must be equal to that of PQ and PR; that is:

$$Ea'/l\sin\theta = 2Ea\sin^2\theta/l$$

whence $$a'/a = 2\sin^3\theta\ .$$

Example 8.8

An axial force of 2 kN is applied to a composite rod in which a copper sheath having a sectional area of 40 mm^2 encloses a steel core having a sectional area of 20 mm^2. Determine the stress in the steel core, given that the values of Young's Modulus for the steel and the copper are 210×10^3, and 110×10^3 N/mm^2, respectively.

Solution

There are two unknown stresses, whereas the parallel system of forces has only one degree of freedom. The problem is therefore statically indeterminate and it will therefore be necessary to satisfy a geometrical condition as well as the laws of equilibrium. Thus, let σ_s, A_s and E_s represent the tensile stress, sectional area and Young's Modulus for the steel, and let σ_c, A_c and E_c represent the corresponding values for the copper.

Equilibrium. The tensile force on any section can be determined by multiplying the tensile stress by the area, and the sum of the tensile forces on the copper and steel can be equated to the given tensile force on the whole.

Thus: $$\sigma_s A_s + \sigma_c A_c = 2 \times 10^3\,\text{N}\ . \qquad \text{(a)}$$

Geometrical Compatibility. Since the steel and the copper have common values of both length and extension, the axial strains in the steel and copper are equal. Thus, the axial strains being determined as the ratio of the stress to Young's Modulus:

$$\sigma_s/E_s = \sigma_c/E_c$$

or $$\sigma_c = (E_c/E_s)\sigma_s \qquad \text{(b)}$$

(b) in (a) $\rightarrow \sigma_s\{A_s + (E_c/E_s)A_c\} = 2 \times 10^3\,\text{N}$

$$\sigma_s = \frac{2 \times 10^3}{20 + (110/210)40}\,\text{N/mm}^2 = 48{\cdot}8\ \text{N/mm}^2\ .$$

Example 8.9

Cylinders of copper and steel are stood on end on a rigid plate, and a shaped rigid yoke is placed across their upper ends, as indicated. Assuming that the

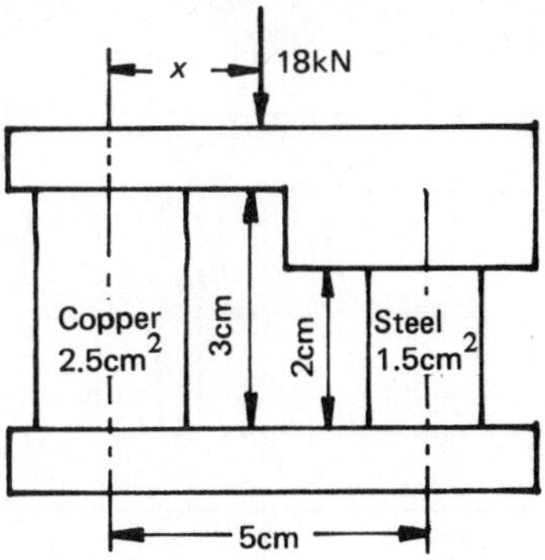

conditions remain elastic, determine at what position, x, a load of 18 kN must be applied in order to retain the level of the upper surface of the yoke.

For steel: $E = 210 \times 10^3$ N/mm^2; for copper: $E = 120 \times 10^3$ N/mm^2.

Solution

By the statement of the problem the extensions of the two cylinders must be equal. Therefore, distinguishing properties of the copper by the suffix 'c', and of the steel by the suffix 's':

$$\sigma_c l_c / E_c = \sigma_s l_s / E_s ,$$

that is: $$\sigma_c / \sigma_s = (l_s / l_c)(E_c / E_s) \qquad \text{(a)}$$

For the equilibrium of the yoke (by taking moments of forces):

$$\sigma_c A_c x = \sigma_s A_s (5 - x)$$

therefore: $$x(\sigma_c A_c + \sigma_s A_s) = 5\sigma_s A_s ,$$

that is: $$x\{(\sigma_c/\sigma_s)(A_c/A_s) + 1\} = 5$$

and substituting for (σ_c/σ_s) from (a):

$$x\{(l_s/l_c)(E_c/E_s)(A_c/A_s) + 1\} = 5$$

that is: $$x\{(2/3)(12/21)(2{\cdot}5/1{\cdot}5) + 1\} = 5,$$

that is: $$x = 5 \times 63/103 = 3{\cdot}05 \text{ cm} .$$

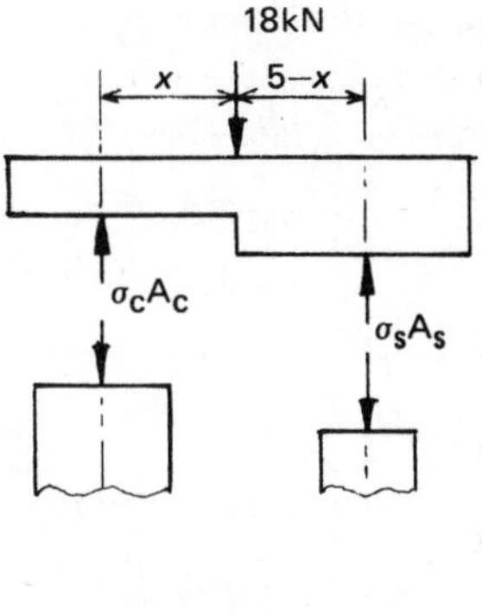

Example 8.10

One cylinder of steel and two of copper, each having a sectional area of 4 cm^2 and a length of 6 cm, are contained symmetrically between rigid end blocks as indicated.

a) Determine the stresses in the cylinders when the temperature increases by 50°C, and compressive axial forces of 60 kN are applied to the rigid end blocks.

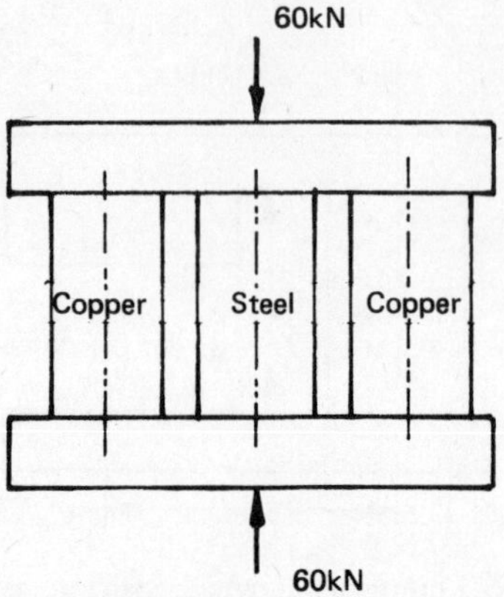

b) Within what range of temperature variation will all three cylinders remain in contact with the two end blocks under the compressive load of 60 kN?

c) Before assembly, the length of the copper cylinders is varied so that when the temperature is raised by 50°C and the compressive load of 60 kN is applied, the stress in the steel is twice the stress in the copper. Determine the necessary variation in the length of the copper cylinders.

For steel: $E_s = 210 \times 10^3\ \text{N/mm}^2$; $\alpha_s = 1{\cdot}0 \times 10^{-5}/°\text{C}$

For copper: $E_c = 120 \times 10^3\ \text{N/mm}^2$; $\alpha_c = 1{\cdot}5 \times 10^{-5}/°\text{C}$.

Solution

a) Let σ_s and σ_c represent the tensile stresses in the steel and copper cylinders respectively.

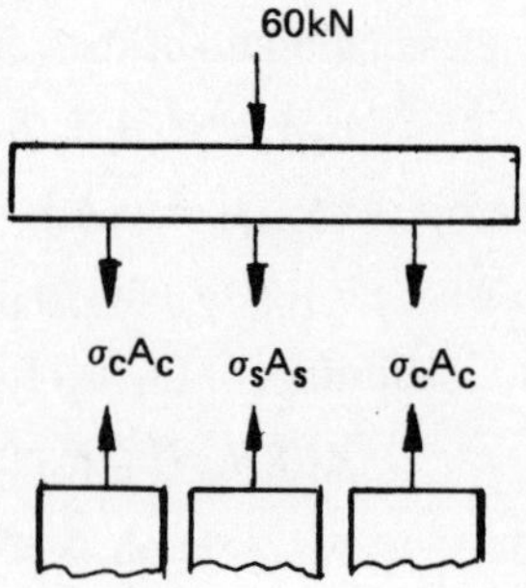

For the equilibrium of the yoke:

$$\sigma_s a_s + 2\sigma_c a_c + 60\,000 = 0$$

$$400\sigma_s + 800\sigma_c + 60\,000 = 0$$

$$\sigma_s = -(2\sigma_c + 150)\ . \qquad \text{(a)}$$

For the equality of strains in the steel and copper cylinders:

$$\sigma_s/E_s + \alpha_s\delta T = \sigma_c/E_c + \alpha_c\delta T$$

$$\sigma_s = (E_s/E_c)\,\sigma_c + E_s(\alpha_c - \alpha_s)\,\delta T$$

$$\sigma_s = 1{\cdot}75\,\sigma_c + 52{\cdot}5 \qquad \text{(b)}$$

Between (a) and (b): $\sigma_c = -54\ \text{N/mm}^2$: $\sigma_s = -42\ \text{N/mm}^2$.

b) When the steel cylinder is on the point of losing contact the load is borne wholly by the copper alone and is given by $\sigma_c = -60\ 000\ \text{N}/(2 \times 400\ \text{mm}^2) = -75\ \text{N/mm}^2$. However, the strains in the steel and the copper are still equal, whence:

$$\sigma_c/E_c + \alpha_c \delta T = \alpha_s \delta T \ ,$$

$$\text{therefore:} \quad \delta T = \frac{-\sigma_c}{E_s(\alpha_c - \alpha_s)} = \frac{75}{1{\cdot}2 \times 0{\cdot}5} = 125°\text{C} \ .$$

Similarly, when the stress in the copper just falls to zero, $\sigma_s = -60\ 000/400\ \text{mm}^2 = -150\ \text{N/mm}^2$, and:

$$\sigma_s/E_s + \alpha_s \delta T = \alpha_c \delta T \ .$$

$$\delta T = \frac{\sigma_s}{E_s(\alpha_c - \alpha_s)} = \frac{-150}{2{\cdot}1 \times 0{\cdot}5} = -142°\text{C} \ .$$

Under the compressive forces of 60 kN, contact between the end pieces and the cylinders is maintained for variations of temperature in the range from −142°C to 125°C.

c) Since the total area of the two copper cylinders is twice that of the single steel cylinder, when the stress in the steel is twice that in the copper, the total load is equally shared between the steel and the copper.

$$\text{Thus} \qquad \sigma_s = 2\sigma_c = -\ 30\ 000\ \text{N}/400\ \text{mm}^2 = -\ 75\ \text{N/mm}^2$$

$$\text{therefore:} \quad \sigma_c - \sigma_s = (\sigma_c/E_c + \alpha_c \delta T) - (\sigma_s/E_s + \alpha_s \delta T)$$

$$= \left(\frac{-37{\cdot}5}{120} - \frac{-75}{210} + 0{\cdot}5 \times 10^{-2} \times 50\right) 10^{-3}$$

$$= (-{\cdot}3125 + {\cdot}3571 + {\cdot}25) \times 10^{-3} = {\cdot}294 \times 10^{-3}$$

Required variation of length $= -0{\cdot}294 \times 10^{-3} \times 60\ \text{mm} = 17{\cdot}4 \times 10^{-3}\ \text{mm}$.

Example 8.11

A cylinder 4 cm dia. and 14 cm long has its diameter turned down to 2 cm over

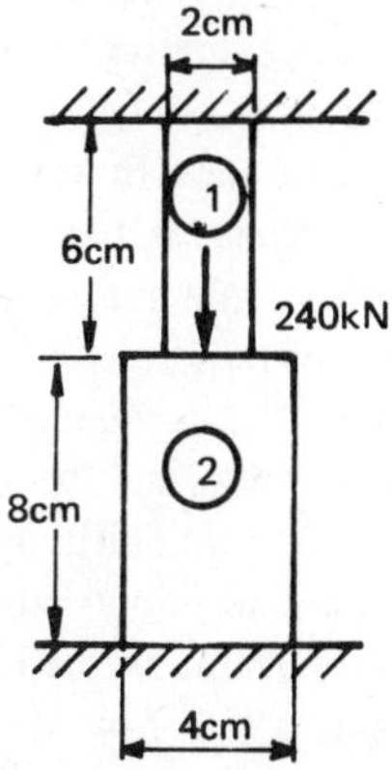

a length of 6 cm, as indicated. Its ends are built into rigidly fixed end plates, and an axial load of 240 kN is applied at the junction section. Given that E for the material is 210×10^3 N/mm^2, and assuming that the stresses in each part are homogeneous, determine the stresses in the two parts.

Solution

By the statement of the problem the extension overall is zero. Therefore, distinguishing quantities that apply to the upper part by the suffix '1', and in the lower part by the suffix '2':

$$\text{Extension overall} = (1/E)(\sigma_1 l_1 + \sigma_2 l_2) = 0$$

therefore:

$$\sigma_1 = -(l_2/l_1)\sigma_2 \ . \qquad \text{(a)}$$

For equilibrium of *forces* in the axial direction:

$$\sigma_1 A_1 = \sigma_2 A_2 + 240 \times 10^3 \ ,$$

that is:

$$\sigma_1 = \sigma_2(A_2/A_1) + 240 \times 10^3/A_1 \ . \qquad \text{(b)}$$

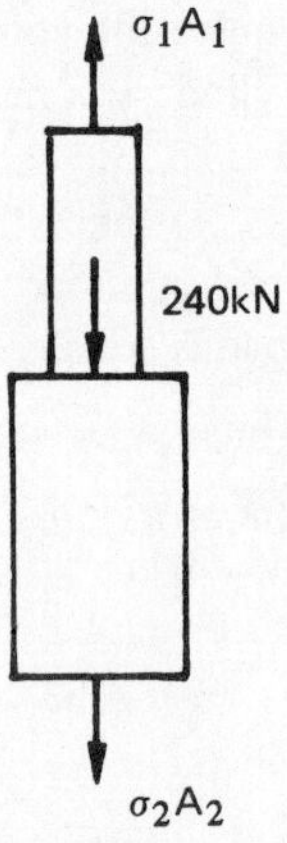

Between (a) and (b)

$$-(l_2/l_1)\sigma_2 = (A_2/A_1)\sigma_2 + 240 \times 10^3/A_1$$

that is:

$$\sigma_2 = -\frac{(240 \times 10^3)/A_1}{\{(l_2/l_1) + (A_2/A_1)\}}$$

$$= -\frac{240 \times 10^3}{\{(8/6) + (4/2)^2\}\,\pi.10^2}\ \text{N/mm}^2$$

$$\sigma_2 = -\ 143{\cdot}2\ \text{N/mm}^2$$

Substituting in (a)

$$\sigma_1 = \quad 191{\cdot}0\ \text{N/mm}^2 \ .$$

8.7 STRESS v STRAIN IN ISOTROPIC HOOKEAN SOLIDS

Having described the stress/strain properties of the common structural solids in simple tension, we now require to generalise the results in a form which applies to states of stress in general. Thus, assuming an elastic solid in which each state of stress is associated with a specific pattern of deformation, let $[\sigma]$ represent a general tensor of a general state of stress, and let $[\epsilon]$ represent the corresponding strain tensor, i.e., the tensor in the same axes as $[\sigma]$, of the small strains of the associated deformation. Then we require to describe the properties of the material in a manner which relates the components of one of the tensors to the components of the other. So considering the deformation of mild steel in simple tension more broadly, we find that although we have so far referred only to the proportionality of the axial and lateral strains, in fact, all the strains are proportional to the tensile stress, and this is true of shear strains as well as linear. Moreover, in more complex states which have three non-zero principal stresses, it is found that the deformations associated with the various principal stresses are sensibly independent of one another. So, for purposes of stress analysis, the properties of mild stell in simple tension within the proportional range are generalised as in the following definition of a hypothetical Hookean material, and this is usually taken to be a reasonable idealisation of the common structtural solids within their normal working ranges.

The Hookean Hypothesis

A Hookean material is postulated as one in which the deformations associated with the various components of any stress tensor are independent of one another, and in each deformation all the strains are proportional to the component of stress with which they are associated. A Hookean solid is therefore one in which each component of either of a corresponding pair of stress and strain tensors can be described as a linear function of the components of the other, and the constant coefficients that appear in such relationships are constants of the material known as elastic constants.

This definition would appear to imply that to describe the stress/strain properties of a Hookean material it will be necessary to specify the 36 coefficients that would be required to describe the six independent parameters of a strain tensor as linear functions of the six independent parameters of a stress tensor. But in practice, symmetries which exist both in the tensors and in the structures of most materials, invariably rob many of the constants of their independence, and ultimately, in isotropic materials, only two elastic constants are required, and only two can be treated as independent.

Isotropy

In general, isotropy implies a property which is uniform in all directions, and this we have, in effect, already assumed, in that we postulated a material in which each each state of stress is associated with a specific pattern of deformation, irrespective of the orientation of the state of stress in the material. However, elastic isotropy is also concerned with the symmetry of the deformation, and a material is said to be elastically isotropic only when its deformation in simple tension is not only independent of the direction of the tensile axis in the material, but is also cylindrically symmetrical about the axis. In part, this implies that the linear strain has a common value in all lateral directions so that transverse planes suffer a uniform dilation from which shears are entirely absent. However it also implies that there is also no relative rotation between the stress axis and any axis at right-angles to it, and it follows that, in an isotropic Hookean solid (in which the deformations associated with the various components of any stress tensor are sensibly independent) a tensile component of any stress tensor causes no shearing of the tensor axes. In particular, the principal stresses cause no shearing of the principal axes, and we can therefore conclude, first, that in an isotropic Hookean solid, the principal axes of the deformation that is associated with any state of stress coincide with the principal axes of the state of stress itself, and second, that in any corresponding pair of tensors of stress and strain, the shear strains are independent of the tensile stresses.

Generalised Stress/Strain Relationships

Recognising that the direct components of any stress tensor affect only the linear components of the corresponding strain tensor, let σ represent the column vector of the direct components of stress, and let ϵ represent the column vector of the linear strains of the corresponding strain tensor. Evidently the strain component ϵ_{11} comprises the axial strain associated with σ_{11}, together with the lateral strains associated with σ_{22} and σ_{33}, and since parallel observations can be made on ϵ_{22} and ϵ_{33}, the column vector ϵ of the strains associated with the column vector σ of stresses is determined as follows:

$$\sigma = \begin{bmatrix} \sigma_{11} \\ \sigma_{22} \\ \sigma_{33} \end{bmatrix} \rightarrow \epsilon = \begin{bmatrix} \epsilon_{11} \\ \epsilon_{22} \\ \epsilon_{33} \end{bmatrix} = \frac{1}{E} \begin{bmatrix} \sigma_{11} - v\sigma_{22} - v\sigma_{33} \\ -v\sigma_{11} + \sigma_{22} - v\sigma_{33} \\ -v\sigma_{11} - v\sigma_{22} + \sigma_{33} \end{bmatrix}$$

$$\sigma = \begin{bmatrix} \sigma_{11} \\ \sigma_{22} \\ \sigma_{33} \end{bmatrix} \rightarrow \epsilon = \begin{bmatrix} \epsilon_{11} \\ \epsilon_{22} \\ \epsilon_{33} \end{bmatrix} = \frac{1}{E} \begin{bmatrix} 1 & -v & -v \\ -v & 1 & -v \\ -v & -v & 1 \end{bmatrix} \cdot \begin{bmatrix} \sigma_{11} \\ \sigma_{22} \\ \sigma_{33} \end{bmatrix} \tag{8.5}$$

In particular, in the principal axes:

$$\mathbf{p} = \begin{bmatrix} p_1 \\ p_2 \\ p_3 \end{bmatrix} \rightarrow \epsilon_p = \begin{bmatrix} \epsilon_{p1} \\ \epsilon_{p2} \\ \epsilon_{p3} \end{bmatrix} = \frac{1}{E} \begin{bmatrix} 1 & -v & -v \\ -v & 1 & -v \\ -v & -v & 1 \end{bmatrix} \begin{bmatrix} p_1 \\ p_2 \\ p_3 \end{bmatrix} \tag{8.5a}$$

Conversely, the direct stresses may be described as linear functions of the associated linear strains as follows:

$$\sigma = \begin{bmatrix} \sigma_{11} \\ \sigma_{22} \\ \sigma_{33} \end{bmatrix} = E \begin{bmatrix} 1 & -v & -v \\ -v & 1 & -v \\ -v & -v & 1 \end{bmatrix}^{-1} \begin{bmatrix} \epsilon_{11} \\ \epsilon_{22} \\ \epsilon_{33} \end{bmatrix}$$

or

$$\sigma = \begin{bmatrix} \sigma_{11} \\ \sigma_{22} \\ \sigma_{33} \end{bmatrix} = \frac{E}{(1-2v)(1+v)} \begin{bmatrix} 1-v & v & v \\ v & 1-v & v \\ v & v & 1-v \end{bmatrix} \begin{bmatrix} \epsilon_{11} \\ \epsilon_{22} \\ \epsilon_{33} \end{bmatrix} \tag{8.5b}$$

The Bulk Modulus K

Taking the sum of the three equations in 8.5a gives

$$\epsilon_{p1} + \epsilon_{p2} + \epsilon_{p3} = \frac{1-2v}{E}(p_1 + p_2 + p_3) \; .$$

But the sum on the left is the invariant which determines the volumetric strain Δ, whilst the sum on the right is the invariant which is equal to three times the hydrostatic stress σ_{00}, and substituting accordingly gives:

$$\Delta = \frac{3(1-2v)}{E}\sigma_{00} = \frac{1}{K}\sigma_{00} \; , \tag{8.6}$$

where $K = E/3(1 - 2v)$ is an elastic constant, called the bulk modulus, which relates the hydrostatic stress σ_{00} to the volumetric strain Δ in just the same way as Young's Modulus relates a tensile stress to the axial strain.

The Shear Modulus G

Now let $\mathbf{p^*}$ represent the column vector of the deviatoric principal stresses of any state and let ϵ_p^* represent the column vector of the associated deviatoric principal strains. The components of $\mathbf{p^*}$ are determined by subtracting the average of the principal stresses from each of the principal stresses in turn, as follows:

$$\mathbf{p^*} = \begin{bmatrix} p_1^* \\ p_2^* \\ p_3^* \end{bmatrix} = \frac{1}{3}\begin{bmatrix} (2p_1 - p_2 - p_3) \\ (-p_1 + 2p_2 - p_3) \\ (-p_1 - p_2 + 2p_3) \end{bmatrix} = \frac{1}{3}\begin{bmatrix} 2 & -1 & -1 \\ -1 & 2 & -1 \\ -1 & -1 & 2 \end{bmatrix}\begin{bmatrix} p_1 \\ p_2 \\ p_3 \end{bmatrix} ,$$

and by substituting accordingly for σ_{11}, σ_{22} and σ_{33} in eqn. (8.5a), the corresponding vector of the deviatoric principal strains ϵ_p^* is determined as follows:

$$\mathbf{p^*} = \frac{1}{3}\begin{bmatrix} 2 & -1 & -1 \\ -1 & 2 & -1 \\ -1 & -1 & 2 \end{bmatrix}\begin{bmatrix} p_1 \\ p_2 \\ p_3 \end{bmatrix} \rightarrow$$

$$\epsilon_p^* = \frac{1}{3E}\begin{bmatrix} 1 & -v & -v \\ -v & 1 & 1 \\ -v & -v & -v \end{bmatrix}\begin{bmatrix} 2 & -1 & -1 \\ -1 & 2 & -1 \\ -1 & -1 & 2 \end{bmatrix}\begin{bmatrix} p_1 \\ p_2 \\ p_3 \end{bmatrix}$$

$$= \frac{(1+v)}{3E}\begin{bmatrix} 2 & -1 & -1 \\ -1 & 2 & -1 \\ -1 & -1 & 2 \end{bmatrix}\begin{bmatrix} p_1 \\ p_2 \\ p_3 \end{bmatrix}$$

or: $$\epsilon_p^* = \frac{(1+v)}{E}\,\mathrm{p}^* = \frac{1}{2G}\,\mathrm{p}^*$$

where $G = E/2(1+v)$ is an elastic constant.

Evidently, the vector ϵ_p^* of deviatoric principal strains can be determined from the vector σ_p of deviatoric principal stresses simply by proportional scaling by the factor $(1+v)/E$, and since it is obvious that this simple proportional relationship would survive the operation for the transformation of the tensors for a rotation of the common axes, it follows that any general deviatoric stress tensor $[\sigma^*]$ would be similarly related to the corresponding deviatoric strain tensor $[\epsilon^*]$ as follows:

$$[\epsilon^*] = \frac{1+v}{E}[\sigma^*] = \frac{1}{2G}[\sigma^*] \ , \tag{8.7}$$

where $G = E/2(1+v)$ is an elastic constant, called the Shear Modulus, which relates any deviatoric stress tensor to the corresponding deviatoric strain tensor, once again in just the same way as Young's Modulus relates a single component of tensile stress to the single component of axial strain.

However, we remember that in any given axes the shear components of a deviatoric tensor are the same as the shear components of the total tensor, and the foregoing result indicates that each component of the deviatoric strain tensor can be described as a simple proportion of the corresponding component of the deviatoric stress tensor. It therefore follows that in any corresponding pair of stress and strain tensors, each of the shear strains can be described as a simple proportion of the corresponding shear stress, as follows:

$$\begin{bmatrix} \gamma_{12} = \gamma_{21} \\ \gamma_{13} = \gamma_{31} \\ \gamma_{23} = \gamma_{32} \end{bmatrix} = 2\begin{bmatrix} \epsilon_{12} = \epsilon_{21} \\ \epsilon_{13} = \epsilon_{31} \\ \epsilon_{23} = \epsilon_{32} \end{bmatrix} = \frac{2(1+v)}{E}\begin{bmatrix} \sigma_{12} \\ \sigma_{13} \\ \sigma_{23} \end{bmatrix} = \frac{1}{G}\begin{bmatrix} \sigma_{12} \\ \sigma_{13} \\ \sigma_{23} \end{bmatrix} \tag{8.8}$$

Lamée's Constant λ

If $[\sigma]$ represents a general tensor of stress, $[\epsilon]$ represents the corresponding strain tensor, σ_{00} represents the hydrostatic stress (which is equal to the average of the tensile components of $[\sigma]$), and Δ represents the volumetric strain (which is equal to the sum of the linear terms of $[\epsilon]$), then eqn. (8.7) can be rewritten by equating the deviatoric stress tensor $[\sigma^*]$ to $[\sigma] - \sigma_{00}[\mathrm{I}]$, and by equating the deviatoric strain tensor $[\epsilon^*]$ to $[\epsilon] - (\Delta/3)[\mathrm{I}]$, as follows:

$$[\sigma] - \sigma_{00}[\mathrm{I}] = 2G([\epsilon] - (\Delta/3)[\mathrm{I}])$$

Finally substituting $\sigma_{00} = K\Delta$ from eqn. (8.6) and transposing gives

$$[\sigma] = 2G\,[\epsilon] + (K - \frac{2G}{3})\,\Delta[\mathrm{I}] \ ,$$

that is:
$$[\sigma] = 2G\,[\epsilon] + \frac{Ev}{(1+v)(1-2v)} \cdot \Delta\,[\mathrm{I}] = 2G\,[\epsilon] + \lambda\,.\,\Delta\,[\mathrm{I}] \tag{8.9}$$

where $\lambda = Ev/\{(1+v)(1-2v)\}$ is an elastic constant known as Lamés Constant.

Summary of Properties

The chief stress/strain properties of an isotropic Hookean solid may be summarised as follows:

a) The principal axes of the deformation that is associated with any state of stress coincide with the principal axes of the state of stress itself.

b) In any pair of corresponding tensors of stress and the associated strain, the shear strains are independent of the direct stresses, and the linear strains are independent of the shear stresses.

i) The linear strains can be determined as linear functions of the tensile stresses as follows:

$$\begin{bmatrix} \epsilon_{11} \\ \epsilon_{22} \\ \epsilon_{33} \end{bmatrix} = \frac{1}{E} \begin{bmatrix} 1 & -v & -v \\ -v & 1 & -v \\ -v & -v & 1 \end{bmatrix} \cdot \begin{bmatrix} \sigma_{11} \\ \sigma_{22} \\ \sigma_{33} \end{bmatrix}$$

ii) Conversely the direct stresses can be determined as linear functions of the associated linear strains as follows:

$$\begin{bmatrix} \sigma_{11} \\ \sigma_{22} \\ \sigma_{33} \end{bmatrix} = \frac{E}{(1-2v)(1+v)} \begin{bmatrix} 1-v & v & v \\ v & 1-v & v \\ v & v & 1-v \end{bmatrix} \begin{bmatrix} \epsilon_{11} \\ \epsilon_{22} \\ \epsilon_{33} \end{bmatrix}$$

iii) Each shear strain depends only on the single corresponding component of shear stress, and the one is related to the other by a simple constant of proportionality (the shear modulus G). Thus:

$$\begin{bmatrix} \gamma_{12} \\ \gamma_{13} \\ \gamma_{23} \end{bmatrix} = 2 \begin{bmatrix} \epsilon_{12} \\ \epsilon_{13} \\ \epsilon_{23} \end{bmatrix} = \frac{1}{G} \begin{bmatrix} \sigma_{12} \\ \sigma_{13} \\ \sigma_{23} \end{bmatrix}$$

where: $G = E/2(1+v)$.

c) Alternatively, the tensors of stress and strain may be considered to comprise the sums of their hydrostatic and deviatoric parts. Thus the hydrostatic term is the invariant average of the three terms on the leading diagonal, whilst the deviatoric tensor is formed simply by subtracting this amount from each term on the leading diagonal of the tensor, when the deviatoric tensors are related by one simple constant of proportionality (the shear modulus G), whilst the hydrostatic stress is related to the volumetric strain by another constant of proportionality (the bulk modulus K). These relationships can be expressed mathematically as follows:

$$\frac{\sigma_{11} + \sigma_{22} + \sigma_{33}}{3} = \sigma_{00} = K\,\Delta = K(\epsilon_{11} + \epsilon_{22} + \epsilon_{33})$$

and: $$[\sigma] - \sigma_{00}[\mathrm{I}] = [\sigma^*] = 2G.\,[\epsilon^*] = 2G([\epsilon] - \frac{\Delta}{3}[\mathrm{I}])$$

where: $G = E/2(1+v)$ and: $K = vE/(1+v)(1-2v)$.

Comparing the alternatives, the relationships in (b) are convenient in so far as the elastic properties are usually determined and specified in terms of E and v. However, to relate the shear terms of a general tensor it is still necessary to determine G, and although this impediment does not arise in dealing with the principal tensors, the necessary relationships are still relatively awkward.

The relationships in (c) require both the determination of the values of G and K, and the preliminary step of dividing the tensors into their hydrostatic and deviatoric parts. However, this is relatively simple, and the necessary relationships then involve only a simple constant of proportionality between each pair of quantities.

Lamé's equation is an adaptation of the relationships in (c) which proves convenient in formal stress analysis.

Example 8.12

Determine the values of Shear Modulus, Bulk Modulus, and Lamé's Constant for a material in which Young's Modulus equals 120×10^3 N/mm², and Poisson's ratio 0·35.

Solution

$$\text{Shear Modulus} = G = \frac{E}{2(1+v)}$$

$$= \frac{120 \times 10^3}{2 \times 1{\cdot}35} = 44{\cdot}44 \text{ N/mm}^2 \ .$$

$$\text{Bulk Modulus} = K = \frac{E}{3(1-2v)}$$

$$= \frac{120 \times 10^3}{0{\cdot}9} = 133{\cdot}3 \times 10^3 \text{ N/mm}^2$$

Lamé's Constant

$$= \lambda = \frac{E}{(1+v)(1-2v)}$$

$$= \frac{0{\cdot}35 \times 120 \times 10^3}{1{\cdot}35 \times 0{\cdot}3} = 103{\cdot}7 \times 10^{-3}\ \text{N/mm}^2\ .$$

Example 8.13
The principal stresses of a state of stress N/mm² are: $p_1 = 80; p_2 = 60; p_3 = -20$. Given that $E = 210 \times 10^3$ N/mm² and that $v = 0{\cdot}4$, determine the principal strains at the point, and the shear strain of the angle between the lines whose direction cosines in the principal axes are $\hat{\text{l}} = (1, 2, 2)/3$ and $\hat{\text{l}}' = (2, 1, -2)/3$.

Solution

$$\begin{bmatrix} \epsilon_{p1} \\ \epsilon_{p2} \\ \epsilon_{p3} \end{bmatrix} = \frac{1}{E}\begin{bmatrix} 1 & -v & -v \\ -v & 1 & -v \\ -v & -v & 1 \end{bmatrix}\begin{bmatrix} p_1 \\ p_2 \\ p_3 \end{bmatrix} = \frac{1}{E}\begin{bmatrix} 10 & -4 & -4 \\ -4 & 10 & -4 \\ -4 & -4 & 10 \end{bmatrix}\begin{bmatrix} 8 \\ 6 \\ -2 \end{bmatrix} = \frac{1}{E}\begin{bmatrix} 64 \\ 36 \\ -76 \end{bmatrix}$$

$$\epsilon_{p1} = 3{\cdot}05 \times 10^{-4};\quad \epsilon_{p2} = 1{\cdot}71 \times 10^{-4};\quad \epsilon_{p3} = -3{\cdot}62 \times 10^{-4}$$

$$\gamma_{ll'} = 2\epsilon_{ll'} = 2\,\hat{\text{l}}\cdot\underline{\epsilon_p}\cdot\hat{\text{l}}'$$

$$\gamma_{ll'} = \frac{2}{9}[1\ \ 2\ \ 2]\begin{bmatrix} 3{\cdot}05 & 0 & 0 \\ 0 & 1{\cdot}71 & 0 \\ 0 & 0 & -3{\cdot}62 \end{bmatrix}\begin{bmatrix} 2 \\ 1 \\ -2 \end{bmatrix}$$

$$= \frac{2}{9}[1\ \ 2\ \ 2]\begin{bmatrix} 7{\cdot}10 \\ 1{\cdot}71 \\ 7{\cdot}24 \end{bmatrix} \times 10^{-4} = 5{\cdot}56 \times 10^{-4}\ .$$

Example 8.14
A tensor of a state of stress in N/mm² has the value $[\sigma]$ shown. Determine the corresponding strain tensor given that $E = 210 \times 10^3$ N/mm², and $v = 0{\cdot}38$.

$$[\sigma] = 10\begin{bmatrix} 8 & -3 & 5 \\ -3 & 4 & 0 \\ 5 & 0 & 3 \end{bmatrix}$$

Solution

Treating the linear and shear effects independently:

$$\begin{bmatrix} \epsilon_{11} \\ \epsilon_{22} \\ \epsilon_{33} \end{bmatrix} = \frac{1}{E} \begin{bmatrix} 1 & -v & -v \\ -v & 1 & -v \\ -v & -v & 1 \end{bmatrix} \begin{bmatrix} \sigma_{11} \\ \sigma_{22} \\ \sigma_{33} \end{bmatrix} = \frac{10}{E} \begin{bmatrix} 1 & -0{\cdot}38 & -0{\cdot}38 \\ -0{\cdot}38 & 1 & -0{\cdot}38 \\ -0{\cdot}38 & -0{\cdot}38 & 1 \end{bmatrix} \begin{bmatrix} 8 \\ 4 \\ 3 \end{bmatrix}$$

$$= 0{\cdot}476 \times 10^{-4} \begin{bmatrix} 5{\cdot}34 \\ -0{\cdot}18 \\ -1{\cdot}56 \end{bmatrix} = \begin{bmatrix} 2{\cdot}54 \\ -0{\cdot}09 \\ -0{\cdot}74 \end{bmatrix} \times 10^{-4}$$

$$1/2G = (1 + v)/E = 0{\cdot}0657 \times 10^{-4}$$

$$\begin{bmatrix} \epsilon_{12} \\ \epsilon_{13} \\ \epsilon_{23} \end{bmatrix} = \frac{1}{2G} \begin{bmatrix} \sigma_{12} \\ \sigma_{13} \\ \sigma_{23} \end{bmatrix} = 0{\cdot}657 \times 10^{-4} \begin{bmatrix} -3 \\ 5 \\ 0 \end{bmatrix} = \begin{bmatrix} -1.97 \\ 3.29 \\ 0 \end{bmatrix}$$

$$[\epsilon] = \begin{bmatrix} 2{\cdot}54 & -1{\cdot}97 & 3{\cdot}29 \\ -1{\cdot}97 & -0{\cdot}09 & 0 \\ 3.29 & 0 & -0{\cdot}74 \end{bmatrix} \times 10^{-4}$$

Alternatively, treating the hydrostatic and deviatoric effects independently:

$$[\epsilon] = \frac{1}{2G}([\sigma] - \sigma_{00}\,[\mathbf{I}]) + \frac{1}{3K}\sigma_{00}\,[\mathbf{I}]$$

where $1/2G = 0{\cdot}0657 \times 10^{-4}$; $1/3K = (1 - 2v)/E = 0{\cdot}0114 \times 10^{-4}$, and $\sigma_{00} = 10(8 + 4 + 3)/3 = 10 \times 5$ N/mm^2.

$$[\epsilon] \times 10^4 = 0{\cdot}657 \begin{bmatrix} 3 & -3 & 5 \\ -3 & -1 & 0 \\ 5 & 0 & -2 \end{bmatrix} + 0{\cdot}114 \times 5\,[\mathbf{I}]$$

$$[\epsilon] \times 10^4 = \begin{bmatrix} 1{\cdot}97 & -1{\cdot}97 & 3{\cdot}29 \\ -1{\cdot}97 & -0{\cdot}66 & 0 \\ 3{\cdot}29 & 0 & -1{\cdot}31 \end{bmatrix} + 0{\cdot}57\,[\mathbf{I}]$$

and adding 0·57 to each term on the leading diagonal of the array gives:

$$[\epsilon] = \begin{bmatrix} 2{\cdot}54 & -1{\cdot}97 & 3{\cdot}29 \\ -1{\cdot}97 & -0{\cdot}09 & 0 \\ 3{\cdot}29 & 0 & -0{\cdot}74 \end{bmatrix} \times 10^{-4}$$

Example 8.15

A flat plate having a uniform thickness of 2 cm is subjected to a regime of plane stress in which the faces are free of stress, and the diagram indicates the stresses on an element isolated by transverse sections. For this point determine the greatest shear strain, the greatest linear strain, and the change of thickness of the plate, given that Young's Modulus E for the material is 210×10^3 N/mm^2, whilst Poisson's Ratio v is equal to 1/3.

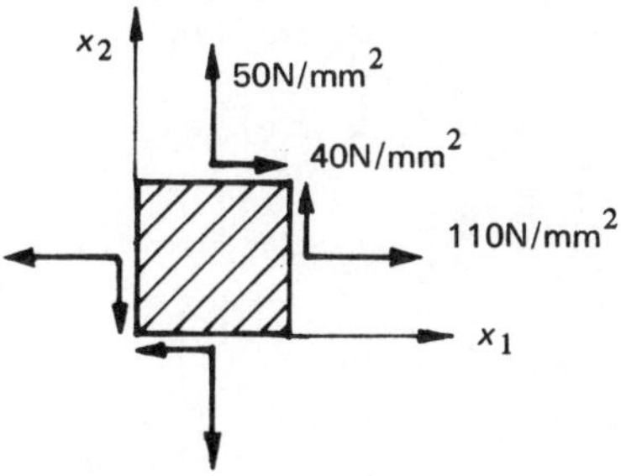

Solution

The greatest linear strain will be one of the principal strains; the greatest shear strain will be equal to the greatest difference between any two principal strains; the direction of the thickness is also a principal direction. In essence the solution of this problem therefore requires only the determination of the principal strains of the state. These may be determined either by first determining the strain tensor in the given axes and then resolving the general strain tensor, or else by first resolving the state of stress to determine the principal stresses, and then determining the principal strains from these. The latter is the simpler because both will require the use of the stress/strain equations for the linear effects, but only the former will also require both the use of the shear relationship, and the determination of the shear modulus G.

Resolving stresses by Mohr's Circle gives:

$$p_1 = 80 + 50 = 130 \text{ N/mm}^2$$

$$p_2 = 80 - 50 = 30 \text{ N/mm}^2$$

$$p_3 = 0 \text{ (given).}$$

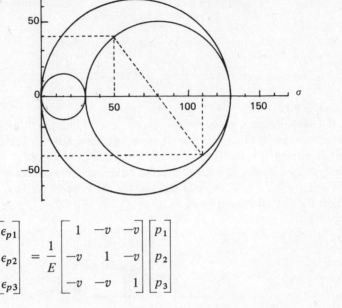

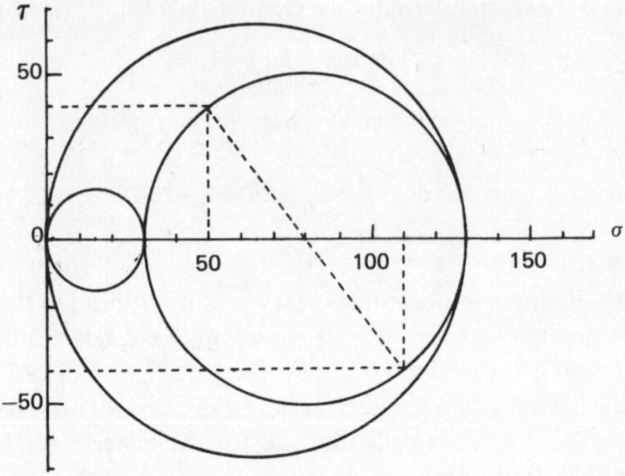

$$\begin{bmatrix} \epsilon_{p1} \\ \epsilon_{p2} \\ \epsilon_{p3} \end{bmatrix} = \frac{1}{E} \begin{bmatrix} 1 & -v & -v \\ -v & 1 & -v \\ -v & -v & 1 \end{bmatrix} \begin{bmatrix} p_1 \\ p_2 \\ p_3 \end{bmatrix}$$

$$= \frac{10}{3E} \begin{bmatrix} 3 & -1 & -1 \\ -1 & 3 & -1 \\ -1 & -1 & 3 \end{bmatrix} \begin{bmatrix} 13 \\ 3 \\ 0 \end{bmatrix} = \frac{10}{3E} \begin{bmatrix} 36 \\ -4 \\ -16 \end{bmatrix} = \begin{bmatrix} 5{\cdot}71 \\ -0{\cdot}63 \\ -2{\cdot}54 \end{bmatrix} \times 10^{-4} .$$

The greatest linear strain is $5{\cdot}71 \times 10^{-4}$.

The greatest shear strain $= 5{\cdot}71 - (-2{\cdot}54) \times 10^{-4} = 8{\cdot}25 \times 10^{-4}$.

The change of thickness $= \epsilon_{p3} \,.\, 20$ mm $= -5{\cdot}08 \times 10^{-3}$ mm.

8.8 THEORIES OF YIELD

The Hookean assumption generalised the proportionality of stress to strain in simple tension in a manner suitable for application to states of stress in general, and now we require similarly to generalise the conditions for the incidence of yield. In other words, the test in simple tension shows that (at moderate rates of strain) yield occurs in a given grade of mild steel at a characteristic value of the simple tensile stress; but other sections are then subject to various combinations of direct stress and shear, and we now require a theory of yield which specifies how the various components of any general state of stress affect the incidence of yield.

But first we note that a given grade of mild steel yields in simple compression at a compressive stress whose value is much the same as that at the tensile yield, and it is usually assumed that the magnitudes of the tensile and compressive yields are equal.

Furthermore, for ductile materials which do not exhibit a true yield it is assumed that the principles which determine the incidence of yield also determine the incidence of a proof stress.

The Hencky-Mises, Octahedral Shear Stress Theory of Yield

We have argued that in crystalline solids large plastic strains can be expected to be independent of the hydrostatic stress of any state, and to depend solely on the deviatoric effect. Furthermore, we have seen that on the faces of a particular octahedral element on which the direct stresses actually have a uniform value equal to the hydrostatic stress, the shear stresses also have a uniform value called the octahedral shear stress, and this suggests that in an isotropic material (where the directional properties of the deviatoric part of any state of stress is irrelevant) the magnitude of the deviatoric effect should be characterised by the octahedral shear stress. Taken together, these observations suggest that the incidence of yield should depend on the value of the octahedral shear stress alone, and this expectation is in good agreement with experiment. So let σ_y represent the simple tensile stress at which yield occurs in a particular grade of material. Since the principal stresses at yield are σ_y, 0 and 0, the octahedral shear stress at yield in simple tension in equal to $\sigma_y\sqrt{2/3}$, and the Hencky-Mises theory postulates that the material will yield at any state of stress having the same value of octahedral shear. Thus if p_1, p_2 and p_3 represent the principal stresses of any yield state, then:

$$(1/3)\sqrt{((p_1-p_2)^2 + (p_1-p_3)^2 + (p_2-p_3)^2)} = \sigma_y\sqrt{2/3}$$

or:

$$(p_1-p_2)^2 + (p_1-p_3)^2 + (p_2-p_3)^2 = 2\,\sigma_y^2 \qquad (8.10)$$

The Tresca, Maximum Shear Stress Theory of Yield

Although the Hencky-Mises theory is the one that best agrees with experiment, its describing equation sometimes proves awkward, and in this case we commonly adopt an approximate alternative. It is attributed to Tresca, and, like the Hencky-Mises theory, it recognises plastic deformation as a manifestation of shear, but it postulates that the magnitude of the shearing effect of any state should be characterised, not by the octahedral shear stress, but by the greatest value of shear on any section. But the greatest shear stress in any state is equal to the greatest algebraic difference between any two of its principal stresses, so the Tresca theory effectively postulates that the incidence of yield is determined by the greatest algebraic difference between any two of the three principal stresses. So again let σ_y represent the simple tensile stress at yield, and let p_1, p_2 and p_3 represent the principal stresses of a general yield state; then, assuming that the magnitude of the yield in simple compression is equal to that in simple tension, depending on which of the principal stresses is the greatest and which the least, the Tresca theory of yield can be expressed formally as follows:

$$\begin{aligned} &\text{when } p_1 > p_2 > p_3, \text{ then: } (p_1 - p_3) = \pm\,\sigma_y \\ &\text{when } p_2 > p_3 > p_1, \text{ then: } (p_2 - p_1) = \pm\,\sigma_y \\ &\text{when } p_3 > p_1 > p_2, \text{ then: } (p_3 - p_2) = \pm\,\sigma_y \end{aligned} \tag{8.11}$$

Yield Surfaces in the Principal Stress Space

If the values of the three principal stresses of any state are adopted as the co-ordinates of a three-dimensional space, each state of stress plots at a unique point, and the space can then be used to visualise various events of interest, including the incidence of yield. For example, the sequence of states of stress that arise at a point during the course of a test can be plotted in the principal stress space as a path of states, and in general such a path might be complex. But tests are commonly so conducted that forces which are applied at given points in given directions, are varied in fixed proportions, and since for a Hookean material this implies that the principal stresses at each point would also vary in fixed proportions, it follows that in such a test the state path for any point would lie on a straight line which radiates from the origin. Thus, on each of the coordinate axes only one of the three principal stresses is non-zero (Fig. 8.6),

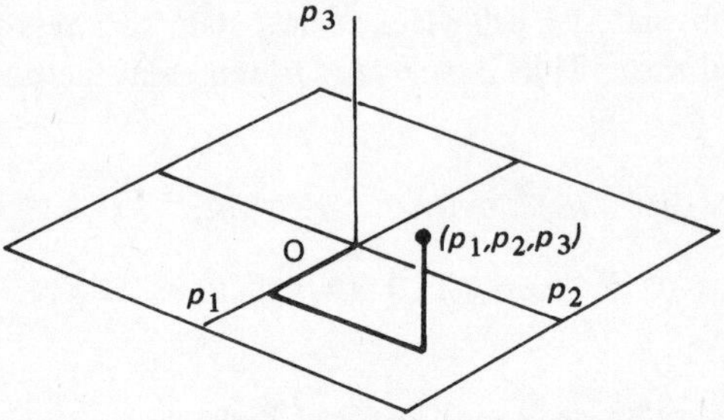

Fig. 8.6

so the coordinate axes themselves can be recognised as the state paths for simple tension or compression in the directions of X_1, X_2 and X_3 in turn. However, the path of particular interest is the one that is equally inclined to the coordinate axes. This is the line of states in which $p_1 = p_2 = p_3$, so this is the path of purely hydrostatic states in which the deviatoric effect is zero.

Consider the state paths of a given point in a body, in a series of tests in which forces applied at given points in given directions are increased from zero up to yield. In each test the forces are increased in fixed proportions which are varied infinitesimally from test to test. The state paths of successive tests would lie along adjacent radial lines, and would lead to yield points which, presumably, would also be adjacent. The points corresponding to all states at which yield occurs are therefore conceived to comprise a continuous surface, and we recognise that in postulating a theory of yield we effectively postulate the shape of the yield surface in the principal stress space. Indeed, if the theory is formulated by drawing an equation between a function of the principal stresses and the yield stress in simple tension, σ_y, this is, in fact, the equation of the yield surface in the principal stress space, in which σ_y determines the scale of a surface whose form is determined by the function of the principal stresses.

Concerning the forms of the yield surfaces for the alternative Hencky-Mises and Tresca theories, we note that both are designed to be consistent with two conditions: first, that mild steel is isotropic; and second, that the yield stress is independent of the hydrostatic stress. Of these, the first requires that the yield surface be equally disposed with respect to the three axes of the principal stress space, whilst the second requires that the intersections of each yield surface with planes normal to the hydrostatic line must have a common form, and each surface therefore takes the form of a prism of uniform section whose longitudinal axis lies along the hydrostatic line.

So, referring to Fig. 8.7a, consider the intersection of the two surfaces with the p_1p_2 plane. This is the plane of all states of stress in which $p_3 = 0$, so this is the plane of states of plane stress in the X_1X_2 plane, and the equations of its

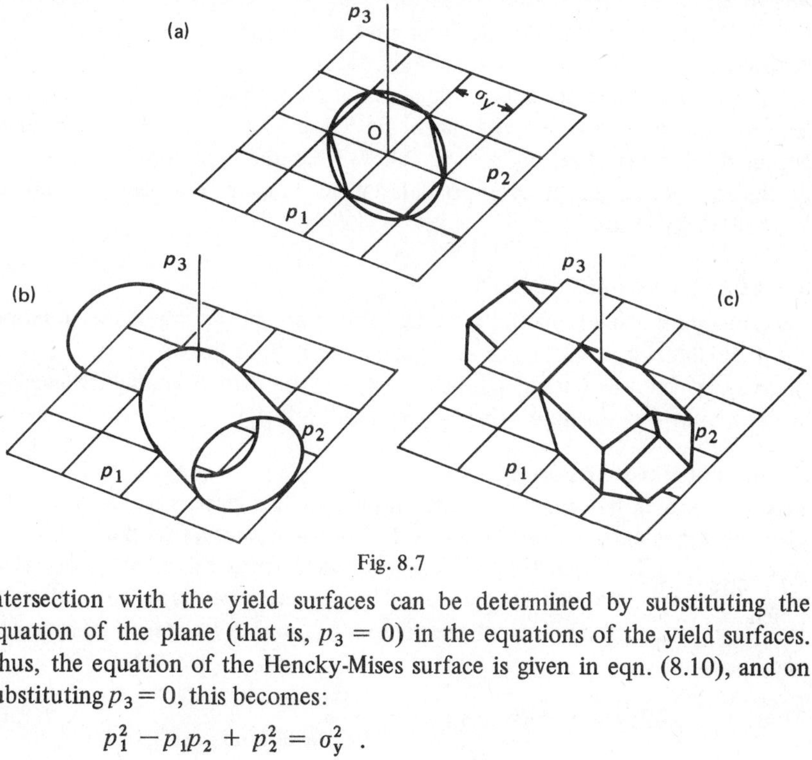

Fig. 8.7

intersection with the yield surfaces can be determined by substituting the equation of the plane (that is, $p_3 = 0$) in the equations of the yield surfaces. Thus, the equation of the Hencky-Mises surface is given in eqn. (8.10), and on substituting $p_3 = 0$, this becomes:

$$p_1^2 - p_1p_2 + p_2^2 = \sigma_y^2 \ .$$

Of course, when $p_1 = 0$, then $p_2 = \pm\sigma_y$, and when $p_2 = 0$, then $p_1 = \pm\sigma_y$. In other words, the yield boundary cuts the axes of simple tension in the values $\pm\sigma_y$, as indeed it must, and it can be shown that the foregoing equation describes an ellipse which passes through these points. The principal axes of the ellipse bisect the angles between the p_1 and p_2 axes, as again, for isotropy, they must, and its semi-major and semi-minor axes are equal to $\sigma_y\sqrt{2}$ and $\sigma_y\sqrt{\frac{2}{3}}$, respectively. We can therefore infer that the Hencky-Mises yield surface is a right-cylinder of radius $\sigma_y\sqrt{\frac{2}{3}}$ whose longitudinal axis lies along the hydrostatic line, and Fig. 8.7b indicates a short length. However, the cylinder really extends indefinitely on both sides of the coordinate plane.

In similarly determining the Tresca boundary in the p_1p_2 plane it is necessary to give separate consideration to the four quadrants. Thus, in the quadrant in which p_1 and p_2 are both positive, $p_3 = 0$ is the algebraic least, and when $p_1 > p_2$ the equation of the boundary is $p_1 = \sigma_y$, whilst when $p_2 > p_1$, the equation is $p_2 = \sigma_y$. Similarly, in the quadrant where p_1 and p_2 are both negative, then $p_3 = 0$ is the algebraic greatest, and the equations of the boundary are $p_1 = -\sigma_y$, and $p_2 = -\sigma_y$. On the other hand, in the two quadrants where the senses of p_1 and p_2 differ, $p_3 = 0$ is the intermediate value, and the equations of the boundaries are, in one case: $p_1 - p_2 = \sigma_y$, and in the other: $p_1 - p_2 = -\sigma_y$. The shape of the Tresca boundary in the p_1p_2 plane is therefore as indicated in Fig. 8.7a, and from this we can infer that the Tresca yield surface is a right hexagonal prism, inscribed in the Hencky-Mises cylinder, as indicated in Fig. 8.7c. This figure therefore gives a visual indication of the order of the discrepancy between the two theories, and since the Tresca surface is wholly contained within the Hencky-Mises cylinder it follows that the approximate theory errs on the safe side in that it never predicts a yield state greater than that predicted by the Hencky-Mises theory.

8.9 ELASTIC v LIMIT DESIGN

Components made of ductile materials are commonly required for situations in which the risk of particular kinds of failure such as brittle fracture, fatigue, or creep can be discounted, and the components are then invariably designed for strength either on elastic principles or on 'limit' principles.

Elastic Design and the Factor of Safety

Elastic design is based on the proposition that it is necessary to ensure that yield does not occur at any point of a component, and a maximum allowable working stress is assigned by dividing the yield stress by a Factor of Safety which is assessed to take account of all the uncertainties and imperfections in the specification of the duty of the component, and in its design, manufacture and operation.

Thus: $$\text{Allowable working stress} = \frac{\text{Yield stress}}{\text{Factor of safety}} \qquad (8.12)$$

For a component which, in operation, is stressed in simple tension, the allowable tensile working stress can be determined, simply by dividing the tensile yield stress by the Factor of Safety. Otherwise the operational states of stress are compared with the tensile yield stress on the basis of an appropriate theory of yield.

Since the method ensures that the stresses are everywhere within the elastic limit, elastic design requires only elastic analysis, usually on the basis of the Hookean stress/strain relationships.

Limit Design and the Load Factor

In some situations there is a tendency for the elastic approach to give way to 'limit' analysis which, in effect, assesses the significance of yield according to the geometry and location of the yield zone, and particularly the extent to which it is contained by parallel, elastic material. Consider a horizontal beam which is cantilevered from a rigid wall, and which carries a load at its free end (Fig. 8.9a). It is intuitively obvious that the section that suffers the greatest bending moment is the one adjacent to the wall, and within this section, it is the outermost material that suffers the greatest stress. So, as the load is increased, yield first occurs in small zones at the outermost edges of the section adjacent to the wall, and as the load is further increased the yield zone gradually spreads both along the length and through the depth until finally the whole of the section adjacent to the wall is at yield. At this point the section has developed its maximum moment of resistance, and to a further increment of the load, the section would then behave as an unresisting hinge. However, this occurs only when the load has reached a value significantly greater than that at which yield first occurs, and this shows that although the stiffness at a point reduces

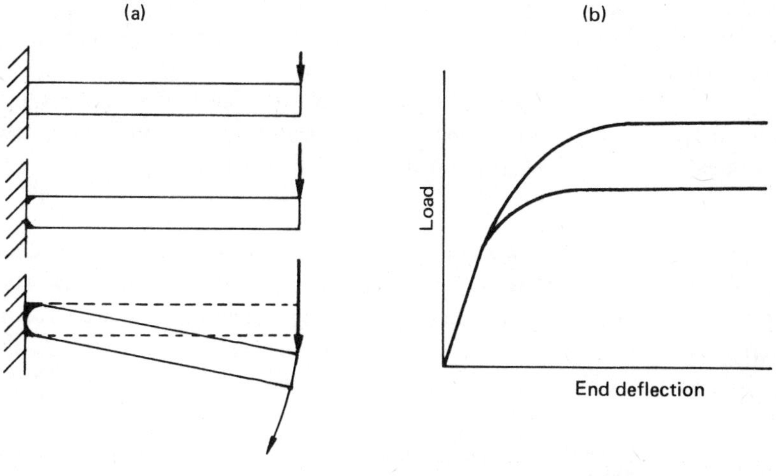

Fig. 8.8

to zero immediately the stress at the point reaches the yield stress, thereafter, the stiffness of the member as a whole only reduces gradually, and it finally becomes zero, only when the yield zones spread in such a manner, that, to further increments of load, the member behaves as a mechanism, and the incremental stiffness of the member falls to zero.

So referring to Fig. 8.8b, consider the load/deflection curves of two such cantilevers of equal length, whose sections, though different, are, by the elastic theory, of equal strength. First yield would then occur in the two at equal values of the load, and up to this point the graphs would be proportional and equal, but thereafter the graphs would develop differently, particularly in respect of the load at which the stiffness falls to zero, and in Limit Analysis, it is proposed that the strength of alternative designs should be compared, not on the basis of stresses, but on the basis of the loads at which the yield zones become so developed that the incremental stiffness falls to zero. The approach therefore ascribes to a component, not an allowable working stress, but an allowable working load, and this is determined by dividing the load at which the incremental stiffness falls to zero by a Load Factor which, like a Factor of Safety, is assessed to take account of all the uncertainties and imperfections in the situation. Thus:

$$\begin{matrix}\text{Allowable}\\ \text{Working load}\end{matrix} = \frac{\text{Load for zero incremental stiffness}}{\text{Load factor}} \tag{8.13}$$

In effect, Limit Analysis allows different Factors of Safety at different positions so as to increase the average effectiveness of the material, and such designs are therefore usually both lighter and more elegant. However, the hoped-for reduction of cost is seldom realised because with Limit Analysis it is important to ensure that the conditions of continuity and stiffness etc. that are necessarily assumed in analysis are realised in the component in practice, and the cost of meeting this condition has so far offset the savings that result from the use of less material.

Evidently, Limit Analysis is more sophisticated and more complex than elastic analysis. In particular it depends upon a study of post-yield behaviour, and a fuller consideration of its principles is therefore deferred.

Example 8.16

If a specimen yields in simple tension at a tensile stress of 240 N/mm^2, determine, for a Factor of Safety of 2.5 in each case, the allowable working values of simple tensile stress, greater shear stress and octahedral shear stress for the material.

Solution

$$\begin{matrix}\text{Allowable tensile}\\ \text{working stress}\end{matrix} = \frac{\text{Tensile yield stress}}{\text{Factor of safety}}$$

Allowable simple tensile stress	$= 240/2{\cdot}5$	$= 96\ \text{N/mm}^2$
Allowable greatest shear stress	$= 96/2$	$= 48\ \text{N/mm}^2$
Allowable octahedral shear stress	$= 96\sqrt{2}/3$	$= 45\ \text{N/mm}^2$.

Example 8.17

The greatest stresses in a component fall at a point on an unloaded surface, and a strain gauge rosette glued to the surface of a $\frac{1}{4}$ scale model at the corresponding position measures the linear strains along three lines q, r and s, and at a load of 1·1 kN their values are as indicated beside the diagram. For the model material,

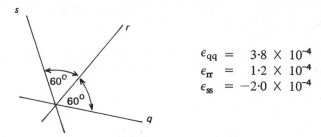

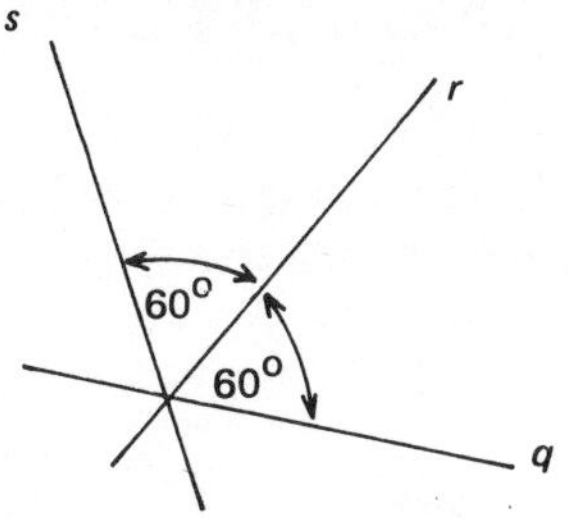

$$\epsilon_{qq} = 3{\cdot}8 \times 10^{-4}$$
$$\epsilon_{rr} = 1{\cdot}2 \times 10^{-4}$$
$$\epsilon_{ss} = -2{\cdot}0 \times 10^{-4}$$

Young's Modulus and Poisson's Ratio are 120×10^3 N/mm^2 and 0·4 respectively. The working load of the full-size component is 20 kN, and the tensile yield stress of the material is 118 N/mm^2. Determine the effective Factor of Safety of the component at its working load.

Solution

To reduce the description of the strains in the plane of the surface to a standard form it is resolved either by drawing Mohr's Circle of Strain to scale (see Ex. 6.8)

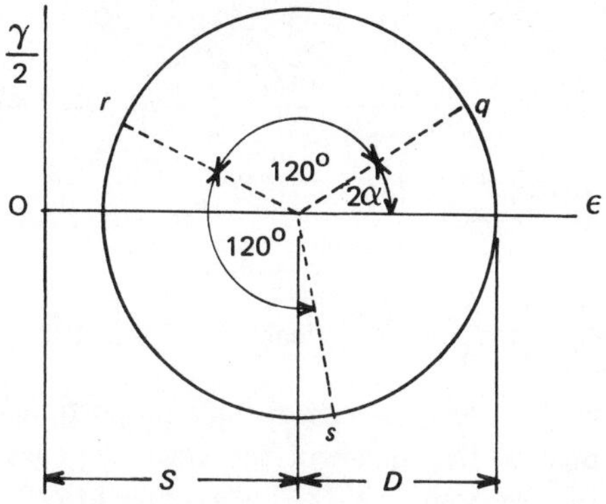

or analytically by letting S and D represent the semi-sum and semi-difference of the principal strains ϵ_{p1} and ϵ_{p2} and by letting α represent the counter-clockwise angle from X_1 to q.

$$\epsilon_{qq} = \quad S + D\cos 2\alpha \tag{a}$$

$$\epsilon_{rr} = S + D\cos 2(\alpha + 60^\circ) = S + \frac{D}{2}(-\cos 2\alpha + \sqrt{3}\sin 2\alpha) \tag{b}$$

$$\epsilon_{ss} = S + D\cos 2(\alpha + 120^\circ) = S + \frac{D}{2}(-\cos 2\alpha - \sqrt{3}\sin 2\alpha) \tag{c}$$

$$(a)+(b)+(c) \rightarrow \quad \epsilon_{qq} + \epsilon_{rr} + \epsilon_{ss} = 3S \tag{d}$$

$$(b)-(c) \rightarrow \quad \epsilon_{rr} - \epsilon_{ss} = \sqrt{3}\,D\sin 2\alpha \tag{e}$$

$$2(a)-(b)-(c) \rightarrow 2\epsilon_{qq} - \epsilon_{rr} - \epsilon_{ss} = 3D\cos 2\alpha \tag{f}$$

$$(e)/(f) \quad \rightarrow 2\alpha = \tan^{-1}\frac{\sqrt{3}(\epsilon_{rr} - \epsilon_{ss})}{2\epsilon_{qq} - \epsilon_{rr} - \epsilon_{ss}} \,. \tag{g}$$

Substituting the given values of ϵ_{qq}, ϵ_{rr} and ϵ_{ss} in (g), (d) and (e), then determines the values of 2α, $2S$ and $2D$ as follows:

$$2\alpha = \tan^{-1}\frac{\sqrt{3}.3\cdot 2}{8\cdot 4} = 33\cdot 42^\circ$$

$$2S = \epsilon_{p1} + \epsilon_{p2} = 2\cdot 0 \times 10^{-4} \tag{h}$$

$$2D = \epsilon_{p1} - \epsilon_{p2} = 6\cdot 71 \times 10^{-4} \,. \tag{j}$$

Solving (h) and (j) simultaneously for ϵ_{p1} and ϵ_{p2} then gives:

$$\epsilon_{p1} = 4\cdot 355 \times 10^{-4}; \quad \epsilon_{p2} = -2\cdot 355 \times 10^{-4} \,.$$

Noting that the third principal stress normal to the unloaded surface is zero, the remaining principal stresses may be determined from the Hookean stress/strain relationships as follows:

$$\epsilon_{p1} = 4\cdot 355 \times 10^{-4} = (\ p_1 - v p_2)/E$$

$$\epsilon_{p2} = -2\cdot 355 \times 10^{-4} = (-v p_1 + \ p_2)/E$$

Whence:

$$p_1 = \frac{(4\cdot 355 - 2\cdot 355\,v)\,.\,E.10^{-4}}{(1 - v^2)} = 48\cdot 7\ \text{N/mm}^2$$

$$p_2 = \frac{-(48\cdot 7 - 4\cdot 355\,.\,E.10^{-4})}{v} = 3\cdot 56\ \text{N/mm}^2$$

therefore: $\tau_0(\text{model}) = \frac{1}{3}\sqrt{(48\cdot 7^2 + 3\cdot 56^2 + 45\cdot 14^2)} = 22\cdot 17\ \text{N/mm}^2$.

In geometrically similar cases, the stresses are proportional to the forces, and inversely proportional to the square of the linear dimensions. The octahedral shear stress in the full-size component at the working load is therefore given by:

$$\tau_0(\text{working}) = 22\cdot 17 \cdot \frac{20}{1\cdot 1}\left(\frac{1}{4}\right)^2 = 25\cdot 2\ \text{N/mm}^2 \,.$$

For the material of the full-size component:

$$\text{Octahedral shear stress at yield} = 118\sqrt{2/3} = 55{\cdot}63 \text{ N/mm}^2 .$$

$$\text{Factor of Safety} = \frac{\text{Octahedral shear stress at yield}}{\text{Working octahedral shear stress}}$$

$$= \frac{55{\cdot}63}{25{\cdot}2} = 2{\cdot}21 .$$

8.10 NEWTONIAN FLUIDS

When the stresses in an isotropic solid are confined to the elastic range, each state of stress is associated with a particular pattern of deformation, and the stresses can therefore be described as functions of the strains. But with some materials it is found that the smallest level of shear stress is sufficient to cause progressive deformation which continues to develop for as long as the stress is applied, and it is such materials that are known as fluids.

Evidently, when the strains associated with a given stress vary with time, the stress can no longer be regarded as a function of the strain; but it could possibly be related to the time-rates of change of the strains, and in the event, it is found that for many common fluids the stresses may be related to rates of strain in much the same way as stresses are related to strains in a Hookean solid. We therefore define a Newtonian fluid as a reasonable idealisation of a range of common fluids, and the definition is such that the components of any stress tensor can be described as linear functions of the time-rates of change of the components of the corresponding strain tensor. In parallel with the elastic constants which relate stresses to strains in a Hookean solid, we therefore now recognise coefficients of viscosity which relate stresses to time-rates of change of strains in a Newtonian fluid.

Throughout, differentiation with respect to time will be indicated in the dot convention, on the understanding that where the operation is applied to a vector or a tensor it is to be applied relative to the frame of reference in which its components are described. It then follows that the operation can be applied to the vector or tensor, simply by its application to each scalar component in turn.

Strain Rates and the Partial Derivatives of the Velocity Field

Although the coefficients of viscosity are defined as coefficients which relate stresses to derivatives of strains with respect to time, these derivatives can alternatively be defined in terms of derivatives of the velocity with respect to the position coordinates. For example, we have seen that if $\mathbf{u}$ represents the displacement of a point in the body in an independent frame of reference, then

$\epsilon_{11} = \partial u_1/\partial x_1$, $\epsilon_{12} = \frac{1}{2}\{(\partial u_1/\partial x_2) + (\partial u_2/\partial x_1)\}$, etc., and since the sequence of successive differentiations is immaterial, we may write

$$\frac{\mathrm{d}\epsilon_{11}}{\mathrm{d}t} = \frac{\mathrm{d}}{\mathrm{d}t}\left(\frac{\partial u_1}{\partial x_1}\right) = \frac{\partial}{\partial x_1}\left(\frac{\mathrm{d}u_1}{\mathrm{d}t}\right) = \frac{\partial v_1}{\partial x_1}$$

where $v_1 = \mathrm{d}u_1/\mathrm{d}t$ represents the scalar component of point velocity parallel to x_1.

So just as ∂u_{11}, ∂u_{12} etc. were used to represent the partial derivatives of the components of the field of point displacements, now let ∂v_{11}, ∂v_{12} etc. represent the partial derivatives of the components of the field of point velocities. Furthermore, just as $[\partial \mathbf{u}]$ was used to represent the tensor of the partial derivatives of the displacements, so now let $[\partial \mathbf{v}]$ represent the tensor of the partial derivatives of the velocities. Then by considering the derivative of each component of $[\partial \mathbf{u}]$ in turn, as above, it may be shown that:

$$\frac{\mathrm{d}}{\mathrm{d}t}([\partial \mathbf{u}]) = [\partial \mathbf{v}] \ ,$$

and since $[\epsilon] = \frac{1}{2}([\partial \mathbf{u}] + [\partial \mathbf{u}^{\mathrm{T}}])$, it follows that:

$$[\dot{\epsilon}] = \frac{\mathrm{d}[\epsilon]}{\mathrm{d}t} = \frac{1}{2}\frac{\mathrm{d}}{\mathrm{d}t}([\partial \mathbf{u}] + [\partial \mathbf{u}]^{\mathrm{T}})$$

or $$[\dot{\epsilon}] = \frac{1}{2}([\partial \mathbf{v}] + [\partial \mathbf{v}]^{\mathrm{T}}) \tag{8.14}$$

Treating the time-rate of change of the volumetric strain similarly gives:

$$\dot{\Delta} = \frac{\mathrm{d}\Delta}{\mathrm{d}t} = \frac{\mathrm{d}}{\mathrm{d}t}(\epsilon_{11} + \epsilon_{22} + \epsilon_{33})$$

that is: $$\dot{\Delta} = \partial v_{11} + \partial v_{22} + \partial v_{33} \tag{8.15a}$$

However, reflection will show that this result can alternatively be described by making use of the partial differential operator ∇ as follows:

$$\frac{\mathrm{d}\Delta}{\mathrm{d}t} = \frac{\partial v_1}{\partial x_1} + \frac{\partial v_2}{\partial x_2} + \frac{\partial v_3}{\partial x_3}$$

that is: $$\dot{\Delta} = \left(\frac{\partial}{\partial x_1} \ \frac{\partial}{\partial x_2} \ \frac{\partial}{\partial x_3}\right) \cdot (v_1 \ v_2 \ v_3)$$

or $$\dot{\Delta} = \nabla \cdot \mathbf{v} \tag{8.15b}$$

The First Coefficient of Viscosity

The Newtonian definition implies that to describe the viscous properties of such a fluid it may generally be necessary to specify many coefficients. But in practice, it is found that the majority of the common fluids, like the majority of the common structural solids, are sensibly isotropic, and again it then falls out that only two of the coefficients can be regarded as independent, and only two are sufficient completely to describe the viscous properties of the material. So again treating the stress and strain tensors of a corresponding pair as the sums of their hydrostatic and deviatoric parts, we define one coefficient to relate the distorting, deviatoric effects, and another to relate the dilational, hydrostatic effects.

First considering the deviatoric effects, we define the First Coefficient of Viscosity, μ, as a simple coefficient of proportionality between each component of any deviatoric stress tensor $[\sigma^*]$, and the time-rate of change of the corresponding component of the corresponding deviatoric strain tensor $[\epsilon^*]$, as follows:

In particular:

$$\sigma_{12} = 2\mu\,\dot{\epsilon}_{12} = \mu\,\dot{\gamma}_{12}, \text{ etc.}$$

and

$$(\sigma_{11} - \sigma_{00}) = 2\mu(\dot{\epsilon}_{11} - \dot{\Delta}/3), \text{ etc.}$$

or in general:

$$[\sigma^*] = 2\mu\,[\dot{\epsilon}^*] \tag{8.16a}$$

But $[\sigma^*] = [\sigma] - \sigma_{00}\,[\mathbf{I}]$, and $[\epsilon^*] = [\epsilon] - (\Delta/3)\,[\mathbf{I}]$.

Therefore:

$$[\sigma] = \sigma_{00}\,[\mathbf{I}] + 2\mu\,[\dot{\epsilon}] - \frac{2}{3}\mu\,\dot{\Delta}\,[\mathbf{I}] \tag{8.16b}$$

or by substituting for $[\dot{\epsilon}]$ and $\dot{\Delta}$ from eqns. (8.14) and (8.15b),

$$[\sigma] = \sigma_{00}\,[\mathbf{I}] + \mu([\partial \mathbf{v}] + [\partial \mathbf{v}]^{\mathrm{T}}) \frac{2}{3}\mu\,(\nabla\cdot\mathbf{v})[\mathbf{I}]\ . \tag{8.16c}$$

To illustrate the case, consider a fluid which is contained between a pair of parallel plates of separation s, when one of the plates is fixed, and the other moves with speed v as shown in Fig. 8.9. The fluid is assumed to be incompressible (so that there are no volumetric effects), and its First Coefficient of Viscosity is represented by μ.

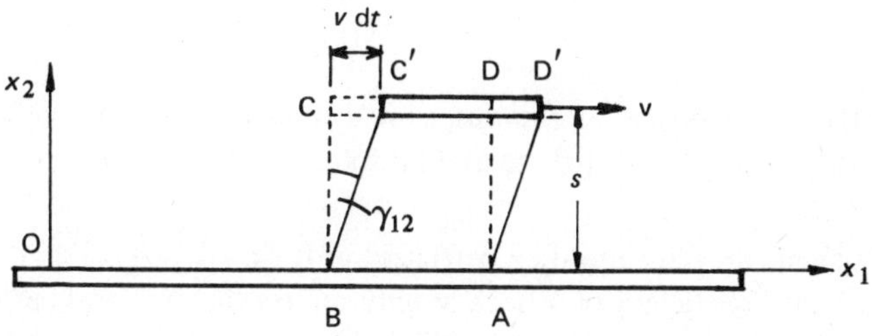

Fig. 8.9

It is an experimental fact that there is no relative motion between any solid surface and the fluid particles with which it is in immediate contact. The fluid in contact with the stationary plate is therefore stationary, whilst the fluid in contact with the moving plate moves with a speed v, and the material being uniform, it is reasonable to suppose, that in between, the speed of the fluid varies in proportion to the distance from the stationary plate. In an infinitesimal interval of time dt, the shear strain of the right-angle $\angle$BAC is $d\gamma_{12} = (v/s)dt$. Therefore $\dot{\gamma}_{12} = d\gamma_{12}/dt = v/s$, and, by definition, this is related to the shear stress σ_{12} by the first coefficient of viscosity as follows:

$$\sigma_{12} = 2\mu\,\dot{\epsilon}_{12} = \mu\,\dot{\gamma}_{12} = \mu(v/s)\ .$$

However, since $\dot{\gamma}_{12}$ is equal to v/s, it is obviously equal to the gradient of the velocity in the direction normal to the flow, so the coefficient of viscosity could alternatively be defined as a constant of proportionality which relates the shear stress of parallel flow to the gradient of the velocity in the direction normal to the flow, and the foregoing result can alternatively be determined from the velocity field by using eqn. (8.16c).

Thus, in the axes $Ox_1x_2x_3$ which are fixed in the stationary plate (it is assumed that conditions are uniform in the direction normal to the x_1x_2 plane, and that there is no motion in this direction) the velocity of a fluid particle at the general position (x_1, x_2, x_3) is given by $\underline{v} = (v/s)x_2\underline{i}$.

$$\text{Therefore}\quad [\partial \mathbf{v}] = \begin{bmatrix} 0 & v/s & 0 \\ 0 & 0 & 0 \\ 0 & 0 & 0 \end{bmatrix};\quad \text{and}\quad [\partial \mathbf{v}] + [\partial \mathbf{v}]^{\mathrm{T}} = \begin{bmatrix} 0 & v/s & 0 \\ v/s & 0 & 0 \\ 0 & 0 & 0 \end{bmatrix}$$

But for an incompressible fluid in which $\Delta = 0$, $\dot{\Delta} = \nabla \cdot \mathbf{v}$ is also zero, and eqn. (8.16c) becomes

$$[\sigma] = \sigma_{00}[\mathrm{I}] + \mu([\partial \mathbf{v}] + [\partial \mathbf{v}]^{\mathrm{T}})$$

$$\text{that is:}\quad [\sigma] = \sigma_{00}[\mathrm{I}] + \mu \begin{bmatrix} 0 & v/s & 0 \\ v/s & 0 & 0 \\ 0 & 0 & 0 \end{bmatrix}$$

Now $\sigma_{00}[\mathrm{I}]$ is the tensor of the unspecified hydrostatic stress, which has no effect on the shear terms of $[\sigma]$, so again it follows that $\sigma_{12} = \mu(v/s)$.

Thermodynamic Pressure and the Coefficient of Bulk Viscosity

Since the First Coefficient of Viscosity only relates the deviatoric effects, it is, of itself, only relevant to an event in which the rate of change of volume is zero, and to take account of the dilational effects we define a Coefficient of Bulk

Viscosity μ_v which relates the hydrostatic stress to the rate of volumetric strain. But fluids include gases and vapours in which it is necessary to take account of the thermodynamic pressure, and the relationship between the hydrostatic stress and the rate of volumetric strain is therefore linear rather than simply proportional.

For we know from elementary thermodynamics, that, because of the thermal motions of its molecules, a given mass of gas can be contained at a constant volume only by the imposition of a hydrostatic pressure (i.e. a negative hydrostatic stress) whose magnitude p_t depends on the volume and temperature.

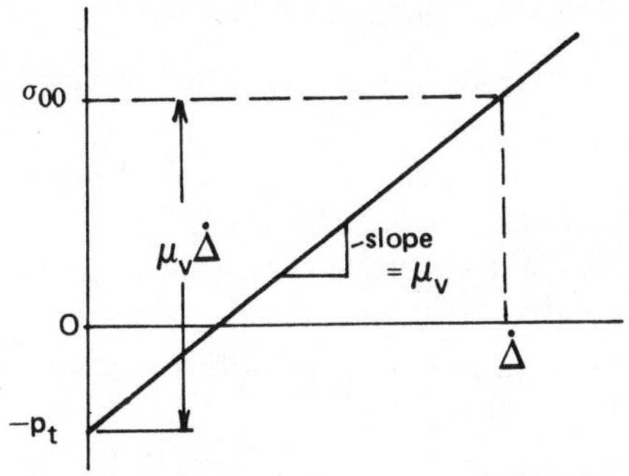

Fig. 8.10

In other words, with a gas or vapour, the rate of volumetric strain is zero only when it is subjected to the hydrostatic stress of thermal equilibrium, $-p_t$, and the coefficient of bulk viscosity is defined as a simple coefficient of proportionality between the rate of volumetric strain, and the amount by which the hydrostatic stress differs from its thermal equilibrium value, as indicated in Fig. 8.10.

Thus. $$\sigma_{00} + p_t = \mu_v\dot{\Delta}$$

or $$\sigma_{00} = -p_t + \mu_v\dot{\Delta} \tag{8.17a}$$

Alternatively, by substituting for $\dot{\Delta}$ from eqn. (8.15b),

$$\sigma_{00} = -p_t + \mu_v(\nabla \cdot \mathbf{v}) \tag{8.17b}$$

We note that the total hydrostatic stress can be regarded as a sum of two parts, in which the stress $u_v\dot{\Delta}$ which is required to produce the rate of volumetric strain $\dot{\Delta}$, is added to the negative stress for thermal equilibrium. However, in an incompressible fluid where $\dot{\Delta}$ is zero, the hydrostatic stress may simply be equated to the negative of the thermodynamic pressure.

Lamé's Equation and the Second Coefficient of Viscosity

Equations (8.16) only relate the distorting effects, whilst eqns. (8.17) only relate the dilating effects, but by substituting for σ_{00} from eqn. (8.17), in eqns. (8.16), we arrive at a relationship which is directly applicable to a case in which both effects occur together. Apart from the complication of the thermodynamic pressure, it is closely parallel with Lamé's equation for a Hookean solid, and it leads to the definition of an alternative, Second Coefficient of Viscosity $\lambda = \mu_v - \frac{2}{3}\mu$ which is comparable with Lamé's Elastic Constant λ. Thus:

Substituting for σ_{00} from eqn. (8.17) in eqns. (8.16) and transposing appropriately, gives

$$[\sigma] = -p_t[\mathrm{I}] + 2\mu\,[\dot{\epsilon}] + \lambda\,\dot{\Delta}[\mathrm{I}] \qquad (8.18a)$$

or

$$[\sigma] = -p_t[\mathrm{I}] + \mu([\partial \mathbf{v} + [\partial \mathbf{v}]^{\mathrm{T}}) + \lambda(\nabla \cdot \mathbf{v})\,[\mathrm{I}]\ . \qquad (8.18b)$$

Stress vs Strain Rate in an Incompressible Fluid

In an incompressible fluid, where the volumetric strain Δ is, by definition, specifically zero, it follows first that $\dot{\Delta} = \nabla \cdot \mathbf{v}$ is also zero, and second, from eqn. (8.15), that $\sigma_{00} = -p_t$, and by substituting accordingly in either of the equations (8.18) gives:

either:

$$[\sigma] = -p_t\,[\mathrm{I}] + 2\mu\,[\dot{\epsilon}] \qquad (8.19a)$$

or:

$$[\sigma] = -p_t\,[\mathrm{I}] + \mu([\partial \mathbf{v}] + [\partial \mathbf{v}]^{\mathrm{T}})\ . \qquad (8.19b)$$

9

The Stress Equations of Small Motion and the Mathematical Theory of Elasticity

9.1 STRESS EQUATIONS OF SMALL MOTIONS

So far we have been concerned only with the relationships between the components of given tensors. In the case of the stress tensor, for example, we have been concerned only with the variation of stress from section to section at a given point, and not with the variation in the state of stress from point to point, and in establishing the rule of resolution it was therefore sufficient to consider the forces on an element only for the case of a uniform field in which every point suffers the same state of stress. In this case, the stress at any point on any section varies only with the orientation of the section, and is independent of both the position of the section and the position of the point within it; and by considering the orders of magnitude of the various terms, we argued that in the limit, the forces generated by the stresses acting on the faces of the element could then be treated as an equilibrium system, irrespective of any body forces, including those associated with the acceleration of the element.

But if the net force generated on an element by the stresses of a uniform field is zero, in a general field, the increments of stress which arise from point to point will give rise to a net force which is commensurate with the body forces, and we now require to determine this force in terms of the spatial derivatives of the stress field.

The Force Generated by the Stresses on the Faces of an Element

The state of stress at each point of a general stress field being described by its tensor in a set of axes parallel to given rectangular coordinates, let $[\sigma]$ represent the tensor for the general point P, described as a function of its coordinates. Then the scalar components of $[\sigma]$ are the scalar functions of the coordinates which describe the variations with position of the components σ_{11}, σ_{12} etc. of the tensors for the various points, and we consider the net force generated by these stresses on the faces of an infinitesimal, rectangular element isolated from the body by two sets of sections parallel to the coordinate planes (Fig. 9.1). One set which intersect in the general point $P = (x_1, x_2, x_3)$ define the three negative faces of the element, whilst the other, which intersect in the adjacent point $Q = (x_1 + dx_1, x_2 + dx_2, x_3 + dx_3)$ define the three positive faces, when it follows that the edges of the element have lengths of dx_1 parallel to x_1, of dx_2 parallel to x_2, and of dx_3 parallel to x_3.

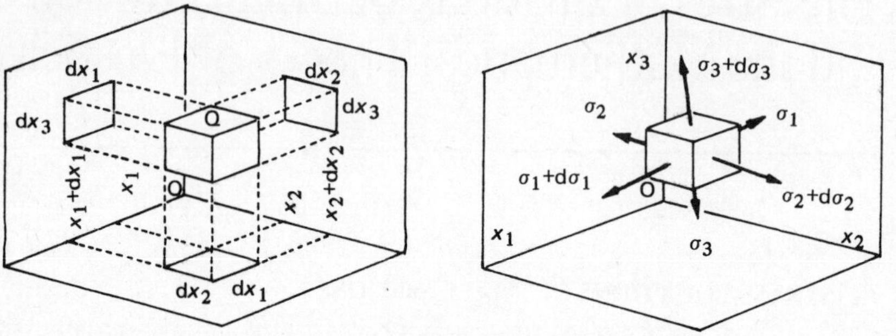

Fig. 9.1

Considering the faces of the element in their three parallel pairs in turn, we note that in a uniform field the stresses on each pair would be equal, and since they would act on faces of equal areas but opposite senses, the net force generated by each pair would be zero. But considering a general field in which the stresses vary from point to point, and dealing first with the pair of faces normal to x_1, let $\bar{\sigma}_1$ represent the average stress on the negative face as a whole. Then the stress on the positive face as a whole would differ by an increment $d\bar{\sigma}_1$, and it follows first, that since the position of the positive face as a whole differs from that of the negative by only the one increment of dx_1 in x_1, the stress increment can be equated to $(\partial\bar{\sigma}_1/\partial x_1)\, dx_1$, and then, that since both faces have an area equal to $dx_2 dx_3$, the net force on the pair of faces may be equated to $(\partial\bar{\sigma}_1/\partial x_1)\, dV$, where $dV = dx_1 dx_2 dx_3$ represents the volume of the element. Thus, for a pair of faces normal to x_1, the force per unit volume may be equated to $(\partial\bar{\sigma}_1/\partial x_1)$, and since parallel conclusions will hold for the pairs of faces normal to x_2 and x_3, we conclude that for an infinitesimal element at any position in a stress field, the force generated by the stresses which act on the surface of the element is proportional to its volume, and the force per unit volume df/dV is given by

$$\frac{df}{dV} = \frac{\partial\bar{\sigma}_1}{\partial x_1} + \frac{\partial\bar{\sigma}_2}{\partial x_2} + \frac{\partial\bar{\sigma}_3}{\partial x_3}\,.$$

But $\bar{\sigma}_1$, $\bar{\sigma}_2$ and $\bar{\sigma}_3$ are the average stresses on sections through P normal to x_1, x_2 and x_3 in turn, and in the limit, as dx_1, dx_2 and dx_3 tend to zero, they tend to equality with the stress vectors σ_1, σ_2 and σ_3 which appear as the rows of the tensor $[\sigma]$ for the point P. Therefore proceeding to the limit by substituting $\bar{\sigma}_1 = [\sigma_{11}\ \sigma_{12}\ \sigma_{13}]$, etc., determines the force per unit volume as follows:

$$\frac{\mathrm{df}}{\mathrm{d}V} = \frac{\partial[\sigma_{11}\ \sigma_{12}\ \sigma_{13}]}{\partial x_1} + \frac{\partial[\sigma_{21}\ \sigma_{22}\ \sigma_{23}]}{\partial x_2} + \frac{\partial[\sigma_{31}\ \sigma_{32}\ \sigma_{33}]}{\partial x_3},$$

and by adding the components in the x_1, x_2 and x_3 directions in turn we determine the force per unit in terms of its scalar components, as follows:

$$\frac{df_1}{dV} = \frac{\partial\sigma_{11}}{\partial x_1} + \frac{\partial\sigma_{21}}{\partial x_2} + \frac{\partial\sigma_{31}}{\partial x_3}$$

$$\frac{df_2}{dV} = \frac{\partial\sigma_{12}}{\partial x_1} + \frac{\partial\sigma_{22}}{\partial x_2} + \frac{\partial\sigma_{32}}{\partial x_3} \qquad (9.1a)$$

$$\frac{df_3}{dV} = \frac{\partial\sigma_{13}}{\partial x_1} + \frac{\partial\sigma_{23}}{\partial x_2} + \frac{\partial\sigma_{33}}{\partial x_3}$$

However, this result may be greatly simplified by again making use of the vector operator ∇ which can be written in row vector form as follows:

$$\nabla = \frac{\partial}{\partial x_1}\mathbf{i} + \frac{\partial}{\partial x_2}\mathbf{j} + \frac{\partial}{\partial x_3}\mathbf{k} = \begin{bmatrix} \frac{\partial}{\partial x_1} & \frac{\partial}{\partial x_2} & \frac{\partial}{\partial x_3} \end{bmatrix}$$

Equations (9.1a) may then be written as:

$$\begin{bmatrix} \frac{df_1}{dV} \\ \frac{df_2}{dV} \\ \frac{df_3}{dV} \end{bmatrix} = \begin{bmatrix} \frac{\partial}{\partial x_1} & \frac{\partial}{\partial x_2} & \frac{\partial}{\partial x_3} \end{bmatrix} \cdot \begin{bmatrix} \sigma_{11} & \sigma_{12} & \sigma_{13} \\ \sigma_{21} & \sigma_{22} & \sigma_{23} \\ \sigma_{31} & \sigma_{32} & \sigma_{33} \end{bmatrix}$$

Or:

$$\frac{\mathrm{df}}{dV} = \nabla \cdot [\sigma] \qquad (9.1b)$$

Example 9.1

When the state of stress at each point in a body is described by its tensor in a given set of axes, the components of the tensors vary with the positions of the points according to:

$$\sigma_{11} = x_1 + 2x_2; \quad \sigma_{22} = 2x_1 + x_2; \quad \sigma_{33} = 4\ ;$$

$$\sigma_{12} = \sigma_{21} = 2x_1^2x_2; \quad \sigma_{13} = \sigma_{31} = \sigma_{23} = \sigma_{32} = 0\ .$$

Detemine the force per unit volume generated by the stresses on the surface of an infinitesimal element as a function of the coordinates of its position.

Solution

$$\frac{\mathrm{d}f}{\mathrm{d}V} = \nabla \cdot [\sigma] = \left[\frac{\partial}{\partial x_1}\ \frac{\partial}{\partial x_2}\ \frac{\partial}{\partial x_3}\right] \begin{bmatrix} (x_1 + 2x_2) & 2x_1^2 x_2 & 0 \\ 2x_1^2 x_2 & (2x_1 + x_2) & 0 \\ 0 & 0 & 4 \end{bmatrix}$$

$$= \left[\left\{\frac{\partial(x_1 + 2x_2)}{\partial x_1} + \frac{\partial(2x_1^2 x_2)}{\partial x_2}\right\} \left\{\frac{\partial(2x_1^2 x_2)}{\partial x_1} + \frac{\partial(2x_1 + x_2)}{\partial x_2}\right\} 0\right]$$

$$= [(1 + 2x_1^2)\quad (4x_1 x_2 + 1)\quad 0]$$

Body Forces

Apart from the 'internal' forces that arise from the stresses which act on its surface, an element may also be liable to 'external' forces that are reacted outside the body, and for elements at the boundary of the body, these may include forces that result from its contact with other bodies. But surface elements are treated as a special case which will be treated later, and for all other elements only body forces are relevant.

These are distributed effects which act on every element, and they are therefore necessarily defined in terms of the force per unity quantity of the appropriate property of the field. Thus, a gravitational force acts on the mass of a body, and a gravitational field is usually specified in terms of the force per unit mass. However, the product of the force per unit mass with the density determines the force per unit volume, and we now suppose that the body forces which act at different positions in the body are described in these terms. Thus the body force per unit volume is represented by a vector function $\mathbf{X} = [X_1 X_2 X_3]$ of the coordinates, when the components represent scalar functions of the coordinates which describe the variation with the position of the scalar elements of the force per unit volume.

The Stress Equations of Small Motions

For any element other than one which falls at the surface of the body, the total force is equal to the sum of the force due to the stresses and the body force, and, by the Second Law of Motion, this may be equated to the mass of the element times its acceleration.

Thus $$(\nabla \cdot [\sigma])\,\mathrm{d}V + X\,.\,\mathrm{d}V = \rho\,\mathrm{d}V\,\mathbf{a}\ ,$$

or $$\nabla \cdot [\sigma] + \mathbf{X} = \rho\,\mathbf{a}\ , \qquad (9.2a)$$

and this may be expanded into its component form as follows:

$$\begin{bmatrix} \dfrac{\partial}{\partial x} & \dfrac{\partial}{\partial y} & \dfrac{\partial}{\partial z} \end{bmatrix} \cdot \begin{bmatrix} \sigma_{11} & \sigma_{12} & \sigma_{13} \\ \sigma_{21} & \sigma_{22} & \sigma_{23} \\ \sigma_{31} & \sigma_{32} & \sigma_{33} \end{bmatrix} + [X_1 \ X_2 \ X_3] = \rho[a_1 \ a_2 \ a_3]$$

or

$$\left.\begin{aligned} \frac{\partial \sigma_{11}}{\partial x_1} + \frac{\partial \sigma_{21}}{\partial x_2} + \frac{\partial \sigma_{31}}{\partial x_3} + X_1 &= \rho\, a_1 \\ \frac{\partial \sigma_{12}}{\partial x_1} + \frac{\partial \sigma_{22}}{\partial x_2} + \frac{\partial \sigma_{32}}{\partial x_3} + X_2 &= \rho\, a_2 \\ \frac{\partial \sigma_{31}}{\partial x_1} + \frac{\partial \sigma_{23}}{\partial x_2} + \frac{\partial \sigma_{33}}{\partial x_3} + X_3 &= \rho\, a_3 \end{aligned}\right\} \qquad (9.2b)$$

These are the stress equations of small motion which specify the limitations which the laws of motion impose on the rates of change of the various components of stress with position.

9.2 THE PRINCIPLE OF STRESS ANALYSIS

Whether it applies to a solid or to a liquid, in a typical problem of stress analysis, a given body of material of known elastic or viscous properties is subjected to a set of external forces and couples applied at known points on known axes. Some of these forces are completely specified as known loads, but for others, the magnitudes may only be implied in a specification of kinematic constraints which are imposed on specified points in the body, and the object of the analysis is to determine the unknown forces, the stresses and the strains that arise at different positions in the body, and, for a solid, the point displacements in the body, or, for a fluid, the point velocities. Thus, in a typical problem, we specify the shape of the body, the boundary forces, the body forces, the kinematic constraints, and the object is to reconcile these conditions within the constraints imposed by the laws of motion, as expressed in the stress equations of small motion (eqns. (9.2)), and the continuity of the body, as expressed, for a solid, in the relationships between the strains and the partial derivatives of the point displacements (eqns. (8.5) and (8.8)), or for a fluid, in the relationships between the rates of change of the strains and the point velocities , (eqns. (8.14 and (8.15)).

Of two approches to this problem, one is mathematically exact, and aims at a formal reconciliation of all the controlling equations in their rigorous differential forms. When applied to a Hookean solid, this approach is pursued under the title of 'The Mathematical Theory of Elasticity', anf by way of illustration, we now briefly outline its principles in application to a particular class of two-dimensional problems.

Specifically, these are cases of plane laminae of an isotropic, Hookean solid in which the faces are 'free', the body forces are negligible, and forces are applied only as a given, equilibrium system of in-plane loads applied to its boundary. In such a case, each point is subjected to a state of plane stress, and the x_1x_2 plane being chosen in the plane of the lamina, the case can be recognised as one in which the body force **X**, the acceleration **a**, and all the components of the stress tensors which bear a suffix 3 are zero.

For application to this class of problems, the various relationships can first be simplified by dropping the zero terms from consideration. Thus, the zero terms of the stress tensors being taken for granted, the non-zero terms may conveniently be presented in a 2 X 2 array as follows.

$$[\sigma] = \begin{bmatrix} \sigma_{11} & \sigma_{12} \\ \sigma_{21} & \sigma_{22} \end{bmatrix};$$

the stress equations of small motion (eqns. (9.2)) reduce to:

$$\left.\begin{aligned} \frac{\partial \sigma_{11}}{\partial x_2} + \frac{\partial \sigma_{12}}{\partial x_1} &= 0 \\ \frac{\partial \sigma_{12}}{\partial x_2} + \frac{\partial \sigma_{22}}{\partial x_1} &= 0 \end{aligned}\right\}; \tag{9.3}$$

and the condition of continuity, as expressed in the relationships between the components of strain and the derivatives of the components of point displacements, (eqns. (7.8)), reduce to:

$$\epsilon_{11} = \frac{\partial u_1}{\partial x_1}; \quad \epsilon_{22} = \frac{\partial u_2}{\partial x_2}; \quad \gamma_{12} = \frac{\partial u_1}{\partial x_2} + \frac{\partial u_2}{\partial x_1}. \tag{9.4}$$

The controlling equations can now be converted to a permissive form by introducing an arbitrary function $\phi(x_1\, x_2)$ of the position coordinates. Thus, dealing first with the stress equations of small motion, consider the stress field whose components are defined in terms of the partial derivatives of any such function, as follows:

$$[\sigma] = \begin{bmatrix} \sigma_{11} & \sigma_{12} \\ \sigma_{12} & \sigma_{22} \end{bmatrix} = \begin{bmatrix} \dfrac{\partial^2 \phi}{\partial x_2^2} & -\dfrac{\partial^2 \phi}{\partial x_1\, \partial x_2} \\ -\dfrac{\partial^2 \phi}{\partial x_1\, \partial x_2} & \dfrac{\partial^2 \phi}{\partial x_1^2} \end{bmatrix} \tag{9.5}$$

Evidently, when the various components of $[\sigma]$ are substituted accordingly in eqns. (9.3), they both reduce to the obvious identity: $0 = 0$, and it follows that any stress field that is generated by eqn. (9.5) from any function $\phi(x_1\, x_2)$ will automatically satisfy the stress equations of small motion, no matter what the form of the function.

As to the question of continuity, this requires that the strains be related to the point displacements as in eqns. (9.4), where the three components of strain are described in terms of only two components of displacement. This implies that the three components of strain cannot be independent, but must rather be liable to a fixed relationship, and by direct substitution for the strains from eqns. (9.4) it can readily be shown that this relationship can be expressed in the equation

$$\frac{\partial^2 \epsilon_{11}}{\partial x_1{}^2} + \frac{\partial^2 \epsilon_{22}}{\partial x_2{}^2} = \frac{\partial^2 \gamma_{12}}{\partial x_1 \, \partial x_2} .$$

But in an isotropic, Hookean solid, the strains are related to the stresses as in eqns. (8.5 and (8.8). For such a material the equation that determines the compatability of the strains with the continuity of the body may be expressed in terms of the stresses by substituting accordingly, and this leads to the result:

$$\frac{\partial^2(\sigma_{11} - v\sigma_{22})}{\partial x_1{}^2} + \frac{\partial^2(\sigma_{22} - v\sigma_{11})}{\partial x_2{}^2} = 2(1+v)\frac{\partial^2 \sigma_{12}}{\partial x_1 \partial x_2} .$$

Finally, for a field of plane stress whose components are determined from a function ϕ, as in eqn. (9.5), the strain compatibility equation can be expressed in terms of ϕ simply by substituting accordingly, and this gives the result:

$$\frac{\partial^4 \phi}{\partial x_1^4} + 2\frac{\partial^4 \phi}{\partial x_1^2 \partial x_2^2} + \frac{\partial^4 \phi}{\partial x_2^4} = 0 \qquad (9.6)$$

We may therefore conclude that, for the class of problems considered here, in any field of plane stress generated by eqn. (9.5), from an arbitrary function $\phi(x_1 \, x_2)$ of the coordinates of position, the forces on each element will automatically satisfy the second law of motion, but where the function ϕ satisfies eqn. (9.6), the associated strains in an isotropic Hookean solid would also be compatible with the physical continuity of the body. Such a function is called an Airy Stress function.

Boundary Conditions

However, such a field does not comprise the solution to a particular problem of stress analysis, because it only describes a field which is boundless. But it does associate a particular state of stress with each point on any arbitrary boundary, and by resolution we can determine what stresses this state requires on the plane which coincides with the boundary. Here, of course, the stresses must arise as externally-reacted contact stresses, and thus we determine what loading must be applied to the boundary for consistency with the stress field within.

We note that the boundary loading is necessarily determined as a distributed effect which is described in terms of its intensity per unit area rather than as point forces, because a point force constitutes a singularity which cannot be accommodated in the calculus of smoothly continuous variables. But from a practical point also, a point force is an unrealistic abstraction in that it implies an infinite stress at the point of application, and in practice all forces are necessarily spread over a finite area. Of course, the area over which a given effect is spread may be small, and even indefinitely small, but within its span each part of the boundary loading must be specified by a smoothly continuous function which describes the boundary force per unit area.

The value of the boundary stress at any point is described in terms of its direct component normal to the boundary σ_N, and its tangential, shearing component σ_T. Note that σ_T is reckoned positive in the sense which bears the

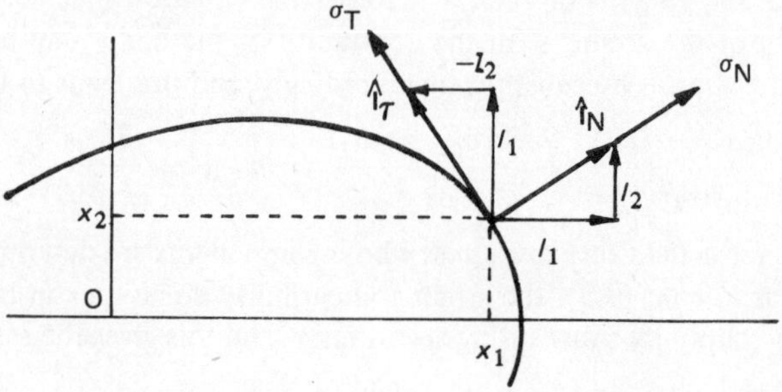

Fig. 9.2

same relationship to σ_N as x_2 bears to x_1. Thus, the convention for boundary stresses is *not* the same as that which is used in the stress tensor, though it is similar to that used in Mohr's circle.

Let $[\sigma]$ represent the state of plane stress that a valid stress function ascribes to any point on an arbitrary boundary, and let $\mathbf{l}_N = (l_1, l_2)$ represent the direction cosines of the outward facing normal at the point. Then, according to the convention for boundary stresses, the direction cosines of the positive tangent are $\mathbf{l}_T = (-l_2, l_1)$, and we can equate σ_N and σ_T to the components of $[\sigma]$ resolved on the plane normal to $\mathbf{l}_N$, first in the direction of $\mathbf{l}_N$, and then in the direction of $\mathbf{l}_T$.

Thus:
$$\sigma_N = \mathbf{l}_N \cdot [\sigma] \cdot \mathbf{l}_N = [l_1 \ \ l_2] \begin{bmatrix} \sigma_{11} & \sigma_{12} \\ \sigma_{21} & \sigma_{22} \end{bmatrix} \begin{bmatrix} l_1 \\ l_2 \end{bmatrix}$$

and
$$\sigma_T = \mathbf{l}_N \cdot [\sigma] \cdot \mathbf{l}_T = [l_1 \ \ l_2] \begin{bmatrix} \sigma_{11} & \sigma_{12} \\ \sigma_{21} & \sigma_{22} \end{bmatrix} \begin{bmatrix} -l_2 \\ l_1 \end{bmatrix}$$

or:
$$\left.\begin{aligned} \sigma_N &= l_1^2\sigma_{11} + l_2^2\sigma_{22} + 2l_1l_2\sigma_{12} \\ \text{and:}\quad \sigma_T &= l_1l_2(\sigma_{22}-\sigma_{11}) + (l_1^2-l_2^2)\sigma_{12} \end{aligned}\right\} . \tag{9.7}$$

These are the equations that relate the boundary stresses to the stress tensor and the direction cosines of the normal to the boundary, for any point on any boundary which contains any field.

9.3 ALTERNATIVE BOUNDARIES TO A GIVEN STRESS FIELD

The field generated by a given stress function can be 'fitted with' a boundary of any form, but in general the required loading on the boundary would be complex so that the problems to which random solutions apply are seldom significant. Nevertheless, for purposes of illustration we will consider the field generated by the function $\phi = x_1^3 x_2 - 3a^2 x_1 x_2 + x_1 x_2^3$, and we will consider the field within boundaries which are circular, square and triangular.

The field generated by $\phi = x_1^3 x_2 - 3a^2 x_1 x_2 + x_1 x_2^3$

It is obvious that this function satisfies the strain compatibility equation because each of the relevant fourth derivatives of each term is zero. Therefore we can generate a valid Hookean field by substituting ϕ in eqn. (9.5) as follows:

$$\begin{bmatrix} \sigma_{11} & \sigma_{12} \\ \sigma_{21} & \sigma_{22} \end{bmatrix} = \begin{bmatrix} \dfrac{\partial^2 \phi}{\partial x_2^2} & -\dfrac{\partial^2 \phi}{\partial x_1 \partial x_2} \\ -\dfrac{\partial^2 \phi}{\partial x_1 \partial x_2} & \dfrac{\partial^2 \phi}{\partial x_1^2} \end{bmatrix}$$

i.e.

$$\begin{bmatrix} \sigma_{11} & \sigma_{12} \\ \sigma_{21} & \sigma_{22} \end{bmatrix} = \begin{bmatrix} 6x_1 x_2 & 3\{a^2 - (x_1^2 + x_2^2)\} \\ 3\{a^2 - (x_1^2 + x_2^2)\} & 6x_1 x_2 \end{bmatrix} \tag{a}$$

Boundary relationships for any boundary

The boundary stresses needed to maintain this field within any given boundary may now be described in terms of the coordinates of a point on the boundary and the direction cosines of the normal to the boundary at the point, by substituting the appropriate values of σ_{11}, σ_{22} and σ_{12} in eqns. (9.7).

Thus

$$\left.\begin{aligned} \sigma_N &= 6(l_1^2 + l_2^2)x_1 x_2 + 6l_1 l_2\{a^2 - (x_1^2 + x_2^2)\} \\ \sigma_T &= 3(l_1^2 - l_2^2)\{a^2 - (x_1^2 + x_2^2)\} \end{aligned}\right\} \tag{b}$$

Boundary loading on a circular boundary

Now consider the loading required on a circular boundary of radius a, having its centre at the origin of coordinates. For any point on this boundary:

$$x_1^2 + x_2^2 = a^2 ,$$

and the direction cosines of the outward-facing normal are:

$$l_1 = x_1/a; \quad l_2 = x_2/a .$$

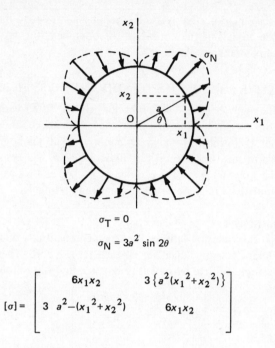

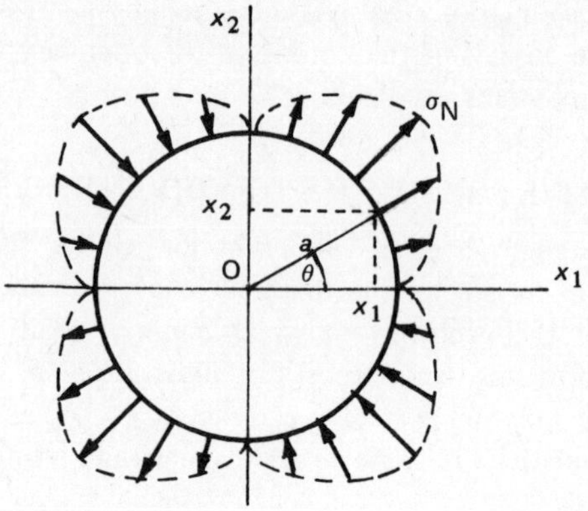

$\sigma_T = 0$

$\sigma_N = 3a^2 \sin 2\theta$

$$[\sigma] = \begin{bmatrix} 6x_1x_2 & 3\{a^2(x_1^2 + x_2^2)\} \\ 3\ a^2 - (x_1^2 + x_2^2) & 6x_1x_2 \end{bmatrix}$$

Therefore substituting accordingly in (a) gives:

$$\sigma_N = 6x_1x_2; \quad \sigma_T = 0 .$$

This result could now be described as a function of only one of the coordinates by substituting for the other from the equation of the boundary, but it is more convenient to express the result in terms of the parameter θ by substituting $x_1/a = \cos\theta$, and $x_2/a = \sin\theta$.

Thus $\quad \sigma_N = 6a^2 \cos\theta \sin\theta; \quad \sigma_T = 0$

or $\quad \sigma_N = 3a^2 \sin 2\theta \quad ; \quad \sigma_T = 0 .$

Thus the circular boundary requires no shear loading, but only normal loading whose intensity varies as indicated in the diagram.

Boundary loading on a square boundary

Alternatively, the same field can be 'fitted to' the square boundary ABCD indicated. Thus considering each edge in turn, we again substitute the equation of each edge, and its direction cosines in equations (b), as in the following table, and again we find that the boundary requires only normal loading which varies as indicated in the diagram.

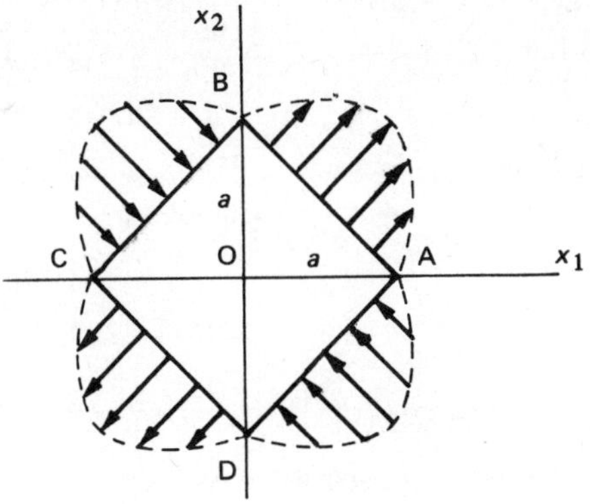

$\sigma_T = 0$; $\sigma_N = 6x_1x_2$

Edge	AB	BC	CD	DA
Equation	$x_1+x_2=a$	$-x_1+x_2=a$	$x_1+x_2=-a$	$x_1-x_2=a$
l_1	$1/\sqrt{2}$	$-1/\sqrt{2}$	$-1/\sqrt{2}$	$1/\sqrt{2}$
l_2	$1/\sqrt{2}$	$1/\sqrt{2}$	$-1/\sqrt{2}$	$-1/\sqrt{2}$
σ_N	$6x_1x_2$	$6x_1x_2$	$6x_1x_2$	$6x_1x_2$
σ_T	0	0	0	0

Boundary loading on a triangular boundary

Following the same procedure for the triangular boundary shows that the rectangular edges require loading in pure shear, and the hypotenuse in simple tension, with interactions which vary as indicated.

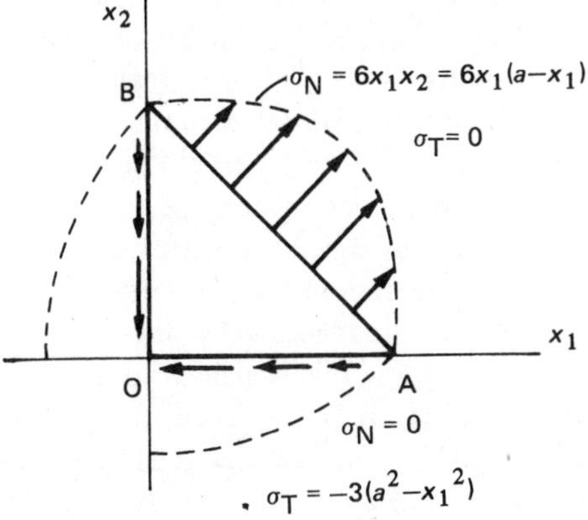

Edge	OA	AB	BO
equation	$x_2 = 0$	$x_1 + x_2 = a$	$x_1 = 0$
l_1	0	$1/\sqrt{2}$	-1
l_2	-1	$1/\sqrt{2}$	0
σ_N	0	$6x_1(a - x_1)$	0
σ_T	$-3(a^2 - x_1^2)$	0	$3(a^2 - x_2^2)$.

9.4 STRENGTH OF MATERIALS

In summary, the formal approach to stress analysis first identifies an arbitrary stress field in which the forces on each element satisfy the laws of motion, and the strains of the elements are consistent with the requirements of physical continuity. Then, for any arbitrary shape of boundary the method determines what externally reacted stresses must be applied to the boundary in order to maintain the arbitrary field of stress within, and although we have not attempted to apply this approach to the solution of a particular problem, it is obvious that the method is not without its difficulties and its limitations. The approach is difficult because it is both devious and complex, and it is limited because it can readily be applied only to situations in which the stress field, the body forces, the boundary loading, and the shape of the boundary itself can all be conveniently described in a common system of coordinates. It is true that the range of problems that can be tackled can be greatly extended by adopting numerical methods rather than algebraic, but this only adds to the complexity and cost of the anlysis, so even though the formal approach is, for some purposes, indispensable, there is an obvious need for a simpler alternative, even though it involves some loss of exactness.

In fact, the exactness of the formal analysis is commonly not a great advantage, because the problem itself can only be expressed in approximate or notional terms. For example, a railway bridge can only be designed for a notional combination of notional trains, or, considering a point of detail, we commonly know only that a certain net force or moment is applied within a given zone of the body, when the precise distribution of the effect within the zone is largely fortuitous. However, in the formal approach it is necessary to specify the conditions in precise detail, so that although the mathematical theory gives an accurate solution to the problem as thus stated, in many cases that statement is in fact no more than a notional idealisation of the physical situation to which it refers.

The alternative approach is pursued under the somewhat misleading title of 'Strength of Materials'. Broadly, this depends on the fact that the common components of engineering are usually designed either to span a distance (beams and transmission shafts), or to cover an area (plates and domes), or to enclose a volume (pressure vessels and shells). Commonly, therefore, the components may have one dimension which is much larger than the other two, or one dimension much smaller than the other two, and/or a high degree of symmetry, and it is significant that in such cases we can often make reasoned assumptions about the deformation of the component which are sufficient to imply a pattern of stress distribution which is, in fact, a reasonable approximation of the truth. Thus from the assumed deformation we infer a description of the variation of stress with position which only involves a couple of unknown constants, and these are determined by applying the controlling conditions in a degenerate integral form.

Here we shall consider the application of this approach to the class of long slender members in which one dimension is much greater than the other two. However, in so doing we shall be particularly concerned with the shape of the section of the member, i.e., with the effect of the distribution or location of the area of the section, and it is therefore convenient first to consider the application of the principle of moments to scalars.

10

Moments of Scalars

10.1 INTRODUCTION

In some respects moments of scalars are similar to moments of vectors, but the differences are no less marked. In particular, whereas with vectors we were concerned only with a single class of moments which were of first degree in distance, with scalars we are concerned with one set of moments of first degree, and another of second degree. Furthermore, with scalars, the moments of first degree are important only in defining the centroids of distributed systems, and it is the moments of second degree that usually determine the effects of their locations.

10.2 MOMENTS OF FIRST DEGREE

Referring to Fig. 10.1, let s represent a scalar quantity which is located at P, and let A represent an arbitrary datum point. Then the position of P relative to the datum is defined by the position vector $\mathbf{r}_{PA}$, and the first moment of s at P

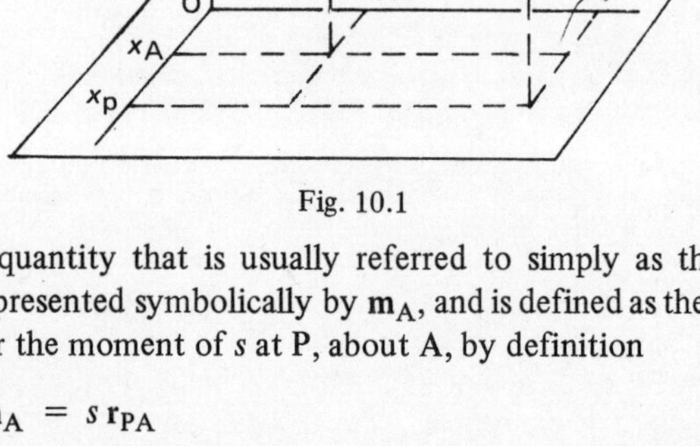

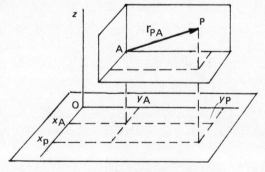

Fig. 10.1

about A (the quantity that is usually referred to simply as the moment of s about A), is represented symbolically by $\mathbf{m}_A$, and is defined as the vector quantity $s\,\mathbf{r}_{PA}$. Thus, for the moment of s at P, about A, by definition

$$\mathbf{m}_A = s\,\mathbf{r}_{PA} \tag{10.1}$$

Scalar First Moments

The first moment being thus defined as a vector quantity, its scalar component in any direction may be regarded as a scalar first moment. In particular, consider the scalar elements of the moment $\mathbf{m}_A = m_{Ax}\mathbf{i} + m_{Ay}\mathbf{j} + m_{Az}\mathbf{k}$ of the scalar s at the point $P = (x_P, y_P, z_P)$, about the datum $A = (x_A, y_A, z_A)$.

By definition:

$$\mathbf{m}_A = s\,\mathbf{r}_{PA}$$

that is $$\mathbf{m}_A = s\{(x_P - x_A)\mathbf{i} + (y_P - y_A)\mathbf{j} + (z_P - z_A)\mathbf{k}\}\ ,$$

Whence $$m_{Ax} = s(x_P - x_A)\ ,$$

$$m_{Ay} = s(y_P - y_A) \qquad (10.2)$$

and $$m_{Az} = s(z_P - z_A)$$

But $(x_P - x_A)$, $(y_P - y_A)$ and $(z_P - z_A)$ can be recognised as the distances of s from planes through A parallel to the coordinate planes. Therefore, whereas a scalar moment of a vector is determined by its distance from an axis, for a scalar it is determined by its distance from a plane, and the scalar moment of a scalar quantity about a given plane is defined as the product of the scalar itself with its perpendicular distance from the datum plane.

However, for points which are contained in a plane, distances from any plane which is normal to the plane of the points are equal to the distances from the line in which the two planes intersect (Fig. 10.2). Therefore for scalars

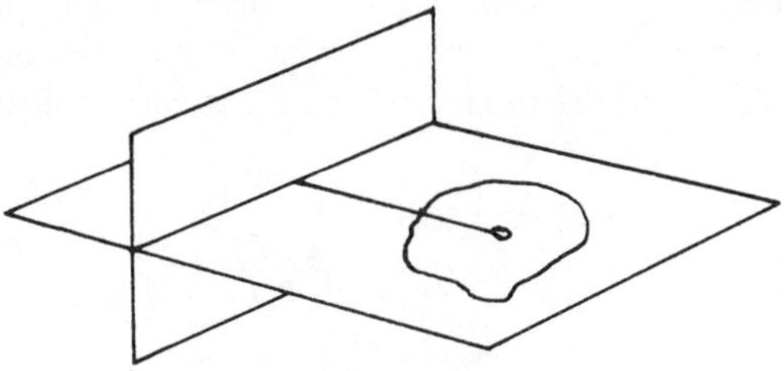

Fig. 10.2

which are confined to a plane, it is possible to define a scalar moment as a moment about an axis, but only for axes that are coplanar with the scalars themselves, and even then, the moment can more properly be considered to be referred, not to the axis itself, but to the plane normal to the plane of the scalar which contains the axis.

In calculating a scalar first moment it is commonly helpful to note that the moment is unaffected by a displacement of the scalar quantity parallel to the datum.

Resultant First Moments

Whether for vector moments or scalar, the resultant moment about a given datum of a set of discrete scalars is defined as the sum of the moments of the several quantities. Thus, if s now represents a typical member of a set of scalars, and if $\mathbf{r}_{PA}$ represents the vector of its position relative to a given datum A, the resultant moment of the set about A is indicated symbolically by Σm_A, and it may be equated, by definition, to the sum of the moments of the several scalars as follows:

$$\Sigma \mathbf{m}_A = \Sigma(s\,\mathbf{r}_{PA}) \ .$$

For a scalar quantity that is continuously distributed over a given domain, the distribution of the scalar is described in terms of its 'density' (i.e. of the amount per unit volume in three dimensions, or per unit area in two), and the moment of the continuum about any datum is determined by first considering the infinitesimal moment d**m** of the element of the scalar that is associated with an infinitesimal element of the domain. The moments for all the elements that make up the domain are then summed by integration. Thus, for a case in three dimensions, if ρ represents the density of the scalar quantity s at a typical position P, the amount of s that is associated with an infinitesimal element of volume dV at P is equal to $\rho\, dV$, its moment about the datum A is, by definition, equal to $\rho\, dV\, \mathbf{r}_{PA}$, and the resultant moment of the whole continuum is determined by integrating over its volume. Thus:

$$\mathbf{m}_A = \int_{Vol} d\mathbf{m}_A = \int_{Vol} \rho\, \mathbf{r}_{PA}\, dV \ .$$

Shift of Datum

Given the resultant (first) moment of a set of located scalars about one datum, we can readily determine the resultant moment about a second datum without recalculating the moments of the several quantities about the new datum. Thus,

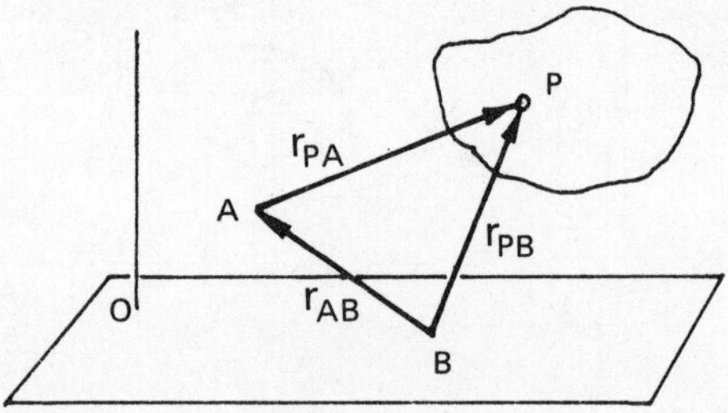

Fig. 10.3

let P (Fig. 10.3) represent the position of a typical scalar s of the set, and let A and B be any pair of alternative datum points. Then, by definition,

$$\Sigma \mathbf{m}_B = \Sigma(s\mathbf{r}_{PB}) = \Sigma\{s(\mathbf{r}_{PA} + \mathbf{r}_{AB})\}$$

Therefore $\Sigma \mathbf{m}_B = \Sigma(s\mathbf{r}_{PA}) + \Sigma(s\mathbf{r}_{AB})$,

whence $\Sigma \mathbf{m}_B = \Sigma \mathbf{m}_A + \mathbf{r}_{AB}\Sigma s$ (10.3)

We therefore conclude, that for the first moments of scalars as for moments of vectors, the resultant (first) moment of a set of located scalars about a second datum can be determined from the resultant moment about a first datum by adding the moment about the second datum of the sum of the scalars lumped at the first.

The Centroid

The centroid of a distributed scalar may be defined as the point about which its resultant moment is zero, when it follows that the resultant moment of the distributed scalar about any datum may be equated to the moment of its sum lumped at its centroid. Thus the resultant moments about any two datum points A and C are related according to:

$$\Sigma \mathbf{m}_C = \Sigma \mathbf{m}_A - \mathbf{r}_{CA} \Sigma s \ .$$

But if C is the centroid, by definition, $\Sigma \mathbf{m}_C$ is zero, and it follows that

$$\Sigma \mathbf{m}_A = \mathbf{r}_{CA} \Sigma s \ , \qquad (10.4)$$

where $\mathbf{r}_{CA} \Sigma s$ is equal to the moment of Σs at C about A.

Example 10.1

The coordinates at four point masses are:

4 kg at (2, −1, 2) m; 3 kg at (2, 0, −1) m

2 kg at (−1, 4, 0) m; 1 kg at (0, 0, 5) m .

Determine: a) The resultant moment of the masses about the origin of the coordinates; b) The coordinates of the centre of mass; and c) The resultant moment of the masses about A = (1, 2, 3) m.

Solution

a) By definition,

$$\Sigma \mathbf{m}_o = \Sigma(s\, \mathbf{r}_{PO})$$

$$\Sigma \mathbf{m}_o = 4(2\mathbf{i} - \mathbf{j} + 2\mathbf{k}) + 3(2\mathbf{i} - \mathbf{k}) + 2(-\mathbf{i} + 4\mathbf{j}) + 1(5\mathbf{k}) \text{ kg m}$$

$$= (8\mathbf{i} - 4\mathbf{j} + 8\mathbf{k}) + (6\mathbf{i} - 3\mathbf{k}) + (-2\mathbf{i} + 8\mathbf{j}) + 5\mathbf{k} \text{ kg m}$$

$$\Sigma \mathbf{m}_o = 12\mathbf{i} + 4\mathbf{j} + 10\mathbf{k} \text{ kg m} \ .$$

b) By the rule for a shift of datum:

$$\Sigma \mathbf{m}_C = \Sigma \mathbf{m}_O - \mathbf{r}_{CO} \Sigma s$$

that is $$0 = (12\mathbf{i} + 4\mathbf{j} + 10\mathbf{k}) - \mathbf{r}_{CO}(4 + 3 + 2 + 1) \ .$$

Therefore: $$\mathbf{r}_{CO} = (1{\cdot}2\mathbf{i} + 0{\cdot}4\mathbf{j} + \mathbf{k}) \text{ kg m} \ ,$$

and the coordinates of C are therefore (1·2, 0·4, 1).

c) For any datum point, the resultant moment of a distributed scalar can be equated to the moment of their sum lumped at the centroid:

$$\Sigma \mathbf{m}_A = (\Sigma s)\mathbf{r}_{CA} = (\Sigma s)(\mathbf{r}_{CO} - \mathbf{r}_{AO})$$

$$\Sigma \mathbf{m}_A = 10\{(1{\cdot}2\mathbf{i} + 0{\cdot}4\mathbf{j} + \mathbf{k}) - (\mathbf{i} + 2\mathbf{j} + 3\mathbf{k})\}\ \text{kg m}$$

$$\Sigma \mathbf{m}_A = (2\mathbf{i} - 16\mathbf{j} - 20\mathbf{k})\ \text{kg m}\ .$$

Example 10.2

The plane section of a machine component has the form shown. Determine the distance y_c of its centre of area from the datum XX.

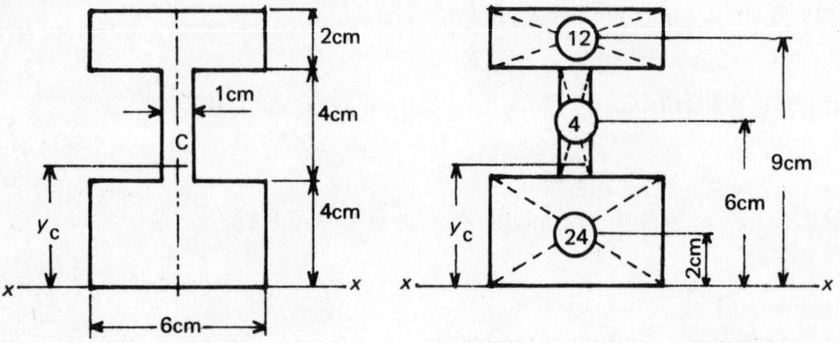

Solution

The resultant moment of the area of the section can be determined as the sum of the moments of three rectangles. But the scalar moment of a scalar can be equated to the total amount of the scalar times the distance of its centroid from the datum, and the centre of area of a rectangle is evidently at its geometrical centre. The total area of the section times y_c can therefore be equated to the sum of the moments of three rectangles whose areas are 24, 4 and 12 cm^2, and whose centres are at 2, 6 and 9 cm from the datum.

Thus $$(24 + 4 + 12)y_c = (24 \times 2) + (4 \times 6) + (12 \times 9)\ \text{cm}^3$$

$$40\,y_c = 48 + 24 + 108 = 180\ \text{cm}^3\ .$$

Whence $$y_c = 4{\cdot}5\ \text{cm}\ .$$

However, the calculation of a moment can often be eased by the choice of the right strategy, and with the moments of second degree (yet to be defined), the easement may be marked. Thus, a scalar moment of a scalar is not affected by a displacement parallel to the datum, so the web being thus displaced, we recognise that for any datum parallel to XX, the moment of the section can be equated to the difference between the moments of only two rectangles. But the moment of the smaller rectangle about any datum that passes through its own centre is zero, so for any such datum, the moment of the section can be equated

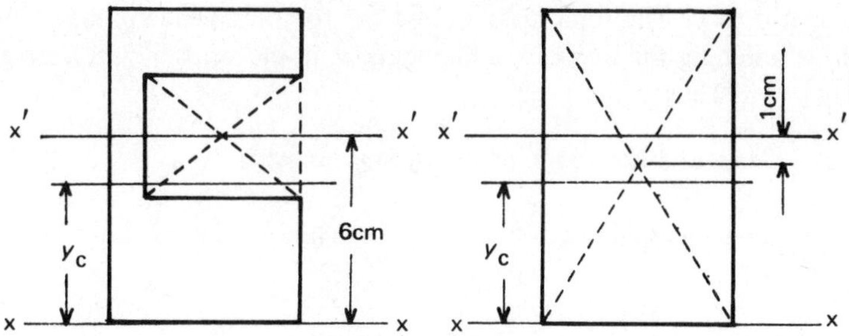

to the moment of the outer rectangle alone. Thus, choosing a datum X′X′ which passes through the centre of the smaller rectangle parallel to XX, we note that the larger rectangle has an area of 60 cm^2 with its centre at 1 cm from X′X′, whilst the section has a total area of 40 cm^2 with its centre at $(6-y_c)$ cm from X′X′, and equating the two descriptions of the moment about X′X′ we have:

$$60 \times 1 = 40(6 - y_c) ,$$

whence $$y_c = 4{\cdot}5 \text{ cm} .$$

Example 10.3

A plane lamina of uniform thickness t has the form of a quadrant of a circle of radius a which lies in the xy plane of rectangular coordinates with its radial edges along the x and y axes. The density ρ varies with y according to $\rho = A + By$, where A and B are constants. Determine the coordinates of the centre of mass.

Solution

Along any strip parallel to the x axis, the density has a constant value, so we consider an elemental strip of width dy at a uniform distance y from the x axis.

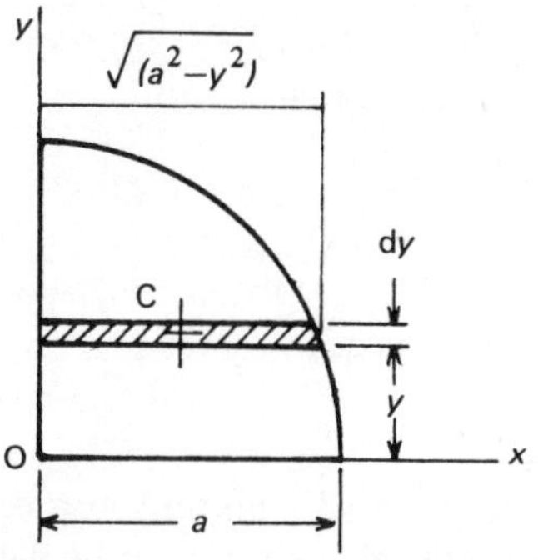

By Pythagoras, this is a strip of length $\sqrt{(a^2 - y^2)}$, and of density $\rho = A + By$ whose centroid falls at the mid point C, whose coordinates are $(\frac{1}{2}\sqrt{(a^2-y^2)}, y)$. The elemental mass d$m$ of the strip is equal to its volume times the density, whilst its elemental moment about the origin dm_o is equal to its mass times the radius vector to C. The total mass m, and the total moment $\mathbf{m}_o$ may be determined by summing the masses and moments of all the strips by integrating from $y = 0$ to $y = a$. Thus

$$\text{Mass of strip AB} = (A + By)(a^2 - y^2)^{\frac{1}{2}} t\,dy \ ,$$

$$\text{Mass of quadrant} = t\int_0^a (A + By)(a^2 - y^2)^{\frac{1}{2}}\,dy$$

$$= \frac{a^2 t}{12}(3\pi A + 4aB) \ .$$

Equating the moment of the strip to its mass times the radius vector to its centroid

$$d\mathbf{m}_o = t(A + By)(a^2 - y^2)^{\frac{1}{2}}\{\tfrac{1}{2}(a^2 - y^2)^{\frac{1}{2}}\mathbf{i} + y\,\mathbf{j}\}\,dy$$
$$= t(A + By)\{\tfrac{1}{2}(a^2 - y^2)\,\mathbf{i} + y(a^2 - y^2)^{\frac{1}{2}}\,\mathbf{j}\}\,dy$$

therefore: $$\mathbf{m}_o = t\int_0^a (A + By)\{\tfrac{1}{2}(a^2 - y^2)\,\mathbf{i} + y(a^2 - y^2)^{\frac{1}{2}}\,\mathbf{j}\}\,dy$$

that is: $$\mathbf{m}_o = \frac{ta^3}{48}\left\{(16A + 6aB)\,\mathbf{i} + (16A + 3\pi a B)\,\mathbf{j}\right\} \ .$$

Since the resultant moment is equal to the total mass times the radius vector to its centroid, it follows that $\mathbf{r}_{co} = \mathbf{m}_o/m$.

Thus $$\mathbf{r}_{co} = \frac{a}{4(3\pi A + 4aB)}\{(16A + 6aB)\mathbf{i} + (16A + 3\pi aB)\mathbf{j}\} \ ,$$

and the coordinates of the centroid are therefore

$$(x_c, y_c) = \frac{a(16A + 6aB)}{4(3\pi A + 4aB)}, \ \frac{a(16A + 3\pi aB)}{4(3\pi A + 4aB)} \ .$$

Example 10.4

The diagram shows a solid three-dimensional body OPQR of uniform density ρ. Determine the coordinates of its centre of mass.

Solution

In the diagram STU is a slice of uniform width dz at a uniform distance z from the xy plane. The mass of the slice dm can be determined as the product of its volume with its constant density, and the moment of the mass about the xy plane can be determined by multiplying its mass dm by its uniform distance z from the xy plane. The mass and moment of the body about the xy plane can then be determined as the sum of the masses and moments of all the slices

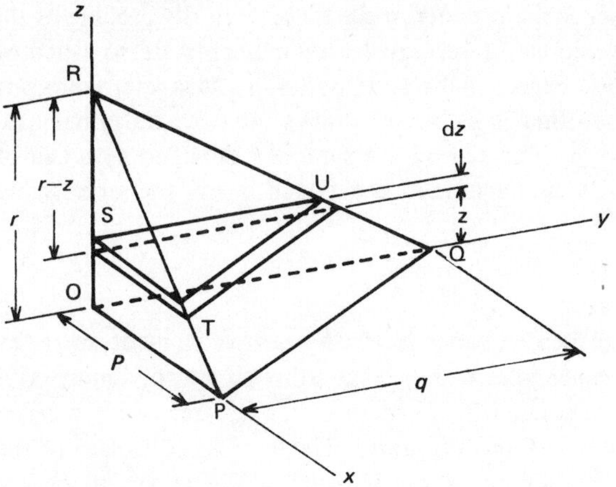

that make up the body by integrating with respect to z from $z = 0$ to $z = z_R = r$. The distance of the centroid from the xy plane is then determined by dividing the moment by the mass, when the x and y coordinates of the centroid follow from symmetry. Thus for the slice STU, by comparing it with the similar face:

$$\text{Area of STU} = \frac{pq}{2}\left(\frac{r - z}{r}\right)^2$$

$$m = \frac{\rho pq}{2r^2}\int_0^r (r - z)^2 \, dz$$

that is: $m = \rho\, p\, q\, r\, /\, 6$

$$m_{xy} = \frac{\rho pq}{2r^2}\int_0^r (r - z)^2 z \, dz$$

that is: $m_{xy} = \rho\, p\, q\, r^2/24$.

Therefore z_c, the distance of the centroid from the xy plane, is determined by the ratio m_{xy}/m, and by symmetry

$$(x_c, y_c, z_c) = (p/4, q/4, r/4) \ .$$

10.3 MOMENTS OF SECOND DEGREE

Whereas a first scalar moment of a scalar is defined as the product of the scalar with its distance from a single datum plane, a moment of second degree is referred to a pair of datum planes, and it is defined, in one case as the product of the scalar with the sum of the squares of its distances from the two planes, and in another as the product of the scalar with the product of the two distances. Thus the two kinds of second degree moments are so called because they are both of second degree in distance, but only one is referred to simply as a second moment. The other is called a product moment. In either case, a moment of second degree is indicated by the symbol I qualified by a pair of suffices which differ for a product moment, but which repeat the same symbol for a second moment.

Second Moments

The datum planes to which a second degree moment are referred are always chosen as a rectangular pair, so that although a second moment is defined as the product of the scalar with the sum of the squares of its distances from the two planes, it follows from Pythagoras' Theorem that the sum of the squares of the distances from the two planes is equal to the square of the distance from the straight line in which they intersect. A second moment (but not a product moment) can therefore by considered to be referred either to a rectangular pair of datum planes, or to the axis in which they intersect, and it may alternatively be defined as the scalar quantity equal in the one case to the product of the scalar with the sum of the squares of its perpendicular distances from the two datum planes, and in the other to the product of the scalar with the square of its perpendicular distance from the datum axis.

A second moment of mass is commonly referred to as a Moment of Inertia.

Radius of Gyration

A second moment being regarded as a moment about an axis, the Radius of Gyration of a distributed scalar is defined as the distance from the axis at which the sum of the scalars would have the same second moment as the distributed scalar. A radius of gyration is indicated by the symbol K, and if I_{xx} represents the resultant second moment about the axis x of a distributed scalar whose sum is represented by Σs, by definition

$$I_{xx} = K_{xx}^2 \Sigma s \tag{10.5}$$

Product Moments

The product moment of a scalar relative to a pair of rectangular datum planes is defined as the scalar quantity equal to the product of the scalar with the product of its perpendicular distances from the two datum planes.

The Set of Second Degree Moments in Rectangular Coordinates

If s represents a located scalar quantity whose coordinates in given rectangular axes are represented by (x, y, z), then x, y and z are the perpendicular distances of the scalar from the rectangular coordinate planes, and for each pair we can define in terms of its coordinates, one second moment of the scalar, and an equal pair of product moments. For example, the distances of s from the xy and xz planes are equal to z and y respectively, and we identify one product moment equal to $s\,y\,z$ which is indicated symbolically by I_{yz}, and another equal to $s\,z\,y$ which is indicated by I_{zy}. However, a product of scalars is commutative, when it follows that I_{yz} is equal to I_{zy}. The second moment with respect to the same pair of planes is defined as the product $s(y^2 + z^2)$, but this can alternatively be regarded as the second moment about the x axis in which the two planes intersect, and it is indicated symbolically by I_{xx}. By thus considering each of the three pairs of coordinate planes in turn we define in the given coordinates a set of nine second degree moments as follows:

$$\begin{array}{ll} \text{Second Moments} & \text{Product Moments} \\ I_{xx} = s(y^2 + z^2) & I_{yz} = I_{zy} = s\,y\,z \\ I_{yy} = s(x^2 + z^2) & I_{xz} = I_{zx} = s\,x\,z \\ I_{zz} = s(x^2 + y^2) & I_{xy} = I_{yx} = s\,x\,y \end{array} \tag{10.6}$$

Example 10.5

For the masses of Example 10.1, determine the product moment about the xy and xz planes, and the moment of inertia and radius of gyration about the x axis.

Solution

By definition

$$I_{yz} = \Sigma(s\,y\,z); \quad I_{xx} = \Sigma(s(y^2 + z^2)); \quad K_{xx} = I_{xx}/\Sigma s.$$

	m	y	z	y^2	z^2	y^2+z^2	$m\,y\,z$	$m(y^2+z^2)$
	4	−1	2	1	4	5	−8	20
	3	0	−1	0	1	1	0	3
	2	4	0	16	0	16	0	32
	1	0	5	0	25	25	0	25
Σ	10						−8	80

$$I_{yz} = -8 \text{ kg m}^2; \quad I_{xx} = 80 \text{ kg m}^2; \quad K_{xx} = 2\sqrt{2} \text{ m}.$$

Example 10.6

For the body of Example 10.4, determine the set of second degree moments in the coordinates indicated.

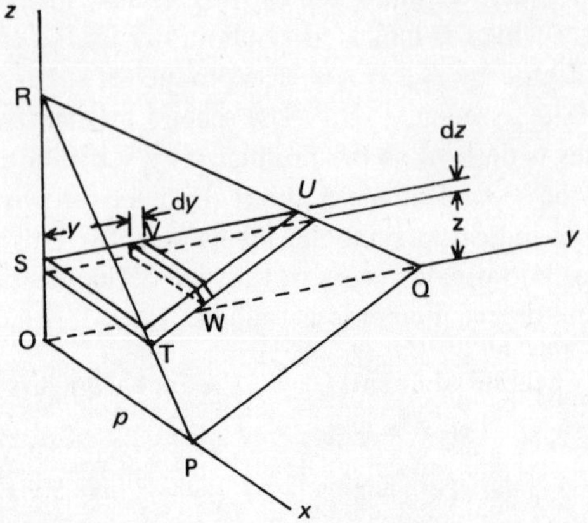

Solution

In the diagram, STU is a slice of uniform thickness dz at a uniform distance z from the xy plane, and VW is a strip of the slice, of uniform width dy at a uniform distance y from the xz plane. Evidently, there is a common value of y and of z for all the elements that go to make up the strip. Furthermore, the density ρ is uniform, so the mass dm of the elemental strip can be determined as the product of its volume with the density, its second moment of mass about the x axis, $\mathrm{d}I_{xx}$, can be equated by $\mathrm{d}m(y^2+z^2)$, and its product moments $\mathrm{d}I_{zy} = \mathrm{d}I_{yz}$ can be equated to $\mathrm{d}m\, y\, z$. Thus:

By similar triangles, length of VW $= \dfrac{p}{qr}(qr - qz - ry)$

therefore: Mass of strip $= \dfrac{\rho p}{qr}(qr - qz - ry)\,\mathrm{d}y\,\mathrm{d}z$

and $$\mathrm{d}I_{xx} = \frac{\rho p}{qr}(qr - qz - ry)(y^2 + z^2)\,\mathrm{d}y\,\mathrm{d}z$$

and $$\mathrm{d}I_{yz} = \frac{\rho p}{qr}(qr - qz - ry)\,y\,z\,\mathrm{d}y\,\mathrm{d}z\ .$$

The moments for the body can now be determined by first integrating with respect to y from $y = y_s = 0$, to $y = y_U = q(r-z)/r$ (whilst treating z and dz as constants), and by then integrating with respect to z from $z = z_0 = 0$ to $z = z_R = r$.

Thus $$I_{xx} = \frac{\rho p}{qr}\int_0^r \int_0^{q(r-z)/r} (qr - qz - ry)(y^2 + z^2)\,dy\,dz$$

that is: $$I_{xx} = \rho pqr(q^2 + r^2)/60 \ .$$

Also $$I_{yz} = \frac{\rho p}{qr}\int_0^r \int_0^{q(r-z)/r} (qr - qz - ry)\,y\,z\,dy\,dz$$

that is: $$I_{yz} = \rho pq^2r^2/120 \ .$$

The remaining second and product moments are then obvious from symmetry, whence:

$$\begin{aligned} I_{xx} &= \rho pqr(q^2 + r^2)/60; & I_{yz} &= \rho pq^2r^2/120 \\ I_{yy} &= \rho pqr(p^2 + r^2)/60; & I_{xz} &= \rho p^2qr^2/120 \\ I_{zz} &= \rho pqr(p^2 + q^2)/60; & I_{xy} &= \rho p^2q^2r/120 \ . \end{aligned}$$

10.4 THE TENSOR OF SECOND DEGREE MOMENTS

It is convenient to present the set of second degree moments in given coordinates in a symmetrical matrix which resembles a stress tensor, and it can be shown that this matrix is, in fact, another tensor of the same kind. To form the matrix, the second moments are entered on the leading diagonal, with the negatives of the equal pairs of product moments placed symmetrically about it.

$$[\mathbf{I}] = \begin{bmatrix} I_{xx} & -I_{xy} & -I_{xz} \\ -I_{yx} & I_{yy} & -I_{yz} \\ -I_{zx} & -I_{zy} & I_{zz} \end{bmatrix} .$$

Transformation for a Rotation of Axes

Again let s represent a scalar that is located in a given frame of reference, and let Oxyz and O$x'y'z'$ represent alternative rectangular coordinates having a common origin O, (Fig. 10.4). The coordinates of s in Oxyz are represented by (x, y, z),

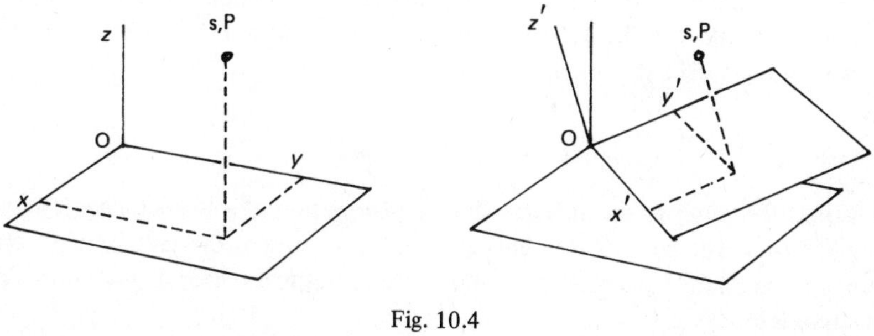

Fig. 10.4

and in $Ox'y'z'$ by (x', y', z'), and representing the second moment of s about x' by I'_{xx}, we now consider the relationship of I'_{xx} to the second moments in $Oxyz$.

By Pythagoras $(x')^2 + (y')^2 + (z')^2 = x^2 + y^2 + z^2$.

Therefore $I'_{xx} = s\{(y')^2 + (z')^2\} = s\{x^2 + y^2 + z^2 - (x')^2\}$.

But if the direction cosines of x' in $Oxyz$ are represented by $l = (l_x\ l_y\ l_z)$, then x' is the component of the vector $[x, y, z]$ resolved in the direction of the unit vector l.

Therefore $x' = l_x x + l_y y + l_z z$,

and $(x')^2 = l_x^2 x^2 + l_y^2 y^2 + l_z^2 z^2 + 2\{l_x l_y xy + l_x l_z xz + l_y l_z yz\}$.

But $l_x^2 = 1 - (l_y^2 + l_z^2);\ l_y^2 = 1 - (l_x^2 + l_z^2);\ \text{and}\ l_z^2 = 1 - (l_x^2 + l_y^2)$,

and substituting accordingly gives

$$(x')^2 = (x^2 + y^2 + z^2) - (l_y^2 + l_z^2)x^2 - (l_x^2 + l_z^2)y^2 - (l_x^2 + l_y^2)z^2$$
$$+ 2\{l_x l_y xy + l_x l_z xz + l_y l_z yz\} .$$

Substituting accordingly in the foregoing expression for I'_{xx} gives

$$I'_{xx} = s\{(l_y^2 + l_z^2)x^2 + (l_x^2 + l_z^2)y^2 + (l_x^2 + l_y^2)z^2$$
$$- 2(l_y l_z yz + l_x l_z xy + l_x l_y xy)\}$$
$$= s\{l_x^2(y^2 + z^2) + l_y^2(x^2 + z^2) + l_z^2(x^2 + y^2)$$
$$- 2(l_y l_z yz + l_x l_z xz + l_x l_y xy)\}$$
$$= s\{l_x^2 I_{xx} + l_y^2 I_{yy} + l_z^2 I_{zz}$$
$$- 2(l_y l_z I_{yz} + l_x l_z I_{xz} + l_y l_z I_{yz})\}$$
$$= [l_x\ l_y\ l_z] \cdot \begin{bmatrix} I_{xx} & -I_{xy} & -I_{xz} \\ -I_{yx} & I_{yy} & -I_{yz} \\ -I_{xz} & -I_{yz} & I_{zz} \end{bmatrix} \cdot \begin{bmatrix} l_x \\ l_y \\ l_z \end{bmatrix} .$$

Since parallel results obtain for each of the second moments in the matrix $[I']$, it follows that if $[l]$ represents the matrix of direction cosines of $Ox'y'z'$ in $Oxyz$, the rule for the transformation of the matrix $[I]$ in $Oxyz$ into the corresponding matrix $[I']$ in $Ox'y'z'$ is

$$[I'] = [l] \cdot [I] \cdot [l]^T \tag{10.7}$$

But this is the same as the rule for the transformation of a stress tensor, and we recognise that, for coordinates having a common origin, the matrix of second degree moments of a given scalar is another symmetrical tensor of the same kind as a stress tensor.

Transformation for a Translation of Axes

But whereas a stress tensor is relevant only to different sets of coordinates at a single point, for the moments, the location of the axes is as relevant as their orientation. We therefore now consider the relationship between the tensors of a given distributed scalar s for sets of axes which are parallel, but which have different origins.

So assuming a given scalar s which has a specific distribution, let Oxyz represent a given set of coordinates, let C$x'y'z'$ represent the parallel set whose origin is chosen at the centroid C of the scalar, and let P identify the position of a typical element ds of the scalar, as indicated in Fig. 10.5. Considering one of

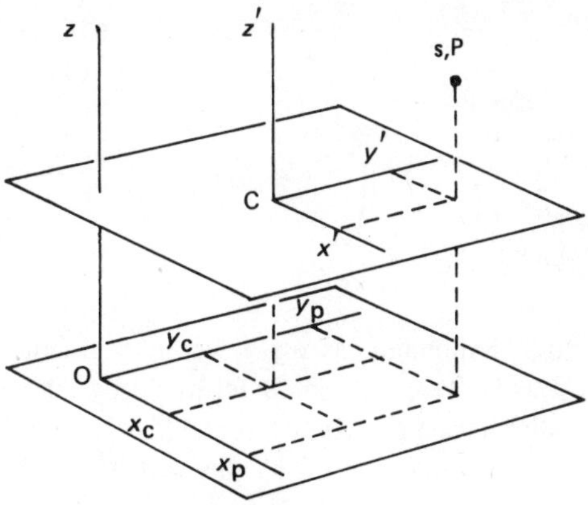

Fig. 10.5

the resultant second moments in Oxyz, say I_{xx}, we note, that since the coordinates of P in Oxyz are represented by (x, y, z), by definition:

$$I_{xx} = \Sigma \mathrm{d}s(y^2 + z^2) = \Sigma \mathrm{d}s\, y^2 + \Sigma \mathrm{d}s\, z^2 \ .$$

But if the coordinates of C in Oxyz are represented by (x_c, y_c, z_c), we may evidently substitute $y = y' + y_c$ and $z = z' + z_c$, when:

$$\begin{aligned} I_{xx} &= \Sigma \mathrm{d}s(y' + y_c)^2 + \Sigma \mathrm{d}s(z' + z_c)^2 \\ &= \Sigma \mathrm{d}s\{(y')^2 + (z')^2\} + \Sigma \mathrm{d}s(y_c^2 + z_c^2) \\ &\quad + 2\Sigma \mathrm{d}s\, y'y_c + 2\Sigma \mathrm{d}s\, z'z_c \ , \end{aligned}$$

or:

$$\begin{aligned} I_{xx} &= \Sigma \mathrm{d}s\{(y')^2 + (z')^2\} + (y_c^2 + z_c^2)\, \Sigma \mathrm{d}s \\ &\quad + 2y_c\, \Sigma \mathrm{d}s\, y' + 2z_c\, \Sigma \mathrm{d}s\, z' \ . \end{aligned}$$

Considering the terms on the right in turn, we note that, by definition, the first determines the resultant second moment of s about Cx', and this may conveniently be represented by I'_{xx}. In the second term, $\Sigma \mathrm{d}s$ may be recognised at the total scalar s, and the term can therefore be recognised as the second moment about Ox, of the total scalar s, lumped at its centroid. In the third and fourth terms $\Sigma ds\, y'$ and $\Sigma ds\, z'$ can be recognised as the resultant first moments of s about planes which pass through the centroid, and they are both therefore zero. We may therefore write

$$\left.\begin{aligned} I_{xx} &= I'_{xx} + (y_c^2 + z_c^2)s \\ \text{Similarly} \quad I_{yy} &= I'_{yy} + (x_c^2 + z_c^2)s \\ \text{and} \quad I_{zz} &= I'_{zz} + (x_c^2 + y_c^2)s \end{aligned}\right\}$$

Similarly, for the product moment I_{xy} in Oxyz, by definition:

$$\begin{aligned} I_{xy} &= \Sigma ds\, xy \\ &= \Sigma ds(x' + x_c)(y' + y_c) \\ &= \Sigma ds\, x'y' + s\, x_c\, y_c \end{aligned}$$

$$\left.\begin{aligned} \text{that is} \quad I_{xy} &= I'_{xy} + s\, x_c\, y_c \\ \text{Similarly} \quad I_{xz} &= I'_{xz} + s\, x_c\, z_c \\ I_{yz} &= I'_{yz} + s\, y_c\, z_c \end{aligned}\right\}.$$

Evidently, for a product moment, as for a second moment, the resultant moment in the general axes Oxyz may be determined by adding to the resultant moment in the parallel axes through the centroid, the simple moment in the general axis of the total scalar lumped at the centroid; and remembering that it is the negatives of the product moments that are entered in the tensor, the relationship between the tensors for the parallel sets of axes through a general point O, and the centroid C may therefore be expressed as follows:

$$\begin{bmatrix} I_{xx} & -I_{xy} & -I_{xz} \\ -I_{yx} & I_{yy} & -I_{yz} \\ -I_{zx} & -I_{zy} & I_{zz} \end{bmatrix} = \begin{bmatrix} I'_{xx} & -I'_{xy} & -I'_{xz} \\ -I'_{yx} & I'_{yy} & -I'_{yz} \\ -I'_{zx} & -I'_{zy} & I'_{zz} \end{bmatrix}$$

$$+ \Sigma s \begin{bmatrix} (y_c^2 + z_c^2) & -x_c y_c & -x_c z_c \\ -y_c x_c & (x_c^2 + z_c^2) & -y_c z_c \\ -x_c z_c & -y_c z_c & (x_c^2 + y_c^2) \end{bmatrix}$$

Plane systems and the Polar Axis Theorem

When the scalar is confined to a plane, it is appropriate to choose a set of coordinates of which one is normal to the plane. For each element the z coordinate is zero, and it follows that the tensor of the second degree moments adopts the same form as a stress tensor of a state of plane stress, as shown. Further-

$$[\mathbf{I}] = \begin{bmatrix} \Sigma sy^2 & \Sigma sxy & 0 \\ \Sigma sxy & \Sigma sx^2 & 0 \\ 0 & 0 & \Sigma s(x^2+y^2) \end{bmatrix}$$

more, we note that the second moment about the 'polar' axis which is normal to the plane of the scalar is equal to the sum of the second moments about the two in-plane axes. This is known as the Polar Axis Theorem.

Summary of Conclusions

In general, a tensor of second degree moments of a given distributed scalar specifies the second moments, and the negatives of the product moments in a set of rectangular coordinates in which the position of the origin as well as the orientation of the axes is arbitrary. With this tensor it is therefore necessary to supplement the standard rule of tensor transformation for a rotation of axes by a rule which provides for a translation of the axes.

a) For any given distributed scalar, the tensor of resultant second degree moment $[\mathbf{I}_{Oxyz}]$ for a set of axes having its origins at a general point O may be determined as a sum of two tensors: one for the distributed scalar in parallel axes $Cxyz$ whose origin falls at the centroid C, and another for the total scalar, lumped at the centroid, in the general axes $Oxyz$.

This rule is applied to the comparison of tensors having arbitrary origins, by comparing each with the tensor for the axes whose origin falls at the centroid.

b) In a rotation of the axes about a given origin, the tensors of second degree moments of a given distributed scalar are resolved or transformed by the standard rule for tensors. Thus, if $Oxyz$ and $Ox'y'z'$ represent alternative coordinates having a common origin O, and if [l] represents the matrix of direction cosines of $Ox'y'z'$ in $Oxyz$, then:

$$[\mathbf{I}_{Ox'y'z'}] = [\mathbf{l}] \cdot [\mathbf{I}_{Oxyz}] \cdot [\mathbf{l}]^T .$$

Alternatively, the tensor may be resolved by Mohr's Circular Diagram, but in either method it is important to note that the off-diagonal terms in the second moment tensors are the negatives of the various products moments, and not the positives. In particular, in comparing the Circular diagrams for moments and stresses, we note that whereas the direct stresses are replaced by the second moments, the shear stresses are replaced by the negatives of the product moments.

c) For each distributed scalar and each origin of coordinates, there is one set of principal axes for which all the product moments are zero, and the second moments about these axes are known as the principal second moments for the given origin. The principal second moments and the direction cosines of the principal axes of second moments, are determined from any general tensor for the given origin in exactly the same way as the principal stresses and the directions of the principal axes of stress are determined from a general tensor of stress.

Example 10.7

In coordinates Oxyz, the second and product moments of a distributed mass have the values (kg m^2):

$$I_{xx} = 50\ ; \quad I_{xy} = I_{yx} = -15$$
$$I_{yy} = 25\ ; \quad I_{xz} = I_{zx} = -30$$
$$I_{zz} = 40\ ; \quad I_{yz} = I_{zy} = 20\ .$$

Determine the tensor of second degree moments of the mass in parallel coordinates Axyz, given that the total mass is 10 kg, and that, in Oxyz, the coordinates of A and of the centroid C of the scalar, in metres, are (0, 1, 2), and (−2, 2, 1), respectively.

Solution

The tensor of second moments of a given distributed scalar in any axes may be equated to the tensor for the distributed scalar in the parallel centroidal axes, plus the tensor in the general axes, of the total scalar lumped at the centroid.

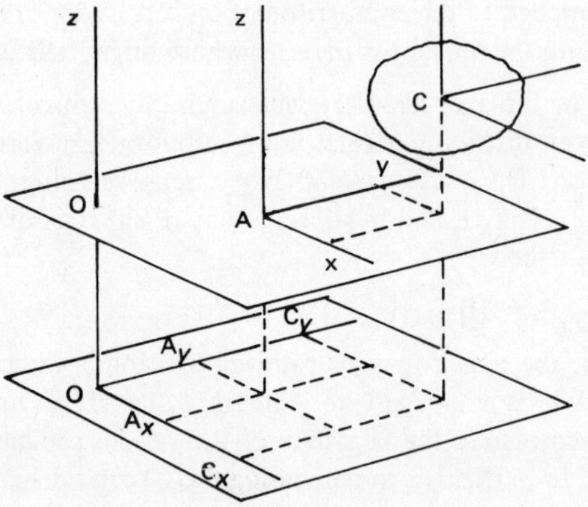

The tensor for the distributed scalar in the centroidal axes can therefore be determined as the difference between the tensors for the distributed and lumped scalars in Oxyz, and the tensor for Axyz can then be determined by adding the tensor in Axyz of the total scalar lumped at the centroid.

Thus, the given second moments, and the negatives of the given product moments being entered appropriately, the tensor $[\mathbf{I}_\text{o}]$ of the distributed scalar in Oxyz is:

$$[\mathbf{I}_O] = \begin{bmatrix} 50 & 15 & 30 \\ 15 & 25 & -20 \\ 30 & -20 & 40 \end{bmatrix} \text{kg m}^2 .$$

The coordinates of C in Oxyz are $(-2, 2, 1)$ m, and in these axes, the tensor for the total scalar lumped at C is:

$$10\begin{bmatrix} (2^2+1^2) & -(-2\,.\,2) & -(-2\,.\,1) \\ -(2.-2) & \{(-2)^2+1^2\} & -(\;2\,.\,1) \\ -(1.-2) & -(1\,.\,2) & \{(-2)^2+2^2\} \end{bmatrix} = \begin{bmatrix} 50 & 40 & 20 \\ 40 & 50 & -20 \\ 20 & -20 & 80 \end{bmatrix}.$$

By subtracting the coordinates of A in Oxyz from the coordinates of C in the same, we determine the coordinates of C in Axyz to be $(-2, 1, -1)$ m, and the tensor for the lumped scalar in Axyz is therefore equal to:

$$10\begin{bmatrix} (1\;+\;1) & -(-2.\;\;1) & -(-2.-1) \\ -(\;1.-2) & (4\;+\;1) & -(\;1.-1) \\ -(-1.-2) & -(-1.\;\;1) & (4\;+\;1) \end{bmatrix} = \begin{bmatrix} 20 & 20 & -20 \\ 20 & 50 & 10 \\ -20 & 10 & 50 \end{bmatrix}.$$

The tensor of the distributed scalar in Axyz is therefore given by:

$$\begin{bmatrix} 50 & 15 & 30 \\ 15 & 25 & -20 \\ 30 & -20 & 40 \end{bmatrix} - \begin{bmatrix} 50 & 40 & 20 \\ 40 & 50 & -20 \\ 20 & -20 & 80 \end{bmatrix} + \begin{bmatrix} 20 & 20 & -20 \\ 20 & 50 & 10 \\ -20 & 10 & 50 \end{bmatrix}$$

$$= \begin{bmatrix} 20 & -5 & -10 \\ -5 & 25 & 10 \\ -10 & 10 & 10 \end{bmatrix}.$$

Example 10.8

Referring to the body of Example 10.6, consider the case when $\rho = 960, p = 1$, $q = 2$, and $r = 2$. Making use of the previous results, determine the tensor of second degree moments of mass of the body relative to a reference CXYZ whose origin falls at the centre of mass, and whose orientation relative to Oxyz is defined by the matrix of direction cosines $[\mathbf{l}]$.

$$[\mathrm{l}] = \begin{bmatrix} 2/3 & -1/3 & 2/3 \\ 3/5 & 0 & 4/5 \\ -4/15 & -2/15 & 1/5 \end{bmatrix} .$$

Solution

Let Cxyz be a reference which is parallel to Oxyz, with its origin at the centre of mass. The coordinates of O in Cxyz are the negatives of the coordinates of C in Oxyz, and from the results of Example 10.4, we have

$$(x_c,\ y_c,\ z_c) = (p/4,\ q/4,\ r/4) = (1/4,\ 1/2,\ 1/2)$$

$$\text{and total mass} = \Sigma m = \rho pqr/6 = 640 \ .$$

From the results of Example 10.7 we also have

$$[\mathbf{I}_{\mathrm{O}xyz}] = \frac{\rho pqr}{120} \begin{bmatrix} 2(q^2+r^2) & -pq & -pr \\ -pq & 2(p^2+r^2) & -qr \\ -pr & -qr & 2(p^2+q^2) \end{bmatrix}$$

$$= 64 \begin{bmatrix} 8 & -1 & -1 \\ -1 & 5 & -2 \\ -1 & -2 & 5 \end{bmatrix} .$$

For the translation between Oxyz and Cxyz:

$$[\mathbf{I}_{\mathrm{O}xyz}] = [\mathbf{I}_{\mathrm{C}xyz}] + \Sigma m \begin{bmatrix} (y_c + z_c) & -x_c y_c & -x_c z_c \\ -x_c y_c & (x_c + z_c) & -y_c z_c \\ -x_c z_c & -y_c z_c & (x_c + y_c) \end{bmatrix}$$

$$[\mathbf{I}_{\mathrm{C}xyz}] = 64 \begin{bmatrix} 8 & -1 & -1 \\ -1 & 5 & -2 \\ -1 & -2 & 5 \end{bmatrix} - \frac{640}{16} \begin{bmatrix} 8 & -2 & -2 \\ -2 & 5 & -4 \\ -2 & -4 & 5 \end{bmatrix}$$

$$= 8 \left\{ 8 \begin{bmatrix} 8 & -1 & -1 \\ -1 & 5 & -2 \\ -1 & -2 & 5 \end{bmatrix} - 5 \begin{bmatrix} 8 & -2 & -2 \\ -2 & 5 & -4 \\ -2 & -4 & 5 \end{bmatrix} \right\}$$

$$[\mathbf{I}_{Cxyz}] = 8 \begin{bmatrix} 24 & 2 & 2 \\ 2 & 12 & 4 \\ 2 & 4 & 15 \end{bmatrix} .$$

For the rotation from $Cxyz$ to $CXYZ$:

$$[\mathbf{I}_{Cxyz}] = [\mathbf{l}]\,[\mathbf{I}_{Cxyz}]\,[\mathbf{l}]^{\mathrm{T}}$$

$$= \frac{8}{15^2} \begin{bmatrix} 10 & -5 & 10 \\ 9 & 0 & 12 \\ -4 & -2 & 3 \end{bmatrix} \begin{bmatrix} 24 & 2 & 2 \\ 2 & 15 & 4 \\ 2 & 4 & 15 \end{bmatrix} \begin{bmatrix} 10 & 9 & -4 \\ -5 & 0 & -2 \\ 10 & 12 & 3 \end{bmatrix}$$

$$= \frac{8}{225} \begin{bmatrix} 10 & -5 & 10 \\ 9 & 0 & 12 \\ -4 & -2 & 3 \end{bmatrix} \begin{bmatrix} 250 & 240 & -94 \\ -15 & 66 & -26 \\ 150 & 198 & 29 \end{bmatrix}$$

$$[\mathbf{I}_{Cxyz}] = \frac{8}{225} \begin{bmatrix} 4075 & 4050 & -520 \\ 4050 & 4436 & -498 \\ -460 & 762 & 515 \end{bmatrix} .$$

Example 10.9
A right, circular cone of uniform density ρ has a base of radius R, and a vertical height h. Determine its Moment of Inertia and Radius of Gyration about its axis.

Solution
Within an elemental disc of thickness dy at a distance y from the base, consider

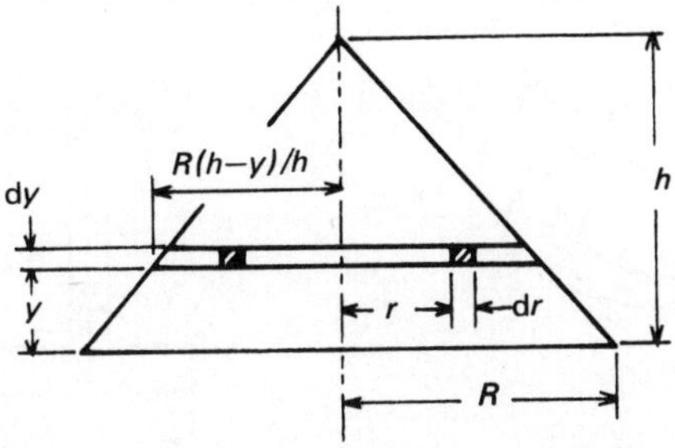

a ring which is concentric with the axis, radius r, and radial width dr. The whole of this ring is at a common distance from the axis, and we may write:

$$\text{Mass of ring} = 2\pi\rho r\,dr\,dy$$

$$\text{Moment of Inertia of ring} = 2\pi\rho r^3 dr\,dy$$

$$\text{moment of inertia of disc} = 2\,dy \int_0^{R(h-y)/h} r^3 dr\,.$$

$$\text{Moment of Inertia of cone} = 2\pi\rho \int_0^h \int_0^{R(h-y)/h} r^3 dr\,dy$$

$$I_{yy} = \frac{\pi\rho R^4}{2h^4} \int_0^h (h-y)^4\,dy$$

$$I_{yy} = \pi\rho h R^4/10$$

$$\text{Mass of cone} = \Sigma m = \pi\rho h R^3/3$$

$$\text{Radius of Gyration} = K_{yy}^2 = I_{yy}/\Sigma m$$

$$K_{yy} = R\sqrt{(0{\cdot}3)}$$

Example 10.10

Treating superficial area as a scalar quantity which is essentially positive:

a) Determine the Second Moments of Area of the following plane figures about the axes indicated:

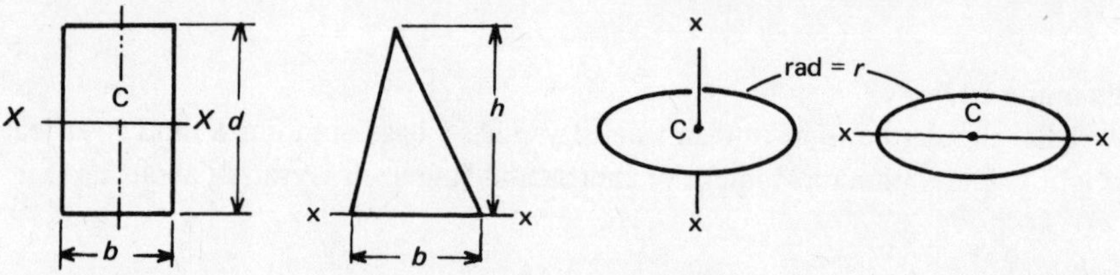

b) Given the foregoing results, and without further integration, determine the second moments of area of the following plane figures about the axes indicated:

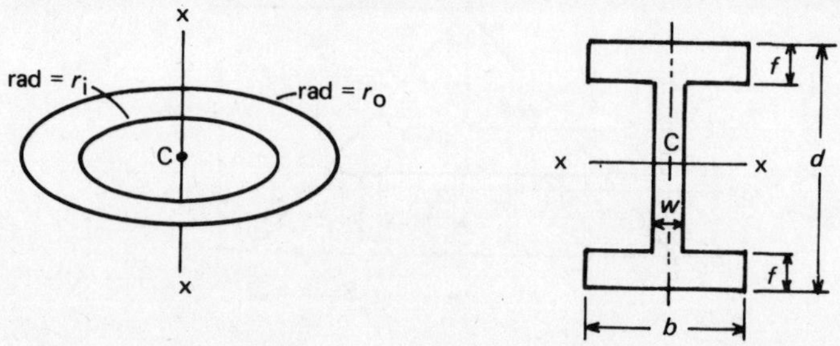

Solution

a) (i) For an elemental strip of width dy at a distance y from XX:

$$\text{Area of strip} = \mathrm{d}A = b\,\mathrm{d}y\ .$$

$$\text{Second Moment} = \mathrm{d}I_{xx} = b\,y^2\mathrm{d}y\ .$$

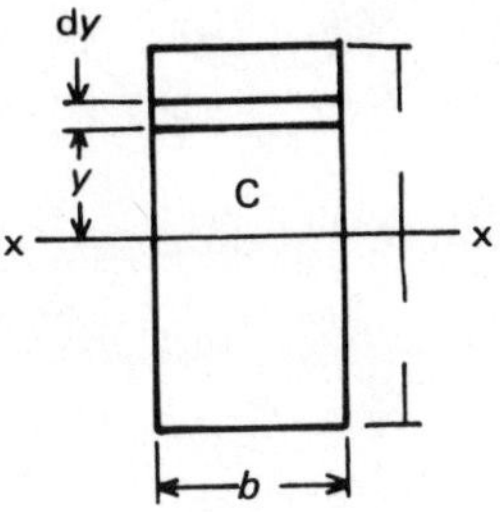

For the complete rectangle:

$$I_{xx} = \int_{-d/2}^{d/2} b\,y^2\mathrm{d}y = b\,\mathrm{d}^3/12\ .$$

a) (ii) For an elemental strip of width dy at a distance y from XX:

$$\text{Area} = \mathrm{d}A = (b/h)\,(h-y)\,\mathrm{d}y$$

$$\text{Second Moment} = \mathrm{d}I_{xx} = (b/h)\,(h-y)y^2\mathrm{d}y\ .$$

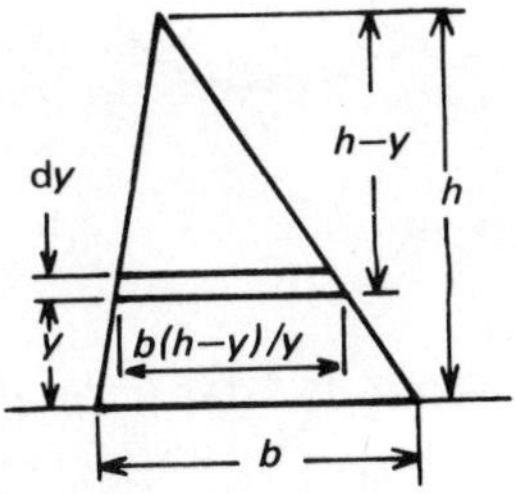

For the complete triangle:

$$I_{xx} = \frac{b}{h}\int_0^h (h-y)\,y^2\mathrm{d}y = bh^3/12\ .$$

a) (iii) For an elemental annulus of width dr at radius r from XX:

$$\text{Area of annulus} = \mathrm{d}A = 2\pi\,r\,\mathrm{d}r$$

$$\text{Second Moment} = \mathrm{d}I_{xx} = 2\pi\,r^3\mathrm{d}r$$

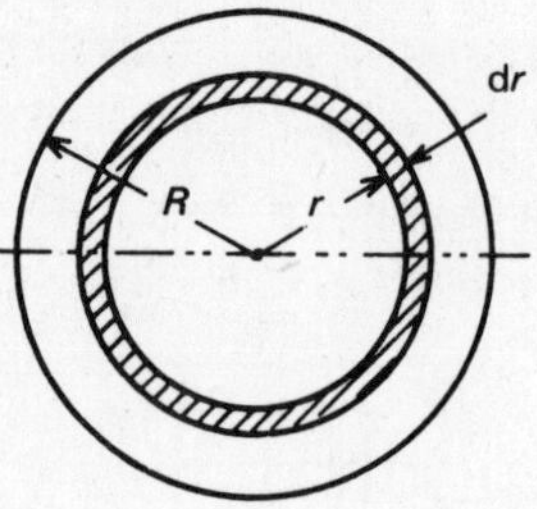

For the complete circle of radius R:

$$I_{xx} = 2\pi \int_0^R r^3 \mathrm{d}r = \pi R^4/2 \ .$$

a) (iv) For an elemental strip of width dy at a distance y from XX:

Area $= \mathrm{d}A = 2\sqrt{(R^2 - y^2)}\, \mathrm{d}y$

Second Moment $= \mathrm{d}I_{xx} = 2\sqrt{(R^2 - y^2)}\, y^2 \mathrm{d}y$

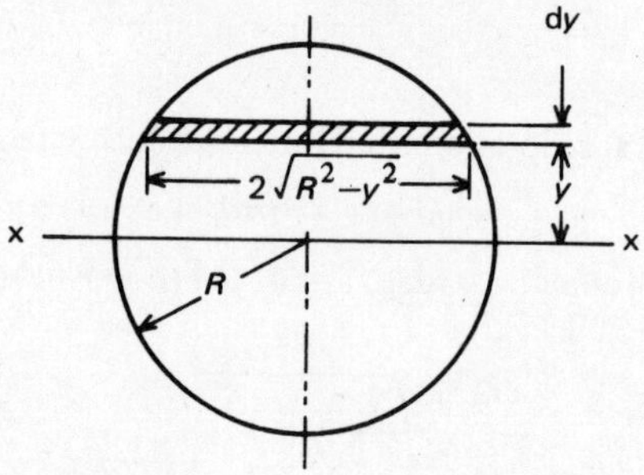

For the complete circle of radius R:

$$I_{xx} = 2 \int_{-R}^{R} \sqrt{(R^2 - y^2)}\, y^2 \, \mathrm{d}y$$

Substitute $y = R \sin\theta$, and $\mathrm{d}y = R\cos\theta \ \mathrm{d}\theta$, noting that when $y = \pm R$, then $\sin\theta = \pm 1$, and $\theta = \pm\pi/2$.

Then: $$I_{xx} = 2R^4 \int_{-\pi/2}^{\pi/2} \sin^2\theta \cos^2\theta \ \mathrm{d}\theta$$

$$= \frac{R^4}{2} \int_{-\pi/2}^{\pi/2} \sin^2 2\theta \ \mathrm{d}\theta = \frac{R^4}{2} \int_{-\pi/2}^{\pi/2} (1 - \cos 4\theta)\, \mathrm{d}\theta$$

Thus $$I_{xx} = \pi R^4/4 \ .$$

Alternatively:

By the Polar Axis Theorem: $I_{zz} = I_{xx} + I_{yy}$.

But by symmetry: $I_{xx} = I_{yy}$.

Therefore: $I_{xx} = I_{yy} = I_{zz}/2$

that is $I_{xx} = I_{yy} = \pi R^4/4$.

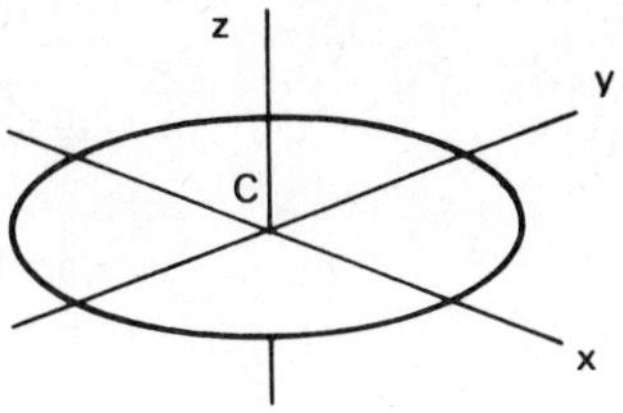

b) (i) Since scalar moments can be added and subtracted as scalars the second

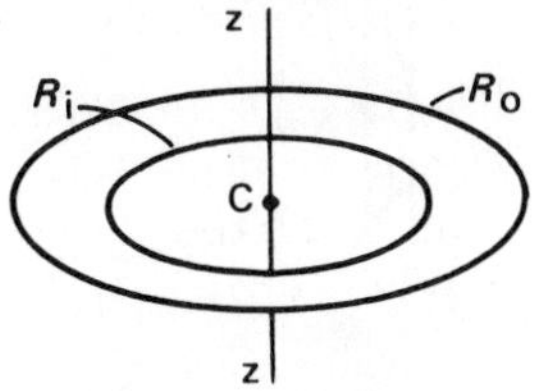

moment of the annulus can be equated to the second moment of the outer circle, less the second moment of the inner circle. Thus:

$$I_{xx} = (R_o^4 - R_i^4)/4 \ .$$

b) (ii) Since a moment of scalar is unaffected by a displacement of any part

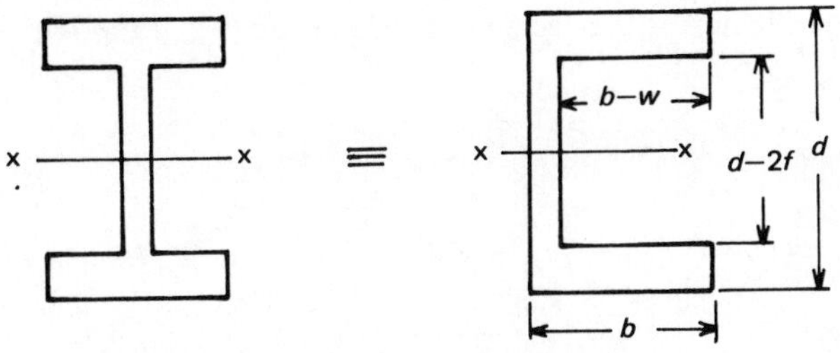

parallel to the datum, second moment of the I section can be determined as the difference in the second moment of two rectangles.

$$I_{xx} = \{b\,d^3 + (b - w)\,(d - 2f)^3\}/12 \ .$$

Example 10.11

The figure indicates the section of a girder, of which C is centre of area, and Y and Z the principal axes.

Determine the second moments of area I_{yy} and I_{zz}, the product of moment of area I_{yz}, the principal second moments of area I_{YY} and I_{ZZ}, and the angle θ.

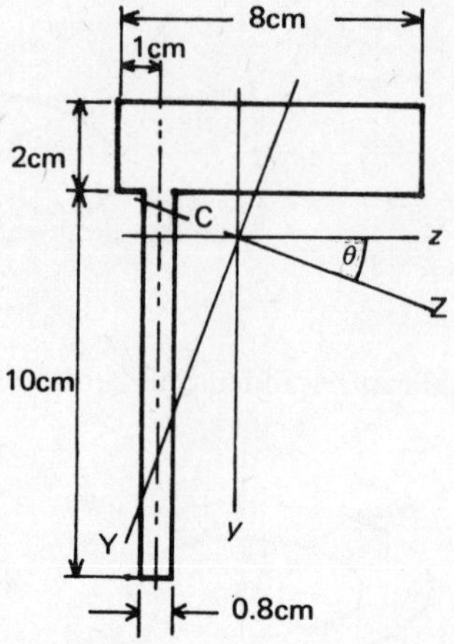

Solution

Each of the moments of the section can be determined as the sum of the moments of the two rectangles of which the section is comprised.

But for first moments about any given datum, the resultant moment of area of a rectangle is equal to the moment of its total area lumped at its centre, and since the resultant moment of the two rectangles about the centroid C is, by definition, zero, it follows that C will divide the straight line between their centres in inverse proportion to their areas, as indicated. Evidently, therefore, the (y, z) coordinates of the rectangles 1 and 2 are $(-2, 1)$ and $(4, -2)$, respectively.

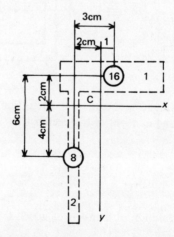

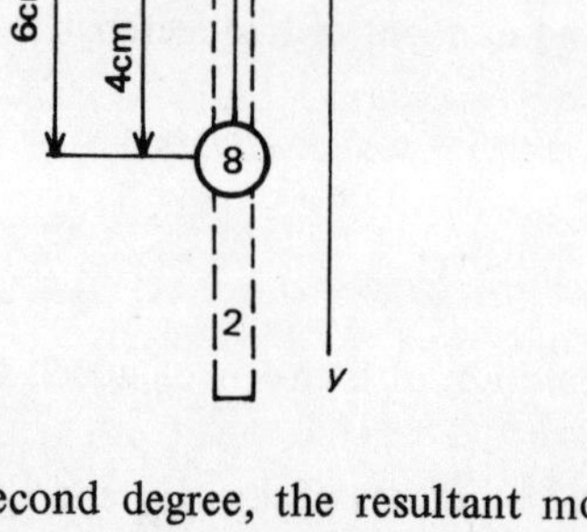

For moments of second degree, the resultant moments of a rectangle in $Cxyz$ can be equated to the resultant moments in the parallel axes which intersect in its centre, plus the moments in $Cxyz$ of the total area lumped at the centre and adding the moments of the two rectangles gives:

$$I_{yy} = (8\times 2^3/12 + 16\times 1^2) + (10\times 0{\cdot}8^3/12 + 8\times 2^2) = 134 \text{ cm}^4$$
$$I_{zz} = (2\times 8^3/12 + 16\times 2^2) + (0{\cdot}8\times 10^3/12 + 8\times 4^2) = 264 \text{ cm}^4$$
$$I_{yz} = (0 + 16\times -2\times 1) + (0 + 8\times 4\times -2) = -96 \text{ cm}^4 .$$

In Mohr's circle for the xy plane, the centre will lie on the axis of second moments at the value $(134 + 264)/2 = 199$, and the radius will be equal to:

$$\sqrt{\{(264 - 134)/2\}^2 + 96^2} = 116$$

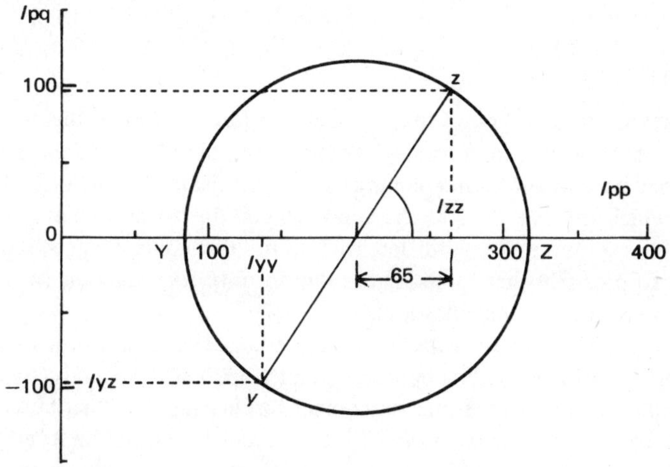

Therefore

$$I_{ZZ} = 199 + 116 = 315 \text{ cm}^4$$
$$I_{YY} = 199 - 116 = 83 \text{ cm}^4$$
$$\theta = \frac{1}{2}\tan^{-1}\frac{96}{65} = 28° .$$

11

Long Slender Members in Strength of Materials

11.1 INTRODUCTION

It has already been suggested that the exactness of the formal theory of stress analysis is seldom of great advantage because the practical situations themselves can seldom be defined with a commensurate precision, but in fact, there is a sense in which the exactness of the approach could even be seen as a weakness, because, being expressed in differential form, the controlling equations are specified so precisely that there is little opportunity for aiding the solution of a problem by recourse to intuition.

To illustrate the point, consider a long, straight bar of uniform section that is bent by a pair of equal but opposite couples applied to its ends (Fig. 11.1a). Such couples might be generated by contact forces which are distributed, in one case over the ends of the bar as in Fig. 11.1b, and in another on its edges as in Fig. 11.1c, so suppose that the stress fields for two such cases were to be determined by the formal theory. Since this ensures that the external forces on each boundary element are exactly balanced by the forces due to the stresses within, it follows that the algebraic forms of the two solutions would be as different as the algebraic descriptions of the boundary loads themselves. But intuition suggests, first, that equal pairs of equal but opposite couples would produce

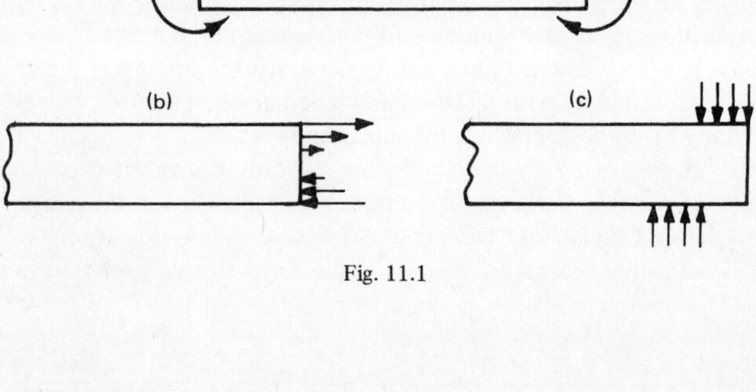

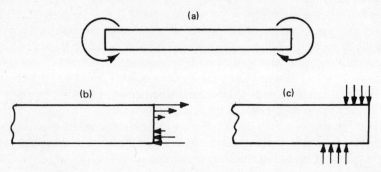

Fig. 11.1

stresses that were nearly equal, irrespective of any difference in the distribution of the boundary loads, and second, that in bending under any pair of equal but opposite couples, plane transverse sections of the bar would remain plane and transverse after bending, and from these two observations we can infer two principles which are crucial to the method of analysis that is pursued under the title of 'Strength of Materials'.

First, it can readily be shown, that if the plane transverse sections of the bar remain plane and transverse after bending, then, the material being Hookean, the stresses on transverse sections will be distributed over the sections in a manner that is very nearly linear, and this is a reasonable approximation to the truth. The example therefore indicates that we can sometimes make intuitive assumptions, particularly about the deformation of the body, which are of themselves sufficient to imply the general form of a stress distribution which is a reasonable approximation to the truth. It is true that conclusions of this kind cannot usefully be introduced into the formal theory where it is necessary to 'guess' a stress function whose stress field is exactly consistent with the equilibrium of every infinitesimal element of the body. However, if the descriptions of the controlling conditions are expressed in less-precise, integral forms, these are not, of themselves, sufficient completely to determine the form of the stress distribution, and they can therefore be combined with intuitive assumptions of the kind referred to. Of course, such an approach is feasible only in those cases for which it is possible to make the necessary assumptions, and the method is therefore usually applicable only to cases which exhibit some strong geometrical characteristic. In particular, this would include bodies in which either one dimension is much greater than the other two, or one dimension is much smaller than the other two, and/or there is a high degree of symmetry, but here we consider only the case of long, slender members in which one dimension is much greater than the other two.

The second principle is generalised by St. Venant as follows.

The Principle of St Venant

Given a body which is in equilibrium under a given set of externally reacted forces and couples applied at known points, suppose that those of the forces and couples that are applied at points which fall within a specific restricted zone are replaced by an alternative set which is statically equivalent to the set removed. In general, the substitution would result in a variation of the stresses throughout the body, but the Principle of St Venant asserts that the amount of the variation would diminish rapidly with distance, and is commonly negligible, except in the near vicinity of the zone in which the substitution is made.

But two sets of forces and couples are statically equivalent when both the resultants of the sets of forces , and the resultant moments of the sets of forces and couples, are equal, and reflection will show that any set of forces can be equated statically to a set in which a single force is associated with a single couple. The single force is equal to the resultant of the actual forces, acting on an arbitrary line of action, and the single couple is equal to the resultant moment of the actual forces and couples about any point which lies on the arbitrary line of action. The Principle of St Venant can therefore be alternatively expressed as follows.

Considering the external forces and couples that are applied to a body at points which fall within a specific, limited domain, let Σf represent the resultant of the forces, and let Σm_A represent the resultant moment of the forces and couples about any point A in the domain. Then, except in the near vicinity

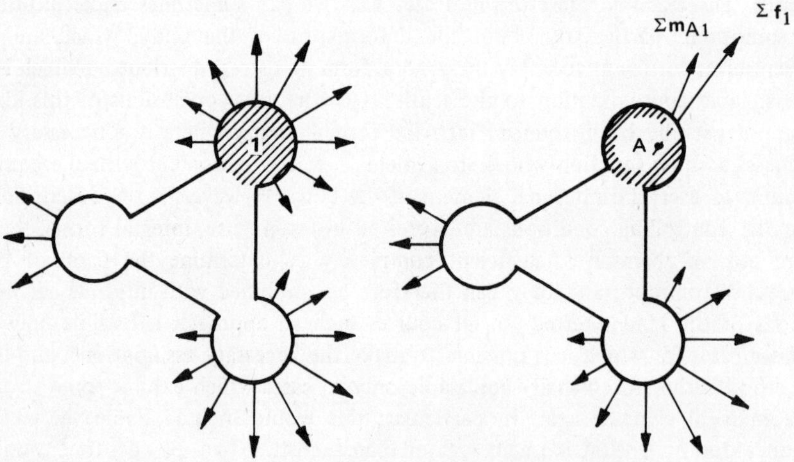

Fig. 11.2

of the zone in which the substitution is made, the stresses in the body would not be materially affected if the actual forces and couples applied within the zone were substituted by a single force equal to Σf, acting through the point A, together with a single couple equal to Σm_A (Fig. 11.2).

Equivalent Resultant Forces and Couples on Plane Sections

So, given a body which supports an equilibrium set of externally reacted forces and couples applied at given points, let the body be considered to comprise the two domains that lie on either side of any plane section that completely divides it. By an extension of the Principle of St Venant we assert that the stresses on the section would not be greatly affected if the forces and couples that are applied in each of the two domains were replaced by a statically equivalent force and couple, and in principle these may be referred to any point in the domain. But a point which lies in the section itself can be considered to lie in both domains, and it is convenient to refer the scalar equivalents for both domains to a common datum point at the centre of area C of the section. Then reflection will show that it follows from the equilibrium of the whole, that the equivalent forces and moments for the two domains will arise as equal but opposite pairs, and in Strength of Materials we therefore postulate as follows.

Irrespective of the distribution of the actual forces and couples that are applied to a body, the stresses on any plane section are assumed to be those that would arise from a pair of equal but opposite forces, together with a pair of equal but opposite couples. The forces are applied to the two domains on a common line of action which passes through the centre of area of the section, and in the equivalent force and couple of each domain, the force is equal to the resultant of the actual external forces, whilst the couple is equal to the resultant moment of the actual external forces and couples about the centre of area of the section (Fig. 11.3).

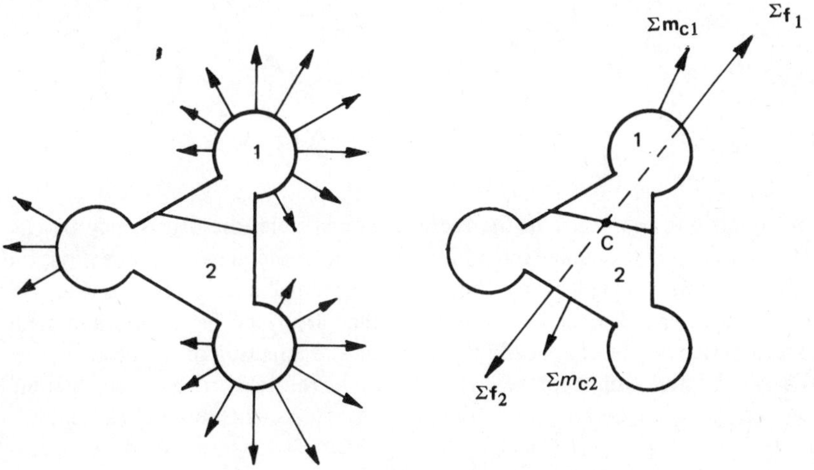

Fig. 11.3

The conventions for the equivalent resultants are defined to agree with the stress conventions, where the positive of the two faces exposed by parting the body on any section is identified as the one on which the outward-facing normal is positive. So, referring to Fig. 11.4, let either direction along the normal to the section be taken as the positive. Then it follows from the equilibrium of the two domains, that the internal forces on the positive face of the negative domain are statically equivalent to the external forces and couples on the positive domain, and the equivalent resultant force and couple on any section are therefore identified either as the resultants of the external forces and couples that are applied within the positive domain, or as the negatives of the resultants of the external forces and couples that are applied within the negative domain.

Furthermore, the resultant force and couple on each section, like the stresses on the section, are described in terms of a set of rectangular directions

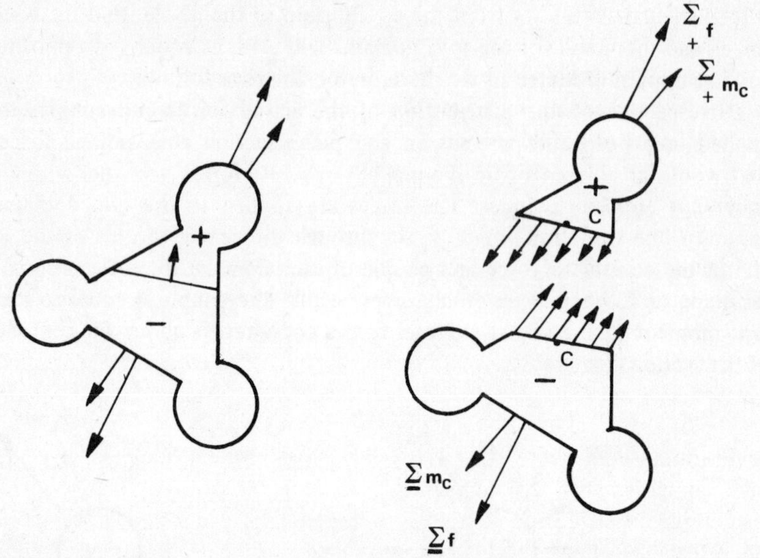

Fig. 11.4

of which one is normal to the section. In principle, the others may be chosen in any pair of rectangular directions which lie in the plane of the section, but in practice it is convenient to choose the principal axes of the section. So let C*xyz* represent the set of coordinates that are fixed in the section, with its origin at the centre of area and the *x* axis in the direction of the positive normal. Then if Σ**f** and Σ**m**$_C$ represent the equivalent resultant force and couple on the section, these are described in terms of their components in C*xyz* as follows:

$$\Sigma \mathbf{f} = F_x\mathbf{i} + F_y\mathbf{j} + F_z\mathbf{k} ,$$

and $$\Sigma \mathbf{m}_C = M_x\mathbf{i} + M_y\mathbf{j} + M_z\mathbf{k} ,$$

and the various scalar components are identified as follows.

The Tensile (or Compressive) Force

The component F_x is one of a pair of equal but opposite forces which are normal to the section. When F_x is positive they tend to stretch the member, and F_x is called the tensile force on the section. When F_x is negative they tend to shorten the member, and F_x is called the compressive force on the section.

Shear Force

Each of the components F_y and F_z represents one of an equal but opposite pair of forces in one of a pair of rectangular directions which lie in the plane of the section. Since they tend to cause the relative slide of shear they are known as Shearing Forces, and the total shear force on the section is the vector sum of the two.

Torsional Moment

The component M_x is one of a pair of equal but opposite couples about the x axis. It describes an effect which tends to twist the member by a relative rotation of succeeding sections about the longitudinal axis, and it is called the torsional moment on the section.

Bending Moment

Each of the components M_y and M_z represents one of a pair of equal but opposite couples about one of a pair of rectangular directions which lie in the plane of the section. Each tends to bend the member by a relative rotation between successive sections about an axis which is coplanar with the section, and they are called the bending moments on the section. The total bending moment on the section is the vector sum of the two.

Equilibrium Condition

Each component of the equivalent resultants on any section, gives rise to a characteristic distribution of stresses on the section, and in the Hookean material the effects of the several components are sensibly independent. They may therefore be treated severally, and in each case the general form of the stress distribution is determined from rationalised or intuitive assumption, particularly about the deformation of the member. The stress at any point in the section is thus determined as a function of position which contains a pair of unknown coefficients, and these are determined from conditions of equilibrium which are expressed in rudimentary, integral form. These conditions are rudimentary in that they ensure the equilibrium, not of all the elements of the body, but only of the two parts into which the section divides the body, and they are formulated by equating the resultant of the internal forces on the positive face of the negative domain to the resultant of the external forces and couples on the positive domain.

Thus let $\Sigma \mathbf{f}$ and $\Sigma \mathbf{m}_C$ represent the equivalent resultant force and couple on a given section. and let σ represent the vector of the stress at a general point P in the section. Then if dA represents the area of an infinitesimal element of the section at P, the force on the element may be equated to $\sigma.\mathrm{d}A$, its moment about the centre of area of the section is equal to $\mathbf{r}_{PC} \times \sigma \, \mathrm{d}A$, and the resultant internal force and moment on the section being determined by integrating over the section, they may be equated to the resultants of the external forces on the section, as follows:

$$\Sigma \mathbf{f} = \int_{\text{Section}} \sigma \,.\, \mathrm{d}A$$

$$\Sigma \mathbf{m}_C = \int_{\text{Section}} \mathbf{r}_{PC} \times \sigma \, \mathrm{d}A \; .$$

However, each of these vector equations can be presented as a set of three scalar equations, and this is the form in which they are invariably used in Strength of Materials.

Continuity

In the formal analysis we introduce a strain compatibility condition which ensures that the strains associated with the stresses in each element are consistent with the physical continuity of the whole. But in Strength of Materials a long slender member is idealised to a line, whose 'stiffness' at each point is determined by the cross-section of the member and we are concerned with the deformation of the member only in respect of the deformation of the idealised line. Thus, attention is focused on transverse sections normal to the line which passes through their centres of area, and it is necessary to assert only the continuity of this centre line.

11.2 EQUATIONS AND DIAGRAMS OF TENSILE AND SHEARING FORCES AND TORSIONAL AND BENDING MOMENTS IN A LONG, SLENDER MEMBER

In applying the foregoing concepts to a long, slender member, attention is focused on both its transverse sections and the centre-line to which they are normal, and the objective is to determine the stresses on the sections, and the deformation of the centre-line. However, the first step is to replace the description of the external forces and couples that are applied to the member with a description of the equivalent resultant forces and couples that they induce on its transverse sections, so let any convenient section be chosen as datum, and let either direction along the centre-line be chosen as positive. Then a general transverse section may be identified by its scalar arc-distance s, measured around the centre-line from the datum, and in each such section we identify a set of rectangular axes, $Cxyz$ whose origin is chosen at the centre of area of the section, with the x axis normal to the section in the sense of s increasing. Then the components of the equivalent resultant forces and couples for the various sections are functions of s which may be determined either as the resultants of the external forces and couples that are applied to the positive side of the section, or as the negatives of the resultant of the external forces and couples that are applied to the negative side of the section, and the functions may be described either algebraically or graphically.

Discontinuities and Singularities

However, the equivalent resultants on any section depend on both the shape of the centre-line of the member and the distribution of the loads, and both are liable to discontinuities or singularities. Consider, for example, a centre-line in which a straight line blends into a circular arc. Although such a curve is smoothly continuous, the algebraic equation of the straight portion is evidently different from that of the circular arc, and it follows that the equations of the components of the equivalent resultants for the two parts will also be different. Similarly for the loads, which may comprise point forces, point couples, or distributed forces whose intensity per unit length of centre-line are defined as functions of s. Each point load, each point couple, and each point at which a distributed load starts, or ends, or suffers a change in its describing equation, constitutes a point of discontinuity at which the equations of the resultants will also vary, and it is therefore necessary to consider the member, bay by bay, as determined by discontinuities in the equation of either the centre line, or the loads.

Superposition

When a given set of forces are applied at specific points in a body, the body is deformed, and sections in the body are displaced relative to the lines of action of the forces. The resultant moments on the sections are therefore modified by the deformations they produce, and in the particular case, either a body which is 'soft' in bending and is subjected to a large axial force, or of a body which is 'soft' in torsion and is subjected to a large shearing force, this mutual interaction may produce effects which are crucial, even to the point of catastrophic instability. But these are special cases that are accorded a separate, special treatment, and otherwise, it can reasonably be assumed that in practice the deformations must generally be limited to values at which their effect on the resultant moments are not significant. So accepting that the special cases will be treated separately, the resultant moments and forces on the transverse sections of a long slender member are otherwise calculated for the undeformed geometry, and reflection will show that when sets of forces are superimposed, their effects on the resultants on the sections of the body (and therefore on the stresses in the body, also) are then superimposed in simple addition.

Relationships between the Loading Intensity and the Components of the Resultant Forces and Moments

Within any one of the bays into which the various discontinuities divide the member the external forces must be smoothly continuous. So, confining attention to one such bay, and identifying an element of the body that lies

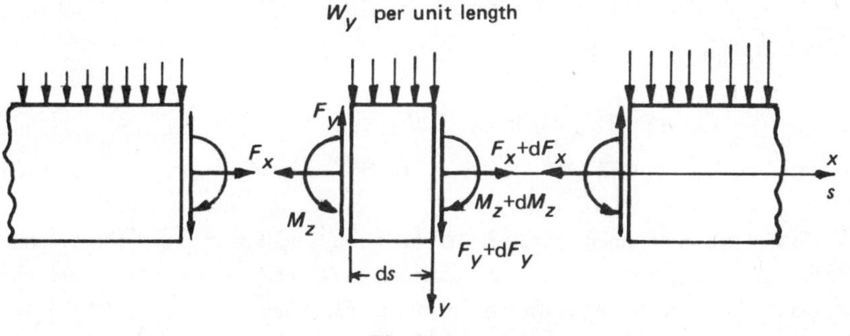

Fig. 11.5

between the transverse sections at the distances s, and $s + \mathrm{d}s$ along the centre-line, let the components of the external loading per unit length on the element be presented by w_x, w_y and w_z, let the components of the resultants on the section at s be represented by F_x, F_y, F_z, M_x, M_y and M_z, and let the corresponding components for the section at $s + \mathrm{d}s$ be represented by $F_x + \mathrm{d}F_x$, $F_y + \mathrm{d}F_y$, $F_z + \mathrm{d}F_z$, $M_x + \mathrm{d}M_x$, $M_y + \mathrm{d}M_y$ and $M_z + \mathrm{d}M_z$, as indicated in Fig. 11.5. Then the components of the external force on the length $\mathrm{d}s$ of the element are equal to $w_x\mathrm{d}s$, $w_y\mathrm{d}s$ and $w_z\mathrm{d}s$, and from the equilibrium of the element in the y direction, we have

$$(F_y + \mathrm{d}F_y) - F_y + w_y\mathrm{d}s = 0 \ ,$$

whence: $\mathrm{d}F_y = -w_y\mathrm{d}s$; or: $w_y = -\mathrm{d}F_y/\mathrm{d}s$.

Similarly, from the equilibrium in the z direction we have:

$$\mathrm{d}F_z = -w_z\mathrm{d}s; \quad \text{or: } w_z = -\mathrm{d}F_z/\mathrm{d}s \ .$$

For the equilibrium of moments about the z axis in the section at $s + \mathrm{d}s$ we have

$$(M_z + \mathrm{d}M_z) - M_z + F_y\mathrm{d}s - w_y\mathrm{d}s\,(\mathrm{d}s/z) = 0$$

or $\mathrm{d}M_z = -F_y\mathrm{d}s + w_y/z\ \mathrm{d}s^2$.

But in the limit, the second order term is negligible compared with the first order terms, whence

$$\mathrm{d}M_z = -F_y\mathrm{d}s; \quad \text{or: } F_y = -\mathrm{d}M_z/\mathrm{d}s \ .$$

Similarly, for the equilibrium about the y axis we have:

$$\mathrm{d}M_y = -F_z\mathrm{d}s; \quad \text{or: } F_z = -\mathrm{d}M_y/\mathrm{d}s \ .$$

We therefore conclude that the relationships between the components of the loading intensity, the shearing force, and the bending moment at any section may be expressed in differential form as follows:

$$\begin{aligned} w_y &= -\mathrm{d}F_y/\mathrm{d}s; \quad & w_z &= -\mathrm{d}F_z/\mathrm{d}s \\ F_y &= -\mathrm{d}M_z/\mathrm{d}s; \quad & F_z &= -\mathrm{d}M_y/\mathrm{d}s \end{aligned} \tag{11.1a}$$

or in integral form as follows:

$$\begin{aligned} F_y &= -\int w_y\,\mathrm{d}s + C_1; \quad & F_z &= -\int w_z\,\mathrm{d}s + K_1 \\ M_z &= -\int F_y\,\mathrm{d}s + C_2; \quad & M_y &= -\int F_z\,\mathrm{d}s + K_2 \end{aligned} \tag{11.1b}$$

Later we shall need to return to these relationships, but for the present it is sufficient to note that, since the stationary values of a function occur where its derivative is zero, the stationary values of M_z occur where F_y is zero, and the stationary values of M_y occur where F_z is zero.

Example 11.1

The diagram illustrates a horizontal beam AB, which is simply supported at A and D, and which is loaded in a vertical plane of symmetry with concentrated

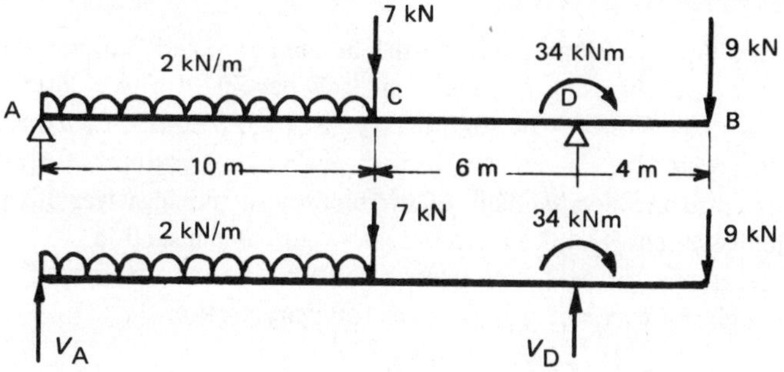

forces of 7 (kN) and 9 (kN) at C and B respectively, with a uniformly distributed load of 2 (kN) per metre between A and C, and with a coplanar couple of 34 (kN m) at D.

Sketch the graphs of Shearing Force and Bending Moment for the beam, marking the salient values, and indicate the equations of Shearing Force and Bending moment for each bay.

Solution

There are two unknown vertical reactions at A and D whose values upwards are represented by V_A and V_D respectively. Since the plane parallel system of forces has two degrees of freedom, the support reactions are statically determinate, and they may be conveniently calculated by taking moments about horizontal axes through A and D. The results may then be checked by substituting in the equation for the equilibrium of forces in the vertical direction. Thus, noting that the moment of the distributed load is equal to the moment of its resultant acting through its centroid:

For moment equilibrium about a horizontal axis through A:

$$(5 \times 20) + (10 \times 7) + (20 \times 9) + 34 - 16\,V_D = 0\ .$$

For moment equilibrium about a horizontal axis through D:

$$34 + (4 \times 9) + (16 \times V_A) - (6 \times 7) - (11 \times 20) = 0\ .$$

For the vertical equilibrium of forces

$$20 + 7 + 9 - V_A - V_D = 0$$

whence $\quad V_A = 12\text{ kN};\quad V_D = 24\text{ kN}\ .$

Let the distance s along the beam be measured from the left-hand end toward the right. For each section the x axis coincides with the centre-line of the beam, and the y axis can conveniently be chosen vertically downward. The z axis is then horizontal and in the diagram is positive downwards from the plane of the paper.

Noting that the singularities in the loading at C and D divide the beam into three bays, and choosing a section in each bay in turn, we determine for the three components of forces in the x, y, and z directions, and the three scalar moments about Cx, Cy and Cz, in turn, either the resultants for all the forces and couples to the positive side of the section, or the negatives of the resultants of all the forces and couples to the negative side of the section.

Since all the forces are parallel to the y axis and have lines of action which pass through the x axis, it follows that, for every section

$$F_x = F_z = M_x = M_y = 0 \ .$$

Thus, only the Shear Force F_y, and the Bending Moment M_z are non-zero, and their equations and diagrams are as follows:

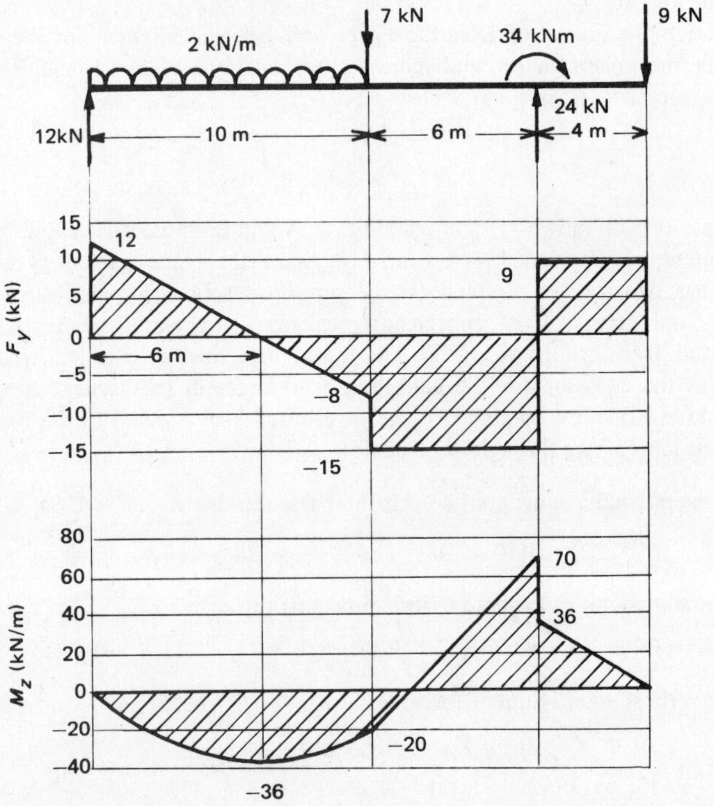

Range DB, $(16 < s < 20)$

$$F_y = 9 \text{ (kN)}$$

$$M_z = (20 - s)\,9 = 180 - 9\,s \text{ (kN m)}$$

Range CD, $(10 < s < 16)$

$$F_y = 9 - 24 \text{ (kN)} = -15 \text{ (kN)}$$

$$M_z = 34 + (20 - s)\,9 - (16 - s)\,24 = -170 + 15s \text{ (kN m)}$$

Range AC, $(0 < s < 10)$

$$F_y = -\{-12 + 2s\} = 12 - 2s \text{ (kN)}$$

$$M_z = -\{s.12 - 2s.\frac{s}{2}\} = -12s + s^2 \text{ (kN m)} \; .$$

Example 11.2

The figure illustrates a long, slender member whose centre-line has the form of

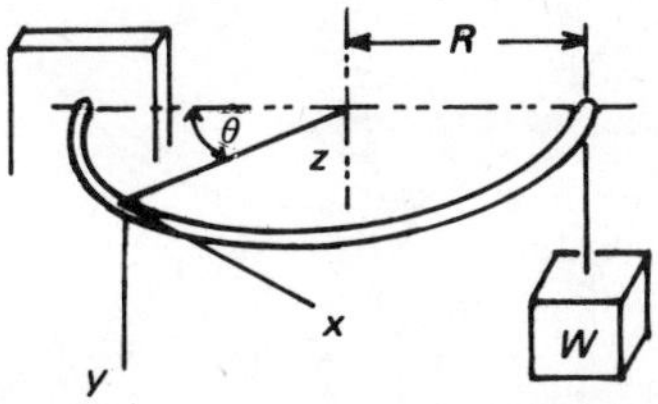

a semi-circle of radius R, which is cantilevered from a rigid, vertical wall so that it lies in a horizontal plane. A load W is suspended from the free end.

A general, transverse section being identified by the angle θ, describe as functions of θ, the components in the axes indicated of the resultant force and moment on the section.

Solution

Only the single force W is applied to the positive side of the section, and this has

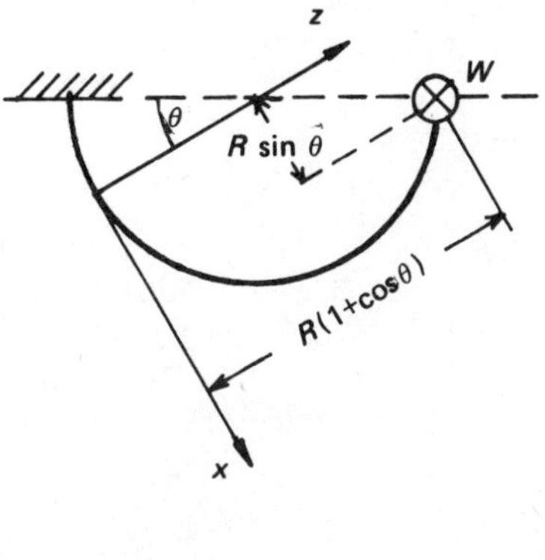

no component in either the x or the y directions. Furthermore, it is parallel to the y axis, therefore:

$$F_x = F_z = M_y = 0$$

and: $$F_y = W; \quad M_z = WR\sin\theta; \quad M_x = -WR\,(1+\cos\theta)$$

Example 11.3

Repeat Example 11.2 for the case when the load W is uniformly distributed over the total length of the arc.

Solution

As in Example 11.2:

$$F_x = F_z = M_y = 0\ .$$

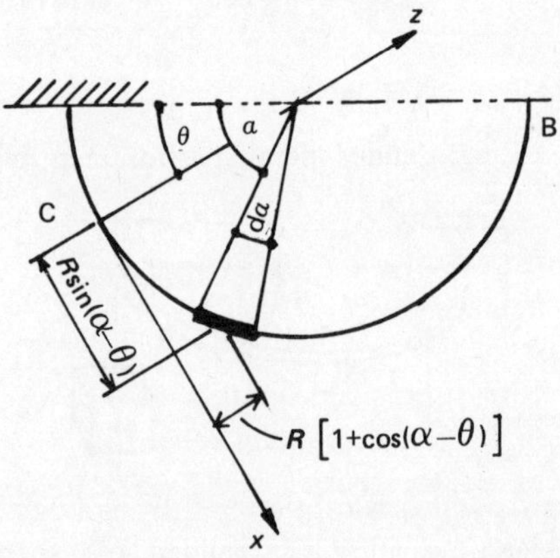

The shear force F_y is equal to the load on the arc from C to B.

that is: $$F_y = W\left(1 - \frac{\theta}{\pi}\right)$$

The moments M_x and M_z are determined by integrating the moments due to the load on the element which subtends the angle $d\alpha$ at α. Thus, the load df on the element is given by:

$$df = (W/\pi)\,d\alpha\ .$$

Therefore $$dM_x = -\frac{WR}{\pi}\{1-\cos(\alpha-\theta)\}\,d\alpha$$

whence $$M_x = -\frac{WR}{\pi}\int_{\theta}^{\pi}\{1-\cos(\alpha-\theta)\}\,d\alpha$$

$$= -\frac{WR}{\pi}\Big[\alpha - \sin(\alpha-\theta)\Big]_{\theta}^{\pi}$$

$$M_x = -\frac{WR}{\pi}\{(\pi-\theta) - \sin(\pi-\theta)\} .$$

Similarly: $$dM_z = +\frac{WR}{\pi}\sin(\alpha-\theta)\,d\alpha$$

$$M_z = \frac{WR}{\pi}\int_{\theta}^{\pi}\sin(\alpha-\theta)\,d\alpha$$

$$M_z = \frac{WR}{\pi}\{1-\cos(\pi-\theta)\} .$$

11.3 STRESSES UNDER THE TENSILE FORCE F_x

Consider a plane section of area A, which lies in the yz plane and is subjected to the uniform simple tension of magnitude σ_{xx}. Then, the force per unit area being uniform, it is obvious, first, that the resultant of the parallel forces that act on the elements of the area has a magnitude equal to $\sigma_{xx}A$, and second, that the centroidal axis of the forces passes through the centre of area of the section. It follows that uniform stresses are statically equivalent to a single force of magnitude $\sigma_{xx}A$ which acts on a line which is normal to the section and passes through its centre of area.

Conversely, therefore, it is assumed that an equivalent resultant tensile force F_x which acts on a section of area A, on a line which passes through its centre of area, induces on the section a state of uniform simple tension of magnitude:

$$\sigma_{xx} = F_x/A \tag{11.2}$$

Since the equivalent resultants on each section are referred to its centre of area, it is assumed that the foregoing conclusion is valid, irrespective of the form of the member, except that the assumptions will obviously break down in the vicinity of any sharp variation in the shape or size of the transverse sections.

11.4 CYLINDRICAL TORSION

Consider a long straight cylinder subjected to pure torsion by equal but opposite couples of magnitude M_x, applied to its ends, as indicated in Fig. 11.6.

It is assumed intuitively that plane transverse sections remain plane and transverse under torsion, so that any tensile stress σ_{xx} that is induced on the transverse section remains parallel to the longitudinal axis of M_x. These stresses would therefore generate no moment about the x axis, and this would imply that the stresses were independent of the moment by which they are induced.

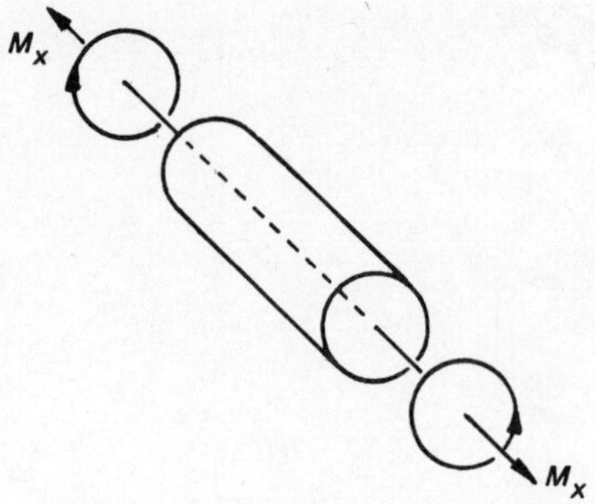

Fig. 11.6

But in a Hookean solid all the stresses are proportional to the loads from which they derive, and we therefore infer, that just as direct stresses on transverse sections generate no moment about the longitudinal axis, so will a moment about the axis generate no direct stresses on the transverse sections.

It is therefore assumed that for circular sections only, the torsional component M_x of the resultant moment on a section induces on the section pure shear only, and from the circular symmetry it is inferred, first, that the total shear stress at each point on the section will lie in the circumferential direction, and second, that the magnitude of the stress at each point will be a function of its radial distance from the axis of the moment only (Fig. 11.7).

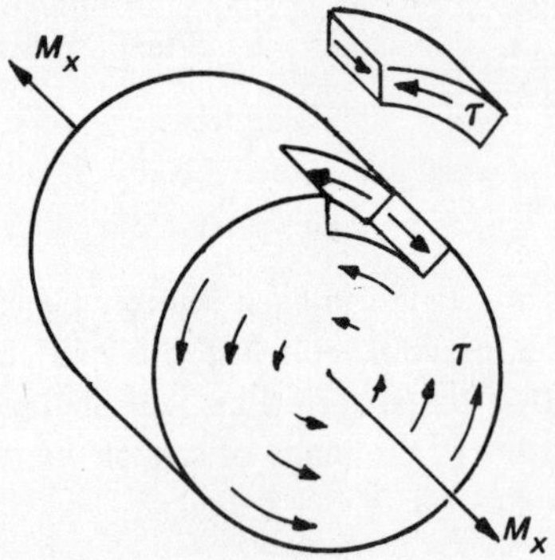

Fig. 11.7

So, again referring to the cylinder under torsion, consider a slice of axial length ds which is isolated between the sections at the distances s and s + ds along the member, and within the slice identify a concentric annulus of radius r, and radial thickness dr, as indicated in Fig. 11.8. Finally, within the annulus we identify a rectangular element abdc which is isolated between a pair of adjacent axial planes. The pure shear stress τ on this element results in a shear

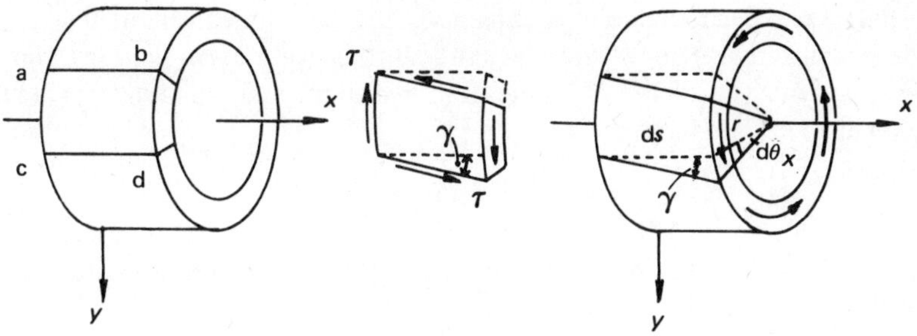

Fig. 11.8

strain γ equal to τ/G, where G is the shear modulus of the material, and the circular ends of the annulus thereby suffer a relative rotation $d\theta_x$ about the x axis, such that:

$$r.d\theta_x = ds\,\gamma = ds\,\tau/G$$

Therefore:
$$\tau = G.\frac{d\theta_x}{ds}.r \tag{a}$$

or
$$\frac{\tau}{r} = G.\frac{d\theta_x}{ds}. \tag{b}$$

Thus, the circumferential shear stress τ at each point in transverse sections is proportional to the radial distance r of the point from the torsional axis, and the constant of proportionality is equal to the product of the shear modulus G with the twist (or relative rotation about the x axis) per unit length ($d\theta_x/ds$).

The stresses and the twist per unit length may now be related to the torsional moment M_x, by invoking the equilibrium condition whereby the equivalent resultant moment M_x on the section may be equated to the resultant moment of the internal forces due to the stresses on the section. Thus, let dA represent the area of an element of the section at a radial distance r from the torsional axis (Fig. 11.9). Then the force on the element has a magnitude equal

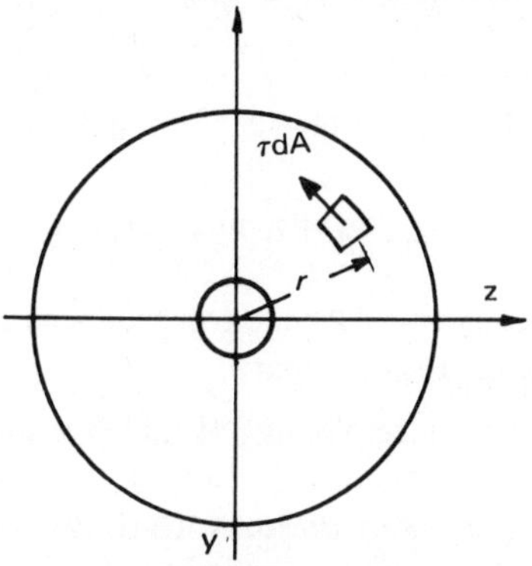

Fig. 11.9

to τ dA in the circumferential direction, and its moment about the torsional axis is equal to $\tau\, r$ dA. Therefore, substituting for τ from eqn. (a), summing the moments for all the elements of the section, and equating this resultant moment to M_x, we have:

$$M_x = \sum_{\text{Section}} (G.\frac{\mathrm{d}\theta_x}{\mathrm{d}s} r^2\, \mathrm{d}A) ,$$

and since G and $\mathrm{d}\theta_x/\mathrm{d}s$ have common values for all elements, this may be written:

$$M_x = G.\frac{\mathrm{d}\theta_x}{\mathrm{d}s} \sum_{\text{Section}}(r^2\, \mathrm{d}A) .$$

But the summation can be recognised as I_{xx}, the second moment of area of the section about the torsional axis, whence:

$$M_x = G\, I_{xx} \frac{\mathrm{d}\theta_x}{\mathrm{d}s}$$

or

$$\frac{M_x}{I_{xx}} = G.\frac{\mathrm{d}\theta_x}{\mathrm{d}s} .$$

Finally, by combining this with eqn. (b), we have

$$\frac{\tau}{r} = \frac{M_x}{I_{xx}} = G\frac{\mathrm{d}\theta_x}{\mathrm{d}s} \tag{11.3}$$

Non-Circular Torsion

The foregoing analysis depended critically on the assumption that plane transverse sections were not liable to warping, and on arguments which depended on circular symmetry. But for other shapes of section, neither of these is justified, so the foregoing results are strictly limited to bodies having circular sections. In particular, it should be noted that if a cylinder is slit along an axial plane, its circular symmetry is entirely destroyed.

Example 11.4

A cylindrical rod of 4 (cm) diameter is tested first in simple tension, and then in pure torsion.

In tension a load of 83 (kN) produces an extension of 65×10^{-3} (mm) on a gauge length of 20 (cm).

In torsion, a couple of 330 (N m) produces a relative rotation of 0.15 degrees between sections 15 (cm) apart.

a) Calculate the values of Young's Modulus E, and Shear Modulus G for the material.

b) Determine the value of Poisson's Ratio for the material, and determine the variation in the diameter of the rod under the tensile load of 83 (kN).

Solution

The properties of the sections of the rod are:

$$\text{Area} = A = \pi r^2 = 400\,\pi\ (\text{mm}^2)$$

$$I_{xx} = \pi r^4/2 = 80\,000\,\pi\ (\text{mm}^4)\ .$$

a) $$\text{Tensile Stress} = \frac{\text{Axial force}}{\text{sectional area}} = \frac{83 \times 10^3}{400\,\pi} = 66\ (\text{N/mm}^2)$$

$$\text{Axial Strain} = \frac{\text{Extension}}{\text{Original length}} = \frac{0{\cdot}065}{200} = 3{\cdot}25 \times 10^{-4}$$

$$E = \frac{\text{Tensile stress}}{\text{Axial strain}} = \frac{66}{3{\cdot}25 \times 10^{-4}} = 203 \times 10^3\ (\text{N/mm}^2)\ .$$

In Torsion

When a rod of uniform section is subjected to a uniform torque along its length, the twist per unit length is also uniform and $(\mathrm{d}\theta_x/\mathrm{d}s) = \theta_x/s$.

Thus $$\frac{M_x}{I_{xx}} = G\frac{\theta_x}{s}\,,$$

therefore $$G = \frac{M_{xx}}{I_{xx}}\frac{s}{\theta_x} = \frac{330 \times 10^3}{80 \times 10^3\pi} \times \frac{150 \times 180}{0{\cdot}15 \times \pi}\,,$$

that is: $$G = 75 \times 10^3\ (\text{N/mm}^2)\ .$$

b) $$G = \frac{E}{2(1+v)}\ .$$

Therefore $$v = \frac{E}{2G} - 1 = \frac{203}{2 \times 75} - 1 = 0{\cdot}35\ .$$

$$\begin{aligned}\text{Change of diameter} &= \text{Lateral strain} \times \text{Original diameter}\\ &= -v \times \text{Longitudinal strain} \times \text{Original diameter}\\ &= -v \times 3{\cdot}25 \times 10^{-4} \times 40\ (\text{mm})\\ &= -4{\cdot}5 \times 10^{-3}\ (\text{mm})\ .\end{aligned}$$

Example 11.5

A solid transmission shaft of 15 cm diameter is to be replaced by a hollow shaft capable of transmitting twice the power at 1·25 times the angular speed. The material of the new shaft will have a proof stress 10% higher than that of the old. The bore diameter of the hollow shaft will be half the outer diameter.

Assuming the same Factor of Safety, determine the necessary outer diameter of the new shaft.

Note: Power is the rate of doing work and is given by:

$$\text{Power (Watts)} = \text{Torque (N m)} \times \text{Angular Speed (rad/s)} .$$

Solution

Quantities for the replacement shaft being indicated by a dash, let r' represent its outer radius:

$$\frac{\text{Power}'}{\text{Power}} = \frac{M_x' w'}{M_x w}$$

$$\frac{M_x'}{M_x} = \frac{\text{Power}'}{\text{Power}} \times \frac{w}{w'} = \frac{2}{1{\cdot}25} = 1{\cdot}6 .$$

In each shaft the greatest shear stress will occur at the outer radius, and the value of greatest shear stress in the new shaft may be 1·1 times its value in the old.

$$\frac{\tau'}{\tau} = \left(\frac{M_x{}'}{I_{xx}'} r'\right) \Big/ \left(\frac{M_x}{I_{xx}} r\right)$$

or

$$\frac{I_{xx}'}{I_{xx}} \frac{r}{r'} = \frac{M_x{}'}{M_x} \cdot \frac{\tau}{\tau'}$$

$$\left(\frac{r'}{r}\right)^3 \left(1 - \frac{1}{16}\right) = 1{\cdot}6 \times \frac{1}{1{\cdot}1} = 1{\cdot}455$$

$$r' = \left(\frac{16}{15} \cdot 1{\cdot}455\right)^{\frac{1}{3}} \cdot 7{\cdot}5 \text{ (cm)} = 8{\cdot}68 \text{ (cm)} .$$

Example 11.6

A composite transmission shaft consists of a solid steel core of 10 (cm) diameter, and a close-fitting bronze sheath of 16 (cm) outer diameter. The core and sheath are rigidly connected so that they twist as one, and the shaft transmits 74·6 (kW) at 240 (r.p.m.). If the shear modulus of the steel is 80×10^3 (N/mm^2), and of the bronze is 40×10^3 (N/mm^2), determine the relative rotation between two transverse sections 1 m apart.

Solution

The torsional couple (or torque) is given by:

$$M_x = \frac{\text{Power}}{\text{Angular speed}} = \frac{74\,600 \times 30}{240\,\pi} = \frac{9325}{\pi} \text{ (N m)} .$$

For the steel core:

$$G_S = 80 \times 10^3 \text{ (N/mm}^2\text{)}$$

and $$I_S = \pi \,.\, 50^4/2 = 3{\cdot}125\,\pi \times 10^6 \text{ (mm}^4\text{)}$$

whence $$G_S I_S = 250\,\pi \times 10^9 \text{ (N mm}^2\text{)} .$$

For the bronze sheath:

$$G_B = 40 \times 10^3 \text{ (N/mm}^2)$$

and $I_B = \pi(80^4 - 50^4)/2 = 17{\cdot}355\,\pi \times 10^6 \text{ (mm}^4)$

whence $G_B I_B = 694{\cdot}2\,\pi \times 10^9 \text{ N/mm}^2$.

For each part , $M_x = G\,I_{xx}(\theta_x/s)$, where θ_x/s is common to both, and the total moment is the sum of the moments in the two parts.

Thus $M_x = (G_S I_S + G_B I_B)(\theta_x/s)$

$$\theta_x = M_x s/(G_S I_S + G_B I_B)$$

$$\theta_x = \frac{9{,}325 \times 10^3 \times 10^3}{\pi^2(250 + 694{\cdot}2) \times 10^9}$$

$$\theta_x = 1{\cdot}0 \times 10^{-3} \text{ rad} \equiv 0{\cdot}057^\circ \ .$$

Example 11.7

A wire of radius r, in the form of a semicircle of radius R, has one end A built into a rigid wall so that the semicircle lies in a horizontal plane. At the other end B, a rigid, radial arm is fixed, and a weight W is suspended from the centre of the semicircle.

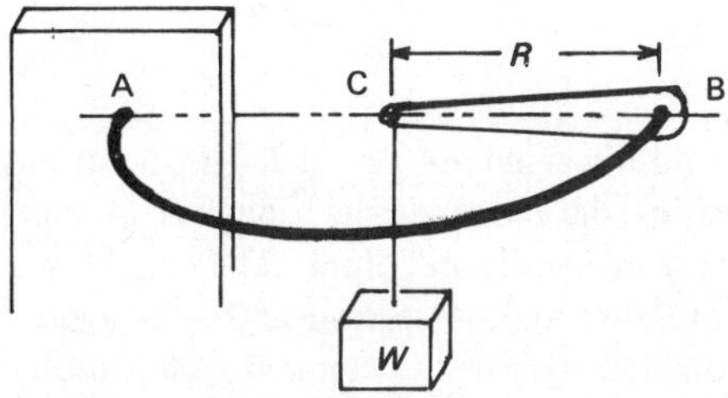

Assuming that the effects of the shear force on the sections of the wire are negligible, and representing the shear modulus of the material by G, show that:

a) The stiffness of the member under the load W, (i.e., the force W per unit displacement of its point of application) is equal to $(Gr^4)/(2R^3)$.

b) The stiffness of the closely-coiled helical spring having N coils of mean radius R, of a wire of uniform radius r is equal to $(Gr^4)/(4NR^3)$.

c) If the material of the spring in (b) has a 0·2% proof stress of 300 (GN/m³), at a Factor of Safety of 2·5 on the maximum shear stress theory, the maximum allowable load is equal to $30\pi r^3/R$ (GN).

Solution

a) For each section only the load W lies to the positive side, and its line of action is parallel to the y axes, and intersects the z axes. Therefore, of the components of the resultants on the sections, only F_y and M_x are non-zero, and F_y being ignored, it remains to consider only the effect of M_x, which has a constant value equal to WR on all sections.

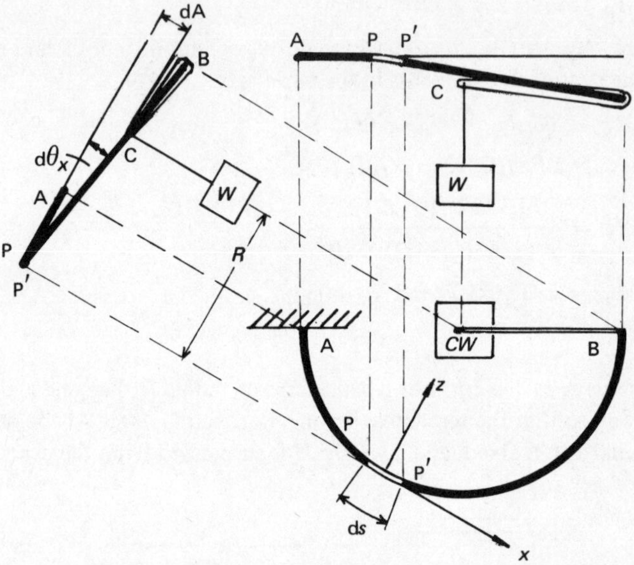

Identifying an infinitesimal arc PP′ of length ds at any position around the semicircle, and treating the remainder as though it were perfectly rigid, consider the effect of twisting in the arc PP′ alone. The arc AP would remain in a horizontal plane, but, relative to the section at P, the section at P′ would rotate about the longitudinal, x axis of the element. As a result, the plane containing the arc P′B and the arm BC would suffer the same rotation, and its amount $\mathrm{d}\theta_x$ can be determined from the relationship:

$$G\frac{\mathrm{d}\theta_x}{\mathrm{d}s} = \frac{M_x}{I_{xx}} = \frac{R\,W}{I_{xx}}$$

From the auxiliary elevation looking directly along the axis of the element, it then follows that the consequential vertical displacement $\mathrm{d}\Delta$ of the load is equal to $R\,\mathrm{d}\theta_x$, and substituting $\mathrm{d}\theta_x = \mathrm{d}\Delta/R$ in the foregoing relationships, and transposing gives:

$$\frac{\mathrm{d}\Delta}{\mathrm{d}s} = \frac{R^2W}{GI_{xx}}\,.$$

Now, R, W, G and I_{xx} have common values throughout, so the foregoing indicates that the displacement of the load per unit length of arc is constant, irrespective of the position of the arc around the circle, and it follows that the displacement of the load due to twisting throughout the length of the semicircle is given by:

$$\Delta = \frac{\pi R\, R^2 W}{G I_{xx}} = \frac{\pi R^3 W}{G I_{xx}} \,.$$

The stiffness of the semicircle under the load W is therefore given by:

$$\frac{W}{\Delta} = \frac{G I_{xx}}{\pi R^3} = \frac{G r^4}{2R^3} \,.$$

b) In a close-coiled helical spring, where the helix angle is small, each half coil is very nearly equal to a plane semicircle, loaded as in (a). The deflection under W will therefore again be proportional to the total length of the coils, so the stiffnesses of coils having different lengths will be inversely proportional to the lengths, and since the stiffness of a half coil is equal to $(Gr^4/2R^3)$, it follows that the stiffness of a spring having N coils is given by:

$$\frac{W}{\Delta} = \frac{1}{2N}\,\frac{G r^4}{2R^3} = \frac{G r^4}{4NR^3} \,.$$

c) The greatest shear stress in the wire occurs at the outside radius, r, and if $\hat{W}$ represents the greatest load, the value of $\hat{\tau}$ is given by

$$\hat{\tau} = \frac{M_x r}{I_{xx}} = \frac{2\,R\,\hat{W}}{\pi\, r^3} \,.$$

But in simple tension, the greatest shear stress is equal to half the tensile stress, so the greatest shear stress at yield is equal to 150 (GN/m^2), and the greatest allowable shear stress is equal to $150/2{\cdot}5 = 60$ (GN/m^2). The maximum load is therefore given by:

$$\frac{2\,R\,\hat{W}}{\pi\, r^3} = 60\ (\text{GN/m}^2)$$

whence
$$\hat{W} = \frac{30\,\pi\, r^3}{R}\ (\text{GN}) \,.$$

Example 11.8

A hollow propeller shaft of outer radius R is required to transmit 8·25 (MW) at 330 rev/min, when the thrust of the propeller is 1·5 (MN).

When a sample of the material is tested in simple tension it yields at a tensile stress of 200 N/mm^2.

Assuming a bore radius equal to half the outer radius, and adopting a Factor of Safety of 2 on the greatest shear stress theory, show that the required value of R is given by:

$$2500\,\pi^4 R^6 - \pi^2 R^2 - 2{\cdot}56 = 0 \,.$$

Solution

The properties of the sections are:

$$\text{Area} = A = \pi R^2(1-0{\cdot}5^2) = 3\pi R^2/4 \ (\text{m}^2)$$

$$I_{xx} = (\pi R^4/2)(1-0{\cdot}5^4) = 15\pi R^4/32 \ (\text{m}^4) \ .$$

The resultants on the various sections comprise a uniform force F_x and a uniform moment M_x, where:

$$F_x = -\ 1{\cdot}5 \times 10^6 \ (\text{N}) \ ,$$

and the moment M_x is determined from

$$M_x = \frac{\text{Power}}{\text{Angular speed}} = \frac{8{\cdot}25 \times 10^6}{11\pi}$$

i.e.: $$M_x = \frac{0{\cdot}75}{\pi} \times 10^6 \ (\text{N m}) \ .$$

The Force F_x induces uniform simple tension, σ_{xx} as follows:

$$\sigma_{xx} = \frac{F_x}{A} = -\frac{1{\cdot}5 \times 10^6}{3\pi R^2/4} = -\frac{2}{\pi R^2} \times 10^6 \ (\text{N/m}^2) \ .$$

The torque M_x induces on transverse sections, circumferential pure shear τ whose magnitude is proportional to the radial distance r from the x axis, as given by

$$\tau = \frac{M_x}{I_{xx}} r = \frac{0{\cdot}75 \times 10^6}{\pi} \ \frac{32}{15\pi R^4} r = \frac{1{\cdot}6 \times 10^6}{\pi^2 R^4} r \ (\text{N/m}^2) \ .$$

The greatest value of τ at the outer radius is therefore:

$$\tau_{(r=R)} = \frac{1{\cdot}6 \times 10^6}{\pi^2 R^3} (\text{N/m}^2) \ .$$

Consider an element of the outer surface, isolated by adjacent pairs of transverse and axial sections. The edges exposed by the transverse sections bear the tensil stress σ_{xx}, and the circumferential shear stress $\tau_{(r=R)}$. The edges exposed by the axial sections bear only the complementary shear stress $\tau_{(r=R)}$, and the faces are free of stress.

Mohr's Circular Diagram for the state of stress at the surface of the shaft is therefore as shown. At any other radius, σ_{xx} would have the same value, but τ would be less, and it is obvious that the diagram for any other radius would be wholly contained within the diagram for the outer radius. The greatest value of shear stress on any section at any point is therefore indicated by the radius of the largest circle in the diagram shown,

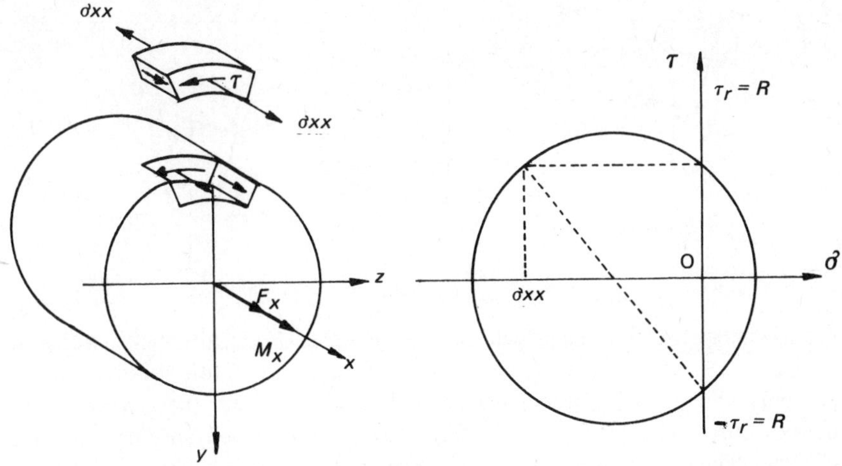

whence: Greatest max. shear $= \sqrt{(\tau^2_{(r=R)} + (\sigma_{xx}/2)^2)}$

$$= \frac{10^6}{\pi^2 R^3}\sqrt{(2{\cdot}56 + \pi^2 R^2)}\ (\mathrm{N/m^2})$$

In simple tension, the maximum shear stress is half the tensile stress, and, by the maximum shear stress theory, the allowable shear stress is equal to the maximum shear stress at yield divided by the Factor of Safety. Therefore

$$\text{Allowable shear} = \frac{200 \times 10^6}{2 \times 2} = 50 \times 10^6\ (\mathrm{N/m^2})\ ,$$

and this being equated to the greatest value of shear stress in the shaft, we have

$$\sqrt{(2{\cdot}56 + \pi^2 R^2)} = 50\,\pi^2 R^3$$

or

$$2500\,\pi^4 R^6 - \pi^2 R^2 - 2{\cdot}56 = 0\ .$$

11.5 PURE BENDING

A section is said to be in pure bending when it is subjected to a bending moment alone, whilst a straight beam is said to be in plane bending when its centre-line is deformed only in a plane, and first we consider the special, but common, case of a member of uniform section which is subjected to pure, plane bending in a plane of symmetry.

Pure, Plane Bending

In a body having a longitudinal plane of symmetry (Fig. 11.10) the intersection of the plane of symmetry with a transverse section is an axis of symmetry of the section. It is therefore a principal axis of the section, and this being chosen as the y axis, consider the case in which the member is loaded in its plane of

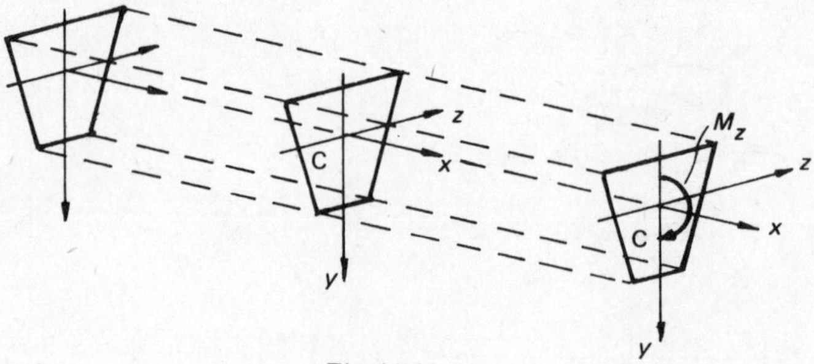

Fig. 11.10

symmetry by a pair of equal but opposite couples of magnitude M_z, which are applied to its ends. Clearly each section is then subjected to the bending moment M_z only. Furthermore it is obvious from symmetry that the centre-line will deform only in the plane of symmetry in which it is loaded, and for such a case of pure plane bending we postulate intuitively as follows:

a) The stresses on longitudinal sections are negligible, and

b) Plane transverse sections remain plane and transverse after bending, so that the member deforms by a simple relative rotation of the plane transverse sections about an axis parallel to the axis z of the applied moment.

So, referring to Fig. 11.11, consider the filaments which span between a pair of adjacent, transverse sections whose initial, infinitesimal separation is represented by ds. The assumptions imply, first, that the filaments are independently

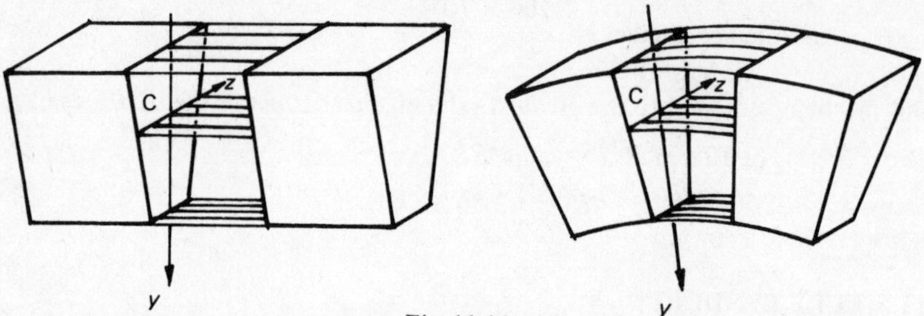

Fig. 11.11

stressed in simple tension or compression, in which case, the tensile stress σ_{xx}, in the filaments is simply proportional to their linear strain ϵ_{xx}, and second, that the extensions, and therefore the strain ϵ_{xx} and the stress σ_{xx}, in the filaments, are independent of the z coordinates of the filaments, and vary linearly with their y coordinates. From the assumptions we therefore infer that pure bending induces only states of simple tension in which the stress σ_{xx} at any position varies linearly with the y coordinate as follows:

$$\sigma_{xx} = a + b \, . \, y \ ,$$

where a and b are constants whose values may be determined by invoking the conditions of equilibrium.

Thus let $\mathrm{d}A$ represent the area of an element of the section at the position whose coordinates are (y, z). The internal force on the element acts in the x direction and has a magnitude $\mathrm{d}f_x$ given by

$$\mathrm{d}f_x = \sigma_{xx}\mathrm{d}A = a\,\mathrm{d}A + b\,y\,\mathrm{d}A \ ,$$

and the resultant of the internal forces in the x direction being equated to the zero value of the resultant external force F_x, we have:

$$\Sigma(a\,\mathrm{d}A) + \Sigma(b\,y\,\mathrm{d}A) = 0 \ .$$

But, a and b, being constants, may be extracted from the summations, when $\Sigma\mathrm{d}A$ may be recognised as the total area A of the section, and we may therefore write:

$$a\,A + b\,\Sigma(y\,\mathrm{d}A) = 0 \ .$$

Now $\Sigma(y\,\mathrm{d}A)$ can be recognised as the first moment of area of the section about the z axis, and since this was chosen through the centre of area, the moment is necessarily zero. However, the area A is not equal to zero, so the constant a must be, and we recognise that σ_{xx} and ϵ_{xx} are in fact simply proportional to the distance y from the z axis.

We may therefore represent the stress on an element by $b.y$, the force on the element by $b\,y\,\mathrm{d}A$, and the moment of this force about the z axis by $-\,b\,y^2\mathrm{d}A$. Thus, the resultant moment of the internal forces about the z axis being equated to the resultant external moment M_z, we may write:

$$-\,b\,\Sigma(y^2\mathrm{d}A) = M_z \ .$$

But $\Sigma(y^2\mathrm{d}A)$ can be recognised as the second moment of area of the section about the z axis, and this being represented by I_{zz}, we have:

$$b = -\frac{M_z}{I_{zz}} \ ,$$

whence $$\sigma_{xx} = -\frac{M_{zz}}{I_{zz}}y \ .$$

Alternatively, the constant of proportionality may be determined in terms of the consequential curvature of the centre-line. Thus, noting that the strain, and therefore the change of length of the centre-line, is zero, let $\mathrm{d}\theta_z$ represent the relative rotation between the pair of sections at an initial separation $\mathrm{d}s$. Then the extension of a filament at the position (y, z) is equal to $-y\,\mathrm{d}\theta_z$, and since the initial length of a filament was $\mathrm{d}s$, its strain is equal to $-y\,\mathrm{d}\theta_z/\mathrm{d}s$.

Therefore $$\sigma_{xx} = E\,\epsilon_{xx} = -\,E\frac{\mathrm{d}\theta_z}{\mathrm{d}s}y \ .$$

Now the angle between the sections is also the angle between the tangents to the centre-line to which they are normal, so $\mathrm{d}\theta_z/\mathrm{d}s$ can be recognised as the curvature of the centre-line, and we therefore conclude as follows.

Pure plane bending under the moment M_z induces only states of simple tension and compression in which the tensile stress σ_{xx} at any position in the section is proportional to its distance from the z-axis of the moment. The constant of proportionality may be equated either to the negative of the ratio of M_z to the second moment of area of the section about the z axis, or to the negative of the product of Young's Modulus E with the consequential curvature of the centre-line.

i.e. $$\sigma_{xx} = -\frac{M_z}{I_{zz}}y = -E\left(\frac{d\theta_z}{ds}\right)y \tag{11.4}$$

Unsymmetrical Bending

Though this is not intuitively apparent, it is found that under a moment about any principal axis, the bending is confined to the plane which is normal to the axis. It therefore follows, that provided that the y and z axes are chosen along the principal axes of the section, comparable relationships may be applied to bending about each. However, when the axes are chosen as a right-handed set, the sign in the relationship for bending about the y axis is positive. Thus Fig. 11.12a indicates that a positive M_z implies compression when y is positive and tension when y is negative, whereas Fig. 11.12b indicates that positive M_y implies tension where z is positive, and compression where z is negative.

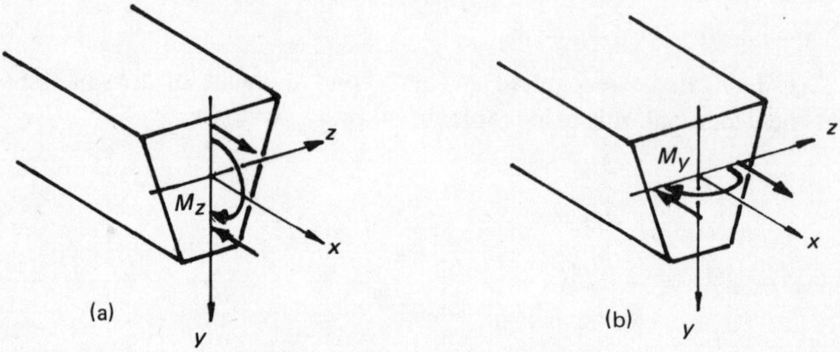

Fig. 11.12

Thus, in pure bending the net bending moment on any section is resolved into its components M_y and M_z about the principal axes of the section. Then the net effect is the simple sum of the effects of M_y and M_z severally, and these are determined as follows:

$$\begin{aligned} \sigma_{xx} &= -\frac{M_z}{I_{zz}}y = -E\left(\frac{d\theta_z}{ds}\right)y \\ \sigma_{xx} &= \frac{M_y}{I_{yy}}z = E\left(\frac{d\theta_y}{ds}\right)z \end{aligned} \tag{11.5}$$

Example 11.9

The diagram illustrates the symmetrical section of an I girder.

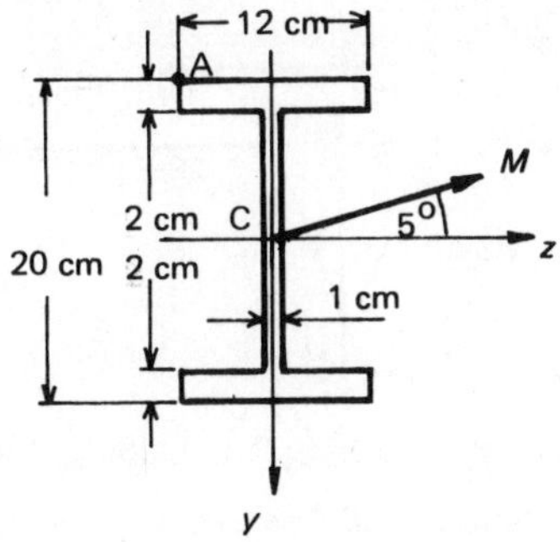

Determine the tensile stress at the point A due to a bending moment $M =$ 40 (kN m) about the axis shown.

Solution

Determining the second moment, of area I_{zz} as the difference of the moments of two rectangles, and of I_{yy} as the sum of the moments of two rectangles gives:

$$I_{zz} = \{(12 \times 20^3) - (11 \times 16^3)\}/12 \text{ (cm}^4\text{)}$$
$$\equiv 42{\cdot}45 \times 10^6 \text{ (mm}^4\text{)} .$$
$$I_{yy} = \{(4 \times 12^3) + (16 \times 1^3)\}/12 \text{ (cm}^4\text{)}$$
$$\equiv 5{\cdot}76 \times 10^6 \text{ (mm}^4\text{)} .$$

Also:

$$(y_A, z_A) = (-100, -60) \text{ (mm)}$$
$$M_z = M \cos 5° = 39{\cdot}85 \times 10^6 \text{ (N mm)}$$
$$M_y = -M \sin 5° = -3{\cdot}49 \times 10^6 \text{ (N mm)} .$$

Since y is an axis of symmetry, y and z are principal axes, and the stress at the point A is given by:

$$\sigma_{xx} = -\frac{M_z}{I_{zz}} y_A + \frac{M_y}{I_{yy}} z_A$$

$$\sigma_{xx} = \frac{39{\cdot}85}{42{\cdot}45} \times 100 + \frac{3{\cdot}49}{5{\cdot}76} \times 60 \text{ (N/mm}^2\text{)}$$
$$= 93{\cdot}9 + 36{\cdot}4$$

that is

$$\sigma_{xx} = 130 \text{ (N/mm}^2\text{)} .$$

Example 11.10

The diagram identifies a point A in the section whose properties are given in Example 10.11.

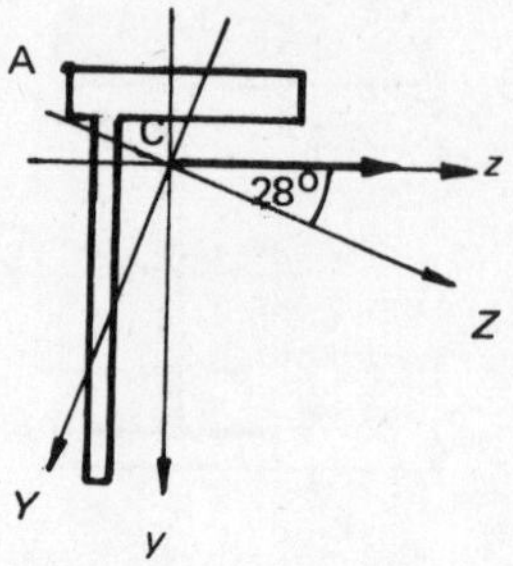

Determine the tensile stress at A due to a moment about the axis z indicated: $M_z = 1$ kN m.

Solution

Since the z axis is not a principal axis, it is necessary to resolve the applied moment along the principal axes, and determine the effect of bending about each.

In Example 10.11, the essential data for the section are determined as follows:

$$\alpha = 28°$$

$$I_{ZZ} = 315\ (\text{cm}^4)$$

$$I_{YY} = 83\ (\text{cm}^4)$$

$$(y_A, z_A) = (-3, -3)\ \text{cm} \ ,$$

therefore
$$\begin{bmatrix} Y_A \\ Z_A \end{bmatrix} = \begin{bmatrix} \cos 28° & -\sin 28° \\ \sin 28° & \cos 28° \end{bmatrix} \begin{bmatrix} -3 \\ -3 \end{bmatrix} = \begin{bmatrix} -1{\cdot}24 \\ -4{\cdot}06 \end{bmatrix} (\text{cm})$$

$$M_Z = 1000 \cos 28° = 883\ (\text{N m})$$

$$M_Y = -1000 \sin 28° = -470\ (\text{N m})$$

$$\sigma_{xx} = \frac{883 \times 10^3}{315 \times 10^4} \times 12{\cdot}4 + \frac{470 \times 10^3}{83 \times 10^4} \times 40{\cdot}6\ (\text{N/mm}^2)$$

$$\sigma_{xx} = 3{\cdot}4 + 23\ (\text{N/mm}^2) = 26{\cdot}4\ (\text{N/mm}^2) \ .$$

Example 11.11

A simple cantilever of uniform $E.I_{zz}$ and length l is loaded in a vertical plane of symmetry with a force W at a distance a from the support. Determine the vertical deflection of the free end of the cantilever:

a) Under the load W at $s = a$,

b) Under an end load W at $s = l$.

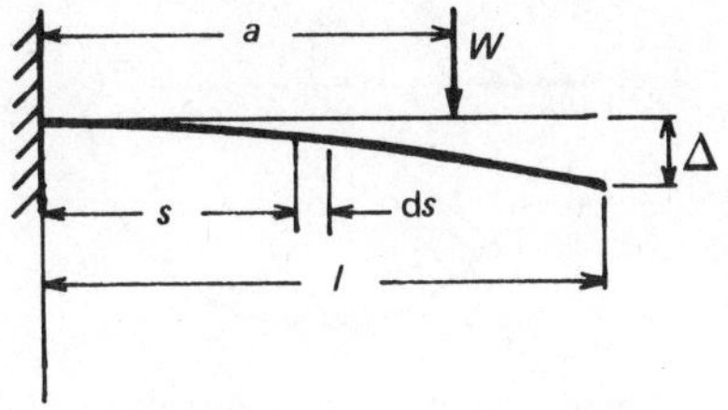

Solution
Within the range $0 < s < a$, the bending moment M_z is equal to $W(a - s)$, but within the range $a < s < l$, it is zero.

Consider the effect of bending in an element of length ds at s, within the range $0 < s < a$, the remainder of the beam being assumed perfectly rigid. Then if $d\theta_z$ represents the relative rotation between the sections at the ends of the element, and dΔ represents the consequential deflection at the end:

$$d\Delta = (l-s)d\theta_z = (l-s)\frac{M_z}{EI_{zz}}ds$$

$$= \frac{W}{EI_{zz}}(l-s)(a-s)\,ds \;.$$

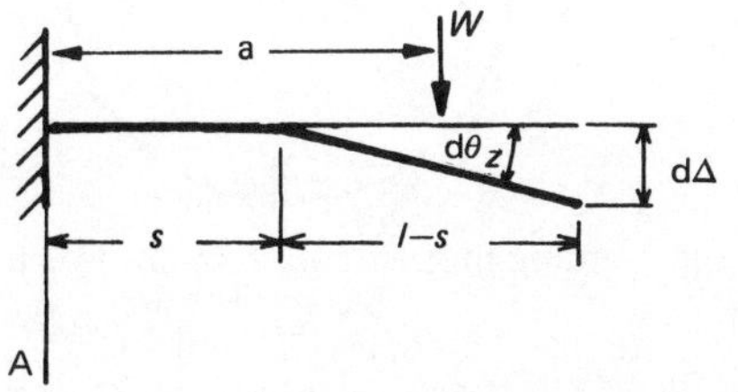

The end deflection due to bending over the range $0 < s < a$ is therefore given by:

$$\Delta = \frac{W}{EI_{zz}}\int_0^a (l-s)(a-s)\,ds \;,$$

that is $$\Delta = \frac{W}{6EI_{zz}}a^2(3l-a) \;.$$

The end deflection under an end load W may be determined by substituting l, for a, when:

$$\Delta = \frac{Wl^3}{3EI_{zz}} \;.$$

Example 11.12
A rod of uniform radius r has the form of a semicircle of radius R. One end is cantilevered from a vertical wall so that the semicircle lies in a horizontal plane.

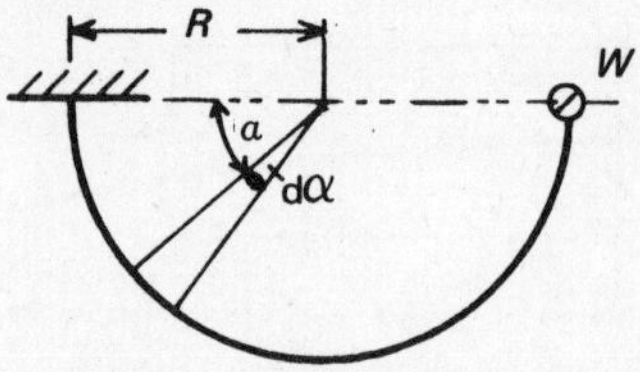

The other carries a vertical load W. Show that the end deflection due to torsion is about eight times that due to bending.

Solution

F_x, F_z and M_y are zero throughout, and F_y being ignored, only the torsional

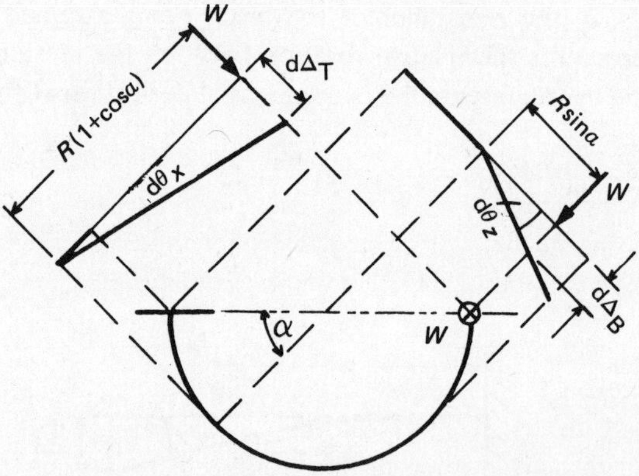

moment M_x, and the bending moment M_z remain. For the element that subtends the angle $d\alpha$ at α:

$$M_x = WR(1 + \cos\alpha); \quad \text{and } M_z = WR\sin\alpha \ .$$

$$\text{In Torsion: } d\theta_x = \frac{M_x R}{G I_{xx}} dx = \frac{WR^2}{G I_{xx}}(1 + \cos\alpha)\, d\alpha$$

$$d\Delta_T = R(1 + \cos\alpha)\, d\theta_x = \frac{WR^3}{G I_{xx}}(1 + \cos\alpha)^2 d\alpha$$

$$\Delta_T = \frac{WR^3}{G I_{xx}} \int_0^{\pi} (1 + \cos\alpha)^2 d\alpha = 3\frac{WR^3}{G r^4} \ .$$

$$\text{In Bending: } d\theta_z = \frac{M_z}{E I_{zz}} R\, d\alpha = \frac{WR^2}{E I_{zz}} \sin\alpha\, d\alpha$$

$$d\Delta_B = R\sin\alpha\, d\theta_z = \frac{WR^3}{E I_{zz}} \sin^2\alpha\, d\alpha$$

$$\Delta_B = \frac{WR^3}{EI_{zz}}\int_0^{\pi} \sin^2\alpha \, d\alpha = \frac{WR^3}{Er^4}$$

$$\frac{\Delta_T}{\Delta_B} = \frac{3E}{G} = 6(1+v) \doteqdot 8 \ .$$

11.6 SHEAR ON AN AXIS OF SYMMETRY

The treatment of the shear components of the equivalent resultants is more complex on two counts.

First, in tension, in cylindrical torsion, or in bending we can readily conceive a simple situation that is both uniform and pure in that the component of interest has a common value on all sections, whilst the remaining components are zero throughout. This is not possible with shear because the shear force is equal to the derivative of the bending moment. This implies that when the sections bear a shear force, they will also be liable to a bending moment which varies from section to section, and in treating the shear force it is necessary to take account of the rate of change of the associated bending moment also.

Second, in shear, the significance that otherwise attaches to the centre of area of the section is transferred to a point called the shear centre.

The Shear Centre

In dealing with the tensile force F_x, for example, we argued that if a section is subjected to a uniform tensile stress, the centroidal axis of the internal forces passes through the centre of area of the section. It is therefore colinear with the resultant external force F_x, and when the tensile stresses are equal to F_x/A, they are therefore statically equivalent to the external force in respect both of force equilibrium, and of moment equilibrium.

But now consider the shear force F_y, recognising that the sections of structural components, like the angle section indicated in Fig. 11.13, can often be

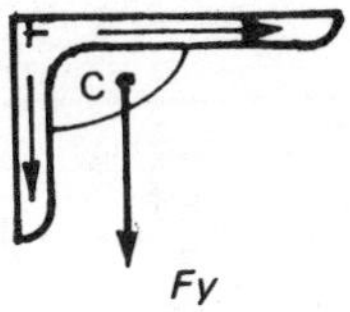

Fig. 11.13

considered to comprise a number of long narrow legs. Then the longitudinal surfaces of the member being free of shear, it follows that the complementary shear stresses on transverse sections are zero also. In other words, at any point on the section adjacent to its boundary, the component of the shear stress normal to the boundary is zero, and this implies that irrespective of either the direction of the external shear force or the location of its line of action, the resultant internal force generated by the shear stresses on each leg will be sensibly colinear with the leg itself. In the case of the angle, for example, this implies that the centroidal axis of the internal forces on both legs will pass through their point of intersection, and this is called the shear centre of the section.

Thus the external force F_y induces shear stresses whose resultant is equal to F_y; but whereas the line of action of F_y passes through the centre of area of the section, the centroidal axis of the internal forces passes through the shear centre. The member is thus subjected to a torsional couple which twists and warps the member in a manner which tends to align the shear centre with the line of action of F_y, as indicated in Fig. 11.14. Evidently, the introduction of

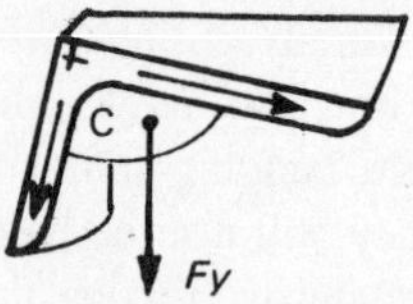

Fig. 11.14

the incidental torsional couple greatly complicates the analysis, but it is also of great practical significance because open sections (as opposed to closed box sections) are particularly susceptible to torsion. However, where a member is loaded in a plane of symmetry, the shear centre, like the centre of area, will lie in the plane of symmetry, where both are colinear with the shear force, so in practice, significant structural members are usually employed only in this manner.

Stresses in Symmetrical Shear

So we now consider the case in which a shear force F_y is applied on the axis of symmetry of a symmetrical section. This induces shear stresses whose component in the direction of F_y is represented by σ_{xy}, and it is assumed that this has a common value along any line, which, like the line AB in Fig. 11.15, is parallel

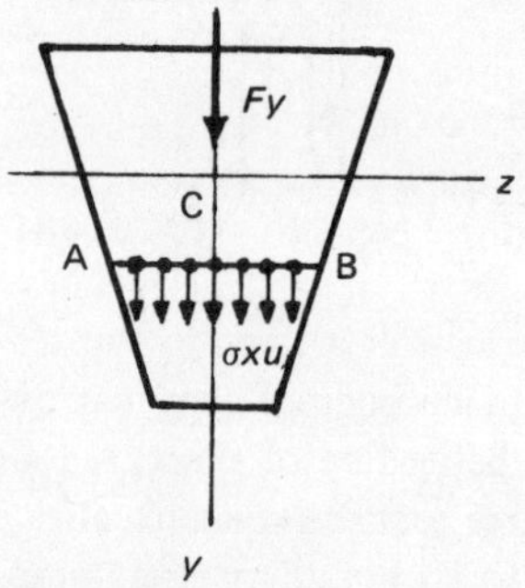

Fig. 11.15

to the z axis, and this value of σ_{xy} is determined by calculating the complementary shear stress σ_{yx} which acts on the longitudinal plane through the line.

Thus, identifying a slice of the member which lies between the transverse sections at the distances s and $s + ds$ along the centre line (Fig. 11.6), let F_y and M_z

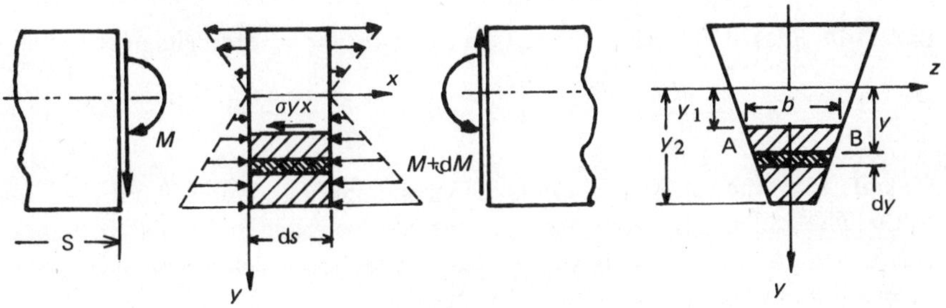

Fig. 11.16

represent the shear force and bending moment on the section at s, let $M_z + dM_z$ represent the bending moment on the section at $s + ds$, and consider the equilibrium in the x direction of the part of the slice isolated by the longitudinal section through the line AB which is parallel to the z axis. The relevent stresses comprise the bending stresses σ_{xx} on the negative of the two faces exposed by the section at s, the bending stresses $\sigma_{xx} + d\sigma_{xx}$ on the positive of the two faces exposed by the section at $s + ds$, and the shear stress σ_{yx} on the negative of the two faces exposed by the longitudinal section, and we first consider the resultant force generated by the bending stresses on the face at s.

For an element of the face area dA at a uniform distance y from the z axis, the force on the element is given by:

$$df'_x = -\sigma_{xx} dA = \frac{M_z}{I_{zz}} y \, dA \ ,$$

and by integration, the force on the whole face is given by

$$f'_x = \frac{M_z}{I_{zz}} \sum_{y_1}^{y_2} y \, dA \ .$$

Similarly, the force on the face at $s + ds$ is given by:

$$f''_x = -\frac{M_z + dM_z}{I_{zz}} \sum_{y_1}^{y_2} y \, dA \ ,$$

and the breadth of the section along the line AB being represented by b, the force on the longitudinal face is given by:

$$f'''_x = -\sigma_{yx} b \, ds \ .$$

Then for the equilibrium of forces in the x direction:

$$f'_x + f''_x + f'''_x = 0 \ ,$$

whence $$\sigma_{yx}\, b\, \mathrm{d}s = -\frac{\mathrm{d}M_z}{I_{zz}} \sum_{y_1}^{y_2} y\, \mathrm{d}A$$

or $$\sigma_{yx} = -\frac{\mathrm{d}M_z}{\mathrm{d}s}\,\frac{1}{b\, I_{zz}} \sum_{y_1}^{y_2} y\, \mathrm{d}A\ .$$

But $-(\mathrm{d}M_z/\mathrm{d}s) = F_y$, and $\sigma_{yx} = \sigma_{xy}$, and substituting accordingly gives

$$\sigma_{xy} = \frac{F_y}{I_{zz}\, b} \sum_{y_1}^{y_2} y\, \mathrm{d}A \qquad (11.6)$$

We note that whereas I_{zz} represents the second moment of area about the z axis of the whole of the transverse section, the sum determines the first moment of area about the z axis of only that part of the section which lies outside the line of length b to which σ_{xy} applies.

We also note, that eqn. (11.6) determines only the component of shear stress in the direction y of the applied force, and there is no simple method of determining the component σ_{xz}. However, we know that, the longitudinal surfaces of the member being free of shear, then at points near the boundary of transverse sections, the component of shear normal to the boundary is zero. In other words, at points near the boundary of transverse sections, σ_{xy} is a component of a total shear stress τ which is necessarily parallel to the boundary.

Example 11.13

Show that when a beam of rectangular section is loaded in a plane of symmetry, the greatest shear stress on transverse sections is 1·5 times the average shear stress.

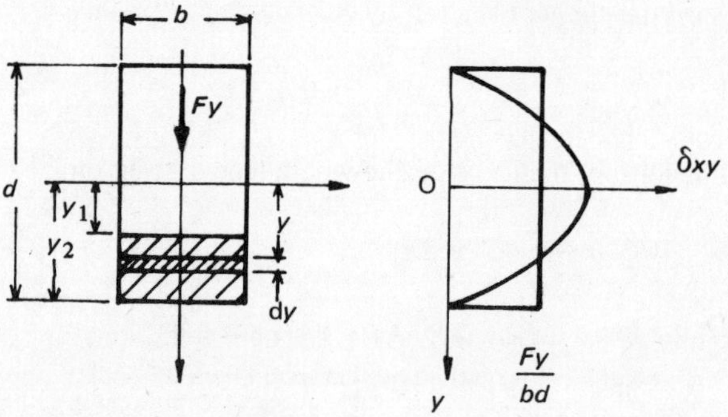

Solution

The breadth and depth of the rectangle being represented by b and d, let the y axis be chosen in the plane of symmetry in which the member is loaded. Then the shear stress σ_{xy} along a line AB at the distance y_1 from the z axis is given by:

$$\sigma_{xy} = \frac{F_y}{I_{zz}\, b}\int_{y_1}^{y_2} y\, \mathrm{d}A = \frac{12\, F_y}{b^2 d^3}\int_{y_1}^{d/2} by\, \mathrm{d}y$$

that is: $$\sigma_{xy} = \frac{3\,F_y}{2\,B\,d^3}(d^2 - y_1^2)\ .$$

This describes a parabola whose greatest value evidently occurs on the z axis where $y_1 = 0$, and its value is

$$\hat{\sigma}_{xy} = 3\,F_y/2\,b\,d\ .$$

But the average shear stress on the section is equal to the ratio of the shear force F_y to the area $b\,d$ of the section, and we therefore recognise that the greatest shear stress on the section occurs along the z axis, and its value is 1·5 times that of the average stress on the section.

Example 11.14

The figure indicates the dimensions of a symmetrical I section which is loaded in the plane of the web.

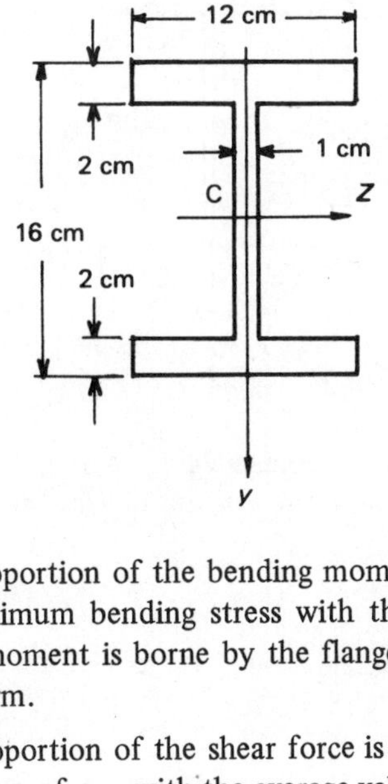

a) Determine what proportion of the bending moment is borne by the flanges, and compare the maximum bending stress with the value determined on the assumption that the moment is borne by the flanges alone, and that the stress on each flange is uniform.

b) Determine what proportion of the shear force is borne by the web, and compare the maximum value of σ_{xy} with the average value, assuming that the whole of the shear force is borne by the web alone.

Solution

Since we are concerned only with proportions it will not be inappropriate to work in centimetre units throughout.

a) The proportions of M_z that are borne by different parts of the section are equal to the proportions of their second moments of area.

For the whole: $I_{zz} = \{12 \times 16^3 - 11 \times 12^3\}/12 = 2512\ (\text{cm}^4)$

For the web: $I_{zz} = (1 \times 12^3)/12 = 144\ (\text{cm}^4)$

For the flanges: $I_{zz} = 2512 - 144 = 2368\ (\text{cm}^4)$

Proportion of M_z borne by flanges $= 2368/2512 \equiv 94\%$.

The greatest bending stress occurs in the fibres most distant from the z axis, and its value is given by:

$$\sigma_{xx} = -\frac{M_z}{I_{zz}}y = \mp \frac{8}{2512} M_z .$$

If the bending stresses on each flange were uniformly of magnitude $\bar{\sigma}_{xx}$, then, since the area of each is equal to 24 cm^2, the stresses on each flange would be

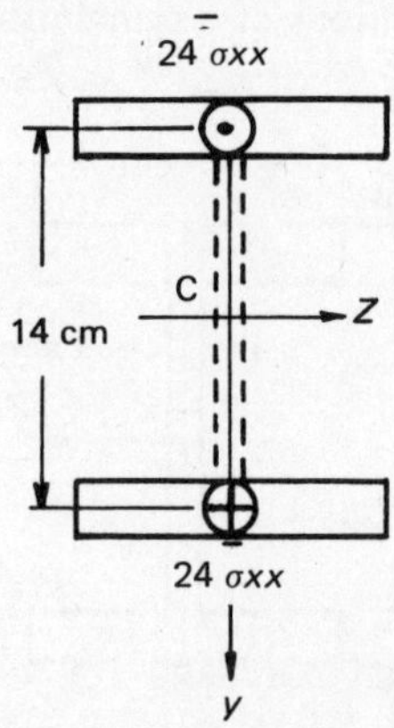

statically equivalent to a force equal to 24 σ_{xx} acting through the centre. Therefore if the whole of M_z were borne by the flanges, the value of $\bar{\sigma}_{xx}$ would be given by:

$$M_z = 14 \times 24\,\bar{\sigma}_{xx}$$

that is $\bar{\sigma}_{xx} = M_z/336$.

Therefore $\bar{\sigma}_{xx}/\sigma_{xx} = 2512/(8 \times 336) \equiv 93{\cdot}5\%$.

Thus the approximate value of σ_{xx} would underestimate the maximum value by about 7%.

b) At the distance y_1 from the z axis:

$$\sum_{y_1}^{y_2} y\,\mathrm{d}A = (6-y_1)(6+y_1)/2 + 24 \times 7 = 186 - y_1^2/2$$

and $$\sigma_{xy} = \frac{F_y}{I_{zz}\,b} \sum_{y_1}^{y_2} y\,\mathrm{d}A = \frac{372 - y_1^2}{5024} \cdot F_y ,$$

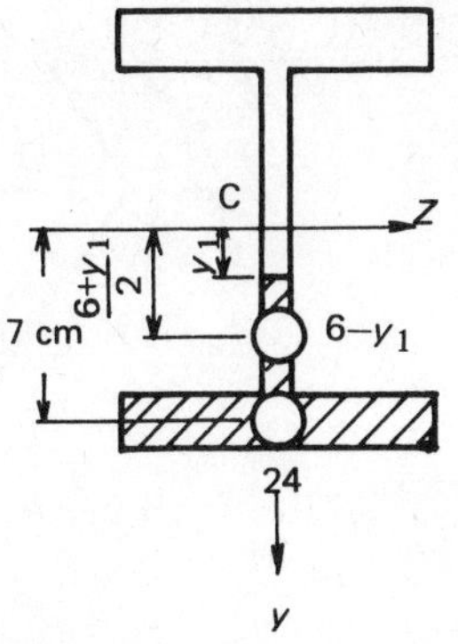

The shear force on a strip of width dy_1 at $y_1 = \sigma_{xy} \, . \, dy_1$. Therefore

$$\text{Force on web} = \frac{2F_y}{5024} \int_0^6 (372 - y_1^2)\, dy_1 \equiv 86\% \, F_y \; .$$

The greatest value of σ_{xy} occurs on the z axis where $y_1 = 0$, and its value σ_{xy} is equal to $(372/5024)\, F_y$.

If it is assumed that the flange bears the whole of F_y, the average value $\bar{\sigma}_{xy}$ would be equal to $F_y/12$.

Therefore $\bar{\sigma}_{xy}/\sigma_{xy} = 5024/(372 \times 12) \equiv 113\%$.

Example 11.15

A shear force F_y acts on a symmetrical section as indicated. Show that the shear stress σ_{xy} has a (maximum) stationary value at $y = s/4$, and derive an expression for the total shear stress τ at the boundary of the section at this distance.

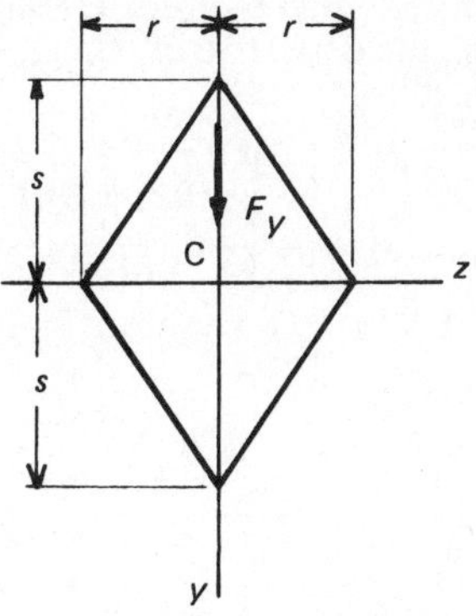

Solution

The second moment of area of a triangle of base b and height h, about its base is given by

$$I_{zz} = \int_0^h \frac{(h-y)}{h} b\, y^2 dy = b\, h^2/12 \; .$$

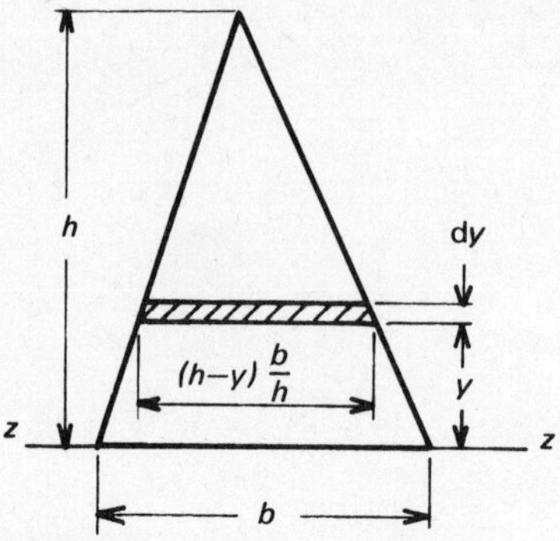

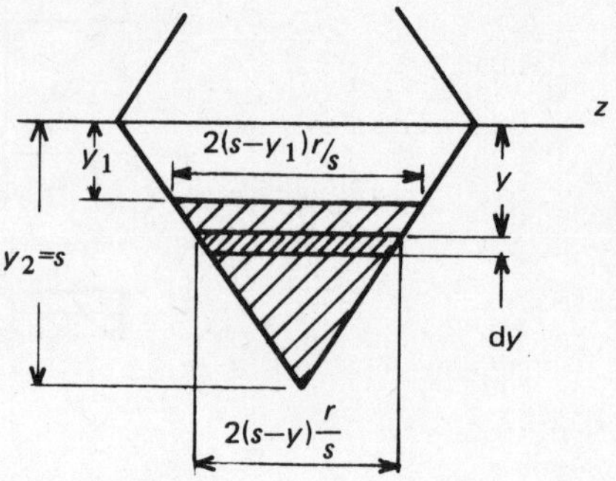

Therefore for the two triangles of base $2r$ and height h:

$$I_{zz} = r\,s^3/3\ .$$

For the shaded area

$$\int_{y=y_1}^{y=y_2} y\,\mathrm{d}A = \frac{2r}{s}\int_{y_1}^{s} y(s-y)\,\mathrm{d}y$$

$$= \frac{r}{3s}(2y_1^3 - 3sy_1^2 + s^3)$$

also $$b = \frac{2r}{s}(s-y_1)\ .$$

Therefore at the distance y_1 from the z axis

$$\sigma_{xy} = \frac{F_y}{I_{zz}\,b}\int_{y=y_1}^{y=y_2} y\,\mathrm{d}A = \frac{(2y_1^3 - 3\,sy_1^2 + s^3)}{2\,r\,s^3(s-y_1)}F_y = \frac{u}{v}F_y\ .$$

For stationary values of σ_{xy}, the derivative of σ_{xy} is zero.

Thus: $$\frac{\mathrm{d}(\sigma_{xy})}{\mathrm{d}y_1} = \frac{\mathrm{d}(u/v)}{\mathrm{d}y_1} = 0; \quad \text{or } u\frac{\mathrm{d}v}{\mathrm{d}y_1} = v\frac{\mathrm{d}u}{\mathrm{d}y_1}\ ,$$

i.e. $$(2y_1^3 - 3sy_1^2 + s^3)(-1) = (s-y_1)(6y_1^2 - 6sy_1)$$

or: $$4\,y_1^3 - 9\,s\,y_1^2 + 6\,s^2y_1 - s^3 = 0$$

$$(4y_1 - s)(y^2 - 2sy_1 - s^2) = 0\ .$$

This has one real solution: $y_1 = s/4$.

The value of σ_{xy} at $y_1 = s/4$ is

$$\sigma_{xy} = \frac{9}{16}\frac{F_y}{rs}$$

$$\tau/\sigma_{xy} = \sqrt{(s^2+r^2)}/s$$

$$\tau = \frac{9\sqrt{(s^2+r^2)}}{16\,r\,s^2}\cdot F_y\ .$$

11.7 BENDING DEFLECTIONS IN STRAIGHT BEAMS

A beam is a member which is designed to support transverse loads over wide spans without excessive deformation, and for this it requires strength and stiffness in both shear and bending. However, apart from closed box sections which introduce problems of their own, strength in bending and transverse shear is not usually to be found in association with torsional stiffness. Beams are therefore usually protected from torsional effects by their being both straight and loaded only in a longitudinal plane of symmetry. We therefore now consider the deformation of straight beams whose sections are subjected to a single component of shear F, and a single component of bending M, both applied in a plane of symmetry, and we consider both the slope θ, and the deflection Δ of the centre-line in the same plane, on the understanding that the slope θ is negligible compared with unity, or, what is equivalent, that θ^2 is negligible compared with θ itself.

Bending Deflection v Shear

Consider, for example, the deflection Δ of the free end of a simple cantilever of uniform section and length l which is subjected to an end load W. Fig. 11. 17 illustrates the effect of the deformation of an elemental slice of infinitesimal

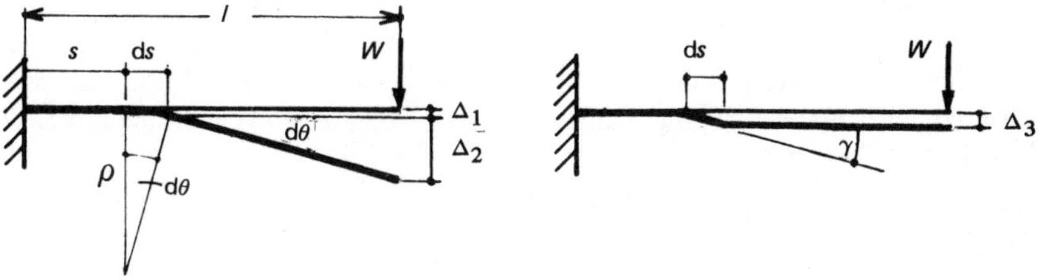

Fig. 11.17

length ds, at a distance s along the beam, and it shows, first the effect of bending under the moment M, and then of shear under the force F. Dealing first with the effect of bending, the diagram shows that the bending deflection can be considered to comprise two parts, of which Δ_1 represents the relative displacement between the sections at s and s + ds, whilst Δ_2 represents the effect of their relative rotation, and to compare the magnitudes of these effects, let ρ represent the consequential radius of curvature of the element. Then:

$$\Delta_1 = \rho(1 - \cos d\theta) \ .$$

But if terms of the fourth and higher orders are ignored, $(1 - \cos d\theta)$ may be equated to $(d\theta)^2/2$. Furthermore, ρ may be equated to the inverse of the curvature, i.e., $\rho = (\mathrm{d}s/\mathrm{d}\theta)$, whilst $\mathrm{d}\theta = (M/EI)\,\mathrm{d}s$, and substituting accordingly gives:

$$\Delta_1 = \frac{M}{2EI}(\mathrm{d}s)^2 \ .$$

Also: $$\Delta_2 = (l-s)\,\mathrm{d}\theta = \frac{M}{EI}(l-s)\,\mathrm{d}s \ .$$

From this it is obvious that Δ_1, which is of second order in $\mathrm{d}s$, is negligible compared with Δ_2, which is of first order.

As to the shear displacement, we note that in shear, parallel lines remain parallel. The shear force F therefore causes no relative rotation between the sections at s and $s + \mathrm{d}s$, and if γ represents the 'average' shear strain of the elemental slice, the associated end displacement Δ_3 is given by: $\Delta_3 = \gamma\,\mathrm{d}s$.

Now, whereas γ, (and therfore Δ_3) depends on F, the bending deflection Δ_2 depends on M, and F and M are independent. It is therefore not possible to compare the magnitudes of Δ_2 and Δ_3 in general. But consider the particular case of the end-loaded cantilever. In this case the shear force F on the elemental slice is equal to the end load W. Therefore if A represents the area of the section of the beam, the average shear stress on the section may be equated to W/A, and it follows, that if G represents the shear modulus of the material, the average shear strain γ in the elemental slice will be approximately equal to W/GA. Furthermore, if k represents the radius of gyration of the section about the bending axis, its second moment of area I may be equated to Ak^2, and since the bending moment M on the element is equal to $W(l - s)$, we may substitute accordingly as follows:

$$\frac{\Delta_3}{\Delta_2} = \frac{EI}{M}\,\frac{\gamma}{(l-s)} = \frac{E}{G}\left(\frac{k}{l-s}\right)^2 .$$

Now reflection will show that the average of the ratio of the radius of gyration of the section k to the length $(l - s)$ will be of the order of the ratio of the cross-sectional dimensions to the length, and if this is of the order of, say 1:10, the shear deflection Δ_3 will be only a few percent of the bending deflection Δ_2.

In the theory of the bending of long beams it is therefore assumed, first, that the deflection due to transverse shear is negligible compared with that due to bending, and second, that the bending deflection depends sensibly on the relative rotations between the succeeding sections rather than their relative displacement.

Slope and Deflection Equations

In a given beam under given loads, the slope θ and the deflection Δ have unique values at each section, and the equations which describe θ and Δ as functions of the distance s along the beam are known as the slope and deflection equations of the beam.

But $$d\theta = \frac{M}{EI} ds$$

Therefore $$\theta = \int \frac{M}{EI} ds + C_3 \;.$$

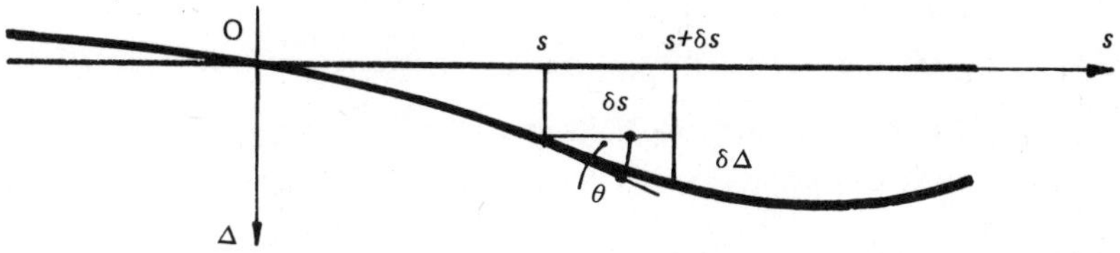

Fig. 11.18

Furthermore it is obvious from Fig. 11.18 that

$$\theta = \tan^{-1}\frac{d\Delta}{ds} = \tan^{-1}\left(\underset{\delta s \to 0}{\text{Limit}}\ \frac{\delta\Delta}{\delta s}\right) .$$

But when θ is negligible compared with unity, the tangent of θ may be equated to the angle θ itself, when:

$$\theta = \frac{d\Delta}{ds}$$

or $$\Delta = \int \theta \, ds + C_4 = \int\int \frac{M}{EI} ds \, . \, ds + C_3 s + C_4 \; .$$

Thus, if M represents the bending moment equation for a beam, and if E and I represent the function which describe the variations of Young's Modulus and the second moment of area of the sections with s, then the slope and deflection equations for a beam may be determined as the first and second derivatives of M/EI, i.e.:

$$\theta = \int \frac{M}{EI} ds + C_3$$
$$\Delta = \int\int \frac{M}{EI} ds \, ds + C_3 s + C_4 \qquad (11.7)$$

Slope and Deflection Equations for a Beam of Uniform *EI*

Commonly both E and I have uniform values at all sections, in which case they can be extracted from under the integrals. Furthermore we have seen in eqns.

(11.1b) that the bending moment equation M can be determined as the negative of the integral of the shear force equation F, and that the negative of F can itself be determined as the integral of the equation of the loading intensity w. For a beam of uniform EI it therefore follows that $EI\theta$ and $EI\Delta$ may be determined either as the first and second integrals of the bending moment equation, or as the negatives of the second and third integrals of the shear force equation, or as the third and fourth integrals of the equation of loading intensity, and we may write:

$$\begin{aligned} -F &= \int w \, ds + C_1 \\ M &= \int -F \, ds + C_2 \\ EI\theta &= \int M \, ds + C_3 \\ EI\Delta &= \int EI\theta \, ds + C_4 \end{aligned} \tag{11.8}$$

Since the equations of loading intensity, shearing force, and bending moment vary from bay to bay, they must be integrated bay by bay, and each bay will therefore yield its quota of constants of integration. To these must be added the unknown support reactions, and the total set of unknowns are determined from the known boundary conditions, together with the known conditions for continuity at each section in which two bays abut.

Boundary Conditions

In the integration of the bending moment and slope equations, the boundary conditions comprise the geometric constraints imposed by the supports. Thus, at an encastré support both $\theta = 0$ and $\Delta = 0$, whereas at a simple support, only $\Delta = 0$. But sometimes we also refer to an elastic prop. This is a support at which the displacement Δ is proportional to the force of reaction R, but the sense of the displacement is opposed to that of the force on the beam. For an elastic prop we may therefore write $\Delta = -f.R$, where f represents the flexibility of the support, and R the force of reaction that it exerts on the beam.

In the integrations of the loading and shear force equations the boundary conditions comprise the conditions for the equilibrium of the elemental slices at the ends of the beam, and reflection will show that these require that at the positive end the internal shear force and bending moment will be equal to the external force and moment, whilst at the negative end, the internal shear force and bending moment will be the negatives of the external force and moment.

Continuity Conditions

As with the boundary conditions, we again distinguish the geometric conditions which apply to the integrations of the bending moment and slope equations, from the equilibrium conditions which apply to the integrations of the loading and shear force equations. Thus, in the immediate vicinity of each section in which two bays abut, the continuity of the slope and deflection of the centre-line requires that the slope and deflection on one side of the section must be equal to the slope and deflection on the other, whilst the equilibrium of an infinitesimal slice which contains the section requires that the shear force and bending moment just to the negative side of the section must exceed the shear force and bending moment just to the positive side by any external point force and point couple that are applied at the section.

Example 11.16

The figure illustrates a straight beam AB, of uniform EI, which is encastré at A and simply supported at B. It is loaded in a plane of symmetry with a distributed load on the range AC whose intensity varies with s according to $\omega = 2s$

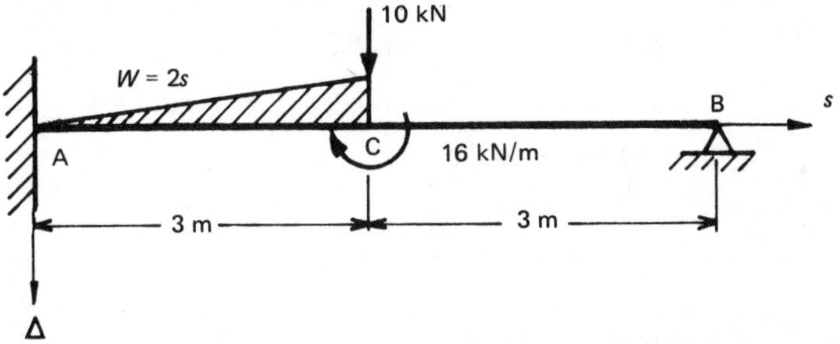

(kN/m), together with a point force $F_c = 10$ (kN), and a point couple $M_c = 16$ (kN m), at C.

Determine its slope and deflection equations.

Solution

The singularities at C divide the beam into two bays, and in the bay AC; $w = 2s$; whilst in the range CB: $w = 0$. The slope and deflection equations for the two bays may therefore be determined by successive integration of the loading equations as follows.

	Range AC	Range CB
w	$2s$	0
$-F$	$s^2 + C_1$	K_1
M	$\frac{s^3}{3} + C_1 s + C_2$	$K_1 s + K_2$
$EI\theta$	$\frac{s^4}{12} + \frac{C_1}{6}s^2 + C_2 s + C_3$	$\frac{K_1}{2}s^2 + K_2 s + K_3$
$EI\Delta$	$\frac{s^5}{60} + \frac{C_1}{6}s^3 + \frac{C_2}{2}s^2 + C_3 s + C_4$	$\frac{K_1}{6}s^3 + \frac{K_2}{2}s^2 + K_3 s + K_4$

The boundary conditions at any point are applied by substituting accordingly in the appropriate equation for the appropriate range. (For example, at A, where $s = 0$, $\Delta = 0$, and substituting accordingly in the slope equation for range AC gives $C_3 = 0$.)

At A, where $s = 0$; $\theta = 0$; therefore $C_3 = 0$ (a)

and $\Delta = 0$; therefore $C_4 = 0$ (b)

and $F = -R_A$; therefore $C_1 = R_A$ (c)

and $M = -M_A$; therefore $C_2 = M_A$ (d)

At B, where $s = 6$: $\Delta = 0$; therefore $36K_1 + 18K_2 + 6K_3 + K_4 = 0$ (e)

and $F = R_B$; therefore $K_1 = -R_B$ (f)

and $M = 0$; therefore $6K_1 + K_2 = 0$. (g)

For continuity at C, where $s = 3$:

$F_{AC} = F_{CB} + 10$; therefore $-C_1 + K_1 = 19$ (h)

$M_{AC} = M_{CB} + 16$; therefore $3C_1 + C_2 - 3K_1 - K_2 = 7$ (j)

$\theta_{AC} = \theta_{CB}$; therefore $\frac{27}{4} + \frac{9}{2}C_1 + 3C_2 + C_3 = \frac{9}{2}K_1 + 3K_2 + K_3$

or: $9C_1 + 6C_2 + 2C_3 - 9K_1 - 6K_2 - 2K_3 = -13{\cdot}5$ (k)

$\Delta_{AC} = \Delta_{CB}$; therefore $\frac{81}{20} + \frac{9}{2}C_1 + \frac{9}{2}C_2 + 3C_3 + C_4 = \frac{9}{2}K_1 + \frac{9}{2}K_2 + 3K_3 + K_4$

or: $9C_1 + 9C_2 + 6C_3 + 2C_4 - 9K_1 - 9K_2 - 6K_3 - 2K_4 = -8{\cdot}1$. (l)

Thus, two of the unknowns are zero, whilst the three equations (c), (d) and (f) constitute an independent set which relate the support reactions to the constants of integration. But the remainder, i.e., equations (e), (g), (h), (j), (k) and (l) form a single set which specify the relationships that must exist between the constants of integration themselves, and they may be presented in matrix form as follows:

$$\begin{bmatrix} 0 & 0 & 36 & 18 & 6 & 1 \\ 0 & 0 & 6 & 1 & 0 & 0 \\ -1 & 0 & 1 & 0 & 0 & 0 \\ 3 & 1 & -3 & -1 & 0 & 0 \\ 9 & 6 & -9 & -6 & -2 & 0 \\ 9 & 9 & -9 & -9 & -6 & -2 \end{bmatrix}^{-1} \begin{bmatrix} 0 \\ 0 \\ 19 \\ 7 \\ -13{\cdot}5 \\ -8{\cdot}1 \end{bmatrix} = \begin{bmatrix} C_1 \\ C_2 \\ K_1 \\ K_2 \\ K_3 \\ K_4 \end{bmatrix} = \begin{bmatrix} -11{\cdot}4125 \\ 18{\cdot}475 \\ 7{\cdot}5875 \\ -45{\cdot}525 \\ 113{\cdot}25 \\ -133{\cdot}2 \end{bmatrix}$$

The back-substitution of the constants then determines the support reactions as follows:

$$M_A = -18{\cdot}5\ (\text{kN m});\quad R_A = -11{\cdot}4\ (\text{kN});\quad R_B = -7{\cdot}6\ (\text{kN})$$

and the slope of and deflection equations for the two ranges as follows:

Range AC	Range CB
$\frac{s^4}{12} - 5{\cdot}71s^2 + 18{\cdot}48s$	$3{\cdot}79s^2 - 45{\cdot}5s + 113$
$\frac{s^5}{60} - 1{\cdot}90s^3 + 9{\cdot}24s^2$	$1{\cdot}26s^3 - 22{\cdot}7s + 113s - 133$

Alternatively, as in the analysis of trusses, the scale of the linear algebra can be reduced by separately considering the equilibrium of the beam as a whole. We then determine the bending moment equation for each bay, when it is necessary to perform only two integrations on each.

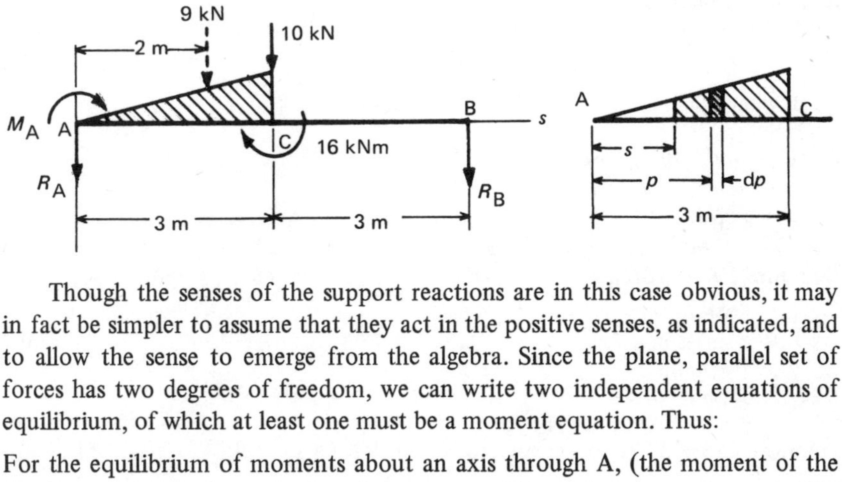

Though the senses of the support reactions are in this case obvious, it may in fact be simpler to assume that they act in the positive senses, as indicated, and to allow the sense to emerge from the algebra. Since the plane, parallel set of forces has two degrees of freedom, we can write two independent equations of equilibrium, of which at least one must be a moment equation. Thus:

For the equilibrium of moments about an axis through A, (the moment of the distributed load is determined as the moment of its total acting on its centroidal axis):

$$M_A + 16 + (2 \times 9) + (3 \times 10) + (6 \times R_B) = 0$$

whence: $M_A = -(6R_B + 64)\ (\text{kN m})$. (a)

For equilibrium in the Δ direction:

$$9 + 10 + R_A + R_B = 0\ ,$$

whence $R_A = -(R_B + 19)\ (\text{kN})$ (b)

Thus both M_A and R_A are now known in terms of R_B.

Bay AC: $$M = 16 + (3-s)10 + (6-s)R_B + \int_s^3 (p-s)2p\,dp$$

or: $$M = (64+6R_B) - (19+R_B)s + \frac{s^3}{3}$$

therefore: $$EI\theta = (64+6R_B)s - \frac{19+R_B}{2}s^2 + \frac{s^4}{12} + C_3 \qquad \text{(c)}$$

and $$EI\Delta = (32+3R_B)s^2 - \frac{19+R_B}{6}s^3 + \frac{s^5}{60} + C_3 s + C_4\ . \qquad \text{(d)}$$

Bay CB: $$M = 6R_B - sR_B$$

therefore $$EI\theta = 6R_B s - \frac{R_B}{2}s^2 + K_3$$

and $$EI\Delta = 3R_B s^2 - \frac{R_B}{6}s^3 + K_3 s + K_4\ .$$

Boundary Conditions

At A, where $s=0$: $\theta = 0$ therefore $C_3 = 0$ (e)

and $\Delta = 0$ therefore $C_4 = 0$. (f)

At B, where $s=6$: $\Delta = 0$ therefore $72R_B + 6K_3 + K_4 = 0$. (g)

For Continuity at C, where $s=3$:

$\theta_{AC} = \theta_{CB}$:

$$\text{therefore } 192 + 18R_B - \frac{171}{2} - \frac{9}{2}R_B + \frac{27}{4} + C_3 = 18R_B - \frac{9}{2}R_B + K_3$$

$$\text{or: } C_3 - K_3 = -113{\cdot}25 \qquad \text{(h)}$$

$\Delta_{AC} = \Delta_{CB}$:

$$\text{therefore } 288 + 27R_B - \frac{171}{2} - \frac{9}{2}R_B + \frac{81}{20} + 3C_3 + C_4 = 27R_B - \frac{9}{2}R_B + 3K_3 + K_4$$

$$\text{or: } 3C_3 + C_4 - 3K_3 - K_4 = -206{\cdot}55\ . \qquad \text{(j)}$$

Thus C_1 and C_2 are zero, and substituting successively in equations (h), (j) and (g) in turn gives:

$$K_3 = -133{\cdot}2; \quad K_4 = 113{\cdot}25; \quad R_B = -7{\cdot}5875\ .$$

Substituting in (a) and (b) then determines the support reactions, and substituting in (c), (d), (e) and (f) then determines the slope and deflection equations for the two bays, as before.

11.8 POINT DEFLECTIONS AND REDUNDANT SUPPORTS

In principle, the determination of the slope and deflection equations of a beam thus requires only the successive integration of the equation of loading intensity, shearing force, or bending moment, but in practice, with up to four additional conditions at each discontinuity, the satisfaction of the boundary and continuity conditions rapidly becomes excessively tedious. However, we are commonly concerned with the slope and deflection only so far as is necessary for the determination of redundant support reactions, and for this it is usually sufficient to determine the slope and deflection at a few specific points only. We therefore now consider alternative approaches that are better suited to this purpose. One may be called the method of sections, and in this the beam itself is considered to be divided along its length. The other, which is called the flexibility method, depends on the principle of superposition, and in this it is the set of external forces and couples that are considered to be divided. Either can conveniently be applied through the concept of flexibility coefficients.

Flexibility Coefficients

Provided that the deformation of a Hookean solid does not sensibly alter its essential geometry, quantities like internal forces and couples, stresses and displacements are linear functions of the external forces and couples. The effects of different loads are then simply additive, and in particular, the effects of an external force or couple that is applied on a given axis are proportional to the force or couple. Thus, when a given beam which is supported in a given manner is loaded on a given axis, quantities like the greatest bending moment in the beam, the shear force or bending moment on a particular section, or the slope and deflection at a given position, are all proportional to the load, and a constant of proportionality which describes the magnitude of such an effect per unit load is called an influence coefficient. However, when the effect is a displacement or a rotation the influence coefficient may be referred to more specifically as a flexibility coefficient.

For example, if one end of a linear spring is fixed, and an axial load is applied at the other, the displacement of the point of action of the force is proportional to the force, and we commonly specify the stiffness of the spring as the force per unit displacement of its point of application. However, we could as well specify the inverse of the stiffness, i.e., the displacement per unit force, and this is called the flexibility of the spring. Furthermore we may be concerned with the displacement of a point other than the point of application of the force. So let 1 and 2 identify two positions along the spring at distances a_1 and a_2 from the fixed end ($a_1 < a_2$), as indicated in Fig. 11.19. The displacement at the position 1 due to a unit force at the position 2 may be indicated by Δ_{12}, whilst the displacement at the position 2 due to a unit force at the position 2 be indicated by Δ_{22}, and it is obvious that Δ_{22} is the inverse of the stiffness of the spring of length a_2, that $\Delta_{12} = (a_1/a_2)\Delta_{22}$, and that the flexibilities of different lengths of a given spring are proportional to their lengths.

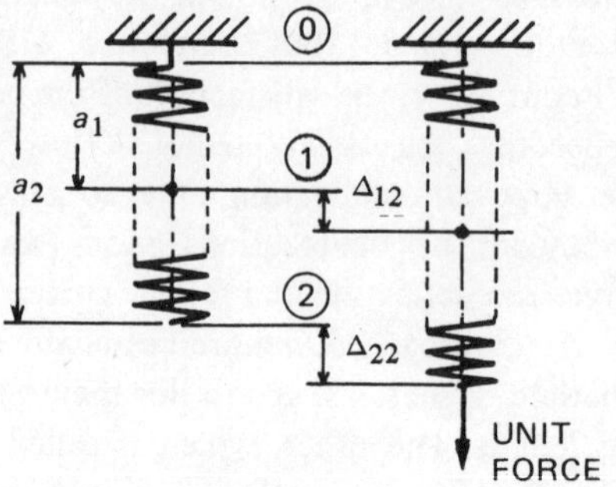

Fig. 11.19

But with beams, we are concerned with both slopes and deflections that result from both forces and couples. We can therefore distinguish four kinds of flexibility coefficient, and these can be distinguished symbolically by indicating slopes and deflections by θ and Δ when they arise from forces, and by α and δ when they arise from couples, as indicated in Fig. 11.20. But a distributed load is a special case in that it is applied over a specific range rather than at a specific

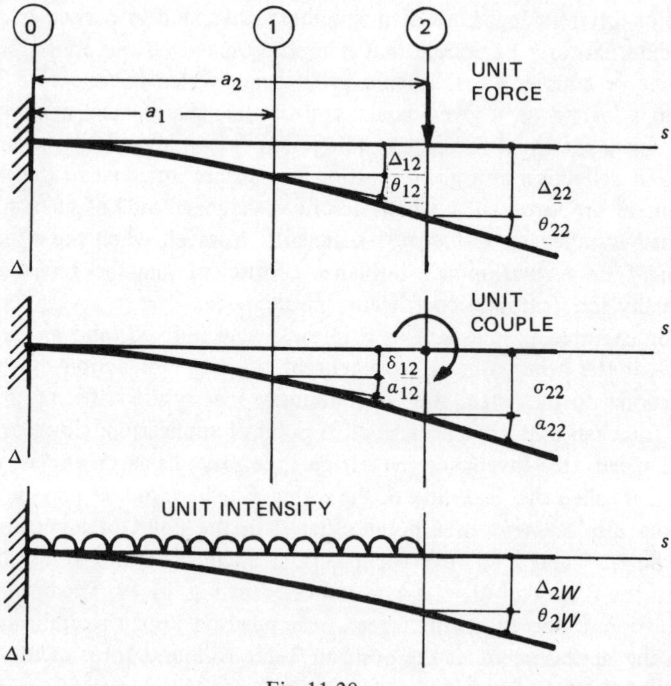

Fig. 11.20

point. On the other hand, we are presently concerned only with the slope and deflection at the termination of a uniformly distributed load which starts at the datum, and these may be represented by θ_{2w} and Δ_{2w}, respectively.

Of immediate interest are the six coefficients which describe the slope and deflection at the point at which the load is applied, of a simple cantilever of uniform EI when it is loaded with a unit force, with a unit couple, and with a uniformly distributed load of unit intensity, and the necessary calculations can conveniently be tabulated as follows.

	UNIT COUPLE 	UNIT FORCE 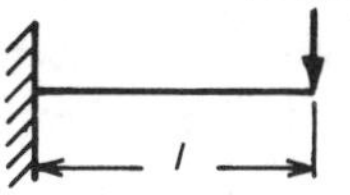	UNIT INTENSITY 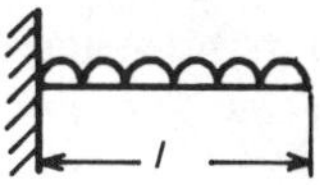
M	1	$(l-s)$	$\dfrac{(l-s)^2}{2}$
$EI\theta$	$s + C_1$	$-\dfrac{(l-s)^2}{2} + C_1$	$-\dfrac{(l-s)^3}{6} + C_1$
$EI\Delta$	$\dfrac{s^2}{2} + C_1 s + C_2$	$\dfrac{(l-s)^3}{6} + C_1 s + C_2$	$\dfrac{(l-s)^4}{24} + C_1 s + C_2$

Since $\theta = 0$, and $\Delta = 0$, when $s = 0$:

C_1	0	$\dfrac{l^2}{2}$	$\dfrac{l^3}{6}$
C_2	0	$-\dfrac{l^3}{3}$	$-\dfrac{l^4}{24}$

Substituting accordingly, and putting $s = l$ then determines the slopes and deflections at the load:

$EI\theta_{ll}$	l	$EI\alpha_{ll}$	$\dfrac{l^2}{2}$	$EI\theta_{lw}$	$\dfrac{l^3}{6}$
$EI\Delta_{ll}$	$\dfrac{l^2}{2}$	$EI\delta_{ll}$	$\dfrac{l^3}{3}$	$EI\Delta_{lw}$	$\dfrac{l^4}{8}$

Given a simple cantilever of uniform EI and length l, consider the slope θ_f and the deflection Δ_f at its free end, when it is subjected to an end couple M, and an end force F, and a uniformly distributed load of intensity w. The effect of each of the loads is proportional to the load, and the effects of the different loads are simply additive, so the net slope and deflection at the end of the cantilever can be determined as the sum of the products of M, F and w with the appropriate coefficients, as follows:

$$\begin{bmatrix} \theta_f \\ \Delta_f \end{bmatrix} = \frac{1}{EI} \begin{bmatrix} l & \frac{l^3}{2} & \frac{l^3}{6} \\ \frac{l^2}{2} & \frac{l^3}{3} & \frac{l^4}{8} \end{bmatrix} \begin{bmatrix} M \\ F \\ w \end{bmatrix} \tag{11.9}$$

But this determines the slope and deflection relative to the tangent to the centre-line at the support, and if this suffers a displacement Δ_i and a rotation

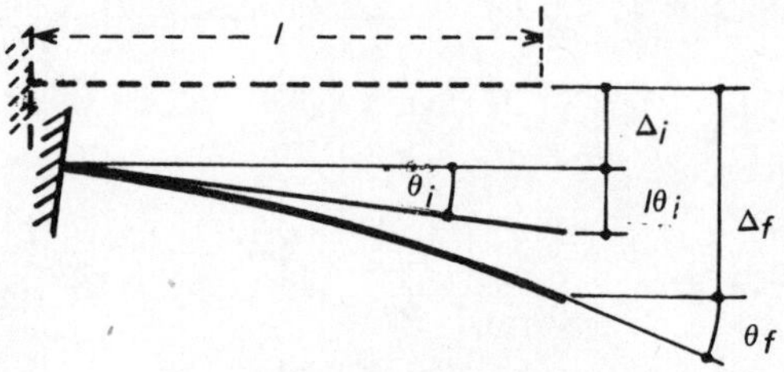

Fig. 11.21

θ_i, it is clear from Fig. 11.21 that the end slope is increased by θ_i, and the end deflection by $\Delta_i + l\theta_i$. Then:

$$\begin{bmatrix} \theta_f \\ \Delta_f \end{bmatrix} = \begin{bmatrix} 1 & 0 \\ l & 1 \end{bmatrix} \begin{bmatrix} \theta_i \\ \Delta_i \end{bmatrix} + \begin{bmatrix} l & \frac{l^2}{2} & \frac{l^3}{6} \\ \frac{l^2}{2} & \frac{l^3}{3} & \frac{l^4}{8} \end{bmatrix} \begin{bmatrix} M \\ F \\ w \end{bmatrix} \tag{11.10}$$

Moment Area Method

Attention being confined to a range of a beam in which the equation of M/EI is smoothly continuous, let two arbitrary sections 1 and 2 be identified by their distances a_1 and a_2 from the datum, and, the beam being treated as an aggregation of elemental slices, consider the effect on the slope and deflection at 2, relative to the tangent to the centre-line at 1, of the deformation of a single, elemental slice of infinitesimal length ds at the distance s from the datum (Fig. 11.22). The relative rotation between the sections at s and $s + ds$ is given by:

$$d\theta = \frac{M}{EI} ds ,$$

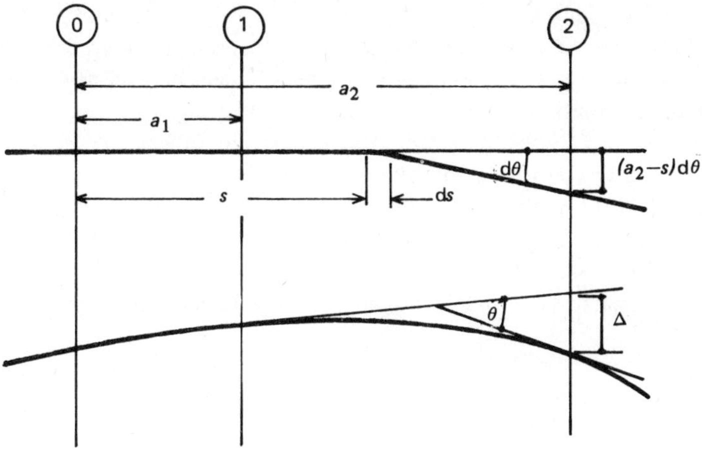

Fig. 11.22

and the total slope at 2 relative to the tangent at 1 can be determined as the sum of the relative rotations between the ends of all the elements between $s = a_1$ and $s = a_2$, as follows:

$$\theta = \int_{a_1}^{a_2} \frac{M}{EI} \mathrm{d}s \tag{11.11}$$

As to the displacement, we have argued that over the infinitesimal length ds, the effect on the displacement at 2 of the relative displacement between the sections at s and $s + \mathrm{d}s$ is negligible compared with that of their relative rotation. Thus, as indicated in Fig. 11.22, the effect of the deformation of the single element on the displacement at 2 is given by:

$$\mathrm{d}\Delta = (a_2 - s)\mathrm{d}\theta = (a_2 - s)\frac{M}{EI}\mathrm{d}s ,$$

and the total displacement at 2 relative to the tangent at 1 is therefore given by

$$\Delta = \int_{a_1}^{a_2} (a_2 - s)\frac{M}{EI}\mathrm{d}s \tag{11.12}$$

We note that the integral in eqn. (11.11) determines the area under the graph of M/EI between $s = a_1$ and $s = a_2$, whilst the integral in eqn. (11.12) determines the moment of this area about $s = a_2$, and it follows that, relative to the tangent to the centre-line at any position $s = a_1$, the slope at any other position $s = a_2$ may be determined as the area under the graph of M/EI between $s = a_1$ and $s = a_2$, whilst the displacement at $s = a_2$ can be determined as the moment of this area about $s = a_2$, and reflection will show that this will still be true even when the equation of M/EI is not smoothly continuous.

Method of Sections

Consider a case in which the equation of M/EI exhibits a number of discontinuities, and in which it is required to determine the slope or deflection at a few specific points only. In practice, these will usually be included among the positions at which the discontinuities occur, but otherwise, any such points are added, and conceiving the beam to comprise the succession of finite bays into which it is thus divided, we imagine the bays to be separated, by parting the member on the transverse sections immediately to the negative side of each position (Fig. 11.23). Each cut exposes two faces on which the shear force and bending moment on the section are revealed as equal but opposite pairs of

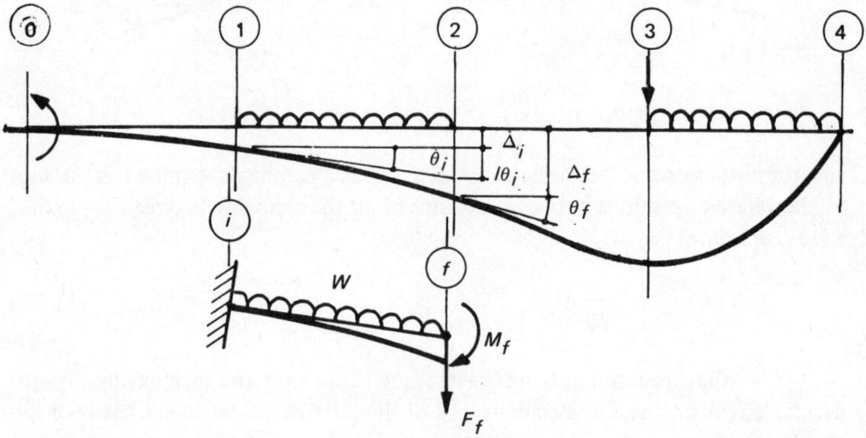

Fig. 11.23

external forces and couples, and if the suffices i and f are used to distinguish the values of quantities that attach to the sections on which the ends of a typical bay are formed, the final end of the bay will be liable to a force equal to the shear force F_f, and a couple equal to the bending moment M_f, whilst the initial end will be liable to a force equal to the shear force F_i and a couple equal to the bending moment M_i. In addition, the initial end will be liable to any singularities in the external loads, but otherwise, the bay will be additionally liable only to a distributed load that is smoothly continuous from end to end. Reflection will show, that each bay can thus be conceived as a simple cantilever which is supported, from its initial section, and that relative to the tangent to the centre-line at its initial point, the deformation of the bay will be that of a simple cantilever which carries an end-force equal to F_f, an end-couple equal to M_f, and a load which is smoothly distributed along its length, and whose intensity may be represented by w.

In principle, both w and EI may be smoothly variable. However, in practice EI is usually constant, and even where a variable EI is used, the variation is usually discontinuous. As to the distributed load, it is seldom realistic to specify a distribution other than uniform because the actual distribution of such a load is commonly largely fortuitous. In practice, therefore, the bays can usually be so chosen that within each bay the values of both w and EI can be specified as constants, and then the slope and deflection of the final point are given by eqn. (11.10) as follows:

$$\begin{bmatrix} \theta_f \\ \Delta_f \end{bmatrix} = \begin{bmatrix} 1 & 0 \\ l & 1 \end{bmatrix} \begin{bmatrix} \theta_i \\ \Delta_i \end{bmatrix} + \begin{bmatrix} l & \dfrac{l^2}{2} & \dfrac{l^3}{6} \\ \dfrac{l^2}{2} & \dfrac{l^3}{3} & \dfrac{l^4}{8} \end{bmatrix} \begin{bmatrix} M_f \\ F_f \\ w \end{bmatrix}$$

Otherwise, for any bay in which w or EI is variable, the corresponding flexibility coefficients may be determined either by formal integration of eqns. (11.11) and (11.12) or by the equivalent moment-area concept.

Evidently for the bay whose initial point lies at the datum, θ_i and Δ_i are zero. At any other section, the final slope and deflection in one bay become the initial slope and deflection of the succeeding bay, and the slope and deflection at the various sections can therefore be determined by adding the effects in the different bays in succession.

Example 11.17

A simple cantilever of uniform EI and length l carries a point load W at the distance a from the support A. Determine the slope and deflection at the end B.

Solution

The bending moment diagram is as indicated, and EI being constant, $EI\theta_B$ can

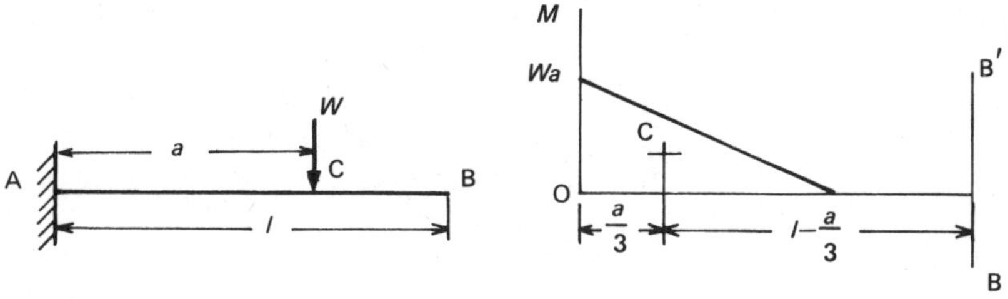

be equated to the area under the diagram between A and B, whilst $EI\Delta_B$ can be equated to the moment of the area about BB′.

Thus: $$EI\,\theta_B = \frac{Wa^2}{2}; \quad EI\,\Delta_B = \frac{Wa^2}{2}\left(l - \frac{a}{3}\right).$$

Example 11.18

The diagram indicates the loads on a simple cantilever in which EI varies from

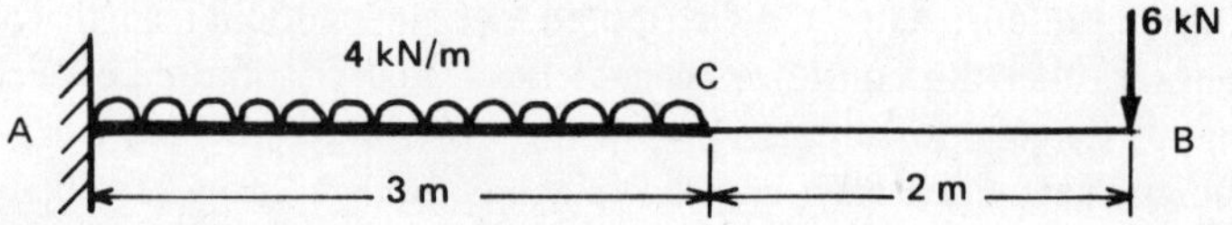

3200 (kN m^2) along AC to 1500 (kN m^2) along CB. Determine the slopes and deflections at C and B.

Solution

The beam is considered to be divided on the section immediately to the negative side of C.

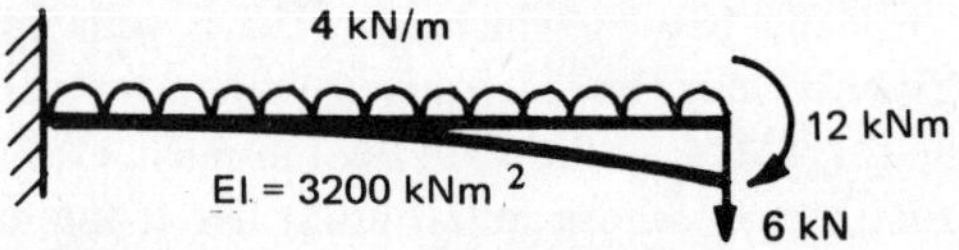

Bay AC: $l = 3$ (m); $EI = 3200$ (N m^2); $\theta_i = \Delta_i = 0$;

$M_f = 12$ (kN m); $F_f = 6$ (kN); $w = 4$ (kN/m)

$$EI\theta_c = M_f l + F_f \frac{l^2}{2} + w\frac{l^3}{6} = 36 + 27 + 18 = 81$$

$$\theta_c = 81/3200 = 0{\cdot}025 \text{ (rad)} .$$

$$EI\Delta_c = M_f\frac{l^2}{2} + F_f\frac{l^3}{3} + w\frac{l^4}{8} = 54 + 54 + 40{\cdot}5$$

$$\Delta_c = 148/3200 = 0{\cdot}046 \text{ (m)} .$$

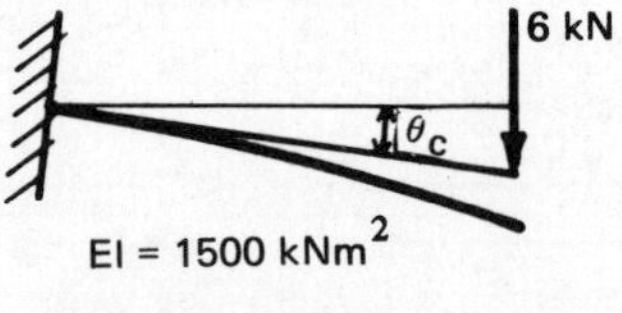

Bay CB: $l = 2$ (m); $EI = 1500$ (N m^2); $\theta_i = 0{\cdot}025$

$\Delta_i = {\cdot}046$ (m); $M_f = 0$; $F_f = 6$ (kN); $w = 0$.

$$\theta_c = \theta_i + \frac{1}{EI}\left(6 \,.\, \frac{l^2}{2}\right) = 0{\cdot}025 + {\cdot}008 = 0{\cdot}033 \text{ (rad)}$$

$$\Delta_c = \Delta_i + l\theta_i + \frac{1}{EI}\left(6 \,.\, \frac{2^3}{3}\right) = 0{\cdot}046 + {\cdot}050 + {\cdot}011 = 0{\cdot}107 \text{ (m)} .$$

Example 11.19

For the cantilever of Example 11.18, determine the deflection at the end B for each of the loading cases indicated in the following diagrams.

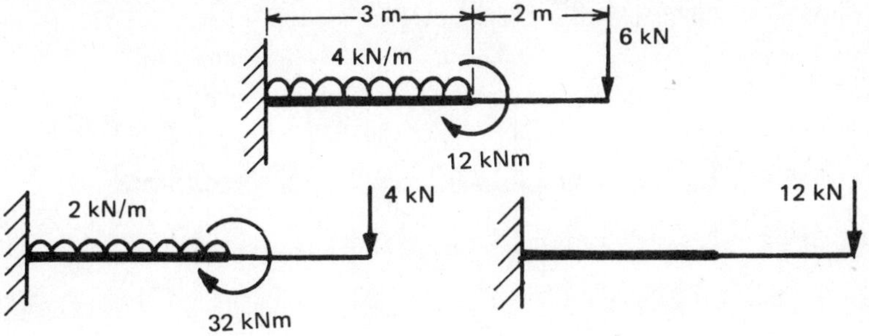

Solution

$$\Delta_B = (\Delta_f - \Delta_i)_{AC} + (\Delta_f - \Delta_i)_{CB} + 2(\theta_f - \theta_i)_{AC} \ .$$

In each bay, a common set of flexibility coefficients apply to all loading cases, and the calculations may conveniently be tabulated as follows.

BAY		AC			CN		
l	m	3			2		
$\begin{bmatrix} l & l^2/2 & l^3/6 \\ l^2/2 & l^3/3 & l^4/8 \end{bmatrix}$		$\begin{bmatrix} 3{\cdot}0 & 4{\cdot}5 & 4{\cdot}5 \\ 4{\cdot}5 & 9{\cdot}0 & 10{\cdot}1 \end{bmatrix}$			$\begin{bmatrix} 2{\cdot}0 & 2{\cdot}0 & 1{\cdot}3 \\ 2{\cdot}0 & 2{\cdot}7 & 2{\cdot}0 \end{bmatrix}$		
LOAD CASE		1	2	3	1	2	3
M_f	kN m	24	40	24	0	0	0
F_f	kN	6	4	12	6	4	12
w	kN/m	4	2	0	0	0	0
$\theta_f - \theta_i$	EI	117	147	126	IRRELEVANT		
$\Delta_f - \Delta_i$	m	202	236	216	16	11	32
EI	kN m^2	3200			1500		
$\theta_f - \theta_i$	10^{-3}	36.6	45.9	39.4	IRRELEVANT		
$\Delta_f - \Delta_i$	mm	63.1	73.8	67.5	10.7	7.3	78.8
$l\theta_i$	mm	0	0	0	73.2	91.8	78.8
Δ_B	mm				147	173	168

Example 11.20

In the cantilever shown in the diagram, EI (kN m^2) varies from 4600 between A and D, to 3000 between D and B. Determine the slope and deflection at each discontinuity.

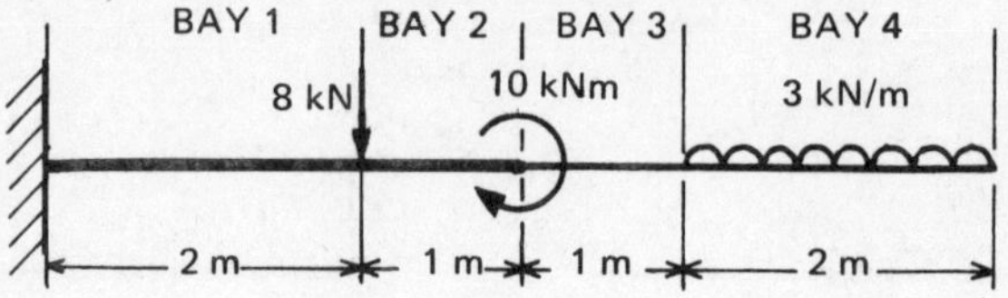

		Bay 1	Bay 2	Bay 3	Bay 4
EI	kN m^2	4600	4600	3000	3000
l	m	2	1	1	2
w	kN/m	0	0	0	3
F	kN	14	6	6	0
M	kN m	28	22	6	0
$wl^3/6$	kN m^2	0	0	0	4
$Fl^2/2$	kN m^2	28	3	3	0
Ml	kN m^2	56	22	6	0
Σ	kN m^2	84	25	9	4
$\dfrac{\Sigma}{EI} = \theta_f - \theta_i$:	10^{-3}	18·3	5·4	3·0	1·3
θ_f	10^{-3}	18·3	23·7	26·7	28·0
$wl^4/8$	kN m^3	0	0	0	6·0
$Fl^3/3$	kN m^3	37·3	2·0	2·0	0
$Ml^2/2$	kN m^3	56·0	11·0	3·0	0
Σ	kN m^3	93·3	13·0	5·0	6·0
$\dfrac{\Sigma}{EI} = \Delta_f - \Delta_i$	mm	20·3	2·8	1·7	2·0
$l\theta_i$	mm	0	18·3	23·7	53·4
$\dfrac{\Sigma}{EI} + l\theta_i$	mm	20·3	21·1	25·4	55·4
Δ	mm	20·3	41·4	66·8	122·2

Example 11.21

A cantilever AC of uniform EI is simply supported at B, and carries an end load of 8 kN, as indicated. Determine the reaction at the support B, and the slope and deflection at the end C.

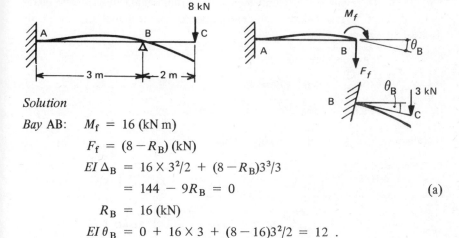

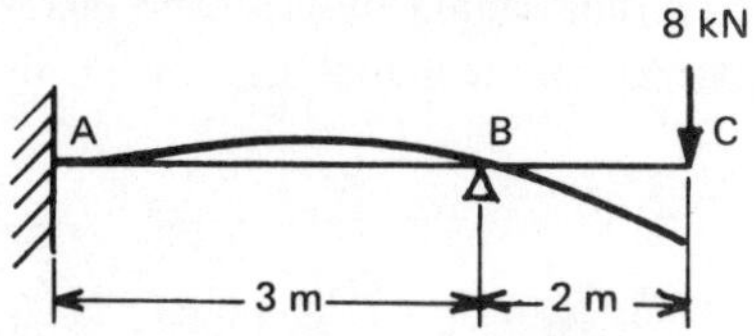

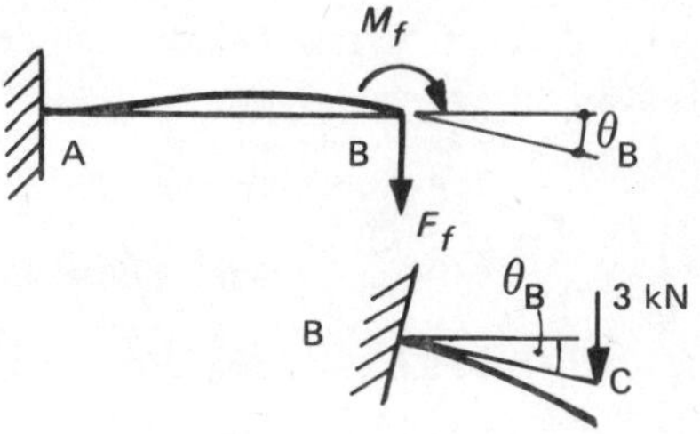

Solution

Bay AB: $M_f = 16$ (kN m)

$F_f = (8 - R_B)$ (kN)

$EI\,\Delta_B = 16 \times 3^2/2 + (8 - R_B)3^3/3$

$= 144 - 9R_B = 0$ (a)

$R_B = 16$ (kN)

$EI\,\theta_B = 0 + 16 \times 3 + (8 - 16)3^2/2 = 12$.

Range BC: $EI\theta_c = 12 + 8 \times 2^2/2 = 28$ (kN m²)

$EI\Delta_c = 12 \times 2 + 8 \times 2^3/3 = 45{\cdot}3$ (kN m³) .

Example 11.22

Repeat Example 11.21 for the case when the support at C is replaced by an elastic prop whose stiffness k (kN/m) is equal to one third the value of EI (kN m²) of the cantilever.

Solution

In the eqn. at (a), the deflection at B must now be equated to $R_B/k = 3R_B/EI$.

Thus: $144 - 9R_B = 3R_B$.

Therefore $R_B = 12$ (kN)

and $EI\theta_B = 0 + 16 \times 3 + (8 - 12)3^2/2 = 30$

Range BC: $EI\theta_c = 30 + 8 \times 2^2/2 = 46$ (kN m²)

$EI\Delta_c = 30 \times 2 + 8 \times 2^3/3 = 81{\cdot}3$ (kN m³) .

Example 11.23

Two positions 1 and 2 along a cantilever of uniform EI are identified by their distances a_1 and a_2 from the support ($a_1 < a_2$). Show that the displacement at 1 per unit load at 2 is equal to the displacement at 2 per unit load at 1, and determine their common value.

Solution

Let the displacement at 1 per unit load at 2 be represented by Δ_{12}. The increment of Δ_{12} that results from the deformation of an elemental slice of length ds at s is given by:

$$\begin{aligned} \mathrm{d}\Delta_{12} &= (a_1 - s)\mathrm{d}\theta \\ EI\mathrm{d}\Delta_{12} &= (a_1 - s)M\,\mathrm{d}s \\ &= (a_1 - s)\,(a_2 - s)\mathrm{d}s \end{aligned}$$

therefore $$EI\Delta_{12} = \int_0^{a_1} (a_1 - s)\,(a_2 - s)\mathrm{d}s\ .$$

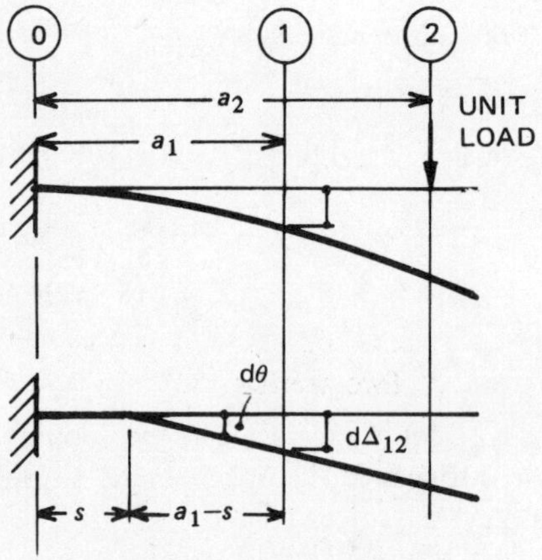

Now consider Δ_{21}, the displacement at 2 per unit load at 1.

$$\begin{aligned} \mathrm{d}\Delta_{21} &= (a_2 - s)\mathrm{d}\theta \\ EI\mathrm{d}\Delta_{21} &= (a_2 - s)M\,\mathrm{d}s \\ EI\Delta_{21} &= \int_0^{a_2} (a_2 - s)M\,\mathrm{d}s \\ &= \int_0^{a_1} (a_2 - s)M\,\mathrm{d}s \quad + \int_{a_1}^{a_2} (a_2 - s)M\,\mathrm{d}s\ . \end{aligned}$$

But for $0 < s < a_1$: $M = (a_1 - s)$,

and for $a_1 < s < a_2$: $M = 0$,

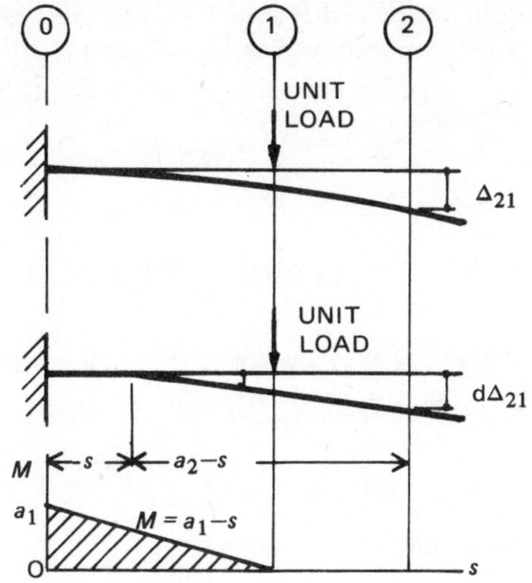

therefore $EI\,\Delta_{21} = EI\,\Delta_{12} = \displaystyle\int_0^{a_1} (a_1 - s)\,(a_2 - s)\,\mathrm{d}s$

$$= \frac{a_1^2(3a_2 - a_1)}{6} \qquad a_1 < a_2 .$$

Example 11.24

The diagram shows a beam of uniform *EI* and length *l* which is simply supported at the sections 0 and 3. Two arbitrary sections 1 and 2 are identified by their

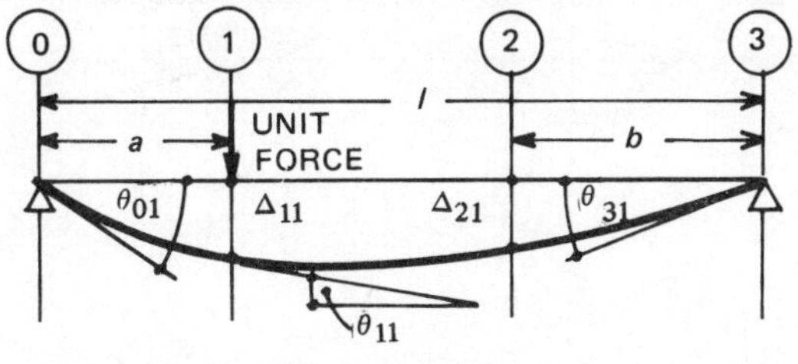

distances *a* and *b* from 0 and 3, $\{(a + b) < l\}$. Determine the following flexibility coefficients for a point force at the section 1:

$$\theta_{11};\ \Delta_{11};\ \theta_{01};\ \theta'_{31}, \text{ and } \Delta_{21} .$$

Also show that $\Delta_{21} = \Delta_{12}$, i.e. show that the displacement at 1 per unit load at 2 is equal to the displacement at 2 per unit load at 1.

Solution

The support reactions are statically determinate and by taking moments about axes in the sections 0 and 3, we obtain

$$R_o = (l - a)/l; \quad R_3 = a/l .$$

Relative to the tangent to the centre line at 1, the two bays comprise a pair of simple cantilevers which support end loads of $R_0 = (l-a)/l$; and $R_3 = a/l$ only, and by comparing the deflection at 1 relative to 0 and 3 we have:

$$\Delta_{11} = a\,.\,\theta_{11} + \frac{(l-a)}{l}\frac{a^3}{3} = -(l-a)\,\theta_{11} + \frac{a}{l}\frac{(l-a)^3}{3}$$

whence:
$$\theta_{11} = \frac{a}{3l}(l-a)(l-2a)$$

and:
$$\Delta_{11} = \frac{a^2}{3l}(l-a)^2$$

$$\theta_{01} = \theta_{11} + \frac{(l-a)}{l}\frac{a^2}{2} = \frac{a}{6l}(l-a)(2l-a)\ .$$

From symmetry (by substituting $(l-a)$ for a):

$$\theta'_{31} = \frac{a}{6l}(l^2-a^2)\ .$$

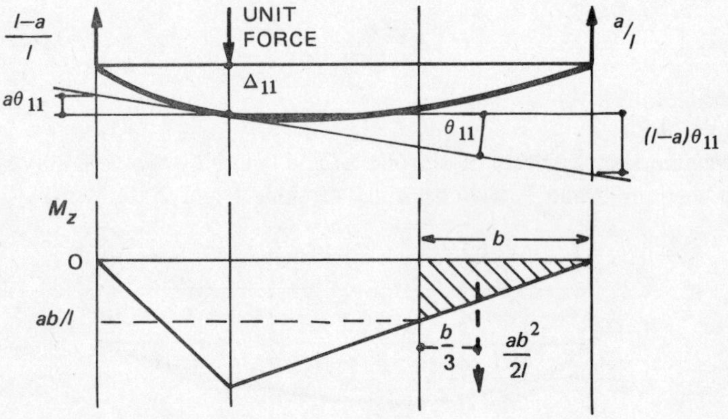

The deflection at 2 relative to the tangent of 3 being determined by the moment area method, we have:

$$\Delta_{21} = b\theta'_{31} - \frac{ab^2}{2l}\cdot\frac{b}{3}$$

$$\Delta_{21} = \frac{ab}{6l}(l^2-a^2-b^2)\ .$$

It is then obvious from symmetry that Δ_{12} (i.e., the deflection at 1 per unit load at 2) can be determined simply by substituting b for a and a for b in Δ_{12}, and from this it is obvious that $\Delta_{12} = \Delta_{21}$.

Example 11.25

A beam of uniform EI and length l is simply supported at its initial section 0 and its final section 2. An arbitrary intermediate section 1 is identified by its distance s from the initial section. Determine the flexibility coefficients for a moment at 0: α_{00}, α'_{20} and δ_{10}.

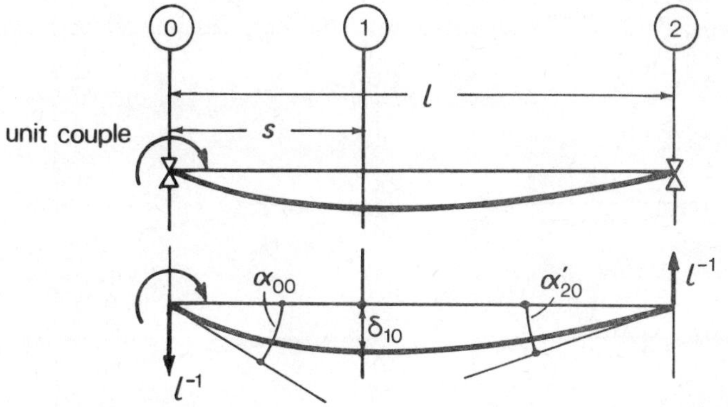

Solution

For equilibrium, the support reactions are both of magnitude l^{-1} in the senses indicated, whence

$$M = -\frac{(l-s)}{l}$$

therefore $$EI\theta = \frac{(l-s)^2}{2l} + C_1$$

and $$EI\Delta = -\frac{(l-s)^3}{6l} + C_1 s + C_2 \; .$$

But $\Delta = 0$, when $s = 0$ and $s = l$.

Therefore: $C_2 = l^2/6; \quad C_1 = -l/6$,

and substituting accordingly gives

$$EI\theta = \frac{(l-s)^2}{2l} - \frac{l}{6} = \frac{(2l^2 - 6ls + 3s^2)}{6l}$$

$$EI\delta_{10} = \frac{-(l-a)^3}{6l} - \frac{l}{6}a + \frac{l^2}{6} = \frac{a}{6l}(l-a)^2 \; .$$

Also, by substituting $s = 0$, and $s = l$ in $EI\theta$ we obtain

$$EI\alpha_{00} = l/3; \text{ and } EI\alpha_{20} = -l/6 \; .$$

Flexibility Method

Fig. 11.24 shows a beam which is encastré at the section O, and simply supported at section 1, and W typifies a plane parallel set of loads which may include point forces, distributed forces, and couples. Since the plane parallel set of forces having only two degrees of freedom includes three unknown components of support reaction, the case is singly redundant, and it follows that if any one of the three support restraints were removed, the beam would then be statically determinate under any coplanar, parallel set of loads. So,

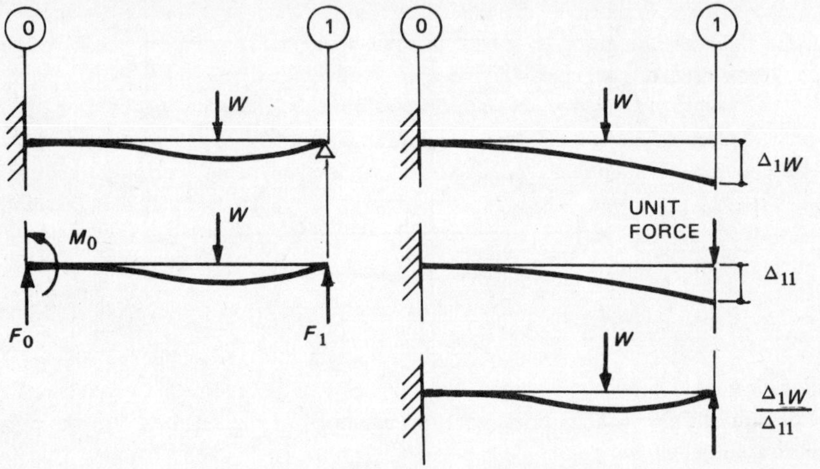

Fig. 11.24

choosing to regard, say, the force of reaction at the simple support as the redundant restraint, let the beam which results from its removal be called the released beam, and suppose that we calculate the deflection of this beam at the section 1 (i.e. at the section from which the redundant restraint was removed), first under the actual loads W, and then under a unit load applied at the position 1. The former may be represented by Δ_{1w}, whilst the latter is given by the flexibility coefficient Δ_{11}, and it follows from the principle of superposition, that the total deflection of the released beam at the section 1 due to the actual loads W, and an additional load F_1 at 1 would be given by $\Delta_1 = \Delta_{1w} + F_1\Delta_{11}$.

But suppose that the magnitude and sense of F_1 were adjusted so as to give Δ_1 the value which satisfies the kinematic constraints imposed by the support. Thus, in an elastic prop, the displacement Δ_1 in the direction opposed to F_1 is proportional to F_1. In other words $\Delta_1 = -F_1 f_1$, where f_1 represents the flexibility of the support itself, and we suppose that F_1 is given the value that satisfies the equation

$$\Delta_1 = -F_1 f_1 = \Delta_{1w} + F_1\Delta_{11} \, ,$$

or:

$$F_1 = -\frac{\Delta_{1w}}{\Delta_{11} + f_1} \tag{11.13}$$

But for a rigid support, both Δ_1 and f_1 are zero, and in that case we suppose that F_1 is given the value:

$$F_1 = -\frac{\Delta_{1w}}{\Delta_{11}} \tag{11.14}$$

Then reflection will show, that when this load F_1 is added to the actual loads W, the boundary conditions on the released beam will be precisely those on the propped cantilever under the actual loads W. The value of F_1 that is thus determined from calculations on the statically determinate, released beam only, must therefore be equal to the support reaction in the propped cantilever, and the remaining support reactions are then statically determinate.

Alternatively, either the linear restraint or the turning restraint at the section O could be treated as the redundant reaction, but in the case of a redundant couple, it is necessary to consider, not the deflection at its point of application, but the slope. Thus, the support couple M_o being treated as the redundant reaction, its removal leaves a released beam which is simply supported at both section O and section 1 (Fig. 11.25), and we consider its slope at O, first

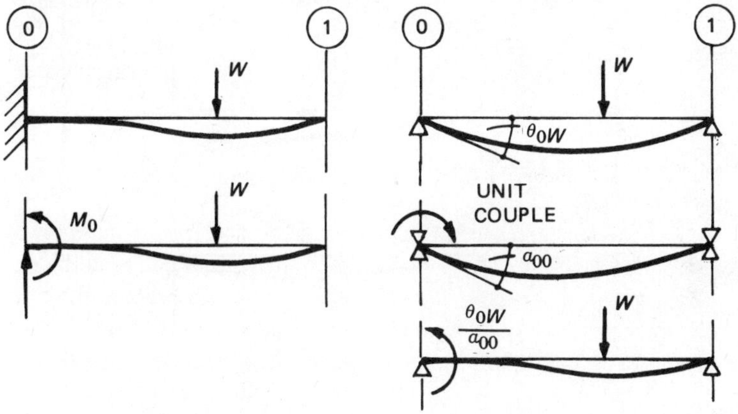

Fig. 11.25

under the loads W, and then under a unit point couple, applied at O. The slope at O due to the loads W may be represented by θ_{ow}, whilst the slope at O per unit couple at O is equal to the flexibility coefficient α_{oo}, and when an additional moment M_o is superimposed on the loads W, the net slope of the released beam at O is given by:

$$\theta_o = \theta_{ow} + M_o \alpha_{oo} \; .$$

But if f_o represents the flexibility of the turning restraint, (i.e. its rotation per unit moment M_o), we may substitute $\theta_o = -M_o f_o$, to obtain:

$$M_o = -\frac{\theta_{ow}}{\alpha_{oo} + f_o} \tag{11.15}$$

and this determines the value of the redundant support moment M_o in the propped cantilever.

Example 11.26

Determine the support reactions in the beam of Example 11.16.

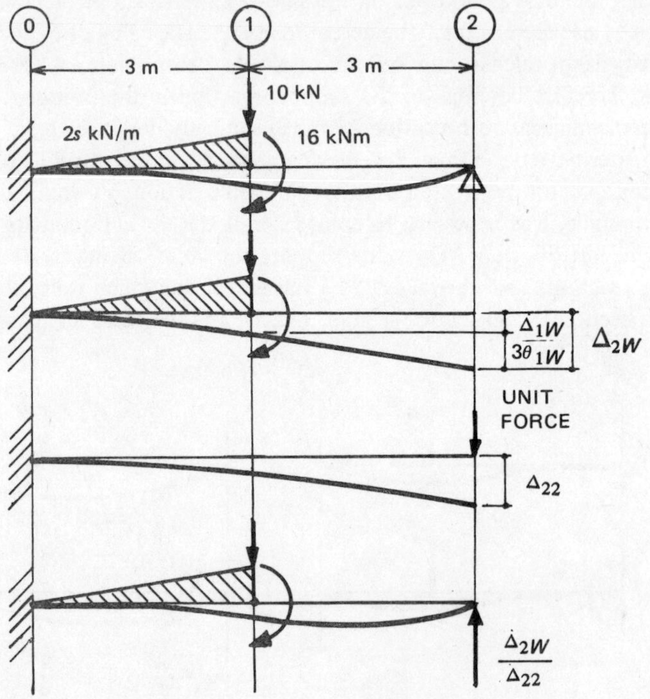

Solution

If the simple support is treated as the redundant, the released beam is a simple cantilever, and the reaction at 2 is given by: $F_2 = -\Delta_{2w}/\Delta_{22}$.

The deflection of the simple cantilever under the loads may be determined by any convenient method. For example:

$$\Delta_{2w} = \Delta_{1w} + 2\theta_{1w} ,$$

where θ_{1w} and Δ_{1w} may be determined by successive integration of the equation of loading for the left-hand bay, as follows:

$$W = 2s$$
$$-F = s^2 + C_1$$

$$M = EI\frac{d\theta}{ds} = \frac{s^3}{3} + C_1 s + C_2$$

$$EI\ \theta = \frac{s^4}{12} + \frac{C_1}{2}s^2 + C_2 s + C_3$$

$$EI\ \Delta = \frac{s^5}{60} + \frac{C_1}{6}s^3 + \frac{C_2}{2}s^2 + C_3 s + C_4\ .$$

Boundary Conditions:

when $s = 0$; then $\theta = 0$; therefore $C_3 = 0$

and $\Delta = 0$; therefore $C_4 = 0$

when $s = 3$; then $F = 10$; therefore $C_1 = -10 - 9 = -19$

and $M = 16$; therefore $C_2 = 16 - 9 + 57 = 64$.

Substituting accordingly, and setting $s = 3$, then determines θ_{1w} and Δ_{1w}, as follows:

$$EI\theta_{1w} = \frac{81}{12} - \frac{19}{2}\times 9 + 64\times 3 = 113{\cdot}25$$

and $$EI\Delta_{1w} = \frac{243}{60} - \frac{19}{6}\times 27 + \frac{64}{2}\times 9 = 207{\cdot}55$$

therefore $EI\Delta_{2w} = 208 + 3\times 113 = 547$

For a simple cantilever, 6 (m) long, $EI\Delta_{22} = l^3/3 = 72$.

But: $EI\Delta_2 = EI\Delta_{2w} + EI.F_2\Delta_{22} = 0$

therefore: $F_2 = -\Delta_{2w}/\Delta_{22} = -547/72 = -7{\cdot}6$ (kN)

and for equilibrium:

$$M_o = -9\times 2 - 3\times 10 - 16 + 7{\cdot}6\times 6 = -18{\cdot}5 \text{ (kN m)}$$

and: $$F_o = -9 - 10 + 7{\cdot}6 = -11{\cdot}4 \text{ (kN)}\ .$$

Example 11.27

A beam of uniform $EI = 3000$ (kN m^2) and of length 5 (m) is encastré at one end, and simply supported at the other. However, the turning restraint at the encastré support is not rigid, and has a flexibility of $(1/3)\times 10^{-3}$ (rad/kN m).

Determine the support moment at the encastré support under a point load of 10 (kN) at 3 (m) from that end.

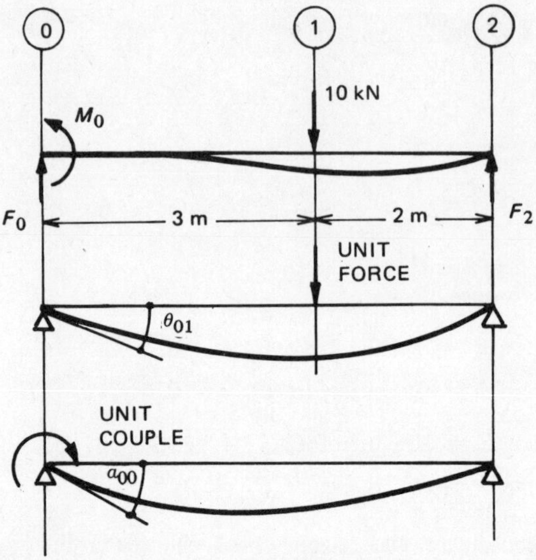

Solution

The support moment at O being treated as the redundant reaction, the released beam is simply supported at each end, and M_o is given by:

$$M_o = \frac{\theta_{ow}}{\alpha_{oo} + f_o} = -\frac{W_1\theta_{o1}}{\alpha_{oo} + f_o} .$$

But from the results of Examples 11.24 and 11.25, the values of θ_{o1} and α_{oo} for a simply supported beam are given by:

$$\theta_{o1} = \frac{a(l-a)(2l-a)}{6EI\,l} = \frac{3 \times 2 \times 7}{30 \times 3000} = \frac{14}{30} \times 10^{-3}$$

and $$\alpha_{oo} = \frac{l}{3EI} = \frac{5}{3 \times 3000} = \frac{5}{9} \times 10^{-3}$$

therefore: $$M_o = -\frac{10(14/30)}{(5/9) + (1/3)} = -5{\cdot}25 \text{ (kN m)} .$$

Multiple Redundancy

In a case of multiple redundancy, it is necessary to consider the slope at the position of each redundant moment, and the deflection at the position of each redundant force. Thus, the beam in Fig. 11.26 is encastré at section O, and simply supported at both section 1 and section 2, and again it supports a plane parallel set of loads, as typified by the forces W_a and W_b. Since the plane parallel set of forces now contains four unknown components of support reaction, the case is doubly redundant, and the two simple supports being taken as the redun-

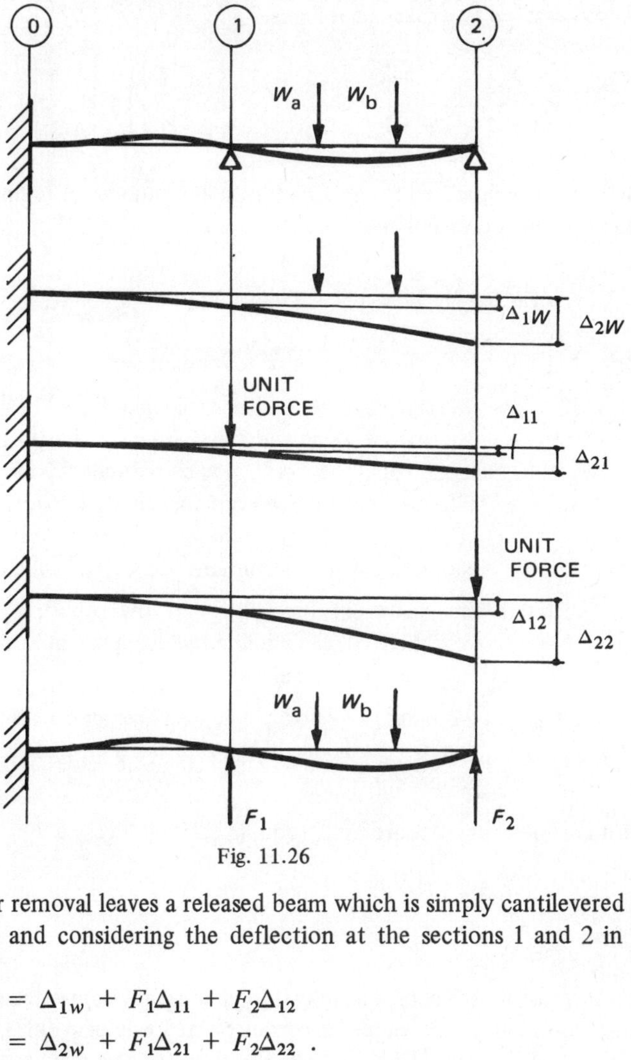

Fig. 11.26

dant pair, their removal leaves a released beam which is simply cantilevered from the section O, and considering the deflection at the sections 1 and 2 in turn, we have:

$$\Delta_1 = \Delta_{1w} + F_1\Delta_{11} + F_2\Delta_{12}$$

and
$$\Delta_2 = \Delta_{2w} + F_1\Delta_{21} + F_2\Delta_{22} \; .$$

Then if f_1 and f_2 represent the flexibilities of the supports at 1 and 2, we require to determine the values of F_1 and F_2 for which both $\Delta_1 = -F_1 f_1$, and $\Delta_2 = -F_2 f_2$, and these are given by the simultaneous solution of the two equations:

$$\Delta_1 = -F_1 f_1 = \Delta_{1w} + F_1\Delta_{11} + F_2\Delta_{12}$$

and
$$\Delta_2 = -F_2 f_2 = \Delta_{2w} + F_1\Delta_{21} + F_2 \text{[illegible]}$$

or:
$$\left\{\begin{bmatrix} \Delta_{11} & \Delta_{12} \\ \Delta_{21} & \Delta_{22} \end{bmatrix} + \begin{bmatrix} f_1 & 0 \\ 0 & f_2 \end{bmatrix}\right\} \begin{bmatrix} F_1 \\ F_2 \end{bmatrix} = - \begin{bmatrix} \Delta_{1w} \\ \Delta_{2w} \end{bmatrix} \qquad (11.16)$$

This result may be re-written as follows:

$$\begin{bmatrix} F_1 \\ F_2 \end{bmatrix} = -\left\{ \begin{bmatrix} \Delta_{11} & \Delta_{12} \\ \Delta_{21} & \Delta_{22} \end{bmatrix} + \begin{bmatrix} f_1 & 0 \\ 0 & f_2 \end{bmatrix} \right\}^{-1} \begin{bmatrix} \Delta_{1w} \\ \Delta_{2w} \end{bmatrix} \tag{11.17}$$

and it may be generalised for application to any number of redundancies, in any combination of forces and couples as follows:

$$[\mathbf{R}] = -\ [[\Delta_{\mathrm{B}}] + [\mathbf{f}]]^{-1}\,[\Delta_w] \tag{11.18}$$

where: $[\mathbf{R}]$ = the column vector of redundant support reactions.

$[\Delta_{\mathrm{B}}]$ = the matrix of flexibilities for the released beam, in which the ith row contains the flexibilities at the position of the ith redundant for loading at each redundancy in turn. When R_i is a force, the ith row contains the deflections at the position of R_i per unit of each redundancy, whilst when R_i is a couple, the ith row contains the slopes at the position of R_i.

$[\mathbf{f}]$ = the matrix of flexibilities of the supports, in which the flexibility of each redundant support in turn is entered on the leading diagonal.

$[\Delta_w]$ = the column vector of the displacements of the released beam due to the loads W, at the position of each redundant reaction in turn.

When all the supports are rigid, (eqn. (11.18) reduces to

$$[\mathbf{R}] = -\ [\Delta_{\mathrm{B}}]^{-1} \cdot [\Delta_w] \tag{11.19}$$

Example 11.28

A cantilever of uniform EI, and 6 (m) long, is simply supported at its 'free' end, and at the middle of its length. It carries point loads of 8 (kN), 4 (kN) and 10 (kN) at 1 (m), 2 (m) and 5 (m) from the encastré support, respectively. Determine the reactions at the simple supports.

Solution

The two props being treated as the redundants, the released beam is a simple cantilever, for which Example 11.23 determines the flexibility coefficient for any two sections at $s = a_1$ and $s = a_2$, $(a_1 < a_2)$, as follows:

$$\Delta_{12} = \Delta_{21} = a_1^2(3a_2 - a_1)/6$$

Then $\quad [\mathbf{f}] \cdot [\mathbf{R}] = -\ [\Delta_w]\ ,$

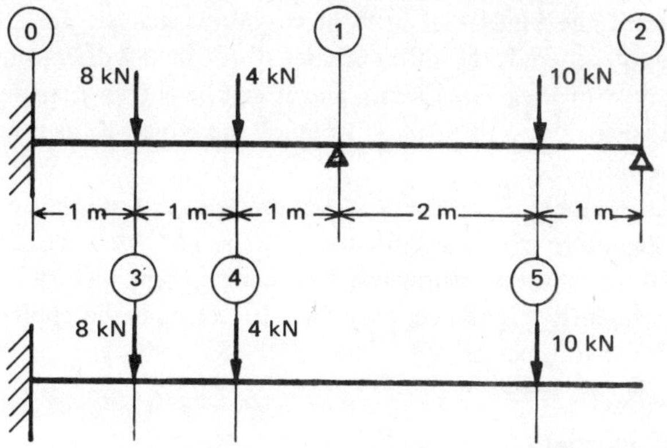

where $[\Delta_w]$, the vector of displacements at the positions of the redundants under the loads W, may be determined by any convenient means. For example, since all the loads, like the two redundants, are specified as point forces, they may in the present case conveniently be determined from the known flexibility coefficient alone. Thus, the sections at which the loads are applied being numbbered 3, 4 and 5 in turn, we have

$$\begin{bmatrix} \Delta_{11} & \Delta_{12} \\ \Delta_{21} & \Delta_{22} \end{bmatrix} \begin{bmatrix} R_1 \\ R_2 \end{bmatrix} = - \begin{bmatrix} \Delta_{13} & \Delta_{14} & \Delta_{15} \\ \Delta_{23} & \Delta_{24} & \Delta_{25} \end{bmatrix} \begin{bmatrix} W_3 \\ W_4 \\ W_5 \end{bmatrix}$$

$$\begin{bmatrix} 9 & 45/2 \\ 45/2 & 72 \end{bmatrix} \begin{bmatrix} R_1 \\ R_2 \end{bmatrix} = - \begin{bmatrix} 4/3 & 14/3 & 18 \\ 17/6 & 32/3 & 325/6 \end{bmatrix} \begin{bmatrix} 8 \\ 4 \\ 10 \end{bmatrix} = - \begin{bmatrix} 209\tfrac{1}{3} \\ 607 \end{bmatrix}$$

Whence $R_1 = -10{\cdot}05$ (kN); $R_2 = -5{\cdot}30$ (kN) .

11.9 SINGULARITY FUNCTIONS IN THE ANALYSIS OF BENDING

Where we are concerned with the deformation only as far as is necessary for the determination of redundant support reactions, the methods of sections and of flexibility thus avoid the inconvenience that results from a multiplicity of constants of integration by considering the slope or deflections only at the positions at which the redundant reactions are applied. However, where an interest in the slope and deflection equations themselves remains, then so does the inconvenience, but it may then be overcome through the concept of singularity functions. These are defined as functions which specifically include discontinuities of the kinds that arise in structural analysis, for although such functions cannot generally be differentiated through the discontinuity, we are concerned only with their integration, and it can be shown that the singularity functions are amenable to integration in much the same way as functions which are smoothly continuous.

The functions include one set of interrupted power functions which may be used to describe interrupted distributed quantities. To these are added two point functions, and since these distinguish between a point quantity and a point couple of the quantity, it is necessary first to consider the application of the concepts of statics to functions.

The Statics of Functions

Consider a given function f of a single scalar s, over a given range of the independent variable from $s = s_1$ to $s = s_2$, as indicated in Fig. 11.27. If the range is divided into a number of small intervals, the value of the function over a typical, infinitesimal interval from s to $s + \mathrm{d}s$ can be indicated by f, and the integral of f, over the range from s_1 to s_2 is defined as the limiting sum of the products $f.\mathrm{d}s$ for all the intervals that make up the range. Then areas in the positive domain being reckoned positive, and areas in the negative domain being reckoned negative, it follows that, between any given limits, the integral of the function may be equated to the area under the graph of the function, and this is regarded as the resultant of the given function between the given limits.

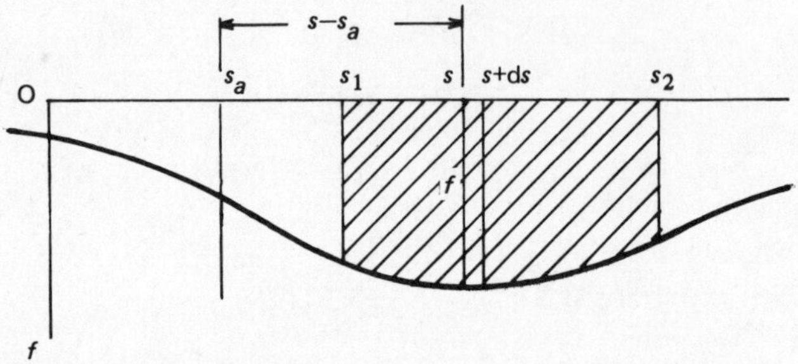

Fig. 11.27

But, identifying a specific value of the independent variable $s = s_a$, now consider the product of $(s - s_a)$ with $f.\mathrm{d}s$. Since $f.\mathrm{d}s$ is the resultant of f over the range $\mathrm{d}s$, and $(s - s_a)$ is the 'distance' of the element from $s = s_a$, the product $(s - s_a) \, . \, f.\mathrm{d}s$ determines the moment of the resultant of the element about $s = s_a$, and between any given limits, it follows, first, that the integral of the elemental moment determines the resultant moment of the function, and second, that this can be equated to the moment of the area under its graph about $s = s_a$.

The Singular Power Functions

Assuming a finite situation, let the zero of the independent variable s be chosen just to the negative side of the position of the first force or couple, so that all the forces and couples fall in the positive domain of s, and consider a set of singularity functions whose members are indicated symbolically by $\langle s - a\rangle^n$, where n may be either zero, or any positive integer. They are defined by first dividing the range of s from $s = 0$ to $s = \infty$ at a specific value $s = a$, when, for values of s less than a the value of $\langle s - a\rangle^n$ is defined to be zero, whilst for all values of s greater than a the value of the singularity function is defined to be equal to that of the smoothly continuous function $(s - a)^n$. Such is illustrated in Fig. 11.28, and it can evidently be regarded as a smoothly continuous power

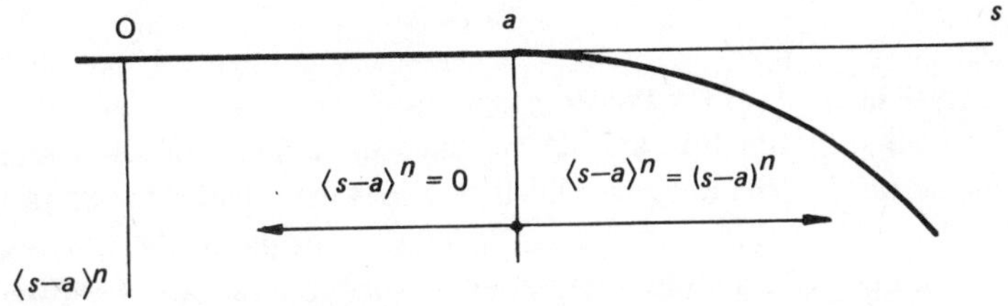

Fig. 11.28

function which is blind to negative values of its argument $(s - a)$, and it may be defined algebraically as follows:

$$\langle s - a\rangle^n = \begin{cases} 0 & \text{when } (s-a) \text{ is negative} \\ (s-a)^n, & \text{when } (s-a) \text{ is positive}\ . \end{cases}$$

The Unit Step Function

In the first function of this series, n is zero, and since $(s - a)^0$ is equal to unity for all positive values of $(s - a)$, it follows that $\langle s - a\rangle^0$ is a function whose value makes a step change from zero, for all negative values of $(s - a)$, to unity for all positive values of $(s - a)$. It is therefore called the unit step function, and its graph is indicated in Fig. 11.29.

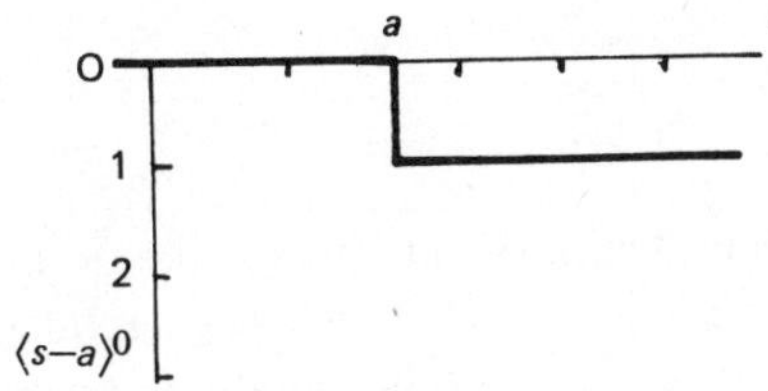

Fig. 11.29

The Unit Ramp Function

In the second function of the series, n is unity, and for all positive values of $(s-a)$, the value of $(s-a)^1$ is equal to that of $(s-a)$ itself. $\langle s-a\rangle^1$ is therefore a function whose graph for positive values of $(s-a)$ is a straight line of unit slope, as indicated in Fig. 11.30, and this is called the unit ramp function.

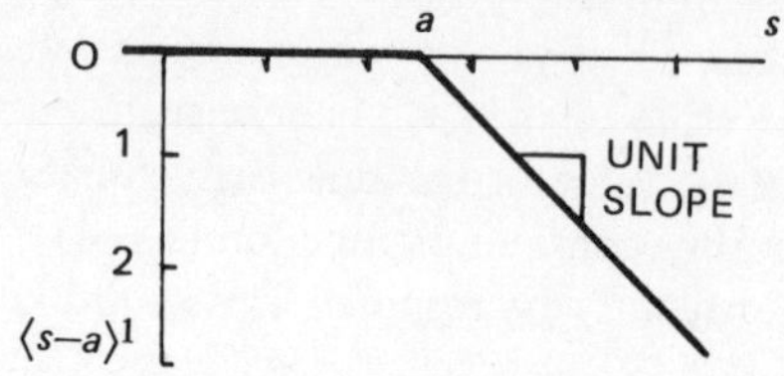

Fig. 11.30

The Indefinite Integrals of the Power Functions

If the integrals of a function are determined from a common lower limit to a variable upper limit s, the function which describes the value of the integrals as a function of the upper limit is called an indefinite integral of the function, and we now consider the indefinite integral of $\langle s-a\rangle^n$ for the specific lower limit at $s = 0$. Since the value of the function is zero for all values of s less than a, the values of the integrals of $\langle s-a\rangle^n$ are also zero. On the other hand, for values of s greater than a, the integral from 0 to s, can be equated to the sum of the integrals from 0 to a, and from a to s. But over the range from 0 to a the value of $\langle s-a\rangle^n$ is zero, whilst for values between a and s the value of $s-a^n$ is equal to that of $(s-a)^n$. The integral of the singularity function from 0 to s, can therefore be equated to that of the smoothly continuous function from a to s, and we have:

$$\int_0^s \langle s-a\rangle^n \mathrm{d}s = \begin{cases} 0 & \text{, when } (s-a) \text{ is negative} \\ \displaystyle\int_a^s (s-a)^n \mathrm{d}s = \frac{(s-a)^{n+1}}{n+1} & \text{, when } (s-a) \text{ is positive .} \end{cases}$$

The integral of $\langle s-a\rangle^n$ is therefore another singularity function of the same kind as $\langle s-a\rangle^n$ itself, and

$$\int_0^s \langle s-a\rangle^n = \frac{\langle s-a\rangle^{n+1}}{n+1} \quad .$$

Thus, the lower limit being set at $s = 0$, the form of the relationship between the singular power functions and their singular integrals is identical with that between the corresponding smoothly continuous power functions and their smoothly continuous integrals.

The Unit Point Function

Now consider two functions $\langle f \rangle$ and $\langle c \rangle$ whose values are non-zero only over a small, variable range ϵ to each side of a specific value, $s = a$, of the independent variable. Within this range, the value of $\langle f \rangle$ first increases at a uniform rate, and then decreases at the same rate, as indicated in Fig. 11.31, and this rate of change is varied in inverse proportion to ϵ, so that at each value of ϵ the value of $\langle f \rangle$ at $s = a$ is equal to $1/\epsilon$. Then it is obvious, that irrespective of the value of ϵ, the area under the graph of $\langle f \rangle$ over the full range from $s = 0$ to $s = \infty$ is equal to unity, whilst the moment of this area about $s = a$ is zero. The complete function is therefore statically equivalent to a unit resultant at $s = a$, and it follows that as ϵ tends to zero, the distributed function $\langle f \rangle$ tends to a unit point function at $s = a$. This limiting function is indicated symbolically by $\langle s - a \rangle_{-1}$.

Though it is not immediately obvious how the concept of integration is to be applied to a function whose value is non-zero only at one specific value of the independent variable, since $\langle s - a \rangle_{-1}$ is defined as a limit of the distributed function $\langle f \rangle$, the integral of $\langle s - a \rangle_{-1}$ can be determined as the limit of the integral of $\langle f \rangle$. Thus, for all values of the upper limit less than $(a - \epsilon)$ the integrals of $\langle f \rangle$ from $s = 0$ are zero, whilst for all values of the upper limit greater than $(a + \epsilon)$ the values of the integrals are equal to the unit area under the graph of $\langle f \rangle$. The graph of the indefinite integral of $\langle f \rangle$ therefore has the form indicated in Fig. 11.31, and from this it is obvious that in the limit, as ϵ tends to zero, and $\langle f \rangle$ tends to $\langle s - a \rangle_{-1}$, then the integral of $\langle f \rangle$ tends to the unit step function $\langle s - a \rangle^0$.

Thus: $$\int_0^s \langle s - a \rangle_{-1} = \langle s - a \rangle^0 .$$

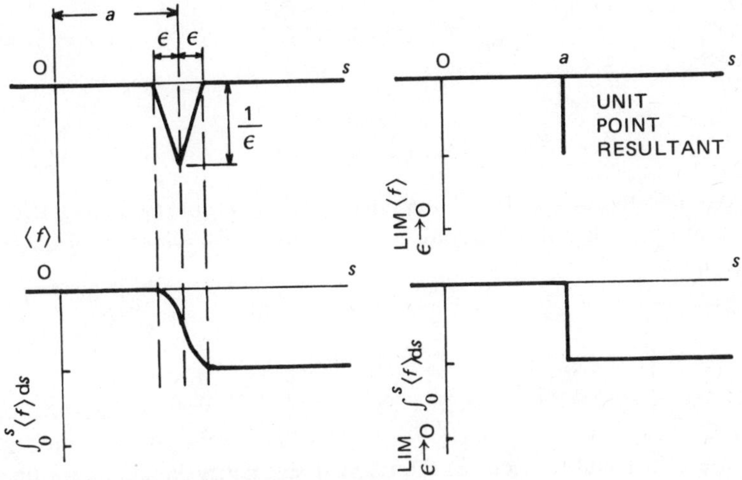

Fig. 11.31

The Unit Couplet

The second function $\langle c \rangle$ is defined as one in which a uniform negative value to the negative side of $s = a$ is matched with a uniform positive value of equal magnitude to the positive side, and in this case the magnitude of the function is varied so that its value at each value of ϵ is equal to $1/\epsilon^2$ (Fig. 11.32). Evidently, whereas the resultant of $\langle c \rangle$ is zero, its resultant moments about different values of s have a common value equal to unity. The function $\langle c \rangle$ is therefore statically equivalent to a unit couple, and in the limit, as ϵ tends to zero, $\langle f \rangle$ tends to a unit pure couple at $s = a$. This limiting function is indicated symbolically by $\langle s = a \rangle_{-2}$.

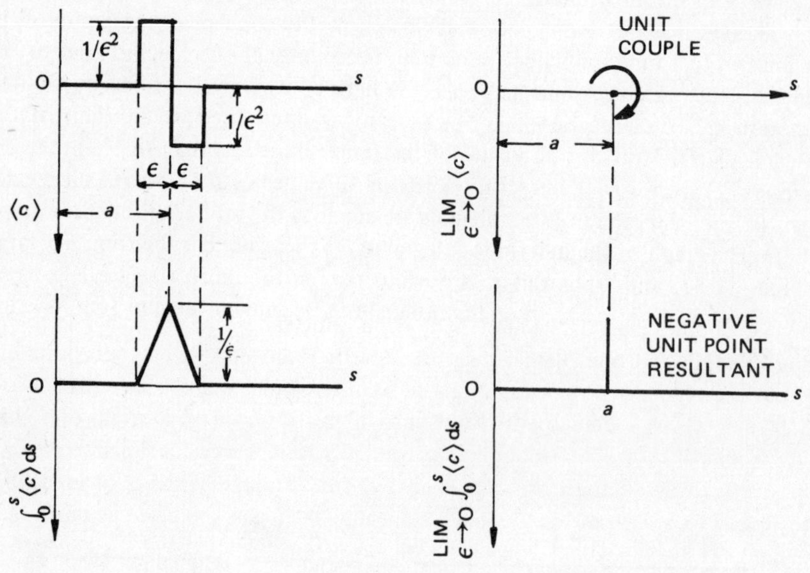

Fig. 11.32

As indicated in Fig. 11.32, the integral of $\langle c \rangle$ is equal to the negative of $\langle f \rangle$. The limit of $\langle f \rangle$ therefore determines the negative of the integral of the limit of $\langle c \rangle$, whence:

$$\int_0^s \langle s - a \rangle_{-2} \, ds = -\langle s - a \rangle_{-1} \, .$$

The values and integrals of the relevant singularity functions are therefore as indicated in Table 11.1.

FUNCTION		SYMBOL	VALUE	$\int_0^s$
UNIT POINT COUPLE	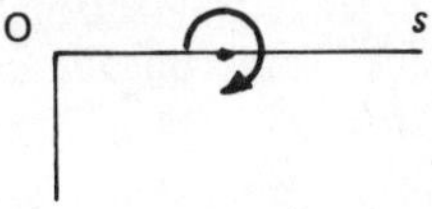	$\langle x-a\rangle_{-2}$	0 WHEN $s \neq a$	$-\langle s-a\rangle_{-1}$
UNIT POINT QUANTITY	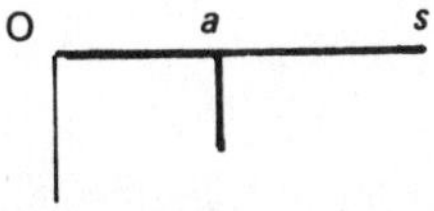	$\langle x-a\rangle_{-1}$	1 WHEN $s = a$	$\langle s-a\rangle^0$
POWER FUNCTION	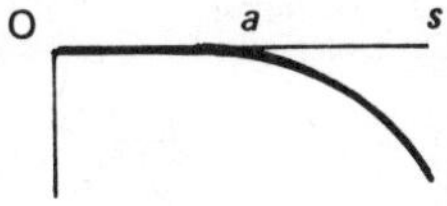	$\langle x-a\rangle^n$	0 WHEN $s < a$ 1 WHEN $s > a$	$\dfrac{\langle s-a\rangle^{n+1}}{n+1}$

Table 11.1

Loading Equations in Terms of Singularity Functions

Evidently, a point couple M at $s=a$ can be represented symbolically by $M\langle s-a\rangle_{-1}$. whilst a point force F at $s = a$ can be represented by $F\langle s-a\rangle_{-1}$. However, a power function remains non-zero from its point of application right through to $s=\infty$. To indicate a distributed load over a finite range, the effects of a power function applied at one point must therefore be removed by the application of appropriate negative power functions at another. For example, as indicated in Fig. 11.33, a distributed load of uniform intensity w per unit length over the range from $s = a$ to $s = b$ is equivalent to a positive uniform load of intensity w from $s = a$ to $s = \infty$, together with a negative uniform load of the same intensity from $s = b$ to $s = \infty$, and the load from $s = a$ to $s = b$ can therefore be indicated by $w\langle s-a\rangle^0 - w\langle s-b\rangle^0$.

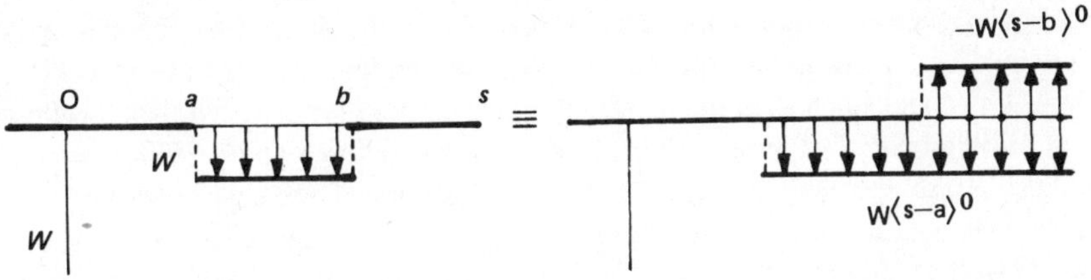

Fig. 11.33

Similarly, as indicated in Fig. 11.34, a distributed load whose intensity per unit length varies linearly from w_a at $s = a$ to w_b at $s = b$ is equivalent to a uniformly distributed load of intensity w_a and a ramp of slope equal to $(w_b - w_a)/(b-a)$ applied from $s=a$ to $s=\infty$, together with a negative ramp of

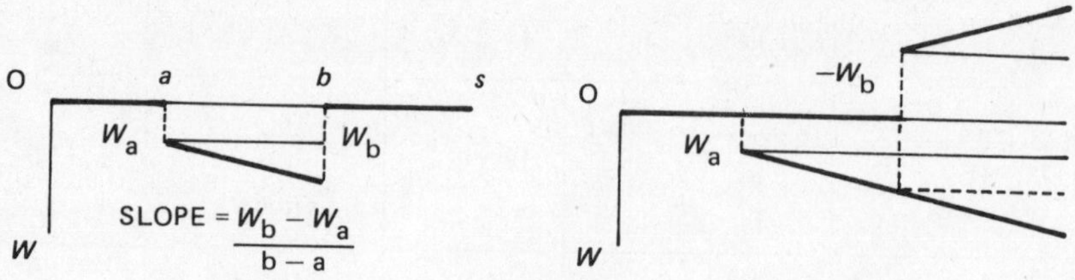

Fig. 11.34

the same slope and a uniformly distributed negative load of intensity w_b per unit length at $s = b$, and the linearly varying load from $s = a$ to $s = b$ can therefore be indicated by

$$w_a\langle s-a\rangle^0 + \frac{(w_b - w_a)}{(b-a)}\langle s-a\rangle^1 - w_b\langle s-b\rangle^0 - \frac{(w_b - w_a)}{(b-a)}\langle s-b\rangle^1 .$$

In principle, more elaborate distributions of continuous loads can be represented by fitting a polynomial of the power functions in the usual way. However, in practice, such an elaboration is seldom justified, and distributed loads can be adequately represented by appropriate combinations of step and ramp functions as above.

The Constants of Integration and the Boundary Conditions

Assuming a beam of uniform EI, let the complete, equilibrium set of forces and couples on the beam, including the unknown support reactions, be described in a function $\langle\omega\rangle$ of singularity functions. Then the equations of shear force, bending moment, $EI\theta$ and $EI\Delta$ may be determined by the successive integration of $\langle\omega\rangle$, and each integration will introduce only a single constant of integration which is equal to the value of the integral at the lower limit. But the lower limit was specifically chosen at $s = 0$, and this origin of s was itself chosen at such a position that all the forces and couples lay to one side only. This implies that both the shear force and the bending moment are zero at the lower limit, and since each of the singularity functions is then zero also, it follows that the constants of integration in both the shear force and bending moment equations are necessarily zero. Thus, the constants of integration in the shear force and bending moment equations being dropped from consideration, only the two that arise in the slope and deflection equations remain, and these, together with the unknown support reactions, may be determined from a combination of the two equations of equilibrium of the forces and couples, with the known conditions of kinematic constraint that are imposed at the supports.

Finally, we note that whereas the shear force and bending moment are liable to step changes, such as arise in the functions $\langle s-a\rangle_{-2}$, $\langle s-a\rangle_{-1}$ and $\langle s-a\rangle^0$, the slope and deflection are not. It follows that in the slope and deflection equations, the distinction between the two points that lie immediately to either side of each discontinuity is no longer of particular significance. Thus, the origin of s being chosen at the position of the first of the forces and couples that act on the beam, the complete set of forces and couples are described by a function $\langle w \rangle$ of singularity functions. Then $EI\theta$ is equated to the third indefinite integral of $\langle w \rangle$ plus a constant of integration, whilst $EI\Delta$ is equated to the integral of $EI\theta$ plus a second constant of integration, and the two constants of integration, and the support reactions are determined from the two equations for the equilibrium of the forces and couples on the beam, together with the kinematic constraints that are imposed at the supports.

Example 11.29

Repeat Example 11.16, by singularity functions.

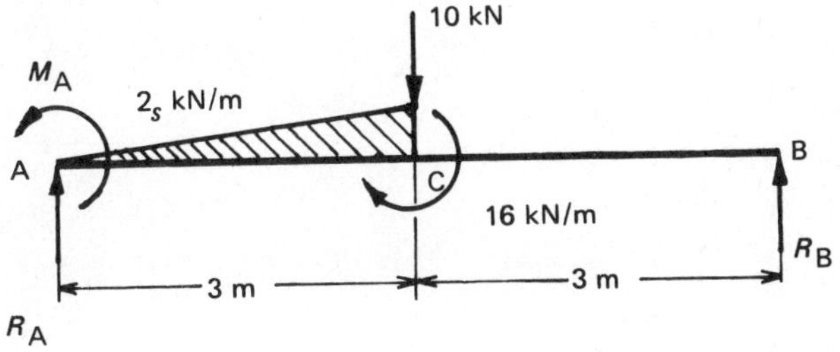

Solution

The support reactions being assumed to cut in the senses indicated in the diagram:

$$\begin{aligned}
\langle w \rangle = {} & - M_A\langle s-0\rangle_{-2} + 16\langle s-3\rangle_{-2} - R_A\langle s-0\rangle_{-1} + 10\langle s-3\rangle_{-1} \\
& - R_B\langle s-6\rangle_{-1} - 6\langle s-3\rangle^0 - 2\langle s-0\rangle^1 - 2\langle s-3\rangle^1 \\
\int\langle w \rangle = {} & M_A\langle s-0\rangle_{-1} - 16\langle s-3\rangle_{-1} - R_A\langle s-0\rangle^0 + 10\langle s-3\rangle^0 \\
& - R_B\langle s-6\rangle^0 - 6\langle s-3\rangle^1 + \langle s-0\rangle^2 - \langle s-3\rangle^2 \\
\iint\langle w \rangle = {} & M_A\langle s-0\rangle^0 - 16\langle s-3\rangle^0 - R_A\langle s-0\rangle^1 + 10\langle s-3\rangle^1 \\
& - R_B\langle s-6\rangle^1 - 3\langle s-3\rangle^2 + \frac{1}{3}\langle s-0\rangle^3 - \frac{1}{3}\langle s-3\rangle^3 \\
EI\theta = {} & M_A\langle s-0\rangle^1 - 16\langle s-3\rangle^1 + \frac{R_A}{2}\langle s-0\rangle^2 + 5\langle s-3\rangle^2 \\
& - \frac{R_B}{2}\langle s-6\rangle^2 - \langle s-3\rangle^3 + \frac{1}{12}\langle s-0\rangle^4 - \frac{1}{12}\langle s-3\rangle^4 + C_1 s \\
EI\Delta = {} & \frac{M_A}{2}\langle s-0\rangle^2 - 8\langle s-3\rangle^2 - \frac{R_A}{6}\langle s-0\rangle^3 + \frac{5}{3}\langle s-3\rangle^3 \\
& - \frac{R_B}{6}\langle s-6\rangle^3 - \frac{1}{4}\langle s-3\rangle^4 + \frac{1}{60}\langle s-0\rangle^5 - \frac{1}{60}\langle s-3\rangle^5 + C_1 s + C_2 .
\end{aligned}$$

Boundary Conditions:

When $s = 0$, then $\theta = 0$; and $\Delta = 0$.

Therefore: $C_2 = 0$; and $C_1 = 0$.

When $s = 6$, then $\Delta = 0$.

Therefore: $M_A - 2R_A = -4{\cdot}35$ (a)

For equilibrium of forces $R_A + R_B = 19$ (b)

of moments about A: $M_A + 6R_B = 64$ (c)

Between (a), (b) and (c):

$$M_A = 18{\cdot}5 \text{ (kN m)};\quad R_A = 11{\cdot}4 \text{ (kN)};\quad R_B = 7{\cdot}6 \text{ (kN)} .$$

Index